C. A. Brebbia J. C. F. Telles
L. C. Wrobel

Boundary Element Techniques

Theory and Applications in Engineering

With 284 Figures

Springer-Verlag
Berlin Heidelberg New York Tokyo 1984

C. A. BREBBIA

Dept. of Civil Engineering
University of Southampton
Southampton SO9 5NH
United Kingdom

J. C. F. TELLES

L. C. WROBEL

COPPE – Univ. Federal do Rio de Janeiro
Programa de Engenharia Civil
Caixa Postal 68506
21944-Rio de Janeiro
Brazil

ISBN 3-540-12484-5 Springer-Verlag Berlin Heidelberg New York Tokyo
ISBN 0-387-12484-5 Springer-Verlag New York Heidelberg Berlin Tokyo

Library of Congress Cataloging in Publication Data

Brebbia, C. A.
Boundary element techniques.
Includes index. 1. Boundary value problems. 2. Engineering mathematics.
I. Telles, J. C. Faria (José Claudio Faria), 1950– . II. Wrobel, L. C. (Luiz Carlos), 1953– . III. Title.
TA347.B69B734 1983 620′.0042 83-4827
ISBN 0-387-12484-5 (U.S.)

Printed in Germany

Typesetting, Offset printing and Bookbinding: Graphischer Betrieb Konrad Triltsch, Würzburg

2061/3020 - 5 4 3 2 1 0

The Preface in a Dialogue

The authors have attempted twice to write an appropriate preface to this book but have on both occasions miserably failed to convey in a brief manner what are the main points of the book. This failure is mainly due to the aversion of the authors to prefaces that promise everything but deliver little. Due to the lack of success the best we can offer is a verbatim report of the two meetings.

Act I

(Our authors start to discuss the writing of this preface. The scene is in Rio de Janeiro beside a swimming pool. The authors are identified by the pseudonyms of Socrates, Plato and Aristotle, not for reasons of vanity but because those illustrious philosophers somewhat characterize their respective points of view).

SOCRATES: I have been reading some of the literature on Integral equations and Boundary elements recently published and feel most unhappy about the lack of a comprehensive text.

PLATO: Yes. I have just been looking at one that is rather written in a hurry I guess to capitalize on the current interest in the topic. The authors failed to comprehend the basic principles of the technique.

ARISTOTLE: That is because people write books without having had a first hand experience in the relevant research topic. I always insist you have to look at the problem and build your theory around it!

PLATO: Well, well, Ari. That may well be the case sometimes but you have to remember that the fundamental mathematical concepts are the essential part of any method.

SOCRATES: I do not think this discussion is leading us anywhere. I propose that we write one book based on our research experience and our fundamental knowledge of approximate and basic techniques, trying to blend our past finite element background with the new method and I think we should define the table of contents and preface right now.

PLATO and ARISTOTLE: Hear! Hear!

SOCRATES: I think that we ought to stress that we will write only about things that we have first hand experience in, in a coherent way that will be useful to engineers and other scientists and stressing the formulation without being too mathematical. We should write with integrity and honesty, giving reference to other authors where reference is due, but avoiding mentioning everybody just to be certain that our book is widely advertised. Above all, the book should be clear and useful.

PLATO: I think we should include a good discussion of fundamental ideas, of how integral equations are formed, pointing out that they are like two dimensional shadows of three dimensional objects, . . .

SOCRATES: Stop there! Remember you are not 'the' Plato!

PLATO: Sorry, I was carried away.

ARISTOTLE: I think that the book should have many applications so that the reader can learn by looking at them how to use the method.

SOCRATES: I agree. But we should be careful. It is easy to include many illustrations and examples in a book in order to disguise its meagre contents. All examples should be relevant.

ARISTOTLE: And we should also include a full computer program to give the reader if so he wishes, a working experience of the technique.

SOCRATES: That is a good idea, provided that the code is well explained and integrates with the theory. Any fool can nowadays attach a computer code to a book but requires work and experience to have it properly related to the theory.

PLATO: I wonder if we will write the book. It seems unlikely.

SOCRATES: Yes it does. Does it not? Well, I am going for a swim.

Act II

(The manuscript is finished and the writers are sitting around it. The scene is now in Southampton in April. A timid ray of sun is coming through a window. The writers are spellbound and looking attentively at the manuscript).

PLATO: I cannot believe it! It is really finished!

SOCRATES: Well, not quite. You will see how the publishers will want us to trim it down. They always do as a matter of principle. 20 to 25% I think.*

ARISTOTLE: But that would be a pity! We have been over the manuscript three times. It is perfect!

---

* Springer-Verlag, to their great credit, accepted the full manuscript. Our apologies – The Authors

PLATO: We can only achieve but pallid reflections of perfection. Still it is a good book.

SOCRATES: You are right. We should fight for it and make them publish the whole work. We have some rights do we not (looks at the contract for a moment and concludes). No, we do not! (meakley) but we can try . . .

ARISTOTLE: And once it is published we have to explain to our colleagues that this is a serious book, a work with applications. We should specially stress (i) that the work has a great unity; (ii) the large range of topics covered in depth; (iii) that it is written by those who have used the method; (iv) that it is well written and clear.

PLATO: How are we going to do that?

ARISTOTLE: (downcast) I do not know . . .

SOCRATES: I know! We should write a Preface (everybody agrees). Well let us start; "Recent new advances and developments in the field of boundary elements . . . etc, etc."

ARISTOTLE and PLATO: Not again!!

The Book

The purpose of this book is to present a comprehensive and up-to-date treatment of the boundary element method (B.E.M.).

The work stresses the non-linear and time-depending applications together with a series of new problems which can now be solved using B.E.M.

The approach followed by the authors is to present the techniques as an outgrowth of the finite element method in a way that is simple for engineers to understand. The mathematical treatment is always subordinate to the applicability of the technique.

The reader will thus find in this definitive monograph a comprehensive treatment of the topic from fundamentals to computer applications, including a fully operational computer program.

The Authors

Contents

Chapter 1

Approximate Methods

1.1. Introduction

Engineers and physical scientists have in recent years become very conversant with numerical techniques of analysis. These techniques are based on the approximate solution of an equation or set of equations describing a physical problem. The first widely known approximate method was finite differences which approximates the governing equations of the problem using local expansions for the variables, generally truncated Taylor series. The technique can be interpreted as a special case of the more general weighted residual methods as shown in Section 1.8.

The finite element method has attracted the attention of the analysts largely due to its property of dividing the continuum into a series of elements, which can be associated with physical parts. The existing literature on finite elements is by now very extensive and encompasses structures [1] as well as fluid flow [2] and other types of problems. The method can sometimes be based on variational principles or more generally on weighted residual expressions. The great interest that the method attracted at the beginning of the 1960s had two important consequences: (i) it originated an impressive amount of work in computational techniques and efficient engineering software, (ii) substantial research into basic physical principles, such as variational techniques and weighted residuals, was stimulated.

The first of the above points comes as a natural consequence of the emergence of new and powerful computers, i.e., second generation machines which were able to solve engineering problems involving large amounts of numerical storage and manipulation. For a while the continuous progress in computer technology including the development of third generation machines distracted scientists from the development of mathematical methods and their basic principles, i.e., point (ii) above. These methods can be traced to precomputer times [3, 4] and involve different ways of solving the governing equations of a problem, i.e., Galerkin, collocation, least squares, line techniques, matrix progression or transfer, the combination of different techniques, etc. Fortunately they were not forgotten and they reappeared in the finite element literature, sometimes with different names, such as Galerkin finite elements, finite element strip method, some time-integration schemes, etc. Another important development in approximate analysis was the investigation of mixed principles and the realization that physical problems can be expressed and solved in many different ways in accordance with the equations that we wish to approximate. These approximations are of fundamental importance for the computer implementation of the different numerical techniques. Mixed methods can be traced to Reissner [5] and more specifically to Pian for finite elements [6]. An excellent exposition of mixed methods in structural mechanics can be found in the book by Washizu [7].

Integral equation techniques were, until recently, considered to be a different type of analytical method, somewhat unrelated to approximate methods. They became known in Europe through the work of a series of Russian authors, such as Muskhelishvili [8], Mikhlin [9], Kupradze [10], and Smirnov [11] but were not very popular with engineers. A predecessor of some of this work was Kellogg [12] who applied integral equations for the solution of Laplace-type problems. Integral equation techniques were mainly used in fluid mechanics and general potential problems and known as the "source" method which is an "indirect" method of analysis; i.e., the unknowns are not the physical variables of the problem. Work on this method continued throughout the 1960s and 1970s in the pioneering work of Jaswon [13], Symm [14], Massonet [15], Hess [16], and many others.

It is difficult to point out precisely who was the first to propose the "direct" method of analysis. It is found in a different form in Kupradze's book [10]. It seems fair, however, from the engineering point of view to consider that the method originated in the work of Cruse and Rizzo [17] in elastostatics. The direct method is the one which will be used mainly in this book as it is the most appealing to engineers and physical scientists.

Since the early 1960s a small research group at Southampton University worked on the applications of integral equations to solve stress analysis problems. Unfortunately, the presentation of the problem, the difficulty of defining the appropriate Green's functions, and the parallel emergence of the finite element method all contributed to minimize the importance of this work. At the beginning of the 1970s recent developments in finite elements started to find their way into the formulation of boundary integral equations and contributed to the development of general curved elements. Still, the question of how effectively one relates the boundary integral equations to other approximate techniques was unresolved. This was done by Brebbia who in the 1970s worked on the relationship of different approximate methods. The work culminated with the first book in 1978 for which the title "Boundary Elements" was used [18]. More recently this work was extended to encompass time-dependent and nonlinear problems [19]. Three important international conferences were held at Southampton University in 1978, 1980, and 1982 and another one in California in 1981. The edited proceedings of these conferences – the only ones so far on this topic – are now standard references [20–23].

In this chapter the common basis of all fundamental techniques will be described, with special reference to boundary elements. This is important to understand the approximations involved when using the techniques and being able to combine boundary elements with other numerical methods.

1.2. Basic Definitions

In this book we will be concerned mainly with the solution of differential equations representing a particular physical problem. These equations can be of elliptic, parabolic, or hyperbolic types. We will consider for the moment only elliptic equations. Assume that we can represent them by an operator $\mathscr{L}$ such that

$$\mathscr{L}(u) = b \qquad \text{in } \Omega. \tag{1.1}$$

The operator $\mathscr{L}$ is defined as a process which when applied to the function u produces another function b. Ω represents the spatial domain, usually represented by coordinates x_i $(i = 1, 2, 3)$, or simply by x in one-dimensional problems.

The operator $\mathscr{L}$ will be considered to be a differential operator such as

$$\mathscr{L}(\,) = \frac{d^2(\,)}{dx^2}$$

or

$$\mathscr{L}(\,) = \frac{d^4(\,)}{dx^4} + \frac{d(\,)}{dx} + (\,) \tag{1.2}$$

or for the two-dimensional Laplace equation

$$\mathscr{L}(\,) = \frac{\partial^2(\,)}{\partial x_1^2} + \frac{\partial^2(\,)}{\partial x_2^2}.$$

When applying the operator to the function u, this function should be substituted inside the parentheses. Although we are presently considering the function u as a scalar, it is important to notice that it could be a vector, such as in solid mechanics, where u can be replaced by a vector $\boldsymbol{u}$, whose components are the three displacements.

We can consider the homogeneous version of (1.1) as

$$\mathscr{L}(u) = 0 \qquad \text{in } \Omega. \tag{1.3}$$

An inner product can be defined such that

$$\int_\Omega \mathscr{L}(u)\, w\, d\Omega = 0. \tag{1.4}$$

These products are sometimes represented by a bracket expression $\langle \mathscr{L}(u), w \rangle$. Definitions of inner products other than (1.4) are also possible, but this is the one which we will apply most of the time. The inner product (1.4) can be integrated by parts until all derivatives in u have been eliminated. This leads to the "transposed" form of the inner product and, as a result of integrating by parts, produces a series of boundary terms. In general we can write

$$\int_\Omega \mathscr{L}(u)\, w\, d\Omega = \int_\Omega u\, \mathscr{L}^*(w)\, d\Omega + \int_\Gamma [S^*(w)\, G(u) - G^*(w)\, S(u)]\, d\Gamma. \tag{1.5}$$

Γ is the exterior surface of the Ω domain and S and G are differential operators due to the integration by parts. By definition $S^*(w)$ contains the w terms resulting from the initial phase of the integration and $S(u)$ contains the corresponding u terms.

The operator $\mathscr{L}^*$ is called the adjoint of $\mathscr{L}$. If $\mathscr{L}^* = \mathscr{L}$, $\mathscr{L}$ is said to be self-adjoint. In this case $G = G^*$ and $S = S^*$ also. Self-adjointness of an operator is analogous to symmetry of a matrix. In addition to determining if the operator is self-adjoint, the integration by parts also generates two different types of boundary conditions. The set $S(u)$ prescribed are called the *essential* boundary conditions and $G(u)$ prescribed are the nonessential or *natural* boundary conditions. One can specify either type of boundary condition on the surface of the domain. However, the essential boundary conditions must be enforced at some point in order for the solution to be unique. Letting Γ_1 and Γ_2 represent complementary positions of the total surface Γ, one can state the boundary conditions for the self-adjoint problem ($\mathscr{L}^* = \mathscr{L}$) as

$$\left.\begin{array}{l} G(u) \text{ prescribed on } \Gamma_2 \\ S(u) \text{ prescribed on } \Gamma_1 \end{array}\right\} \Gamma = \Gamma_1 + \Gamma_2 .$$

A self-adjoint operator is also positive definite if

$$\int_{\Omega} (\mathscr{L}(u))\, u \, d\Omega \geqq 0 \tag{1.6}$$

for all u and only equal to zero for the trivial case $u \equiv 0$. To determine if $\mathscr{L}$ is positive definite we can integrate the inner product by parts until it contains derivatives of the same order. This operation is the midpoint of the transformation of $\mathscr{L}$ into $\mathscr{L}^*$. Positive definiteness is an extremely valuable property in establishing solution schemes and also in constructing variational statements.

Example 1.1. Consider the equation

$$\mathscr{L}(u) = \frac{d^2 u}{dx^2} - \lambda^2 u = 0 \tag{a}$$

with $0 < x < 1$.

We can propose the inner product

$$\int_0^1 \mathscr{L}(u)\, w \, dx = \int_0^1 \left\{ \frac{d^2 u}{dx^2} - \lambda^2 u \right\} w \, dx = 0 . \tag{b}$$

Next we integrate by parts once which gives

$$\int_0^1 \mathscr{L}(u)\, w \, dx = \int_0^1 \left(- \frac{du}{dx} \frac{dw}{dx} - \lambda^2 u \, w \right) dx + \left[\frac{du}{dx} w \right]_0^1 . \tag{c}$$

Integrating again,

$$\int_0^1 \mathscr{L}(u)\, w \, dx = \int_0^1 \left(u \frac{d^2 w}{dx^2} - \lambda^2 u \, w \right) dx + \left[\frac{du}{dx} w \right]_0^1 - \left[u \frac{dw}{dx} \right]_0^1 . \tag{d}$$

Notice that the first term on the right-hand side can be written as a function of the $\mathscr{L}(\,)$ operator, giving

$$\int_0^1 \mathscr{L}(u)\, w\, dx = \int_0^1 \mathscr{L}(w)\, u\, dx + \left[S(w)\, G(u)\right]_0^1 - \left[S(u)\, G(w)\right]_0^1 . \tag{e}$$

The operator is then *self-adjoint* and the boundary operators are

$$S(\,) = (\,),$$

$$G(\,) = \frac{d(\,)}{dx}.$$

Notice that the operator is also negative definite, i.e., if we take $u = w$ and the homogeneous boundary conditions, the first integration of (a) gives

$$\int_0^1 \mathscr{L}(u)\, u\, dx = -\int_0^1 \left\{\left(\frac{du}{dx}\right)^2 + \lambda^2 u^2\right\} dx .$$

Negative definite operators can be easily converted into positive definite by multiplying by -1.

Example 1.2. Consider now an operator with a first-order derivative as well, i.e.,

$$\mathscr{L}(u) = \frac{d^2u}{dx^2} + \frac{du}{dx} + u \tag{a}$$

with $0 < x < 1$. We can now write the inner product and integrate by parts twice which gives

$$\int_0^1 \left\{\frac{d^2u}{dx^2} + \frac{du}{dx} + u\right\} w\, dx = \int_0^1 \left\{\frac{d^2w}{dx^2} - \frac{dw}{dx} + w\right\} u\, dx + \left[w\left(\frac{du}{dx} + u\right) - u\left(\frac{dw}{dx}\right)\right]_0^1 . \tag{b}$$

Notice that now the adjoint operator is

$$\mathscr{L}^*(w) = \frac{d^2w}{dx^2} - \frac{dw}{dx} + w \neq \mathscr{L}(w) \tag{c}$$

and the boundary operators are

$$\begin{aligned} G(u) &= \frac{du}{dx} + u, & S(u) &= u, \\ G^*(w) &= \frac{dw}{dx}, & S(w) &= w. \end{aligned} \tag{d}$$

Example 1.3. Another important operator in engineering applications is the fourth-order (beam-type) operator

$$\mathscr{L}(u) = \frac{d^4 u}{dx^4}. \tag{a}$$

In this case the inner product is

$$\int_0^1 \mathscr{L}(u) w \, dx = \int_0^1 \frac{d^4 u}{dx^4} w \, dx. \tag{b}$$

Integrating by parts twice we have

$$\int_0^1 \mathscr{L}(u)\, w \, dx = \left[w \frac{d^3 u}{dx^3} \right]_0^1 - \left[\frac{dw}{dx} \frac{d^2 u}{dx^2} \right]_0^1 + \int_0^1 \frac{d^2 w}{dx^2} \frac{d^2 u}{dx^2} \, dx. \tag{c}$$

Integrating twice again by parts, we obtain

$$\begin{aligned} \int_0^1 \mathscr{L}(u)\, w \, dx = & \left[w \frac{d^3 u}{dx^3} \right]_0^1 - \left[\frac{dw}{dx} \frac{d^2 u}{dx^2} \right]_0^1 \\ & + \left[\frac{d^2 w}{dx^2} \frac{du}{dx} \right]_0^1 - \left[\frac{d^3 w}{dx^3} u \right]_0^1 + \int_0^1 \frac{d^4 w}{dx^4} u \, dx. \end{aligned} \tag{d}$$

Notice that in this case we have two types of boundary operators, i.e.,

$$\begin{aligned} \int_0^1 \mathscr{L}(u)\, w \, dx = & \int_0^1 \mathscr{L}^*(w)\, u \, dx + [S_1(w)\, G_1(u) - S_2(w)\, G_2(u)]_0^1 \\ & + [G_2(w)\, S_2(u) - G_1(w)\, S_1(u)]_0^1. \end{aligned} \tag{e}$$

Hence the essential boundary conditions are

$$S_1(u) = u, \qquad S_2(u) = \frac{du}{dx} \tag{f}$$

and the natural boundary conditions are

$$G_1(u) = \frac{d^3 u}{dx^3}, \qquad G_2(u) = \frac{d^2 u}{dx^2}. \tag{g}$$

The operator $\mathscr{L}^*$ is self-adjoint, i.e.,

$$\mathscr{L}^*(\,) = \mathscr{L}(\,) = \frac{d^4(\,)}{dx^4}. \tag{h}$$

If we put $w = u$ in equation (c) with homogeneous boundary conditions, one obtains

$$\int_0^1 \mathscr{L}(u)\, u\, dx = \int_0^1 \left(\frac{d^2u}{dx^2}\right)^2 dx \geqq 0\,. \tag{i}$$

Notice that this integral is essentially positive definite and only zero for $u = \alpha x + \beta$; hence in order to be positive for all u, one must prescribe at least two values of u or one of u and the other of du/dx. For a beam problem this means that all rigid-body-type movements need to be suppressed.

1.3. Approximate Solutions

Most engineering problems which are expressed in a differential form can only be solved in an approximate manner due to their complexity. The best known techniques are the finite difference and the finite element methods. These two techniques although apparently different achieve similar objectives by reducing the infinite degrees of freedom of a continuous system to a finite set. By doing this the problem can be solved numerically and becomes amenable to computer solving. The finite difference technique defines a series of nodes at which the discrete version of the differential equation is satisfied. In finite elements the differential equation or its inner product version is satisfied in an average sense over a region or element. The two techniques discretize the domain as well as the boundaries of the region under consideration. However, the boundary element method is based on the discretization of the exterior boundary only. The three techniques are closely related if one focuses on the approximation involved. To do so let us first define the following set of equations:

$$\mathscr{L}(u_0) = b \qquad \text{in } \Omega \tag{1.7}$$

with boundary conditions

$$\begin{aligned} S(u_0) &= s \qquad \text{on } \Gamma_1\,, \\ G(u_0) &= g \qquad \text{on } \Gamma_2\,. \end{aligned} \tag{1.8}$$

u_0 represents the *exact* solution of the problem which is usually impossible to find. The function u_0 can be approximated by a set of functions $\varphi_k(x)$ such that

$$u = \sum_{k=1}^{n} \alpha_k\, \varphi_k + \alpha_0\,. \tag{1.9}$$

(The α_0 coefficient is included to satisfy the nonhomogeneous part of the boundary conditions.). α_k are undetermined parameters and φ_k are linearly independent functions taken from a complete sequence of functions such as

$$\varphi_1(x),\, \varphi_2(x),\, \ldots,\, \varphi_n(x)\,; \tag{1.10}$$

x represents the spatial coordinates in the Ω domain.

These functions are usually chosen to satisfy certain given conditions, called admissibility conditions, relating to the boundary conditions and the degree of continuity.

A sequence of functions such as the ones defined in (1.10) is said to be linearly independent if

$$\alpha_1 \varphi_1 + \alpha_2 \varphi_2 + \cdots + \alpha_n \varphi_n = 0 \tag{1.11}$$

is true only when all α_i are zero.

A sequence of linearly independent functions is said to be complete if a number n of terms and a corresponding set of constants α_k can be found such that, given an admissible but otherwise arbitrary function u_0, we have that the square root of the inner product of the difference between u_0 and its approximation can be made as small as one requires, i.e.,

$$\{\int (u_0 - u)^2 dx\}^{1/2} \leqq \beta,$$

where β is a small positive quantity.

Returning to Eq. (1.7) we will initially require that the approximate functions satisfy all the boundary conditions of the problem and have the necessary degree of continuity to make the left-hand side of Eq. (1.7) different from zero (i.e., $\mathscr{L}(u) \neq 0$).

Substituting the approximation for u_0 into Eq. (1.7) produces a "residual" or "error" function R such that

$$R = \mathscr{L}(u) - b \neq 0. \tag{1.12}$$

These errors will be forced to be zero in certain average sense, and this is done differently for each particular method.

If the function u does not satisfy *all* boundary conditions one may have two other types of error, or residual functions:

(i) error in the essential boundary conditions

$$R_1 = S(u) - s \neq 0 \quad \text{on } \Gamma_1; \tag{1.13}$$

(ii) error in the natural conditions

$$R_2 = G(u) - g \neq 0 \quad \text{on } \Gamma_2.$$

Our aim is now to make the errors as small as possible over the domain and on the boundary. In order to do so the errors can be distributed and the way in which this distribution is carried out produces different types of approximate methods.

Example 1.4. Let us consider the following equation

$$\mathscr{L}(u) = \frac{d^2 u}{dx^2} = b, \qquad 0 < x < 1, \tag{a}$$

where b is assumed constant from 0 to 1 and the boundary conditions are homogeneous, $u = 0$ at $x = 0$ and $x = 1$. The exact solution can be easily obtained by integrating equation (a) and gives

$$u_0 = \tfrac{1}{2}\, b\, x\,(x-1)\,. \tag{b}$$

Consider now that a sinusoidal approximation is proposed such that as a first try one can take

$$u = \alpha_1\, \varphi_1 = \alpha \sin \pi\, x\,. \tag{c}$$

Notice that this function satisfies the – homogeneous – boundary conditions for this problem.

The residual function is

$$R = \frac{d^2 u}{dx^2} - b = -\,\alpha\, \pi^2 \sin \pi\, x - b\,. \tag{d}$$

In this case we will determine the α parameter simply by setting $R \equiv 0$ at $x = \frac{1}{2}$. (This is a very simple application of the collocation method.)

$$\begin{aligned} -\,\alpha\, \pi^2 &= b\,, \\ \alpha &= -\frac{b}{\pi^2}\,. \end{aligned} \tag{e}$$

The approximate solution is

$$u = -\frac{b}{\pi^2} \sin \pi\, x\,. \tag{f}$$

The results for the exact and approximate solutions and the residual function are tabulated in Table 1.1 for $b = 1$ and the error plotted in Fig. 1.1,

$$R = b \sin \pi\, x - b\,. \tag{g}$$

Notice the difference between the errors of the approximate function and the residual function itself, which is directly related to the differential equation.

Table 1.1

x	u_0 (exact)	u (approx.)	$R = \sin \pi\, x - 1$
0.10	– 0.045000	– 0.031309	– 0.690983
0.20	– 0.080000	– 0.059555	– 0.412214
0.30	– 0.105000	– 0.081970	– 0.190983
0.40	– 0.120000	– 0.096362	– 0.048943
0.50	– 0.125000	– 0.101321	0.000000

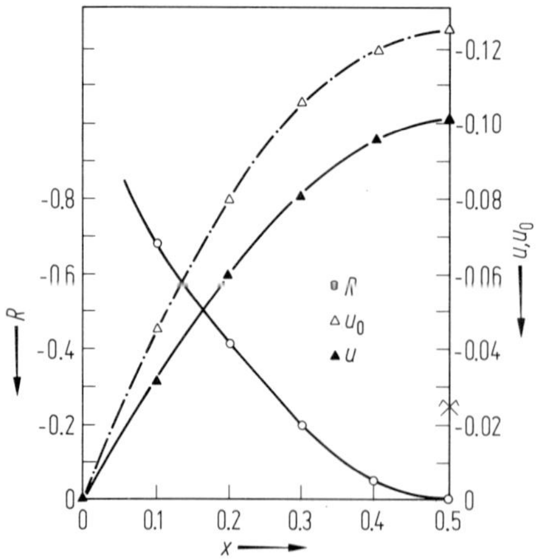

Fig. 1.1. Exact, approximate, and residual functions (first approximation)

Another possibility is to make the residual zero at $x = \frac{1}{4}$ instead of $x = \frac{1}{2}$. For this case we have

$$R_{1/4} = [-\alpha\,\pi^2 \sin \pi\, x - b]_{1/4} = 0\,, \qquad \alpha = -\frac{b}{\pi^2}\sqrt{2}\,. \tag{h}$$

The new values for the approximation to u for $b = 1$, i.e.,

$$u = -\frac{1}{\pi^2}\sqrt{2}\,\sin \pi\, x \tag{i}$$

and the residual function

$$R = \sqrt{2}\,\sin \pi\, x - 1 \tag{j}$$

are tabulated in Table 1.2 and shown in Fig. 1.2. Notice that the error has decreased.

Table 1.2

x	u_0 (exact)	u (approx.)	$R = \sqrt{2}\,\sin \pi\, x - 1$
0.10	− 0.045000	− 0.044278	− 0.562983
0.20	− 0.080000	− 0.084223	− 0.168746
0.30	− 0.105000	− 0.115923	+ 0.144122
0.40	− 0.120000	− 0.136276	+ 0.344997
0.50	− 0.125000	− 0.143289	+ 0.414213

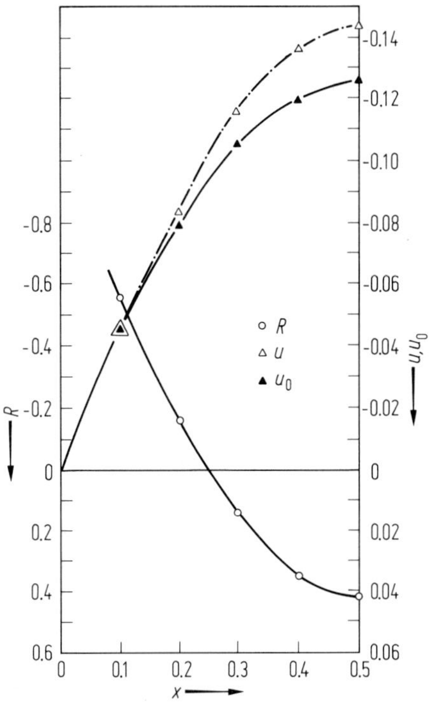

Fig. 1.2. Exact, approximate, and residual functions (second trial)

Example 1.5. Let us consider the following equation in the $0 < x < 1$ domain:

$$\mathscr{L}(u_0) - b = \frac{d^2 u_0}{dx^2} + u_0 + x = 0 \tag{a}$$

with boundary conditions $u = 0$ at $x = 0$ and 1.

We can now propose an approximate set of functions such as

$$\varphi_1 = 1, \qquad \varphi_2 = x, \qquad \varphi_3 = x^2, \qquad \varphi_4 = x^3, \; \ldots \, . \tag{b}$$

The approximate function u can be written as

$$\begin{aligned} u &= a_1\,\varphi_1 + a_2\,\varphi_2 + a_3\,\varphi_3 + a_4\,\varphi_4 + \cdots, \\ &= a_1\,1 + a_2\,x + a_3\,x^2 + a_4\,x^3 + \cdots. \end{aligned} \tag{c}$$

Let us first consider three terms and try to make u satisfy the boundary conditions of the problem.

$$\begin{aligned} &\text{at} \quad x = 0 \rightarrow u = a_1 = 0\,, \\ &\text{at} \quad x = 1 \rightarrow u = a_1 + a_2 + a_3 = 0\,. \end{aligned} \tag{d}$$

Hence $a_2 = -a_3$ and the approximate function which satisfies the boundary conditions is

$$u = \alpha x - \alpha x^2 = \alpha x(1 - x), \tag{e}$$

where $\alpha = a_2$.

If we take four terms, we have

$$u = a_1 1 + a_2 x + a_3 x^2 + a_4 x^3, \tag{f}$$

and after satisfying the homogeneous boundary conditions we find that

$$u = x(1 - x)(\alpha_1 + \alpha_2 x). \tag{g}$$

In general,

$$u = x(1 - x)(\alpha_1 + \alpha_2 x + \alpha_3 x^2 + \cdots). \tag{h}$$

The exact solution of equation (a) with the above boundary condition is

$$u_0 = \frac{\sin x}{\sin 1} - x. \tag{i}$$

Note that in this case *admissibility* is satisfied by equation (h) functions which fulfilled the boundary conditions and have the necessary degree of continuity. *Completeness* is also satisfied as the norm of $(u - u_0)$ will decrease when the number of terms in equation (h) increases.

1.4. Method of Weighted Residuals

If one assumes that the residual or error functions are

$$\begin{array}{ll} R & \text{in } \Omega \\ R_1 & \text{on } \Gamma_1 \\ R_2 & \text{on } \Gamma_2 \end{array} \Bigg\} \; \Gamma_1 + \Gamma_2 = \Gamma, \tag{1.14}$$

one can now propose a way of distributing these errors such that they are zero in an average sense. Let us define another set of linearly independent functions ψ_i such as

$$\psi_1(x), \quad \psi_2(x), \quad \psi_3(x), \ldots, \psi_k(x).$$

These functions are sometimes assumed to satisfy the homogeneous form of the boundary conditions although this, as we will see, is not always necessary. We can now define a set of arbitrary coefficients β_i which allow us to write the set ψ_i

in a more compact form as a w function;

$$w = \beta_1 \psi_1 + \beta_2 \psi_2 + \beta_3 \psi_3 + \cdots . \tag{1.15}$$

Let us assume for the moment that the approximate function u identically satisfies all the boundary conditions of the problem. In this case $R_1 = R_2 \equiv 0$ and only $R \neq 0$. We can now *distribute* the error R in Ω by multiplying it by w which is now called a weighting function, i.e.,

$$\langle R, w \rangle = \int_{\Omega} R w \, d\Omega = 0 . \tag{1.16}$$

This ensures that the error R is distributed according with the functions in w. As the β_i coefficients are arbitrary, Eq. (1.16) implies

$$\int_{\Omega} R \psi_i \, d\Omega = 0 \qquad \text{for } i = 1, 2, \ldots, k . \tag{1.17}$$

Several well-known approximate techniques can now be reexamined in the light of these definitions.

1.4.1. The Collocation Method

As seen in Example 1.4, instead of satisfying the equations in an "average" form, we try to satisfy them at only a series of chosen points. These points are usually, but not necessarily, evenly distributed in the domain.

For the case of homogeneous boundary conditions we can define an approximating function

$$u = \sum_{k=1}^{n} \alpha_k \varphi_k , \tag{1.18}$$

where the φ_k functions satisfy the homogeneous boundary conditions.

We have a residual function

$$R = \mathscr{L}(u) - b = 0 \qquad \text{in } \Omega \tag{1.19}$$

which has to be satisfied at n points in the domain Ω. Note that in principle the number of α_k has to be the same as the number of collocation points chosen. We can express this condition in a weighted residual form by defining the ψ_i as Dirac delta functions. The Dirac delta function is a generalized function which can be defined as the limit of a normal function (Fig. 1.3). In the limit this function is zero at every point of the domain except where the argument is zero at which point it is infinite (i.e., when $x = \xi$, ξ being the coordinate of the point i of application of the Dirac function). Hence,

$$\Delta(x - \xi) = \Delta_i .$$

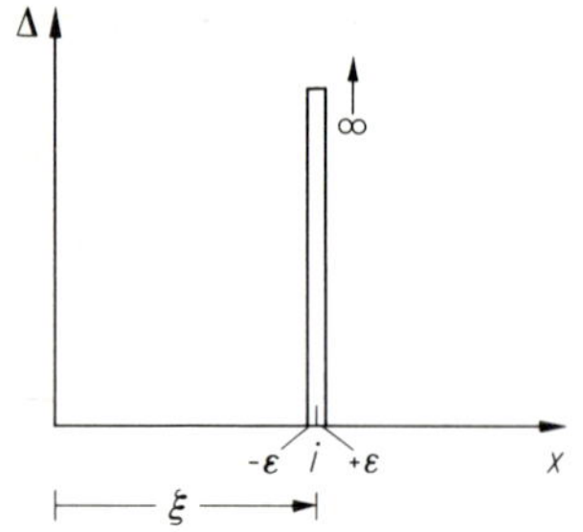

Fig. 1.3. Dirac delta function

The Dirac delta function has a property very useful in engineering:

$$\int_{-\infty}^{\infty} \Delta(x-\xi)\,dx = \int_{\xi_i-\varepsilon}^{\xi_i+\varepsilon} \Delta(x-\xi)\,dx = 1\,,$$

where ε is any positive number. We also have that for any function $f(x)$ continuous at x_i we can write

$$\int_{-\infty}^{\infty} f(x)\,\Delta(x-\xi)\,dx = \int_{\xi-\varepsilon}^{\xi+\varepsilon} f(x)\,\Delta(x-\xi)\,dx = f(\xi) = f_i\,, \tag{1.20}$$

where f_i or $f(\xi)$ is the value of the function f at the point i. In what follows where it is not necessary to represent the arguments of the functions explicitly, the shorthand notation f_i and Δ_i will be used for simplicity.

The collocation method at a series of points i – or point collocation method – can now be represented by Eq. (1.16), i.e.,

$$\int_{\Omega} R\,w\,d\Omega = 0\,, \tag{1.21}$$

where

$$w = \beta_1\Delta_1 + \beta_2\Delta_2 + \beta_3\Delta_3 + \cdots + \beta_k\Delta_k\,. \tag{1.22}$$

Δ_i represents the Dirac delta functions at the collocation points $i = 1, 2, \ldots, k$.

Example 1.6. Consider now the following equation in the $0 < x < 1$ domain:

$$\mathscr{L}(u) - b = \frac{d^2u}{dx^2} + u + x = 0 \tag{a}$$

with the homogeneous boundary conditions

$$u = 0 \quad \text{at } x = 0\,,$$

$$u = 0 \quad \text{at } x = 1\,.$$

Let us take (Example 1.5) an approximate function with only two unknown parameters, i.e.,

$$u = x(1-x)(\alpha_1 + \alpha_2 x)\,. \tag{b}$$

Substituting this approximate u in the equilibrium equation (a) we obtain the residual function, i.e.,

$$R = \frac{d^2 u}{dx^2} + u + x = (-2 + x - x^2)\,\alpha_1 + (2 - 6x + x^2 - x^3)\,\alpha_2 + x\,. \tag{c}$$

We can now propose a weighting function in terms of Dirac delta functions applied at $\xi = \frac{1}{4}$, $\xi = \frac{1}{2}$, i.e.,

$$w = \beta_1 \Delta_1 (x - \tfrac{1}{4}) + \beta_2 \Delta_2 (x - \tfrac{1}{2})\,. \tag{d}$$

This means that

$$\int_0^1 R w\, dx = 0 \quad \Rightarrow \beta_1 R(\tfrac{1}{4}) + \beta_2 R(\tfrac{1}{2}) = 0 \tag{e}$$

becomes

$$R = 0 \quad \text{at } x = \tfrac{1}{4} \text{ and } x = \tfrac{1}{2}\,. \quad \because \beta_i \text{ is arbitrary}, i = 1, 2,$$

This gives two equations which can be written in matrix form as follows:

$$\begin{bmatrix} \dfrac{29}{16} & -\dfrac{35}{64} \\[2ex] \dfrac{7}{4} & \dfrac{7}{8} \end{bmatrix} \begin{Bmatrix} \alpha_1 \\[2ex] \alpha_2 \end{Bmatrix} = \begin{Bmatrix} \dfrac{1}{4} \\[2ex] \dfrac{1}{2} \end{Bmatrix}. \tag{f}$$

Solving this system we obtain

$$\alpha_1 = \frac{6}{31}, \qquad \alpha_2 = \frac{40}{217} \tag{g}$$

which gives

$$u = \frac{x(1-x)}{217}(42 + 40\,x)\,. \tag{h}$$

Notice that the error function becomes

$$\begin{aligned} R &= \frac{6}{31}(-2 + x - x^2) + \frac{40}{217}(2 - 6\,x + x^2 - x^3) + x \\ &= \frac{1}{217}(-4 + 19\,x - 2\,x^2 - 40\,x^3)\,. \end{aligned} \tag{i}$$

These results can be seen in tabulated and graphic form in Table 1.3 and Fig. 1.4.

Table 1.3

x	u_0 (exact)	u (approx.)	R
0.10	0.018641	0.019078	− 0.009953
0.20	0.036097	0.036866	− 0.002764
0.30	0.051194	0.052258	+ 0.002027
0.40	0.062782	0.064147	+ 0.003317
0.50	0.069746	0.071428	+ 0.000000
0.60	0.071018	0.072995	− 0.009032
0.70	0.065585	0.065806	− 0.024884
0.80	0.052502	0.054562	− 0.048663
0.90	0.030901	0.032350	− 0.081474

Note: $R = 0$ at $x = 0.25$ and $x = 0.5$.

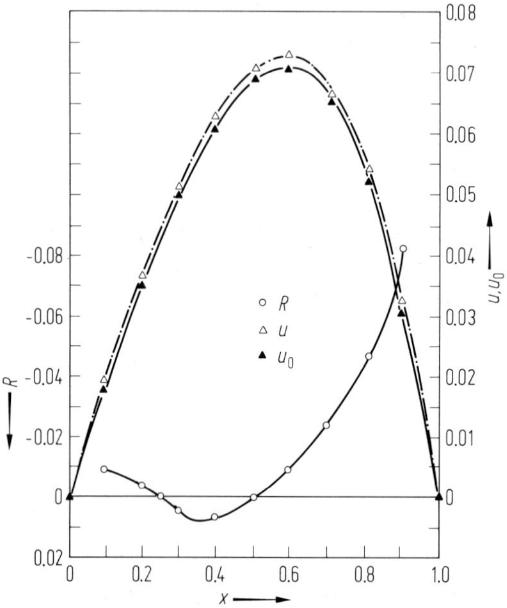

Fig. 1.4. Exact, approximate, and residual functions (Example 1.6)

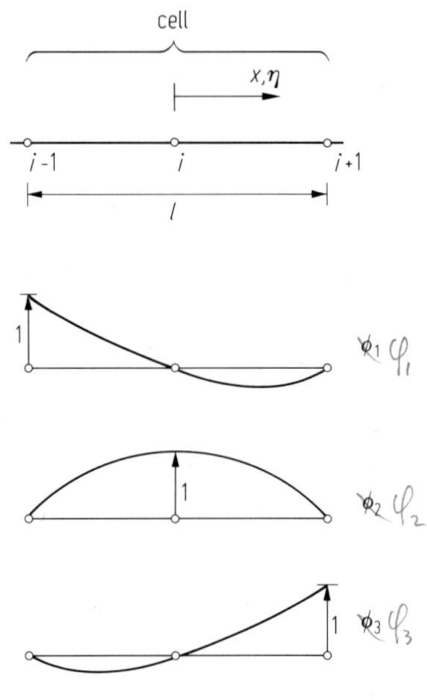

Fig. 1.5. Cell collocation

Example 1.7. It is interesting to point out that a special type of collocation method gives the finite difference method. Consider for instance a region or "cell" (Fig. 1.5) around the node i under consideration. We can propose a "local" approximating function over each "cell" as follows:

$$u = u_{i-1}\,\varphi_1 + u_i\,\varphi_2 + u_{i+1}\,\varphi_3; \tag{a}$$

u_i, etc., are the values of the function at the nodes of the finite difference grid. The functions φ_i are quadratic functions such that referring to the dimensionless system of coordinates η one can write

$$\varphi_1 = \tfrac{1}{2}\,\eta(\eta - 1)\,,$$
$$\varphi_2 = (1 - \eta)(1 + \eta)\,, \qquad \text{(b)}$$
$$\varphi_3 = \tfrac{1}{2}\,\eta(1 + \eta)\,.$$

These shape functions are plotted in Fig. 1.5 for reference, $\eta = 2\,x/l$.

If we differentiate functions (a) taking into consideration the above shape functions and collocate at i we can write the residual

$$R_i = \frac{d^2u}{dx^2} + u + x = \frac{4}{l^2}\,(u_{i-1} - 2u_i + u_{i+1}) + u_i + x_i = 0\,. \qquad \text{(c)}$$

This expression is the same as the one obtained using central finite differences.

The above technique which is discussed in detail in Ref. [24] permits us to find many different finite difference molecules and also facilitates the use of curvilinear systems of coordinates for this method.

1.4.2. Method of Collocation by Subregions

This method is also a collocation technique, but now instead of requiring that the error function be zero at certain points, one seeks that the integral of the error function over different regions will be zero. Then one takes

$$\int_{\Omega_i} R\, d\Omega = 0 \qquad \text{over a subregion } \Omega_i\,. \qquad (1.23)$$

Example 1.8. Consider again the equation

$$\frac{d^2u}{dx^2} + u + x = 0 \qquad \text{(a)}$$

with boundary conditions $u = 0$ at $x = 0, 1$. Let the approximate function be

$$u = x(1 - x)(\alpha_1 + \alpha_2 x + \cdots)\,. \qquad \text{(b)}$$

For the first approximation (i.e., α_1 only) one can take the entire region as the subdomain with

$$R = (-2 + x - x^2)\,\alpha_1 + x\,. \qquad \text{(c)}$$

This gives

$$\int_0^1 \{(-2 + x - x^2)\,\alpha_1 + x\}\,dx = -\frac{11}{6}\,\alpha_1 + \frac{1}{2} = 0 \qquad \text{(d)}$$

$$\therefore \alpha_1 = \frac{3}{11}\,.$$

Hence the first approximation gives

$$u^{(1)} = \frac{3}{11} x(1-x)\,. \tag{e}$$

and so take 2 values of α_i into consideration.

For the second approximation one can take two regions $0 < x < \frac{1}{2}$ and $0 < x < 1$. Notice that one of them superimposes the other. The residual function is the same as for Example 1.6, i.e.,

$$R = (-2 + x - x^2)\,\alpha_1 + (2 - 6\,x + x^2 - x^3)\,\alpha_2 + x\,. \tag{f}$$

After integrating

$$\int_0^{1/2} R\,dx = 0 \qquad \text{and} \qquad \int_0^1 R\,dx = 0\,, \tag{g}$$

one obtains

$$\begin{aligned} \frac{1}{8} - \frac{11}{12}\alpha_1 + \frac{53}{192}\alpha_2 &= 0\,, \\ \frac{1}{2} - \frac{11}{6}\alpha_1 - \frac{11}{12}\alpha_2 &= 0 \end{aligned} \tag{h}$$

which gives

$$\alpha_1 = \frac{97}{517}\,, \qquad \alpha_2 = \frac{24}{141}\,; \tag{i}$$

the second approximate solution becomes

$$u^{(2)} = x(1-x)\left(\frac{291 + 264\,x}{1551}\right). \tag{j}$$

Table 1.4

x	u_0	$u^{(1)}_{(\text{approx})}$ (α_1 only)	$u^{(2)}_{(\text{approx})}$ (α_1 and α_2)	$R^{(1)}$	$R^{(2)}$
0.1	0.018641	0.024545	0.018417	− 0.420909	− 0.018526
0.2	0.036097	0.043636	0.035466	− 0.301818	− 0.003605
0.3	0.051194	0.057272	0.050123	− 0.188181	+ 0.008924
0.4	0.062782	0.065454	0.061369	− 0.080000	+ 0.018042
0.5	0.069746	0.068181	0.068181	+ 0.022727	+ 0.022727
0.6	0.071018	0.065454	0.069539	+ 0.120000	+ 0.021957
0.7	0.065585	0.057272	0.064421	+ 0.211818	+ 0.014711
0.8	0.052502	0.043636	0.051806	+ 0.298181	− 0.000030
0.9	0.030901	0.024545	0.030673	+ 0.379090	− 0.023292

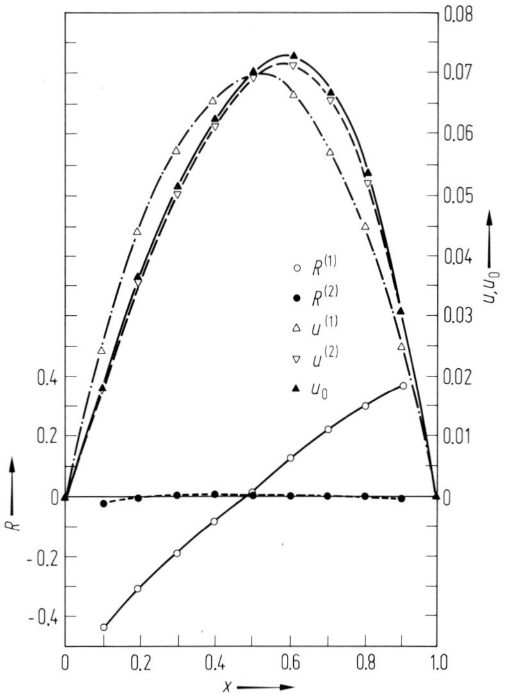

Fig. 1.6. Exact, approximate, and residual functions (Example 1.8)

The values of u for the two approximations and their residuals are compared in Table 1.4 and Fig. 1.6 with the exact solution. Notice that the results have improved dramatically in the second case.

Although at first sight the method of subregions would appear to be advantageous compared with point collocation, in practice the advantage of averaging the error function to zero over set regions is often offset by the fact that the error function is likely to change signs several times in the region and thus the method covers up quite large errors.

Example 1.9. A very interesting application of the subregion method is the case where the weighting function is a step-type function as shown in Fig. 1.7. Notice

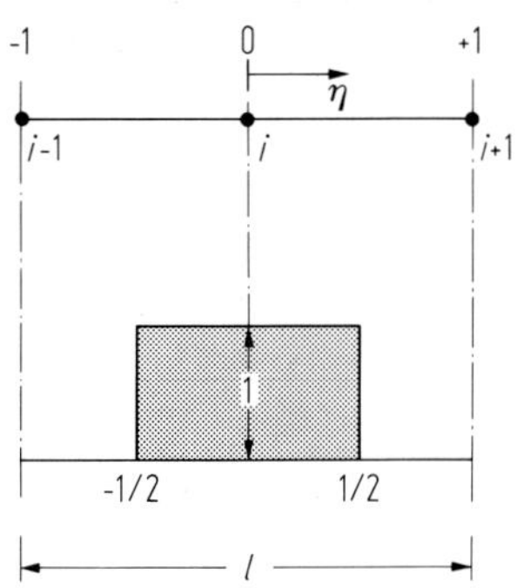

Fig. 1.7. Collocation by subregions in a finite difference or finite element grid

that here we are considering an element similar to a finite element, with shape functions

$$\varphi_1 = \tfrac{1}{2}\,\eta(\eta - 1)\,, \qquad \varphi_2 = (1-\eta)(1+\eta)\,, \qquad \varphi_3 = \tfrac{1}{2}\,\eta(1+\eta)\,. \tag{a}$$

Hence,

$$u = u_{i-1}\,\varphi_1 + u_i\,\varphi_2 + u_{i+1}\,\varphi_3\,, \tag{b}$$

where u_{i-1}, u_i, and u_{i+1} are the nodal values of the function at $i-1$, i, and $i+1$, respectively.

Consider the finite difference model for an equation such as

$$\frac{d^2u}{dx^2} = b\,. \tag{c}$$

It is given by (Example 1.7)

$$\frac{4}{l^2}\,(u_{i-1} - 2u_i + u_{i+1}) = b_i\,. \tag{d}$$

Let us first consider the weighted residual statement

$$\int \left(\frac{d^2u}{dx^2} - b\right) w\, dx = 0 \tag{e}$$

and apply it between $-\frac{l}{4}$ and $\frac{l}{4}$.
Integrating by parts we have

$$\int_{-l/4}^{l/4} \left(\frac{du}{dx}\frac{dw}{dx}\right) dx = -\int_{-l/4}^{l/4} b\, w\, dx + \left[\frac{du}{dx}\, w\right]_{-l/4}^{l/4}. \tag{f}$$

But as w is a step function, we are left only with

$$\int_{-l/4}^{l/4} b\, dx = \left[\frac{du}{dx}\right]_{l/4} - \left[\frac{du}{dx}\right]_{-l/4}. \tag{g}$$

If b is constant, we have the following relationship for node i:

$$\frac{b_i l}{2} = \frac{2}{l}\,[u_{i-1} - 2u_i + u_{i+1}] \tag{h}$$

or

$$\frac{4}{l^2}\,[u_{i-1} - 2u_i + u_{i+1}] - b_i = 0\,. \tag{i}$$

Notice that this is exactly the same molecule as in Example 1.7 but now we have obtained it by taking the operator to the boundary using an integration over subregions.

It is easy through these examples to realize the potentialities of the collocation method which is extensively applied in boundary element formulations as we will see in later chapters.

Example 1.10. Equation (1.16) can be used to generate a wide variety of weighted residual techniques such as the method of moments, the Galerkin technique, (see Section 1.5) as well as the collocation methods already seen.

The method of moments consists in taking moments of the error function in much the same way as it is done in probabilistic analysis. In one-dimensional problems for instance, the error function R is weighted with respect to

$$1, \quad x, \quad x^2, \quad x^3, \ldots . \tag{a}$$

Hence,

$$w = \beta_1 1 + \beta_2 x + \beta_3 x^2 + \cdots . \tag{b}$$

Let us illustrate this case by considering the same equation as previously, i.e.,

$$\frac{d^2 u}{dx^2} + u + x = 0 \tag{c}$$

with the approximating function

$$u = \alpha_1 \varphi_1 + \alpha_2 \varphi_2 + \cdots = \alpha_1 x (1 - x) + \alpha_2 x^2 (1 - x) + \cdots . \tag{d}$$

The residual for two terms is then

$$R = x + (-2 + x - x^2)\, \alpha_1 + (2 - 6x + x^2 - x^3)\, \alpha_2 . \tag{e}$$

This function needs to be orthogonalized with respect to 1 and x, i.e., we take the following two moments:

$$\int_0^1 R\, 1\, dx = 0, \qquad \int_0^1 R\, x\, dx = 0. \tag{f}$$

Integrating these equations from 0 to 1 one finds a 2×2 system of equations, i.e.,

$$\begin{bmatrix} \frac{11}{6} & \frac{11}{12} \\ \frac{11}{12} & \frac{19}{20} \end{bmatrix} \begin{Bmatrix} \alpha_1 \\ \alpha_2 \end{Bmatrix} = \begin{Bmatrix} \frac{1}{2} \\ \frac{1}{3} \end{Bmatrix}.$$

Its solution gives

$$\alpha_1 = \frac{122}{649}, \qquad \alpha_2 = \frac{110}{649},$$

whence

$$u = x(1-x)\left\{\frac{122}{649} + \frac{110}{649}x\right\}$$

and the residual becomes

$$R = x + (-2 + x - x^2)\frac{122}{649} + (2 - 6x + x^2 - x^3)\frac{110}{649}.$$

The results are shown in Table 1.5 and Fig. 1.8.

Table 1.5

x	u_0	u	R
0.1	0.018641	0.018443	− 0.020231
0.2	0.036097	0.035500	− 0.004869
0.3	0.051194	0.050154	+ 0.008089
0.4	0.062782	0.061386	+ 0.017627
0.5	0.069746	0.068181	+ 0.022727
0.6	0.071018	0.069522	+ 0.022372
0.7	0.065585	0.064391	+ 0.015546
0.8	0.052502	0.051771	+ 0.001232
0.9	0.030901	0.030647	− 0.021587

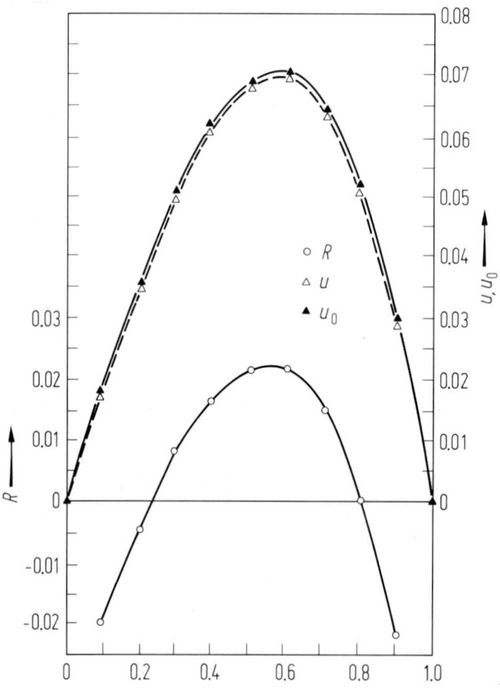

Fig. 1.8. Results for method of moments

1.5. Method of Galerkin

Galerkin's method is a particular weighted residual method for which the weighting functions belong to the same set as the approximating or trial functions. Given the system

$$\mathscr{L}(u_0) = b \qquad \text{in } \Omega \tag{1.24}$$

with homogeneous boundary conditions we can propose an approximate function which satisfies these conditions and such that

$$u_0 \cong u = \sum_{k=1}^{n} \alpha_k \varphi_k, \tag{1.25}$$

producing a residual

$$R = \mathscr{L}(u) - b \neq 0 \tag{1.26}$$

which in Galerkin's method is then orthogonalized with respect to the same functions φ_k used in the approximation, i.e.,

$$\int_{\Omega} R\, w\, d\Omega = 0, \tag{1.27}$$

where

$$w = \beta_1 \varphi_1 + \beta_2 \varphi_2 + \beta_3 \varphi_3 + \cdots. \tag{1.28}$$

For a linear operator $\mathscr{L}(\,)$, Eq. (1.27) produces a linear system of algebraic equations from which the β_i coefficients can be calculated.

As the same functions are used for u and w and the β's are arbitrary it is common to write the w function as a variation of u, i.e.,

$$w = \delta u = \delta\alpha_1 \varphi_1 + \delta\alpha_2 \varphi_2 + \delta\alpha_3 \varphi_3 + \cdots, \tag{1.29}$$

where $\delta\alpha_i \equiv \beta_i$. These variations can be associated with virtual quantities such as virtual displacements or velocities.

The property of having the same functions for the weighting and approximating functions is important in engineering practice as it produces symmetrical coefficients in many cases. Most finite element models are based on Galerkin-type techniques.

Example 1.11. Let us return to our original equation and try to solve it using Galerkin's method. We can choose the same approximating functions as previously, i.e.,

$$u = \alpha_1 \varphi_1 + \alpha_2 \varphi_2 = \alpha_1 x(1-x) + \alpha_2 x^2(1-x). \tag{a}$$

Note that, as previously, α_1 and α_2 are not nodal values of u but unknown generalized coefficients.

The weighted residual statement is

$$\int_0^1 R\, w\, dx = 0 \tag{b}$$

which produces the following two equations:

$$\int_0^1 R\, \varphi_1\, dx = 0 \quad \text{and} \quad \int_0^1 R\, \varphi_2\, dx = 0. \tag{c}$$

The R function is the same as in Example 1.10. Hence,

$$\begin{aligned} &\int_0^1 [x + (-2 + x - x^2)\,\alpha_1 + (2 - 6x + x^2 - x^3)\,\alpha_2]\,[x\,(1-x)]\,dx = 0,\\ &\int_0^1 [x + (-2 + x - x^2)\,\alpha_1 + (2 - 6x + x^2 - x^3)\,\alpha_2][x^2\,(1-x)]\,dx = 0 \end{aligned} \tag{d}$$

which, after integration, gives the following system:

$$\begin{bmatrix} \frac{3}{10} & \frac{3}{20} \\ \frac{3}{20} & \frac{13}{105} \end{bmatrix} \begin{Bmatrix} \alpha_1 \\ \alpha_2 \end{Bmatrix} = \begin{Bmatrix} \frac{1}{12} \\ \frac{1}{20} \end{Bmatrix}. \tag{e}$$

Note that the matrix is symmetric because the approximate and weighting functions are the same.

The α's are

$$\alpha_1 = \frac{71}{369}, \qquad \alpha_2 = \frac{7}{41}. \tag{f}$$

The approximate solution is

$$u = x\,(1-x)\left(\frac{71}{369} + \frac{7}{41}\,x\right) \tag{g}$$

which compares well with the exact solution (Table 1.6 and Fig. 1.9).

Table 1.6. Results for Example 1.11.

x	u_0	u	R
0.1	0.018641	0.018853	− 0.026945
0.2	0.036097	0.036249	− 0.011989
0.3	0.051194	0.051162	+ 0.000485
0.4	0.062782	0.062569	+ 0.009452
0.5	0.069746	0.069444	+ 0.013888
0.6	0.071018	0.070764	+ 0.012769
0.7	0.065582	0.065504	+ 0.005070
0.8	0.052502	0.052639	− 0.010233
0.9	0.030901	0.031146	− 0.034165

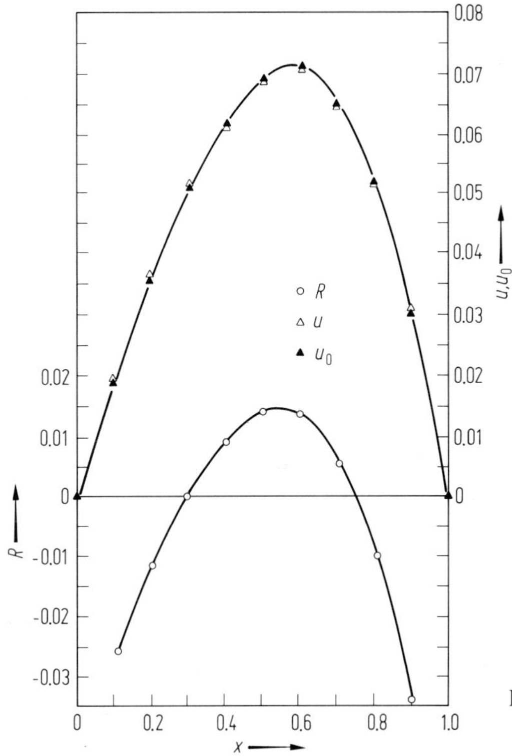

Fig. 1.9. Results for Galerkin's method

1.6. Weak Formulations

The examples treated in the previous sections were restricted to the case of approximate functions identically satisfying the boundary conditions of the problem. Weighted residuals can also be used to approximate the boundary conditions and in this way allow only for their partial satisfaction and, more significantly, the use of a set of functions having relaxed requirements.

If we consider for simplicity the second-order equation used in some of the examples, i.e.,

$$\mathscr{L}(u) - b = \frac{d^2u}{dx^2} + u + x = 0 \qquad \text{in } 0 < x < 1, \tag{1.30}$$

and its weighted residual statement

$$\int_0^1 \left(\frac{d^2u}{dx^2} + u + x \right) w \, dx = 0, \tag{1.31}$$

it is clear that a different order of continuity is required for u than for w. To define this continuity requirements we need to introduce a classification for the degree of continuity of a function. Let us assume that a function f is discontinuous at discrete points but is finite throughout the region (Fig. 1.10); its norm satisfying the following condition

$$\int f^2\, dx < \infty . \tag{1.32}$$

The function f is then said to be *square integrable.*

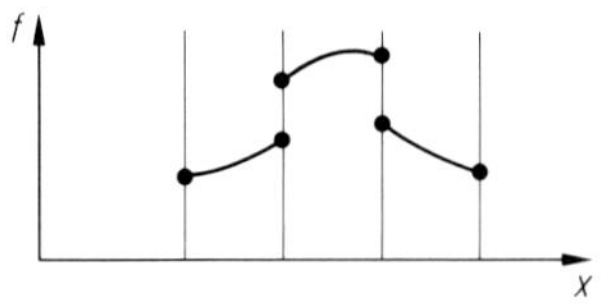

Fig. 1.10. Square integrable function

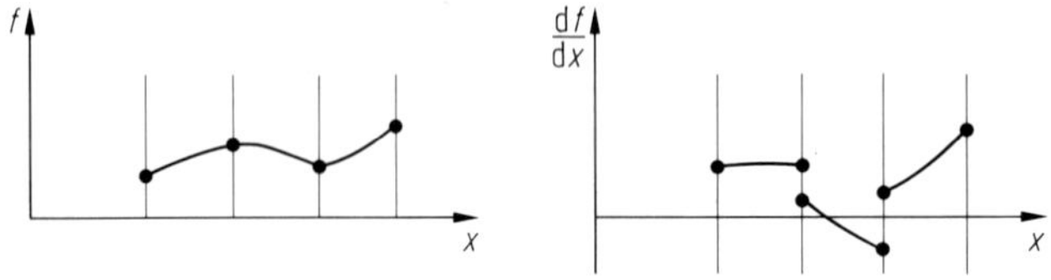

Fig. 1.11. First derivative square integrable function

If we impose conditions on the first derivative (Fig. 1.11), the function is said to be a *first derivative square integrable* function and the following norm has to be bounded:

$$\int \left\{ f^2 + \left(\frac{df}{dx} \right)^2 \right\} dx < \infty . \tag{1.33}$$

We can continue defining higher-order continuity. For example, the functions whose *second derivative* is *square integrable* (Fig. 1.12) have the following norm:

$$\int \left\{ f^2 + \left(\frac{df}{dx} \right)^2 + \left(\frac{d^2 f}{dx^2} \right)^2 \right\} dx < \infty . \tag{1.34}$$

The above definitions can be extended to two- and three-dimensional problems by replacing the scalar products with vector products.

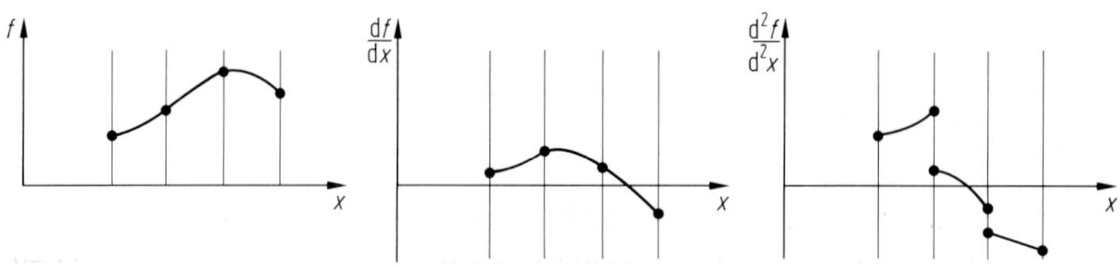

Fig. 1.12. Second derivative square integrable function

It is clear now that the approximating function u in Eq. (1.31) needs to be "second derivative square integrable" (Fig. 1.12), while the weighting function w is required only to be "square integrable." This was demonstrated in several examples and in particular in Example 1.9 where u was taken as a second-order polynomial and w as a constant.

In many cases we prefer to reduce the order of the continuity required for u and this can be done by integrating by parts. If we consider Eq. (1.31) and integrate the first term by parts, one finds

$$\int_0^1 \left(\frac{d^2u}{dx^2} + u + x\right) w\, dx = -\int_0^1 \frac{du}{dx}\frac{dw}{dx}\, dx + \int_0^1 (u+x)\, w\, dx + \left[q\, w\right]_0^1 = 0\,, \tag{1.35}$$

where $q = du/dx$. Note that now both functions u and w need to be continuous (or piecewise continuous) up to their first derivatives, i.e., we can take a set of "first derivative square integrable" basis functions for u and w:

$$\begin{aligned} u &= u_1\, \varphi_1 + u_2\, \varphi_2 + \cdots, \\ w &= \delta u = \delta u_1\, \varphi_1 + \delta u_2\, \varphi_2 + \cdots. \end{aligned} \tag{1.36}$$

(If the boundary conditions for u are nonhomogeneous, terms are usually added to the u expression to satisfy them.)

The term involving the function q in Eq. (1.35) is associated with the boundary conditions, i.e.,

$$\begin{aligned} &\text{Essential} \qquad u = \bar{u} \\ \text{or} \quad & \\ &\text{Natural} \qquad q = \bar{q} \end{aligned} \tag{1.37}$$

where the terms with overbars indicate known values of u or q. If the essential conditions are given and the approximate functions satisfy them, the functions w also identically satisfy their homogeneous versions. Let us assume for the sake of argument that the boundary conditions for Eq. (1.35) are

$$\begin{aligned} u &= 0 \qquad \text{at } x = 0 \qquad (\text{essential})\,, \\ q &= \bar{q} \qquad \text{at } x = 1 \qquad (\text{natural})\,. \end{aligned} \tag{1.38}$$

It is tempting to substitute $\bar{q}$ for q in Eq. (1.35) and write

$$-\int_0^1 \frac{du}{dx}\frac{dw}{dx}\, dx + \int_0^1 (u+x)\, w\, dx + [\bar{q}\, w]_{x=1} = 0\,. \tag{1.39}$$

There is, however, a subtle point to be considered in that by substituting q for $\bar{q}$ we are making an approximation. This is better explained by trying to retrieve Eq. (1.31) starting with Eq. (1.39). If we integrate the first term in Eq. (1.39) by parts, we obtain

$$\int_0^1 \left(\frac{d^2u}{dx^2} + u + x\right) w\, dx + [(\bar{q} - q)\, w]_{x=1} = 0\,. \tag{1.40}$$

Notice that the boundary term $(\bar{q}-q)$ is not necessarily zero as we are satisfying identically only the essential boundary conditions and the natural ones are being approximated. Hence Eq. (1.40) is the correct starting expression for the case for which we want to approximate the governing equation *and* the natural boundary condition.

To illustrate this point even better consider the case of the Laplace equation

$$\nabla^2 u_0 = 0 \qquad \text{in } \Omega \tag{1.41}$$

with boundary conditions

$$\left.\begin{array}{lll} \text{Essential} & u_0 = \bar{u} & \text{on } \Gamma_1 \\ \text{Natural} & q_0 = \bar{q} & \text{on } \Gamma_2 \end{array}\right\} \Gamma = \Gamma_1 + \Gamma_2 .$$

If we approximate u_0 by u and q_0 by q $(q = \partial u/\partial n)$, it is correct to start by writing

$$\int_\Omega (\nabla^2 u)\, w\, d\Omega = \int_{\Gamma_2} (q-\bar{q})\, w\, d\Gamma . \tag{1.42}$$

(The essential boundary conditions are identically satisfied.)

Integrating by parts the Laplacian (Gauss theorem) we find

$$\int_\Omega \frac{\partial u}{\partial x_k}\frac{\partial w}{\partial x_k}\, d\Omega = \int_{\Gamma_2} \bar{q}\, w\, d\Gamma , \tag{1.43}$$

where the indicial notation has been used, with Einstein's summation convention. Formula (1.43) is the correct starting expression for the Laplace equation finite element models.

We can now generalize the above by defining two residual, or error, functions:

$$\begin{array}{ll} R = \mathscr{L}(u) - b & \text{in the domain } \Omega, \\ R_2 = G(u) - g & \text{on the } \Gamma_2 \text{ part of the boundary.} \end{array} \tag{1.44}$$

Both residuals can now be weighted by the function w,

$$\int_\Omega (\mathscr{L}(u) - b)\, w\, d\Omega = \int_{\Gamma_2} (G(u) - g)\, w\, d\Gamma . \tag{1.45}$$

We can then proceed by integrating by parts the operator $\mathscr{L}(\,)$. If the operator is self-adjoint, it will produce two equal operators of reduced order which can be called $\mathscr{D}(\,)$, i.e.,

$$\int_\Omega \mathscr{D}(u)\,\mathscr{D}(w)\, d\Omega + \int_\Omega b\, w\, d\Omega = \int_{\Gamma_2} g\, w\, d\Gamma . \tag{1.46}$$

For the two-dimensional Laplace equation,

$$\mathscr{D}(\,) = \left[\frac{\partial(\,)}{\partial x_1}, \frac{\partial(\,)}{\partial x_2}\right]$$

and

$$g = \bar{q} . \tag{1.47}$$

Note that expression (1.43) could also have been obtained by considering

$$\int_{\Omega} (\nabla^2 u)\, w \, d\Omega = 0, \tag{1.48}$$

integrating by parts,

$$\int_{\Omega} \frac{\partial u}{\partial x_k} \frac{\partial w}{\partial x_k} \, d\Omega = \int_{\Gamma} \frac{\partial u}{\partial n} w \, d\Gamma . \tag{1.49}$$

and replacing $\partial u/\partial n$ by $\bar{q}$ on Γ_2 (on Γ_1, $w \equiv 0$). This gives

$$\int_{\Omega} \frac{\partial u}{\partial x_k} \frac{\partial w}{\partial x_k} \, d\Omega = \int_{\Gamma_2} \bar{q} \, w \, d\Gamma . \tag{1.50}$$

Once again this is valid only if we are aware that an approximation has been introduced when $q = \partial u/\partial n$ is replaced by $\bar{q}$ on Γ_2.

Example 1.12. Consider again the second-order equation

$$\frac{d^2 u_0}{dx^2} + u_0 + x = 0 \tag{a}$$

with boundary conditions $u_0 = 0$ at $x = 0$, but at $x = 1$ we impose the condition

$$q = \frac{du_0}{dx} = \bar{q} \tag{b}$$

where $\bar{q}$ is known.

Let us start with the following approximation:

$$u = \alpha_1 + \alpha_2 x + \alpha_3 x^2 \tag{c}$$

and try to satisfy the *two boundary conditions*

$$\begin{aligned} &\text{at } x = 0 \qquad u = \alpha_1 = 0, \\ &\text{at } x = 1 \qquad q = \frac{du}{dx} = 2\alpha_3 + \alpha_2 = \bar{q} . \end{aligned} \tag{d}$$

Hence,

$$\alpha_2 = \bar{q} - 2\alpha_3 . \tag{e}$$

Calling $\alpha_3 = \alpha$ we can rewrite equation (c) as

$$u = \alpha x^2 + (\bar{q} - 2\alpha) x = \bar{q} x - 2\alpha x + \alpha x^2 . \tag{f}$$

The residual for this case is

$$R = \frac{d^2 u}{dx^2} + u + x = 2\alpha + \bar{q} x - 2\alpha x + \alpha x^2 + x . \tag{g}$$

This error can be distributed by orthogonalizing it with respect to the function $(x^2 - 2x)$ only, i.e.,

$$w = \delta u = \delta\alpha\,(x^2 - 2x). \tag{h}$$

(The part in $\bar{q}$ in (f) is needed to satisfy the nonhomogeneous part of the boundary conditions and consequently should not appear in w.) This gives the following weighted residual statement:

$$\int_0^1 (2\alpha + \bar{q}\,x - 2\alpha\,x + \alpha\,x^2 + x)(x^2 - 2x)\,dx = 0,$$

$$\frac{24}{5}\alpha + \frac{5}{2}\bar{q} + \frac{5}{2} = 0, \tag{i}$$

$$\alpha = -\,0.5208\,(\bar{q} + 1).$$

The value of u is given by

$$u = \bar{q}\,x - 0.5208\,(\bar{q} + 1)(x^2 - 2x). \tag{j}$$

This result has been obtained satisfying both boundary conditions. Let us now consider what happens if the condition $(u = 0)$ is satisfied but the other $(q = \bar{q})$ is approximated. Now we have

$$\int_0^1 \left(\frac{d^2u}{dx^2} + u + x\right)\delta u\,dx = \left[\left(\frac{du}{dx} - \bar{q}\right)\delta u\right]_{x=1}. \tag{k}$$

Integrating the first term by parts one obtains

$$\int_0^1 \left(\frac{du}{dx}\frac{d\delta u}{dx} - (u + x)\,\delta u\right)dx = [\bar{q}\,\delta u]_{x=1}. \tag{l}$$

We can take the following approximation:

$$u = \alpha_1 x^2 + \alpha_2 x \tag{m}$$

which identically satisfies $u = 0$ at $x = 0$, but not $q = \bar{q}$ at $x = 1$. Integral (l) becomes

$$\int_0^1 [(2\alpha_1 x + \alpha_2)(2\delta\alpha_1 x + \delta\alpha_2) - (\alpha_1 x^2 + \alpha_2 x + x)(\delta\alpha_1 x^2 + \delta\alpha_2 x)]\,dx = \bar{q}\,(\delta\alpha_1 + \delta\alpha_2). \tag{n}$$

After integration one finds the following system:

$$\begin{bmatrix} \frac{17}{15} & \frac{3}{4} \\ \frac{3}{4} & \frac{2}{3} \end{bmatrix} \begin{Bmatrix} \alpha_1 \\ \alpha_2 \end{Bmatrix} = \begin{Bmatrix} \bar{q} + \frac{1}{4} \\ \bar{q} + \frac{1}{3} \end{Bmatrix}. \tag{o}$$

This system gives

$$\begin{aligned} \alpha_1 &= -0.4319 - 0.4322\,\bar{q}, \\ \alpha_2 &= 0.9859 + 1.9864\,\bar{q}. \end{aligned} \tag{p}$$

Let us now compute the value of q at $x = 1$:

$$q = \left.\frac{du}{dx}\right|_{x=1} = 2\,\alpha_1 + \alpha_2 = 0.1221 + 1.122\,\bar{q}. \tag{q}$$

Hence the value of q at $x = 1$ is not equal to $\bar{q}$ but is an approximation.
At the center point $x = \frac{1}{2}$ we have

$$u_c = \tfrac{1}{4}\,\alpha_1 + \tfrac{1}{2}\,\alpha_2 = 0.3850 + 0.8852\,\bar{q}. \tag{r}$$

Note that the first solution (equation (j)) gives

$$u_c = 0.3906 + 0.8906\,\bar{q}. \tag{s}$$

The values of u for the two approximations are given in Table 1.7 and shown in Fig. 1.13 for the case $\bar{q} = 0$. The reader is left to calculate the residuals to see how the imposition of the natural boundary conditions affects their distribution.

Table 1.7. u function for Example 1.12.

x	All boundary conditions are identically satisfied u	Only essential boundary condition is identically satisfied u
0.1	0.098958	0.094271
0.2	0.187500	0.179904
0.3	0.265625	0.256899
0.4	0.333333	0.325256
0.5	0.390625	0.384975
0.6	0.437500	0.436056
0.7	0.473958	0.478499
0.8	0.500000	0.512304
0.9	0.515625	0.537471
1.0	0.520833	0.554000

Example 1.13. Consider the case of a beam on elastic foundation (Fig. 1.14) which behaviour is governed by

$$\mathrm{EI}\,\frac{d^4 u}{dx^4} + k\,u = b; \tag{a}$$

u are the vertical displacements, EI is the beam rigidity, k is the spring constant for the foundation, and b is the load; notice $b = b\,(x)$.

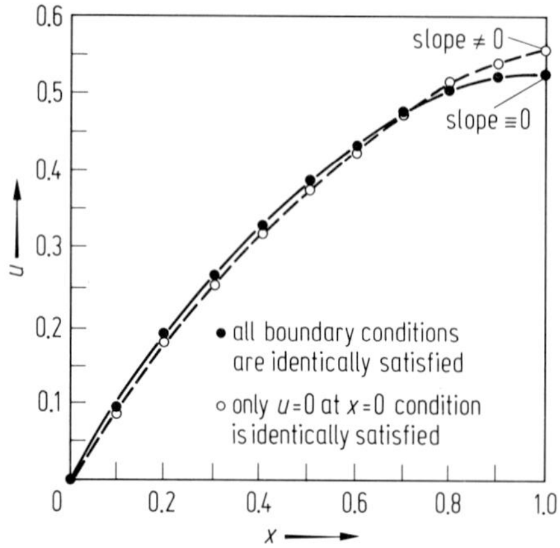

Fig. 1.13. u functions for Example 1.12

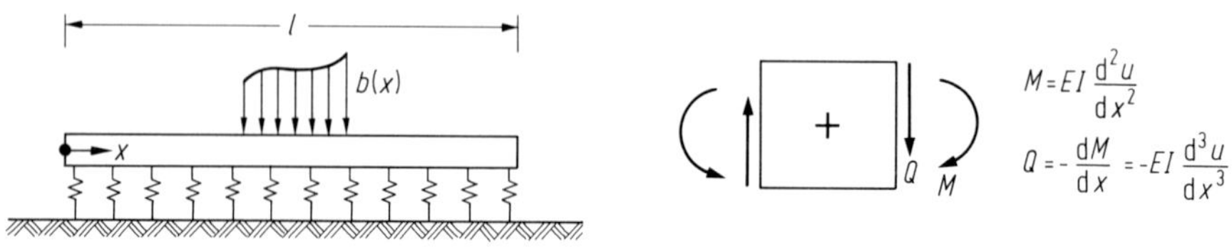

Fig. 1.14. Beam on elastic foundation

We can write the following weighted residual statement:

$$\int_0^l \left(\mathrm{EI} \frac{d^4 u}{dx^4} + k\,u - b \right) w\,dx = 0. \tag{b}$$

Integrating once by parts, expression (b) becomes

$$\int_0^l \left(-\,\mathrm{EI} \frac{d^3 u}{dx^3} \frac{dw}{dx} + k\,u\,w - b\,w \right) dx + \left[\mathrm{EI} \frac{d^3 u}{dx^3}\, w \right]_{x=0}^{x=l} = 0. \tag{c}$$

Integrating again we obtain

$$\int_0^l \left(\mathrm{EI} \frac{d^2 u}{dx^2} \frac{d^2 w}{dx^2} + k\,u\,w - b\,w \right) dx + \left[-\,Q\,w - M \frac{dw}{dx} \right]_{x=0}^{x=l} = 0, \tag{d}$$

where $Q = -\,\mathrm{EI}\dfrac{d^3 u}{dx^3}$ and $M = \mathrm{EI}\dfrac{d^2 u}{dx^2}$. Note that this is the starting point for finite element beam models.

Integrating once more,

$$\int_0^l \left(-\,\mathrm{EI} \frac{du}{dx} \frac{d^3 w}{dx^3} + k\,u\,w - b\,w \right) dx + \left[-\,Q\,w - M \frac{dw}{dx} + \theta\,\mathrm{EI} \frac{d^2 w}{dx^2} \right]_{x=0}^{x=l} = 0, \tag{e}$$

where $\theta = \dfrac{du}{dx}$ is the rotation. A fourth and final integration gives

$$\int_0^l \left(\mathrm{EI}\, u \frac{d^4 w}{dx^4} + k\, u\, w - b\, w \right) dx \\ - \left[Q\, w + M \frac{dw}{dx} - \theta\, \mathrm{EI} \frac{d^2 w}{dx^2} + u\, \mathrm{EI} \frac{d^3 w}{dx^3} \right]_{x=0}^{x=l} = 0. \tag{f}$$

Notice that by integrating by parts we can reduce the order of the functions required for u but increase those needed for w. The boundary conditions required for the beam can be classified as follows:

$$\begin{array}{ll} \text{Natural conditions} & M \text{ and } Q, \\ \text{Essential conditions} & u \text{ and } \theta\,. \end{array} \tag{g}$$

Example 1.14. Let us consider the beam shown in Fig. 1.15. The governing equation is

$$\mathscr{L}(u) = \mathrm{EI} \frac{d^4 u}{dx^4} = 0, \tag{a}$$

where u is the vertical displacement. EI is the beam rigidity (assumed to be constant along the beam). The boundary conditions described in Fig. 1.15 are

$$\begin{array}{ll} \text{Essential conditions} & \begin{cases} u = \dfrac{du}{dx} = 0 & \text{at } x = 0, \\ u = 0 & \text{at } x = l; \end{cases} \\ \text{Natural conditions} & \left\{ M = \mathrm{EI} \dfrac{d^2 u}{dx^2} = \bar{M} \quad \text{and} \quad Q = -\,\mathrm{EI} \dfrac{d^3 u}{dx^3} = \bar{Q} \quad \text{at } x = 2l. \right. \end{array} \tag{b}$$

We can start with the following weighted residual statement (Example 1.13).

$$\int_0^{2l} \left\{ \mathrm{EI} \frac{d^4 u}{dx^4} \right\} w\, dx = \left[\left(\mathrm{EI} \frac{d^3 u}{dx^3} + \bar{Q} \right) w - \left(\mathrm{EI} \frac{d^2 u}{dx^2} - \bar{M} \right) \frac{dw}{dx} \right]_{x=2l}. \tag{c}$$

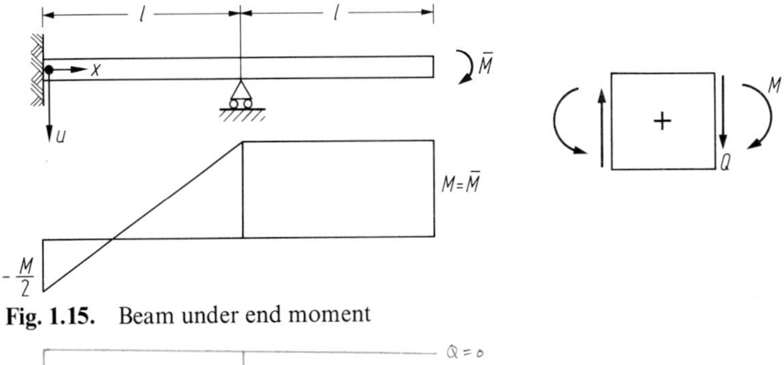

Fig. 1.15. Beam under end moment

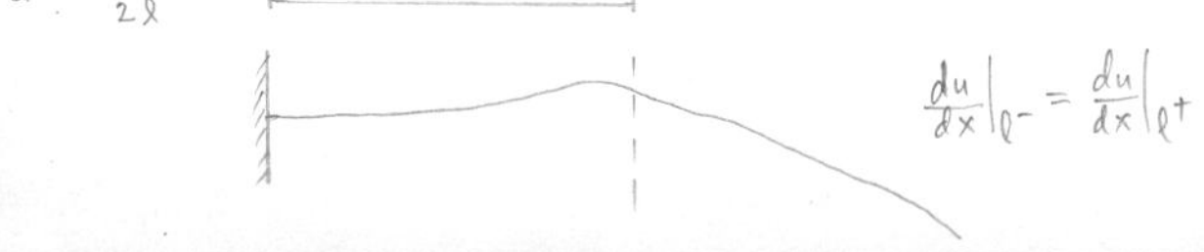

In principle the functions for u need to be square integrable up to their fourth derivative. Integrating equation (c) by parts twice one can reduce the order of the functions for u, obtaining

$$\int_0^{2l} \mathrm{EI} \frac{d^2 u}{dx^2} \frac{d^2 w}{dx^2} dx = \left[\bar{Q}\, w + \bar{M} \frac{dw}{dx} \right]_{x=2l}. \tag{d}$$

This new expression allows us to approximate the natural boundary conditions and use second derivative square integrable functions for our approximation.

The simplest choice for the u function is

$$u^{(1)} = \alpha_1^{(1)} \varphi_1 = \alpha_1^{(1)} x^2 (x - l). \qquad W^{(1)} = \beta_1^{(1)} \varphi_1 \tag{e}$$

Notice that this function identically satisfies the essential boundary conditions of the problem but not the natural conditions.

Substituting this into equation (d), taking into consideration $\bar{Q} = 0$, gives

$$\mathrm{EI} \int_0^{2l} \frac{d^2}{dx^2} (\alpha_1^{(1)} \varphi_1) \frac{d^2}{dx^2} (\beta_1^{(1)} \varphi_1)\, dx = \left[\bar{M} \frac{d}{dx} (\beta_1^{(1)} \varphi_1) \right]_{x=2l},$$
$$\mathrm{EI} \int_0^{2l} \alpha_1^{(1)} (6x - 2l)^2\, dx = 8 l^2 \bar{M}. \tag{f}$$

After integration,

$$56\, \mathrm{EI}\, \alpha_1^{(1)} l^3 = 8 l^2 \bar{M},$$
$$\therefore \alpha_1^{(1)} = \left(\frac{\bar{M}}{\mathrm{EI}\, l} \right) \frac{1}{7}. \tag{g}$$

The rotation and values of M and Q at $x = 2l$ can now be calculated. They are

$$\left(\frac{du}{dx} \right)_{x=2l} = 8 \alpha_1^{(1)} l^2 = \left(\frac{\bar{M} l}{\mathrm{EI}} \right) \frac{8}{7},$$
$$M = \mathrm{EI} \left(\frac{d^2 u}{dx^2} \right)_{x=2l} = \frac{10}{7} \bar{M}, \qquad Q = -\,\mathrm{EI} \left(\frac{d^3 u}{dx^3} \right)_{x=2l} = -6 \frac{\bar{M}}{7l}. \tag{h}$$

We can see that the values of Q and M are quite different from the applied $\bar{M}$ and $\bar{Q} = 0$ forces. The rotation, however, is not too far off from the exact value:

$$\left(\frac{du}{dx} \right)_{x=2l}^{\text{exact}} = 1.250 \left(\frac{\bar{M} l}{\mathrm{EI}} \right). \tag{i}$$

We can now improve upon these results by taking a second approximation for u such that

$$u^{(2)} = \alpha_1^{(2)} \varphi_1 + \alpha_2^{(2)} \varphi_2, \tag{j}$$

where

$$\varphi_1 = x^2 (x - l), \qquad \varphi_2 = x^3 (x - l). \tag{k}$$

We can substitute these values into equation (d) and obtain

$$\begin{aligned} \mathrm{EI} \int_0^{2l} \frac{d^2}{dx^2} (\alpha_1^{(2)} \varphi_1 + \alpha_2^{(2)} \varphi_2) \frac{d^2}{dx^2} (\beta_1^{(2)} \varphi_1 + \beta_2^{(2)} \varphi_2)\, dx \\ = \left[\bar{M} \frac{d}{dx} (\beta_1^{(2)} \varphi_1 + \beta_2^{(2)} \varphi_2) \right]_{x=2l}, \end{aligned} \tag{l}$$

where $\beta_1^{(1)}$ and $\beta_2^{(2)}$ are arbitrary. After integration, each of them gives one independent equation:

$$\begin{aligned} &\text{for } \beta_1^{(2)} \rightarrow 56\, \alpha_1^{(2)} + 152\, l\, \alpha_2^{(2)} = \frac{8 \bar{M}}{\mathrm{EI}\, l}; \\ &\text{for } \beta_2^{(2)} \rightarrow 152\, \alpha_1^{(2)} + 441.6\, l\, \alpha_2^{(2)} = \frac{20 \bar{M}}{\mathrm{EI}\, l}. \end{aligned} \tag{m}$$

Thus,

$$\alpha_1^{(2)} = 0.30315 \frac{\bar{M}}{\mathrm{EI}\, l}, \qquad l\, \alpha_2^{(2)} = -\, 0.05905 \frac{\bar{M}}{\mathrm{EI}\, l}. \tag{n}$$

Hence the rotation and end forces at $x = 2l$ can be computed, giving

$$\begin{aligned} &\left(\frac{du}{dx} \right)_{x=2l} = 1.244 \left(\frac{\bar{M} l}{\mathrm{EI}} \right), \\ &[M]_{x=2l} - 0.9055\, \bar{M}, \qquad [Q]_{x=2l} = -\, 0.6614 \frac{\bar{M}}{l}. \end{aligned} \tag{o}$$

Notice that this second trial solution gives excellent results for the rotations but convergence of the natural (forces) boundary conditions at $x = 2l$ is not as rapid.

1.7. Inverse Problem and Boundary Solutions

Up to now we have considered functions which satisfy the boundary conditions (or some of them) and are approximate in the domain, that is, not satisfying exactly the governing equations. Conversely, we could propose using functions which identically satisfy the governing equations and only approximately satisfy the boundary conditions. This will give origin to boundary formulations and, in particular, to the singular boundary integral technique.

We can differentiate two cases:

(i) when the approximate functions u are chosen in such a way that they identically satisfy the governing equations, i.e., $\mathscr{L}(u) \equiv 0$. Note that if the same functions are used for u and w, this implies that also the weighting function satisfies the governing equations, i.e., $\mathscr{L}(w) = 0$.

By $\int_{\Omega} (w\nabla^2 u - u\nabla^2 w)\,d\Omega = \int_{\Gamma} \left(w\frac{\partial u}{\partial n} - u\frac{\partial w}{\partial n}\right) d\Gamma$ (Green's Theorem)

(1.51) ⇒ (1.53)

(ii) when the weighting function satisfies the governing equations $\mathscr{L}(w)$ (self-adjoint problem) or their adjoint version $\mathscr{L}^*(w)$ (non-self-adjoint case).

To formulate these ideas in a clearer way, consider the case of the self-adjoint operator ∇^2 with boundary conditions operators $\frac{\partial(\,)}{\partial n}$ and ( ) on Γ_2 and Γ_1 parts of the boundary. We can start by writing the following weighted residual expression:

$$\int_{\Omega} (\nabla^2 u - b)\, w\, d\Omega = 0\,. \tag{1.51}$$

Integrating by parts once we have

$$\int_{\Omega} \left(\frac{\partial u}{\partial x_k}\frac{\partial w}{\partial x_k} + b\, w\right) d\Omega = \int_{\Gamma} q\, w\, d\Gamma\,, \tag{1.52}$$

where $q = \frac{\partial u}{\partial n}$. Integrating again,

$$\int_{\Omega} (\nabla^2 w)\, u\, d\Omega = \int_{\Gamma} u \frac{\partial w}{\partial n}\, d\Gamma - \int_{\Gamma} q\, w\, d\Gamma + \int_{\Omega} b\, w\, d\Omega\,. \tag{1.53}$$

If we now introduce the boundary conditions

$$u = \bar{u} \quad \text{on } \Gamma_1\,,$$
$$q = \bar{q} \quad \text{on } \Gamma_2\,,$$

Eq. (1.53) becomes

$$\int_{\Omega} (\nabla^2 w)\, u\, d\Omega = \int_{\Gamma_1} \bar{u} \frac{\partial w}{\partial n}\, d\Gamma + \int_{\Gamma_2} u \frac{\partial w}{\partial n}\, d\Gamma - \int_{\Gamma_1} q\, w\, d\Gamma - \int_{\Gamma_2} \bar{q}\, w\, d\Gamma + \int_{\Omega} b\, w\, d\Omega\,. \tag{1.54}$$

By introducing the boundary conditions in this way we are making an approximation which can be demonstrated by integrating by parts Eq. (1.54) twice to retrieve the operator $\nabla^2(\,)$ on u. This gives

$$\int_{\Omega} (\nabla^2 u - b)\, w\, d\Omega = \int_{\Gamma_2} (q - \bar{q})\, w\, d\Gamma - \int_{\Gamma_1} (u - \bar{u}) \frac{\partial w}{\partial n}\, d\Gamma\,. \tag{1.55}$$

Notice that expression (1.55) implies that we have now three errors or residual functions. Equation (1.55) can also be represented as

$$\int_{\Omega} R\, w\, d\Omega = \int_{\Gamma_2} R_2 w\, d\Gamma - \int_{\Gamma_1} R_1 \frac{\partial w}{\partial n}\, d\Gamma\,. \tag{1.56}$$

This new expression is a generalization of the integral expression (1.42) used in finite elements and other techniques. In Eq. (1.42) one approximates the governing equation and natural boundary conditions of the problem. In Eq. (1.56) the essential boundary conditions are approximated as well; hence the latter can be seen as a generalization of all the previous statements. In elastostatics for instance, equations such as Eq. (1.42) can be interpreted as an application of virtual displacements to find the approximate solution (virtual displacements being those which satisfy the homogeneous essential or kinematic boundary conditions but not the natural or mechanical ones). Equations of the type (1.56) can be seen as an application of virtual work, for which the virtual displacements are generalized as those which do not need to satisfy any boundary condition. Unfortunately much confusion still remains regarding these definitions but they are of fundamental importance not only to understand the boundary element method but also many hybrid, equilibrium, or mixed finite element models as well as being able to apply variational techniques in engineering.

We will now consider some examples to illustrate how a boundary solution can be obtained using approximate or weighting functions which satisfy the governing equations.

Let us first consider the case for which the functions for u satisfy the governing equations; in this case $\nabla^2 u \equiv 0$. The problem is governed by the integral equation (1.55) with $b = 0$, i.e.,

$$\int_{\Omega} (\nabla^2 u)\, w \, d\Omega = \int_{\Gamma_2} (q - \bar{q})\, w \, d\Gamma - \int_{\Gamma_1} (u - \bar{u}) \frac{\partial w}{\partial n} \, d\Gamma \,. \tag{1.57}$$

As $\nabla^2 u = 0$, we have to satisfy

$$\int_{\Gamma_2} (q - \bar{q})\, w \, d\Gamma = \int_{\Gamma_1} (u - \bar{u}) \frac{\partial w}{\partial n} \, d\Gamma \,. \tag{1.58}$$

If the same functions are chosen for u and w, one has the method of Trefftz [25]. Note that from Green's theorem,

$$\int_{\Omega} (u \nabla^2 w - w \nabla^2 u) \, d\Omega = \int_{\Gamma} \left(\frac{\partial u}{\partial n} w - u \frac{\partial w}{\partial n} \right) d\Gamma \,, \tag{1.59}$$

where $\Gamma = \Gamma_1 + \Gamma_2$.

When the same functions are chosen for u and w, we have $\nabla^2 u \equiv \nabla^2 w \equiv 0$ and can write $w = \delta u$. We can write Eq. (1.59) as

$$\int_{\Gamma} \frac{\partial u}{\partial n} \delta u \, d\Gamma = \int_{\Gamma} u \frac{\partial \delta u}{\partial n} \, d\Gamma \,. \tag{1.60}$$

Applying boundary conditions we can write Eq. (1.60) as

$$\int_{\Gamma_1} q \, \delta u \, d\Gamma + \int_{\Gamma_2} \bar{q} \, \delta u \, d\Gamma = \int_{\Gamma_1} \bar{u} \frac{\partial \delta u}{\partial n} \, d\Gamma + \int_{\Gamma_2} u \frac{\partial \delta u}{\partial n} \, d\Gamma \,. \tag{1.61}$$

We will illustrate the application of this integral in Examples 1.15 and 1.16 and the case with functions w (but not u) which satisfy the Laplacian operator in Examples 1.17 and 1.18. It is important to point out that for the Trefftz method boundary solutions are possible without using the inverse relationship, but Eq. (1.55) presents an elegant way of introducing the boundary element method.

Example 1.15 (Trefftz Technique). Consider the following Poisson's equation:

$$\nabla^2 u - b = 0 \tag{a}$$

with homogeneous boundary conditions $u = 0$ at $x = \pm 1$ and $y = \pm 1$.

We can first reduce equation (a) to a Laplace equation by defining a new function v such that

$$u = \tfrac{1}{4} b (x^2 + y^2) + v \,. \tag{b}$$

Substituting equation (b) into equation (a) we find that

$$\nabla^2 u - b = \nabla^2 v = 0 \tag{c}$$

with $v = -\tfrac{1}{4} b (x^2 + y^2)$ on the boundaries $x = \pm 1$ and $y = \pm 1$.

We can now approximate v by a trial function that satisfies Laplace's equation. For instance,

$$v = (x^4 - 6 x^2 y^2 + y^4)\, \alpha \,. \tag{d}$$

Because of symmetry we only need to consider a part of the boundary, say $x = 1, 0 < y < 1$.

Notice that for this case the boundary is all of the Γ_1 type. Equation (1.60) becomes

$$\int_{\Gamma_1} \bar{v} \frac{\partial \delta v}{\partial n} d\Gamma = \int_{\Gamma_1} \frac{\partial v}{\partial n} \delta v \, d\Gamma \,. \tag{e}$$

Taking into consideration symmetry this reduces to

$$\int_0^1 \left(\bar{v} \frac{\partial \delta v}{\partial x} - \frac{\partial v}{\partial x} \delta v \right) dy = 0 \,. \tag{f}$$

Substituting equation (d) into equation (f) we obtain the following equation:

$$\int_0^1 [\tfrac{1}{4} b (1 + y^2)(4 - 12 y^2)\, \delta\alpha + \alpha (4 - 12 y^2)(1 - 6 y^2 + y^4) \delta\alpha] \, dy = 0 \,. \tag{g}$$

Hence,

$$\int_0^1 [\tfrac{1}{4} b (1 + y^2) + \alpha (1 - 6 y^2 + y^4)](4 - 12 y^2) \, dy = 0 \,. \tag{h}$$

This gives, after integration,

$$-\frac{1}{5} b + \alpha \frac{144}{35} = 0 \,, \tag{i}$$

$$\therefore \alpha = \frac{7}{144} b. \tag{j}$$

Hence the value of v becomes

$$v = (x^4 - 6x^2y^2 + y^4)\,\alpha = \frac{7}{144} b\,(x^4 - 6x^2y^2 + y^4) \tag{k}$$

and u is given by formula (b). Note that the magnitude of u or v may be displaced from the correct solution by a constant value. In order to determine this value one can carry out the following integration over the $x = 1$, $0 < y < 1$ side:

$$\int_0^1 (\bar{v} - v - c)\, dy = 0 \tag{l}$$

or

$$\int_0^1 [\tfrac{1}{4} b(1 + y^2) + \alpha(1 - 6y^2 + y^4) + c]\, dy = 0\,. \tag{m}$$

Thus,

$$c = -\frac{53}{180} b\,. \tag{n}$$

Example 1.16. Let us now solve the same problem as in Example 1.15 but using boundary collocation and starting with Eq. (1.58) instead of Eq. (1.60). We can take as an approximate solution the same function as in the previous example, i.e.,

$$u = \tfrac{1}{4} b(x^2 + y^2) + \alpha_1 + \alpha_2(x^4 - 6x^2y^2 + y^4) \tag{a}$$

and collocate it over the boundary $x = 1$, $0 < y < 1$. The starting expression is

$$\int_{\Gamma_1} (u - \bar{u}) \frac{\partial \delta u}{\partial n}\, d\Gamma = 0 \tag{b}$$

or

$$\int_0^1 (u - \bar{u}) \frac{\partial \delta u}{\partial x}\, dy = 0 \tag{c}$$

which for two collocation points on the boundary reduces to

$$\int_0^1 (u - \bar{u})\, \Delta_i\, dy = 0\,, \qquad i = 1, 2\,. \tag{d}$$

Let us take the two points on $x = 1$, one at $y = 0.25$ and the other at $y = 0.75$. This will give the following system of equations:

$$\begin{bmatrix} 1 & 0.6289 \\ 1 & -2.0586 \end{bmatrix} \begin{Bmatrix} \alpha_1 \\ \alpha_2 \end{Bmatrix} = -\begin{Bmatrix} 0.2656 \\ 0.3906 \end{Bmatrix} b \tag{e}$$

with

$$\alpha_1 = -\,0.2949\,b\,, \qquad \alpha_2 = 0.0465\,b\,. \tag{f}$$

This gives

$$u = \tfrac{1}{4}\,b\,(x^2 + y^2) - 0.2949\,b + 0.0465\,b\,(x^4 - 6\,x^2\,y^2 + y^4)\,. \tag{g}$$

Equation (g) gives a u value similar to the one obtained in Example 1.15. For instance, if we compare the values of u at the center $(x = y = 0)$ for the two examples, we have

$$\begin{array}{ll} \text{Previous example} & u_c = -\,0.2944\ \text{b}\,, \\ \text{This example} & u_c = -\,0.2949\ \text{b}\,. \end{array} \tag{h}$$

Example 1.17 (Boundary method using $\nabla^2 w = 0$ solution). Consider again the equation

$$\frac{d^2 u_0}{dx^2} + u_0 + x = 0 \qquad \text{in } \Omega \tag{a}$$

with boundary conditions $u_0 = 0$ at $x = 0$ and at $x = 1$.

The statement corresponding to Eq. (1.57) can be written (note all the boundary is of the Γ_1 type) as

$$\int_0^1 \left(\frac{d^2 u}{dx^2} + u + x\right) w\,dx + \left[(u - \bar{u})\,\frac{dw}{dx}\right]_{x=0}^{x=1} = 0\,. \tag{b}$$

Integrating by parts twice we obtain

$$\int_0^1 \left(\frac{d^2 w}{dx^2} + w\right) u\,dx + \int_0^1 x\,w\,dx + [q\,w]_{x=0}^{x=1} - \left[\bar{u}\,\frac{dw}{dx}\right]_0^1 = 0\,. \tag{c}$$

Note that now we need to find a solution that satisfies the equation

$$\frac{d^2 w}{dx^2} + w = 0\,. \tag{d}$$

The general solution is

$$w = \beta_1 \cos x + \beta_2 \sin x \tag{e}$$

with

$$\frac{dw}{dx} = -\,\beta_1 \sin x + \beta_2 \cos x\,. \tag{f}$$

Note that the solution for u and w can be of any form and that these functions are defined only on the boundary. In this particular example the boundary reduces to two points; hence we have as unknowns q_0 and q_1 (i.e., du/dx at 0 and 1).

Consider now equation (c) and substitute into it the weighting function (e). This gives

$$\int_0^1 x\, w\, dx + [q\, w]_1 - [q\, w]_0 = 0\,,$$
$$\int_0^1 x(\beta_1 \cos x + \beta_2 \sin x)\, dx + q_1(\beta_1 \cos 1 + \beta_2 \sin 1) - q_0(\beta_1) = 0\,. \tag{g}$$

This gives the following equations (note that β_1 and β_2 are arbitrary):

$$\int_0^1 x \cos x\, dx = -(q_1 \cos 1 - q_0)\,,$$
$$\int_0^1 x \sin x\, dx = -\, q_1 \sin 1\,, \tag{h}$$

$$q_1 = \frac{\cos 1}{\sin 1} - 1\,, \qquad q_0 = \frac{1}{\sin 1} - 1\,. \tag{i}$$

These are the exact boundary fluxes du/dx at $x = 1$ and du/dx at $x = 0$.

Example 1.18 (Use of the fundamental solution). Considering again the same equation as seen in Example 1.17, we can take as our starting expression

$$\int_0^1 \left(\frac{d^2 w}{dx^2} + w\right) u\, dx + \int_0^1 x\, w\, dx + [q\, w]_{x=0}^{x=1} - \left[\bar{u}\, \frac{dw}{dx}\right]_{x=0}^{x=1} = 0\,. \tag{a}$$

It is usual – although not necessary – in boundary elements to choose the fundamental solution as the weighting function w. This solution will be indicated by an asterisk, i.e., w^*, to stress its special character and is the solution of

$$\frac{d^2 w^*}{dx^2} + w^* = \Delta_i\,, \tag{b}$$

where Δ_i indicates a Dirac delta function which is different from zero at the point i of coordinate ξ (Fig. 1.16) but zero everywhere else. The Dirac delta function is such that

$$\int_0^1 \left(\frac{d^2 w^*}{dx^2} + w^*\right) u\, dx = \int_0^1 \Delta_i\, u\, dx = u_i\,. \tag{c}$$

Hence the first integral in Eq. (a) can be replaced by the value of the function at the point i where the Dirac delta function is applied. Taking into account the

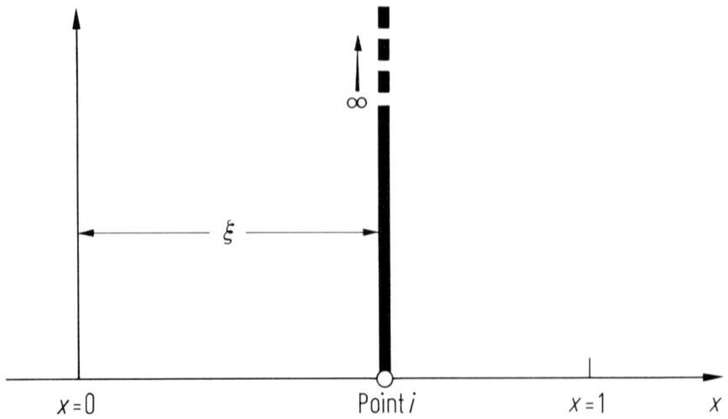

Fig. 1.16. Dirac delta function

boundary condition $\bar{u}=0$, expression (a) becomes

$$u_i = -\int_0^1 x\, w^*\, dx - [q\, w^*]_{x=0}^{x=1} . \tag{d}$$

The function w^* which satisfies equation (b) is

$$w^* = \tfrac{1}{2} \sin r , \tag{e}$$

where $r = |x - \xi|$.

One can now substitute the values of w^* into equation (d) to obtain a system of equations (one at $x=0$, the other at $x=1$) from which the two values of q at $x=0$, $x=1$ can be found. As previously, the results are the exact ones, i.e.,

$$q_0 = \frac{1}{\sin 1} - 1 , \qquad q_1 = \frac{\cos 1}{\sin 1} - 1 . \tag{f}$$

Note that equation (d) can be used to calculate the values of the u function at any internal point. If we choose ξ to be at the midpoint of our internal domain we obtain the value of u at $x = \frac{1}{2}$ as

$$u(\tfrac{1}{2}) = -\tfrac{1}{2} \int_0^{1/2} x \sin(\tfrac{1}{2} - x)\, dx - \tfrac{1}{2} \int_{1/2}^{1} x \sin(x - \tfrac{1}{2})\, dx - q_1 \sin \tfrac{1}{2} + q_0 \sin \tfrac{1}{2} , \tag{g}$$

in which

$$r = \begin{cases} \tfrac{1}{2} - x & \text{for } 0 < x < \tfrac{1}{2} , \\ x - \tfrac{1}{2} & \text{for } \tfrac{1}{2} < x < 1 . \end{cases}$$

Thus we obtain

$$\begin{aligned} u(\tfrac{1}{2}) &= -\tfrac{1}{2}[(\cos 1 - 1)/\sin 1] \sin \tfrac{1}{2} - \tfrac{1}{2}(\tfrac{1}{2} - \sin \tfrac{1}{2}) - \tfrac{1}{2}(\tfrac{1}{2} + \sin \tfrac{1}{2} - \cos \tfrac{1}{2}) \\ &= 0.069746964 \end{aligned} \tag{h}$$

which is the same as the exact solution, i.e.,

$$u\left(\tfrac{1}{2}\right) = \frac{\sin \frac{1}{2}}{\sin 1} - \tfrac{1}{2} = 0.069746964\,. \qquad \text{(i)}$$

1.8. Classification of Approximate Methods

Notice that the weighted residual statements written for the Poisson equation for instance can be classified as follows:

(i) Original statement

$$\int_{\Omega} (\nabla^2 u - b)\, w\, d\Omega = \int_{\Gamma_2} (q - \bar{q})\, w\, d\Gamma - \int_{\Gamma_1} (u - \bar{u}) \frac{\partial w}{\partial n}\, d\Gamma\,; \qquad (1.62)$$

(ii) Weak statement

$$\int_{\Omega} \frac{\partial u}{\partial x_k} \frac{\partial w}{\partial x_k}\, d\Omega + \int_{\Omega} b\, w\, d\Omega = \int_{\Gamma_2} \bar{q}\, w\, d\Gamma + \int_{\Gamma_1} q\, w\, d\Gamma + \int_{\Gamma_1} (u - \bar{u}) \frac{\partial w}{\partial n}\, d\Gamma\,; \qquad (1.63)$$

(iii) Inverse statement

$$\int_{\Omega} (\nabla^2 w)\, u\, d\Omega - \int_{\Omega} b\, w\, d\Omega = -\int_{\Gamma_2} \bar{q}\, \mathrm{w}\, \mathrm{d}\Gamma - \int_{\Gamma_1} q\, w\, d\Gamma + \int_{\Gamma_2} u \frac{\partial w}{\partial n}\, d\Gamma + \int_{\Gamma_1} \bar{u} \frac{\partial w}{\partial n}\, d\Gamma\,. \qquad (1.64)$$

This classification can be extended to other self-adjoint and even non-self-adjoint operators and allows us to differentiate between alternative approximate methods.

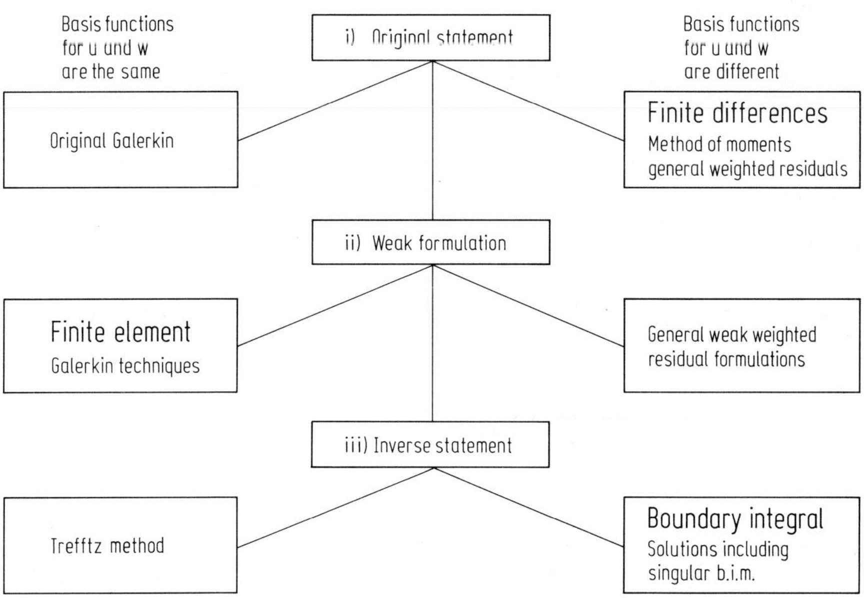

Fig. 1.17. Classification of different approximate techniques

Another essential difference between techniques refers to the type of basis function used for the approximation u and for the weighting w. We can divide the numerical methods according to those for which the same basis functions are used for u and w and those for which they are different.

A rational classification is attempted in Fig. 1.17 where we see that the main engineering methods can be differentiated as follows:

(1) **Finite differences.** Here one normally has different basis functions for u and w, the latter being taken in the form of Dirac delta functions (Example 1.7). Most finite difference schemes are based on statement (i) although some (like energy schemes) use statement (ii).
(2) **Finite elements.** It is usual in this technique to take the same basis functions for u and w to obtain symmetric matrices. Finite element schemes are based on weak formulations (type (ii)).
(3) **Boundary elements.** Boundary element schemes are generally based on the inverse relationship (iii) as shown in Examples 1.17 and 1.18. For the weighting function w, they use a set of basis functions which eliminates the domain integrals and reduces the problem to a boundary-only problem. These functions can be singular functions produced by applying a Dirac delta at a particular point or can be regular as in the case of those resulting from the homogeneous equation solution.

Other numerical techniques such as moments, the original Galerkin method (i.e., without weak formulation), and the Trefftz method (discussed in Example 1.15) can be easily fitted within the diagram shown in Fig. 1.17. Other techniques such as finite difference "energy" techniques, "upwind" finite elements such as those used in convective problems, mixed finite element formulations, etc., can also be classified as above.

References

1. Brebbia, C. A., and Connor, J. J., *Fundamentals of Finite Element Techniques for Structural Engineers*, Butterworths, London, 1973.
2. Connor, J. J., and Brebbia, C. A., *Finite Element Techniques for Fluid Flow*, Butterworths, London, 1976.
3. Kantorovich, L. V., and Krylov, V. I., *Approximate Methods of Higher Analysis*, Noordhoff, Groningen, 1958.
4. Courant, R., and Hilbert, D., *Methods of Mathematical Physics*, Interscience, New York, 1953.
5. Reissner, E., A note on variational principles in elasticity, Int. J. Solids Structures **1**, 93–95 (1965).
6. Pian, T. H. H., and Tong, P., Basis of finite element method for solid continua, Int. J. Numerical Methods Engng. **1**, 3–28 (1969).
7. Washizu, K., *Variational Methods in Elasticity and Plasticity*, 2nd ed., Pergamon, New York, 1975.
8. Muskhelishvili, N. I., *Some Basic Problems of the Mathematical Theory of Elasticity*, Noordhoff, Groningen, 1953.
9. Mikhlin, S. G., *Integral Equations*, Pergamon, New York, 1957.
10. Kupradze, O. D., *Potential Methods in the Theory of Elasticity*, Daniel Davey & Co., New York, 1965.
11. Smirnov, V. J., Integral equations and partial differential equations, in *A Course in Higher Mathematics*, Vol. IV, Addison-Wesley, London, 1964.

12. Kellogg, O. D., *Foundations of Potential Theory,* Dover, New York, 1953.
13. Jaswon, M. A., Integral equation methods in potential theory, I, Proc. Roy. Soc. Ser. A **275,** 23–32 (1963).
14. Symm, G. T., Integral equation methods in potential theory, II, Proc. Roy. Soc. Ser. A **275,** 33–46 (1963).
15. Massonnet, C. E., Numerical Use of Integral Procedures, in *Stress Analysis* (O. C. Zienkiewicz and G. S. Holister, Eds.), Wiley, London, 1966.
16. Hess, J. L., and Smith, A. M. O., Calculation of potential flow about arbitrary bodies, Progress in Aeronautical Sciences Vol. 8 (D. Küchemann, Ed.), Pergamon, London, 1967.
17. Cruse, T. A., and Rizzo, F. J., A direct formulation and numerical solution of the general transient elasto-dynamic problem, I, J. Math. Anal. Appl. **22,** 244–259 (1968).
18. Brebbia, C. A., *The Boundary Element Method for Engineers,* Pentech Press, London; Halstead Press, New York, 1978.
19. Brebbia, C. A., and Walker, S., *Boundary Element Techniques in Engineering,* Newnes-Butterworths, London, 1980.
20. Brebbia, C. A. (Ed.), *Recent Advances in Boundary Element Methods,* Proc. 1st Int. Conference Boundary Element Methods, Southampton University, 1978, Pentech Press, London, 1978.
21. Brebbia, C. A. (Ed.), *New Developments in Boundary Element Methods,* Proc. 2nd Int. Conference Boundary Element Methods, Southampton University, 1980, CML Publications, Southampton and Butterworths, London, 1980.
22. Brebbia, C. A. (Ed.), *Boundary Element Methods,* Proc. 3rd Int. Conf. Boundary Element Methods, Irvine, California, 1981; Springer-Verlag, Berlin, 1981.
23. Brebbia, C. A. (Ed.), *Boundary Element Methods in Engineering,* Proc. 4th Int. Conference Boundary Element Methods, Southampton University, 1982, Springer Verlag, Berlin, 1982.
24. Lau, P., and Brebbia, C. A., The cell collocation method, Int. J. Mech. Sci. **20,** 83–95 (1978).
25. Trefftz, E., Ein Gegenstück zum Ritzschen Verfahren, Proc. 2nd Int. Congress Appl. Mech., Zurich, 1926.

Bibliography

Biezeno, C. B., and Koch, J. J., Over een nieuwe methode ter brerkening van vlokke platen met toepassing op. Eukele Voor de Technick Belangrijke Belastingsgevallen, Ing. Grav. **38,** 25–36 (1923).

Biezeno, C. B., and Grammel, R., *Engineering Dynamics,* Vol. 1, *Theory of Elasticity,* Boekie & Son, London, 1955.

Brebbia, C. A., and Tottenham, H. (Eds.), *Variational Methods in Engineering* (2 vols), Southampton University Press, England, 1973; reprinted 1975.

Brebbia, C. A., and Dominguez, J., The boundary element method for potential problems, Appl. Math. Modelling, **1,** 372–378 (1977).

Brebbia, C. A., Fundamentals of boundary elements, in *New Developments in Boundary Element Methods,* Butterworths, London, 1980.

Brebbia, C. A., Basis of boundary elements, in *Progress in Boundary Elements,* Vol. 1, Pentech Press, London; Wiley, New York, 1981.

Collatz, L., *The Numerical Treatment of Differential Equations,* 3rd Ed., Springer-Verlag, Berlin, 1967.

Finlayson, B. A., *The Method of Weighted Residuals and Variational Principles,* Academic Press, New York, 1972.

Galerkin, B., Contribution a la solution generale du probleme de la theorie de l'elasticite dans le cas de trois dimensions, Comptes Rendus **190,** 1047–1048 (1930). Comptes Rendus (Dokl.) Acad. Sci. URSS Ser. A **14,** 353 (1930).

Hellinger, E., Der Allgemeine Ansatz der Mechanik der Kontinus, in *Encyclopädie der Mathematischen Wissenschaften*, Vol. 4, Part 4, 1974.

Hildebrand, F. B., *Methods of Applied Mathematics*, Prentice-Hall, Englewood Cliffs, N.J., 1965.

Hutton, S. G., and Anderson, D. L., Finite element methods: A Galerkin approach, J. Engng. Mech. Div. ASCE **7**, 1503–1520 (1971).

Kantorovich, L. V., and Krylov, V. I., *Approximate Methods of Higher Analysis*, Noordhoff, Groningen, 1958.

Lanczos, C., *The Variational Principles of Mechanics*, University of Toronto Press, Toronto, 1949.

Mikhlin, S. G., *Variational Methods in Mathematical Physics*, MacMillan, New York, 1964.

Morse, P., and Feshbach, H., *Methods of Theoretical Physics*, Parts I and II, McGraw-Hill, New York, 1953.

Oliveira, E. R. A., Theoretical foundation of the finite element method, Int. J. Solids Structures, **4**, 929–952 (1968).

Reissner, E., On variational principles in elasticity, in *Proc. of Symposium on Applied Mathematics*, Vol. 8, 1–6, McGraw-Hill, New York, 1958.

Wendland, W. L., Asymptotic accuracy and convergence, in *Progress in Boundary Elements*, Vol. 1, Pentech Press, London; Wiley, New York, 1981.

Chapter 2

Potential Problems

2.1. Introduction

The boundary element method is now firmly established as an important alternative technique to the prevailing numerical methods of analysis in continuum mechanics. One of the most important applications is for the solution of a range of problems such as temperature diffusion, some types of fluid flow motion, flow in porous media, electrostatics, and many others which can be written as a function of a potential and whose governing equation is the classical Laplace or Poisson equation. All the cases are potential problems and they can generally be efficiently and economically analyzed using boundary elements.

The technique consists in the transformation of the partial differential equation describing the behavior of the unknown inside and on the boundary of the domain into an integral equation relating only boundary values, and then finding out the numerical solution of this equation. If values at internal points are required, they are calculated afterwards from the boundary data. Since all numerical approximations take place only at the boundaries, the dimensionality of the problem is reduced by one and a smaller system of equations obtained in comparison with those achieved through differential methods.

The present chapter is concerned with the application of the boundary element method for the solution of steady potential flow problems. By steady potential problems we mean those governed by the Laplace (or Poisson) equation. Two-dimensional, axisymmetric, and fully three-dimensional problems are considered.

Historically, the application of integral equations to formulate the fundamental boundary-value problems of potential theory dates back to 1903 when Fredholm [1] demonstrated the existence of solutions to such equations, on the basis of a discretization procedure. Due to the difficulty of finding analytical solutions, the use of integral equations has, to a great extent, been limited to theoretical investigations of existence and uniqueness to solutions of problems of mathematical physics. However the advent of high-speed computers made it possible to implement discretization procedures analytically and enabled numerical solutions to be readily achieved.

Fredholm integral equations follow from the representation of harmonic potentials by single-layer or double-layer potentials and set up the foundations of the so-called indirect boundary element method. Vector integral equations analogous to the Fredholm integral equations of potential theory were introduced by Kupradze [2] in the context of the theory of elasticity.

Integral equations for linear problems can alternatively be formulated through the application of Green's third identity [3], which represents a harmonic function as the superposition of a single-layer and a double-layer potential. Taking the field point to the boundary, an integral equation relating only boundary values and normal derivatives of the harmonic function is obtained. Its counterpart in elasticity is Somigliana's identity [4], and its use gave rise to the direct boundary element method. More recently, it has been demonstrated that the same integral relationships can be obtained through weighted residual considerations [5] as shown in Chapter 1. In this way, it became easier to relate and combine the boundary element method with other numerical techniques, such as the finite element method, as well as to extend it for the analysis of problems governed by more complex partial differential equations, including nonlinearities. In what follows the weighted residual formulation will be used in preference to the others as it is more general and more in line with the approximate methods familiar to engineers and other analysts.

Although integral equations have been extensively employed to formulate boundary-value problems of potential theory, analytical solutions to such equations are limited to very simple geometries and are obtained using the Green's function for the geometry which satisfies the prescribed boundary conditions of the problem [6–8]. The Green's function method of solving boundary-value problems is most directly applicable to elliptic partial differential equations. In fact, the concept of Green's function grew out of a detailed study of such boundary-value problems, but the method can also be extended to solve parabolic and hyperbolic partial differential equations as shown in Chapters 4 and 10. For general problems with complex geometry and boundary conditions, however, it may be assumed that no exact Green's function, or any other analytical treatment, is available.

In 1963, Jaswon [9] and Symm [10] presented a numerical technique to solve Fredholm boundary integral equations. The technique consists of discretizing the boundary into a series of small segments (elements), assuming that the source density remains constant within each segment. By using the method of collocation, the discretized equation is applied to a number of particular points (nodes) in each element, and the influence coefficients are computed approximately using Simpson's rule. Exception is made for the singular coefficients resulting from the self-influence of each element, which are computed either analytically (for Dirichlet problems or problems for which all conditions are essential) or by the summation of the off-diagonal coefficients plus the free term (for Neumann problems or problems for which all conditions are natural). This produces a system of linear algebraic equations which can be solved computationally by the Gauss elimination method.

Applying such techniques, Jaswon and Symm obtained accurate solutions for simple two-dimensional Neumann and Dirichlet problems. They also proposed a more general numerical formulation for solving Cauchy boundary-value problems (also called mixed because the boundary conditions are essential on parts of the boundary and natural on other parts) through the application of Green's third identity, which yields a boundary integral equation where boundary values and normal derivatives of the physical variable play the role of the fictitious source densities. Results using this formulation are reported by Symm [10] and Jaswon and Ponter [11].

Hess and Smith [12] developed a parallel work for the solution of Neumann-type boundary-value problems, more specifically, the problem of potential flow around arbitrary bodies, Applying basically the same (indirect) technique they computed the quantities of interest (potential and velocity) from the source density distribution by using direct quadratures of the corresponding equations. They extended the method to analyze a variety of body shapes: two-dimensional, axisymmetric and fully three-dimensional results were presented. The influence coefficients were all computed analytically for the two- and three-dimensional cases although in the latter, in order to improve the computer efficiency, multipole expansions were employed to calculate the influence of elements located far from the actual node and the system of equations was solved iteratively by the Gauss–Seidel method. For axisymmetric problems, the influence coefficients were computed numerically using Simpson's rule but the number of subelements was scaled in such a way that the farther the element lies from the actual node, the fewer the number of subelements used in the calculation. The singular (self-influence) coefficients were computed analytically by means of series expansions.

Harrington *et al.* [13] applied the technique to solve some two-dimensional electrical engineering problems with the more general impedance boundary condition, which is of the Robin type, i.e., it prescribes a linear relation between the potential and its normal derivative. They also proposed a piecewise variation for the source density. Mautz and Harrington [14] solved axisymmetric electrical engineering problems with Dirichlet boundary conditions, again employing the indirect formulation and assuming the source density to remain constant within each element. Some of their numerical considerations were later discussed by Jaswon and Symm [15].

The present chapter starts by showing how a problem governed by Laplace's equation can be recast into an integral equation which, through a limiting process, produces a boundary integral equation relating only boundary values. Both the indirect and the direct formulations of the boundary element method are discussed. The weighted residual technique is then employed to formulate integral equations. It is shown how several features such as internal sources, nonhomogeneity, orthotropy and anisotropy can be included in the formulation. Two-dimensional, axisymmetric, and fully three-dimensional problems are treated and results are presented.

2.2. Elements of Potential Theory

Some basic elements of classical potential theory will now be briefly reviewed. We shall introduce only the concepts that are of importance to our subject matter and in doing so we follow Cruse [16] and Jaswon and Symm [15]. For a more formal mathematical treatment, including all necessary and relevant rigorous proofs, see for instance Kellogg [3], Courant and Hilbert [17], and Sternberg and Smith [18].

If a particle of unit mass, subjected only to the force of a specific field F, is moved from a point ξ to a point x in space, the work done on the particle by the field during the motion is given by

$$W = \int_{\xi}^{x} \boldsymbol{F}\, d\boldsymbol{r}\,, \tag{2.1}$$

where $\boldsymbol{F}$ is the force field vector and $d\boldsymbol{r}$ is the differential motion of the particle on the path from ξ to x.

The work is, in general, dependent not merely on the position of the points, but also on the path of the particle between them. If the field is such that the work is independent of the path, i.e., it has the same value when taken over any two paths connecting ξ and x which can be continuously deformed one into the other, the field is called conservative.

Considering the point ξ as fixed and x as variable, the integral (2.1) represents a function of x alone. This scalar function,

$$u(x) = \int_{\xi}^{x} \boldsymbol{F}\, d\boldsymbol{r}\,, \tag{2.2}$$

is called the potential of the field F.

When the field is gravitational, the potential is a Newtonian one. The Newtonian potential generated by two particles of masses m_1 and m_2, located at points ξ (fixed) and x (variable), respectively, is of the form

$$u(x) = \int_{\xi}^{x} G\, m_1\, m_2\, \boldsymbol{\nabla}\left(\frac{1}{r}\right) d\boldsymbol{r} = G\, m_1\, m_2 \frac{1}{r} + \text{constant}\,, \tag{2.3}$$

where G is the gravitational constant and r is the distance between ξ and x, that is,

$$\begin{aligned} r(\xi, x) &= |\xi - x| \\ &= \{[x_1(\xi) - x_1(x)]^2 + [x_2(\xi) - x_2(x)]^2 + [x_3(\xi) - x_3(x)]^2\}^{1/2}, \end{aligned} \tag{2.4}$$

where x_i are the coordinates of the point under consideration.

Attractional forces of the same character as those occurring in gravitation also act between electric charges, and between the poles of magnets. For generality, we will then refer to sources rather than masses throughout this work and state that a unit simple source, located at a source point ξ in space (Fig. 2.1), generates at a field point x the Newtonian potential

$$\frac{1}{r(\xi, x)}\,. \tag{2.5}$$

This potential is a continuous function of x, differentiable to all orders, everywhere except at the source point ξ.

Similarly, a discrete distribution of simple sources of intensities $\sigma_1, \sigma_2, \ldots, \sigma_N$ located at points $\xi_1, \xi_2, \ldots, \xi_N$, respectively, generates the Newtonian potential

$$u(x) = \sum_{n=1}^{N} \sigma(\xi_n) \frac{1}{r(\xi_n, x)} \tag{2.6}$$

at point x. Again, this potential is a continuous function of x, together with its derivatives of all orders, everywhere except when x is coincident with one of the source points ξ_n.

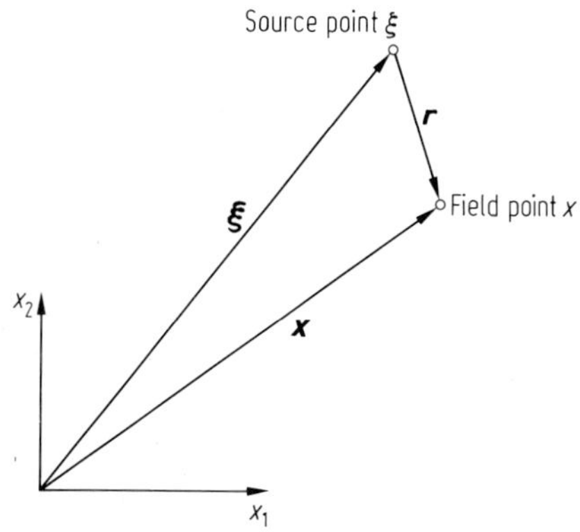

Fig. 2.1. Notation for source and field points

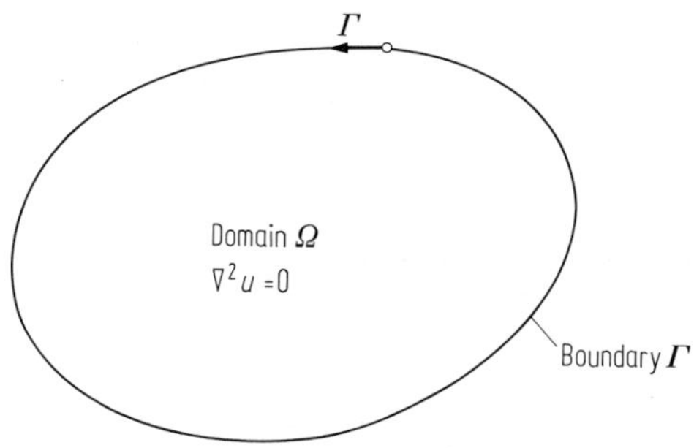

Fig. 2.2. Notation for potential problems

Now consider a continuous distribution of simple sources of volume density ϱ throughout the domain Ω (Fig. 2.2). The potential associated with this force field is a volume potential, obtained by the integration

$$u(x) = \int_{\Omega} \varrho(\xi) \frac{1}{r(\xi, x)} \, d\Omega(\xi) . \tag{2.7}$$

This volume potential is a continuous function of x, differentiable to all orders, at all points of free space, that is, points located outside the attracting domain Ω. When the field point x lies inside the domain Ω, the integrand in Eq. (2.7) contains a singularity. However, if the density ϱ is bounded throughout Ω, the potential $u(x)$ exists at all points $x \in \Omega$ and is everywhere continuous and differentiable throughout space [3]. This amounts to saying that the derivatives of the first order of u may be obtained by differentiating under the sign of integration as

$$\frac{\partial u(x)}{\partial x_i(x)} = \int_{\Omega} \varrho(\xi) \frac{\partial}{\partial x_i(x)} \left(\frac{1}{r(\xi, x)} \right) d\Omega(\xi) . \tag{2.8}$$

The same is not valid for the derivatives of second order. In fact, the mere continuity of the density does not suffice to ensure the existence of these derivatives. Therefore, it is necessary to impose that the density $\varrho(\xi)$ satisfies a Hölder condition [3] at x,

$$|\varrho(\xi) - \varrho(x)| \leqq A \, r(\xi, x)^{\alpha} , \tag{2.9}$$

where A and α are positive constants.

In order to investigate the partial derivatives of u of second order, we can start by integrating Eq. (2.8) by parts, obtaining

$$\frac{\partial u(x)}{\partial x_i(x)} = -\int_\Gamma \varrho(\xi) \frac{1}{r(\xi, x)} n_i(\xi)\, d\Gamma(\xi) + \int_\Omega \frac{\partial \varrho(\xi)}{\partial x_i(\xi)} \frac{1}{r(\xi, x)}\, d\Omega(\xi) \tag{2.10}$$

through the application of the divergence theorem, and noting that

$$\frac{\partial}{\partial x_i(x)}\left(\frac{1}{r(\xi, x)}\right) = -\frac{\partial}{\partial x_i(\xi)}\left(\frac{1}{r(\xi, x)}\right). \tag{2.11}$$

Making use of the identity

$$\frac{\partial}{\partial x_i(\xi)} \varrho(\xi) = \frac{\partial}{\partial x_i(\xi)} [\varrho(\xi) - \varrho(x)] \tag{2.12}$$

and taking the second derivative of Eq. (2.10) yields the relation

$$\frac{\partial^2 u(x)}{\partial x_i^2(x)} = -\int_\Gamma \varrho(\xi) \frac{\partial}{\partial x_i(x)}\left(\frac{1}{r(\xi, x)}\right) n_i(\xi)\, d\Gamma(\xi)$$

$$+ \int_\Omega \frac{\partial}{\partial x_i(\xi)} [\varrho(\xi) - \varrho(x)] \frac{\partial}{\partial x_i(x)}\left(\frac{1}{r(\xi, x)}\right) d\Omega(\xi). \tag{2.13}$$

The second integral in Eq. (2.13) may be integrated by parts with respect to $x_i(\xi)$ and subjected to the divergence theorem once more to obtain

$$\frac{\partial^2 u(x)}{\partial x_i^2(x)} = -\int_\Gamma \varrho(\xi) \frac{\partial}{\partial x_i(x)}\left(\frac{1}{r(\xi, x)}\right) n_i(\xi)\, d\Gamma(\xi)$$

$$+ \int_\Gamma [\varrho(\xi) - \varrho(x)] \frac{\partial}{\partial x_i(x)}\left(\frac{1}{r(\xi, x)}\right) n_i(\xi)\, d\Gamma(\xi)$$

$$+ \int_\Omega [\varrho(\xi) - \varrho(x)] \frac{\partial^2}{\partial x_i^2(x)}\left(\frac{1}{r(\xi, x)}\right) d\Omega(\xi), \tag{2.14}$$

which reduces to

$$\frac{\partial^2 u(x)}{\partial x_i^2(x)} = \varrho(x) \int_\Gamma \frac{\partial}{\partial x_i(\xi)}\left(\frac{1}{r(\xi, x)}\right) n_i(\xi)\, d\Gamma(\xi)$$

$$+ \int_\Omega [\varrho(\xi) - \varrho(x)] \frac{\partial^2}{\partial x_i^2(x)}\left(\frac{1}{r(\xi, x)}\right) d\Omega(\xi). \tag{2.15}$$

Thus, adding up the three second-order derivatives (Eq. (2.15) for $i = 1, 2, 3$) yields the Laplacian of u,

$$\nabla^2 u(x) = \varrho(x) \int_\Gamma \frac{\partial}{\partial n(\xi)} \left(\frac{1}{r(\xi, x)} \right) d\Gamma(\xi)$$

$$+ \int_\Omega [\varrho(\xi) - \varrho(x)] \nabla^2 \left(\frac{1}{r(\xi, x)} \right) d\Omega(\xi) . \tag{2.16}$$

In the second integral in Eq. (2.16), the volume may be divided into two parts: one is a small sphere of radius ε surrounding the point x; the other is the entirety of the remaining volume, denoted by Ω_ε. As x is exterior to Ω_ε and $1/r$ is a harmonic function (as will be demonstrated later), the Laplacian term equals zero throughout this domain. The integral over the sphere also approaches zero with ε since $\varrho(\xi)$ satisfies a Hölder condition at x. Thus there remains the surface integral to be evaluated (Fig. 2.3).

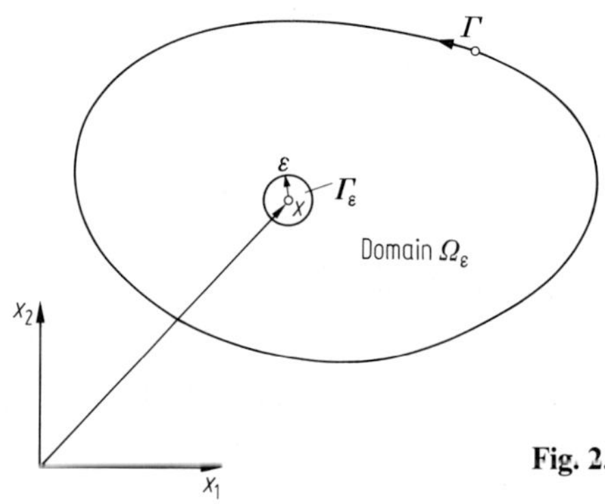

Fig. 2.3. Integration around a sphere

Again, consider a small sphere of radius ε around x, with surface Γ_ε. Integrating Eq. (2.16) around this surface gives

$$\int_{\Gamma_\varepsilon} \frac{\partial}{\partial n(\xi)} \left(\frac{1}{r(\xi, x)} \right) d\Gamma(\xi) = -\frac{1}{\varepsilon^2} \int_{\Gamma_\varepsilon} d\Gamma = -4\pi . \tag{2.17}$$

Since there are no sources in the domain between Γ_ε and Γ, the Newtonian field is solenoidal, i.e., there is no flux out of this region. Numerically, we can write

$$\int_{\Gamma_\varepsilon} \frac{\partial}{\partial n} \left(\frac{1}{r} \right) d\Gamma_\varepsilon + \int_\Gamma \frac{\partial}{\partial n} \left(\frac{1}{r} \right) d\Gamma = 0 \tag{2.18}$$

where the normal is outward on Γ, but inward on Γ_ε. Combining Eq. (2.17) and (2.18) and noticing the reversal of the normal on Γ_ε gives

$$\int_\Gamma \frac{\partial}{\partial n(\xi)} \left(\frac{1}{r(\xi, x)} \right) d\Gamma(\xi) = -4\pi . \tag{2.19}$$

Inserting Eq. (2.19) into Eq. (2.16) produces the result

$$\nabla^2 u(x) = -4\pi \varrho(x). \tag{2.20}$$

Other Newtonian potentials can be generated, including surface potentials. In particular, two of them are of importance to what follows and will be defined next. The first is the potential associated with a continuous distribution of simple sources extending over a surface Γ and of surface density σ, which is of the form

$$u(x) = \int_\Gamma \sigma(\xi) \frac{1}{r(\xi, x)} d\Gamma(\xi) \tag{2.21}$$

and is called a *single-layer potential.*

Let us now consider two surfaces, $\Gamma(\xi)$ and $\Gamma(\xi')$, separated by a small distance $h(\xi, \xi')$, carrying distributions of attraction of magnitude $\sigma(\xi)$ and $\sigma(\xi')$, respectively. These distributions are such that, for corresponding area elements,

$$\sigma(\xi)\, d\Gamma(\xi) = -\sigma(\xi')\, d\Gamma(\xi'). \tag{2.22}$$

The potential due to the two surfaces is then

$$\begin{aligned} u(x) &= \int_\Gamma \sigma(\xi) \frac{1}{r(\xi, x)} d\Gamma(\xi) + \int_{\Gamma'} \sigma(\xi') \frac{1}{r(\xi', x)} d\Gamma(\xi'). \\ &= \int_\Gamma \sigma(\xi)\, h(\xi, \xi') \left[\frac{1}{h(\xi, \xi')} \left\{\frac{1}{r(\xi, x)} - \frac{1}{r(\xi', x)}\right\}\right] d\Gamma(\xi). \end{aligned} \tag{2.23}$$

If we let $h \to 0$ and $\sigma \to \infty$ so that $\sigma h \to \mu$ everywhere uniformly on Γ and also compute the limit of the term in brackets,

$$\lim_{h(\xi, \xi') \to 0} \left[\frac{1}{h(\xi, \xi')} \left\{\frac{1}{r(\xi, x)} - \frac{1}{r(\xi', x)}\right\}\right] = \frac{\partial}{\partial n(\xi)} \left(\frac{1}{r(\xi, x)}\right), \tag{2.24}$$

the potential

$$u(x) = \int_\Gamma \mu(\xi) \frac{\partial}{\partial n(\xi)} \left(\frac{1}{r(\xi, x)}\right) d\Gamma(\xi) \tag{2.25}$$

obtained as the limit of the potential of two single layers of opposite signs that approach coincidence is called a *double-layer potential.* The function μ is the surface density, or moment, of the double layer.

The potentials in Eqs. (2.21) and (2.25) are continuous functions of x, differentiable to all orders everywhere except at $x \in \Gamma$, where the integrands in these equations contain singularities. In order to investigate the behavior of these surface potentials near the singularity, the boundary Γ is broken into two surfaces: one is a small disc tangent to the surface at a point x, noting that it was assumed that the surface possesses a unique tangent plane at any point; the other is the entirety of the remaining surface and contains no singularity, as $x \neq \xi$.

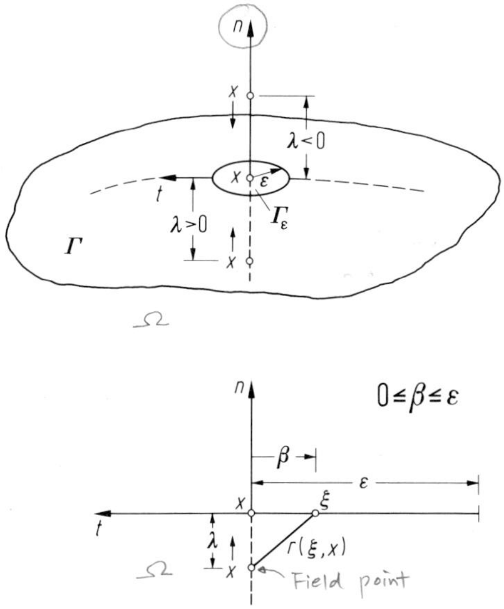

Fig. 2.4. Discontinuity of three-dimensional double-layer potential

According to Fig. 2.4, the field point x, originally located inside Ω, moves along the normal to Γ until it is on the Γ surface. The disc centered at x has a radius ε and is denoted Γ_ε; the remainder of the surface is denoted $\Gamma-\Gamma_\varepsilon$. The point x is originally located at a distance λ from the surface such that $|\lambda| \ll \varepsilon$ and such that $\lambda < 0$ if x is originally outside Ω and $\lambda > 0$ if x is inside Ω. The integrals in Eqs. (2.21) and (2.25) can then be separated as

$$u(x) = \lim_{\varepsilon \to 0} \left\{ \int_{\Gamma-\Gamma_\varepsilon} \sigma(\xi) \frac{1}{r(\xi, x)} d\Gamma(\xi) + \int_{\Gamma_\varepsilon} \sigma(\xi) \frac{1}{r(\xi, x)} d\Gamma(\xi) \right\}, \tag{2.26}$$

$$u(x) = \lim_{\varepsilon \to 0} \left\{ \int_{\Gamma-\Gamma_\varepsilon} \mu(\xi) \frac{\partial}{\partial n(\xi)} \left(\frac{1}{r(\xi, x)} \right) d\Gamma(\xi) + \int_{\Gamma_\varepsilon} \mu(\xi) \frac{\partial}{\partial n(\xi)} \left(\frac{1}{r(\xi, x)} \right) d\Gamma(\xi) \right\}. \tag{2.27}$$

It is clear that the integrals over $\Gamma-\Gamma_\varepsilon$ are continuous as the field point x passes through the surface and will again produce the integrals in Eqs. (2.21) and (2.25) when the limit is taken.

The integral over Γ_ε in Eq. (2.26) contains a weak singularity and is also continuous as the field point passes through the surface, provided the density σ is bounded at all points along Γ. This statement does not hold for the second integral in Eq. (2.27) which, because of the normal derivative term, contains a singularity of higher order. This integral can be written

$$\int_{\Gamma_\varepsilon} \mu(\xi) \frac{\partial}{\partial n(\xi)} \left(\frac{1}{r(\xi, x)} \right) d\Gamma(\xi) = \int_{\Gamma_\varepsilon} [\mu(\xi) - \mu(x)] \frac{\partial}{\partial n(\xi)} \left(\frac{1}{r(\xi, x)} \right) d\Gamma(\xi) + \mu(x) \int_{\Gamma_\varepsilon} \frac{\partial}{\partial n(\xi)} \left(\frac{1}{r(\xi, x)} \right) d\Gamma(\xi). \tag{2.28}$$

constant w.r.t. ξ

This means that the potential of a surface whose density is continuous at x is the sum of the potentials of a surface whose density vanishes at x and of a surface with constant density, equal to that at x. If the density $\mu(\xi)$ satisfies a Hölder condition at x, then the first integral on the right-hand side of Eq. (2.28) is continuous as the field point passes through the surface. The second integral becomes (refer to Fig. 2.4 for notation)

$$\int_{\Gamma_\varepsilon} \frac{\partial}{\partial n}\left(\frac{1}{r}\right) d\Gamma = -\int_0^\varepsilon \frac{\lambda}{r^3}\, 2\pi\beta\, d\beta . \tag{2.29}$$

Since $\beta\, d\beta = r\, dr$ for a given $|\lambda| \ll \varepsilon$, an interchange of variables produces

$$-\int_0^\varepsilon \frac{\lambda}{r^3}\, 2\pi\beta\, d\beta = 2\pi\lambda \int_{|\lambda|}^{\varepsilon} -\frac{dr}{r^2} = 2\pi \left[\frac{\lambda}{r}\right]_{|\lambda|}^{\varepsilon} = 2\pi\frac{\lambda}{\varepsilon} - 2\pi\,\mathrm{sgn}(\lambda), \tag{2.30}$$

where $\mathrm{sgn}(\lambda)$ takes the sign of λ.

Taking the limit as $\varepsilon \to 0$ (noting that $\lambda \to 0$ much quicker) (2.28) gives

$$\lim_{\varepsilon \to 0} \left\{ \int_{\Gamma_\varepsilon} \mu(\xi) \frac{\partial}{\partial n(\xi)} \left(\frac{1}{r(\xi, x)}\right) d\Gamma(\xi) \right\} = -2\pi\,\mathrm{sgn}(\lambda)\,\mu(x). \tag{2.31}$$

Thus the limiting form of Eq. (2.25) as x goes to the surface Γ from the inside can be written as

$$u^+(x) = -2\pi\mu(x) + \int_\Gamma \mu(\xi) \frac{\partial}{\partial n(\xi)} \left(\frac{1}{r(\xi, x)}\right) d\Gamma(\xi) \tag{2.32}$$

and from the outside as

$$u^-(x) = 2\pi\mu(x) + \int_\Gamma \mu(\xi) \frac{\partial}{\partial n(\xi)} \left(\frac{1}{r(\xi, x)}\right) d\Gamma(\xi). \tag{2.33}$$

The three-dimensional double-layer potential is then said to have a discontinuity or jump of $-4\pi\mu(x)$ as the point x passes from outside to inside the region, that is,

$$u^+(x) - u^-(x) = -4\pi\mu(x). \tag{2.34}$$

All concepts presented thus far are also valid for two-dimensional problems, where the equivalent of the Newtonian potential is the logarithmic potential

$$\ln \frac{1}{r(\xi, x)} \tag{2.35}$$

in which $r(\xi, x)$ is now

$$r(\xi, x) = |\xi - x| = \{[x_1(\xi) - x_1(x)]^2 + [x_2(\xi) - x_2(x)]^2\}^{1/2}. \tag{2.36}$$

The logarithmic potential can be derived either by starting with two-dimensional force fields acting on a line source or by integrating the Newtonian potential for a line source at ξ [3, 17, 18].

The two-dimensional volume potential

$$u(x) = \int_{\Omega} \varrho(\xi) \ln \frac{1}{r(\xi, x)} d\Omega(\xi) \tag{2.37}$$

satisfies Poisson's equation

$$\nabla^2 u(x) = -2\pi \varrho(x) \tag{2.38}$$

for every $x \in \Omega$ by an analogy to the Newtonian volume potential, noting that

$$\int_{\Gamma_\varepsilon} \frac{\partial}{\partial n(\xi)} \left(\ln \frac{1}{r(\xi, x)} \right) d\Gamma(\xi) = -\frac{1}{\varepsilon} \int_{\Gamma_\varepsilon} d\Gamma = -2\pi \tag{2.39}$$

where Γ_ε is now a curve in the plane region.

The single-layer potential for two-dimensional problems is given by

$$u(x) = \int_{\Gamma} \sigma(\xi) \ln \frac{1}{r(\xi, x)} d\Gamma(\xi) \tag{2.40}$$

and, as in the three-dimensional case, is continuous as the field point passes through the surface for a density σ which is bounded at all surface points.

The two-dimensional double-layer potential is of the form

$$u(x) = \int_{\Gamma} \mu(\xi) \frac{\partial}{\partial n(\xi)} \left(\ln \frac{1}{r(\xi, x)} \right) d\Gamma(\xi) \tag{2.41}$$

and contains a discontinuity which can be investigated in a manner similar to the three-dimensional case.

As before, the bounding curve Γ is divided into $\Gamma - \Gamma_\varepsilon$ and Γ_ε, the latter being a short straight line centered at point x (Fig. 2.5), where it was assumed that the

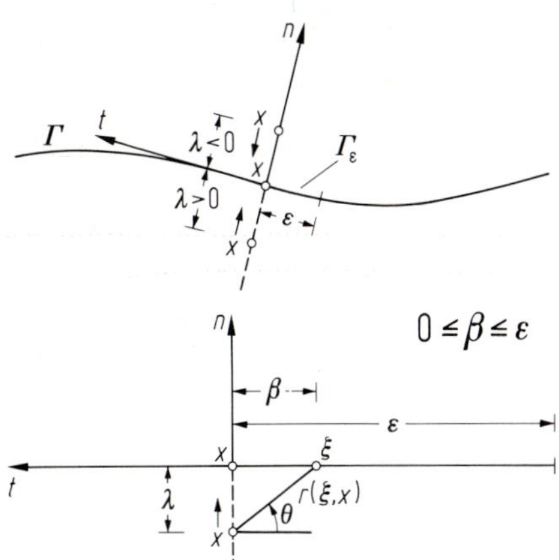

Fig. 2.5. Discontinuity of two-dimensional double-layer potential

surface possesses a smooth contour. The point x is taken to the boundary along the normal to the surface that passes through the original position of x and the distance λ between the two points is taken to be much less than 2ε, the length of Γ_ε. Dividing the integral in Eq. (2.41) as was done in Eq. (2.28) and assuming that $\mu(\xi)$ satisfies a Hölder condition at x, the discontinuity is given by

$$\lim_{\varepsilon \to 0} \left\{ \mu(x) \int_{\Gamma_\varepsilon} \frac{\partial}{\partial n(\xi)} \left(\ln \frac{1}{r(\xi, x)} \right) d\Gamma(\xi) \right\}. \tag{2.42}$$

This integral contains a perfect differential since for θ defined as in Fig. 2.5 its integrand can be written

$$\frac{\partial}{\partial n} \left(\ln \frac{1}{r} \right) d\Gamma = -\frac{\lambda}{r^2} \left(\frac{-\lambda}{\sin^2 \theta} \right) d\theta = d\theta. \tag{2.43}$$

Thus, evaluating the limit in Eq. (2.42) and noting that $|\lambda| \ll \varepsilon$ gives

$$\lim_{\varepsilon \to 0} \left\{ \mu(x) \int_{\Gamma_\varepsilon} \frac{\partial}{\partial n(\xi)} \left(\ln \frac{1}{r(\xi, x)} \right) d\Gamma(\xi) \right\} = -\pi \operatorname{sgn}(\lambda)\, \mu(x). \tag{2.44}$$

The limiting form of Eq. (2.41) as the point x approaches the boundary from the inside becomes

$$u^+(x) = -\pi \mu(x) + \int_\Gamma \mu(\xi) \frac{\partial}{\partial n(\xi)} \left(\ln \frac{1}{r(\xi, x)} \right) d\Gamma(\xi), \tag{2.45}$$

and from the outside it becomes

$$u^-(x) = \pi \mu(x) + \int_\Gamma \mu(\xi) \frac{\partial}{\partial n(\xi)} \left(\ln \frac{1}{r(\xi, x)} \right) d\Gamma(\xi); \tag{2.46}$$

the jump in the integral is now

$$u^+(x) - u^-(x) = -2\pi \mu(x). \tag{2.47}$$

2.3. Indirect Formulation

In this section, we study solutions to Laplace's equation

$$\nabla^2 u(x) = 0 \qquad \text{in } \Omega \ (x \in \Omega) \tag{2.48}$$

with boundary conditions of the Dirichlet type, i.e.,

$$u(x) = \bar{u}(x) \qquad \text{on } \Gamma_1 \ (x \in \Gamma_1), \tag{2.49}$$

or the Neumann type

$$q(x) = \frac{\partial u(x)}{\partial n} = \bar{q}(x) \qquad \text{on } \Gamma_2 \, (x \in \Gamma_2), \tag{2.50}$$

where n is the unit outward normal to surface Γ, and $\bar{u}$ and $\bar{q}$ are prescribed values of the function and its normal derivative over the boundary Γ. Notice that $\Gamma = \Gamma_1 + \Gamma_2$ (Fig. 2.6).

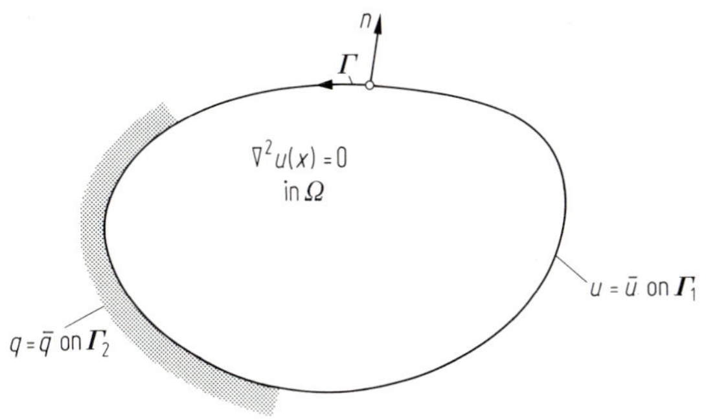

Fig. 2.6. Notation

A function u is said to be *harmonic* within the domain Ω, bounded by a closed surface Γ, if it satisfies the following conditions:

(i) u is continuous in Ω and Γ;
(ii) u is differentiable to at least the second order in Ω;
(iii) u satisfies Laplace's equation in Ω.

Any harmonic function can be represented by a potential distribution and, conversely, every potential is a harmonic function [3, 18].

Thus, an effective method of formulating the boundary-value problems of potential theory is to represent the harmonic function by a single-layer or a double-layer potential generated by continuous source distributions over Γ, provided these potentials satisfy the boundary conditions prescribed for u. This procedure leads to the formulation of integral equations which define the source densities concerned. These equations can be discretized and solved numerically, and values of u at internal points can be computed afterwards from the boundary data by using numerical quadratures as will be shown later.

To obtain an integral equation for the solution of the Neumann problem, we can assume that the unknown function u may be expressed solely as a single-layer potential with unknown density σ,

$$u(x) = \int_\Gamma \sigma(\xi) \, u^*(\xi, x) \, d\Gamma(\xi). \tag{2.51}$$

Here, $x \in \Omega$

The function $u^*(\xi, x)$ is the Newtonian potential (2.5) for three-dimensional problems or the logarithmic potential (2.35) for two-dimensional problems and is called the fundamental solution to Laplace's equation.

Taking the derivative of Eq. (2.51) in the direction of the outward normal to Γ as x is taken to Γ yields the boundary relation

$$q(x) = -\alpha\pi\sigma(x) + \int_\Gamma \sigma(\xi)\frac{\partial u^*(\xi, x)}{\partial n(x)}\, d\Gamma(\xi), \tag{2.52}$$

where $\alpha = 1$ for two-dimensional problems and $\alpha = 2$ for three-dimensional problems. This constitutes a Fredholm equation of the second kind for σ in terms of q, as the unknown appears both inside and outside the integral. After solving the system of corresponding algebraic equations, values of u at any interior or boundary point can be calculated by using Eq. (2.51) since $u^*(\xi, x)$ is continuous as x is taken to Γ.

It is important to note that Eq. (2.52) has a solution only if the Gauss condition [3]

$$\int_\Gamma q(x)\, d\Gamma(x) = 0 \tag{2.53}$$

holds and that this solution is unique only to within an arbitrary additive constant. However, a unique solution of Eq. (2.52) can be obtained by imposing some extra "normalizing" condition [15].

The above method was extensively employed by Hess and Smith [12] to solve a series of fluid flow problems, including flow past hydrofoils, cascades, and lifting aerofoils. Numerical results can also be found in [10, 11, 15].

To obtain an integral equation for the solution of the Dirichlet problem, the classical approach is to assume that the unknown function u may be expressed solely as a double-layer potential with unknown density μ,

$$u(x) = \int_\Gamma \mu(\xi)\frac{\partial u^*(\xi, x)}{\partial n(\xi)}\, d\Gamma(\xi). \tag{2.54}$$

Taking into account the jump in the double-layer potential, the limit of Eq. (2.54) may be taken as

$$u(x) = -\alpha\pi\mu(x) + \int_\Gamma \mu(\xi)\frac{\partial u^*(\xi, x)}{\partial n(\xi)}\, d\Gamma(\xi). \tag{2.55}$$

As $u(x)$ is known for the Dirichlet problem, the source density μ is the only unknown. Again, Eq. (2.55) constitutes a Fredholm equation of the second kind which, after being solved, enables us to compute $u(x)$ everywhere in Ω using Eq. (2.54). Numerical results using this formulation were obtained, for instance, by Kantorovich and Krylov [19].

Since $u^*(\xi, x) = u^*(x, \xi)$, the integral equation (2.55) is said to contain the adjoint kernel of Eq. (2.52). The kernel is the function of (ξ, x) multiplying the density under the integral sign in the integral equations. For scalar kernels, the adjoint is obtained by interchanging x and ξ.

An alternative approach to obtaining an integral equation for the solution of the Dirichlet problem is to assume that the unknown function u may be expressed solely as a single-layer potential with unknown density σ,

$$u(x) = \int_{\Gamma} \sigma(\xi)\, u^*(\xi, x)\, d\Gamma(\xi), \qquad x \in \Omega. \tag{2.56}$$

Since the kernel in this equation is continuous as x passes through the surface, the limit of Eq. (2.56) as x is taken to Γ gives

$$u(x) = \int_{\Gamma} \sigma(\xi)\, u^*(\xi, x)\, d\Gamma(\xi), \qquad x \in \Gamma \tag{2.57}$$

and, as $u(x)$ is known, the source density σ is the only unknown in the equation.

Equation (2.57) is a Fredholm equation of the first kind, as the unknown appears only inside the integral. For many Dirichlet problems, formulations using such equations have proven to be more illuminating physically and more convenient mathematically than using equations of the second kind.

Regarding the numerical solution of the system of corresponding algebraic equations obtained by discretization, the presence of the term outside the integral, for equations of the second kind, ensures that the system matrix will always be diagonally dominant. An equation of the first kind with a nonsingular kernel can be very difficult to solve, being essentially ill-conditioned [20]; however, in the present case, the singularity of the kernel ensures diagonal dominance in the system matrix and the problem is in general well conditioned. For numerical solutions of Eq. (2.57), see, for instance, [10, 11, 13–15].

2.4. Direct Formulation

A conceptual disadvantage of single-layer and double-layer potentials is the introduction of formal source densities which usually bear no physical relation to the problem. This can be overcome by using the direct formulation of the boundary element method, where values of the function and its normal derivative over Γ play the role of the source densities in generating u throughout Ω. This formulation can be deduced through Green's third identity, Betti's or similar theorems or principles such as virtual work. Alternatively we can follow the theory described in Chapter 1 and formulate it through weighted residual considerations [5]. The advantage of using a weighted residual technique is its generality: it permits a straightforward extension of the method to solve more complex partial differential equations; since it can also be employed to formulate other numerical techniques such as the finite element method, it becomes easier to relate and combine the boundary element method with more classical numerical methods.

We are seeking an approximate solution to the problem governed by

$$\nabla^2 u(x) = 0 \qquad \text{in } \Omega \tag{2.58}$$

with boundary conditions

$$\begin{aligned} u(x) &= \bar{u}(x) \qquad \text{on } \Gamma_1, \\ q(x) &= \bar{q}(x) \qquad \text{on } \Gamma_2. \end{aligned} \tag{2.59}$$

The error introduced by replacing u and q by an approximate solution can be minimized by writing the following weighted residual statement:

$$\int_\Omega \nabla^2 u(x)\, u^*(\xi, x)\, d\Omega(x) = \int_{\Gamma_2} [q(x) - \bar{q}(x)]\, u^*(\xi, x)\, d\Gamma(x) - \int_{\Gamma_1} [u(x) - \bar{u}(x)]\, q^*(\xi, x)\, d\Gamma(x), \tag{2.60}$$

where u^* is interpreted as a weighting function and

$$q^*(\xi, x) = \frac{\partial u^*(\xi, x)}{\partial n(x)}. \tag{2.61}$$

The integration of Eq. (2.60) by parts with respect to x_i gives

$$-\int_\Omega \frac{\partial u(x)}{\partial x_i(x)} \frac{\partial u^*(\xi, x)}{\partial x_i(x)}\, d\Omega(x) = -\int_{\Gamma_1} q(x)\, u^*(\xi, x)\, d\Gamma(x) - \int_{\Gamma_2} \bar{q}(x)\, u^*(\xi, x)\, d\Gamma(x) - \int_{\Gamma_1} [u(x) - \bar{u}(x)]\, q^*(\xi, x)\, d\Gamma(x), \tag{2.62}$$

where $i = 1, 2, 3$ and Einstein's summation convention for repeated indices is implied. Integrating by parts once more,

$$\int_\Omega \nabla^2 u^*(\xi, x)\, u(x)\, \mathrm{d}\Omega(x) = -\int_{\Gamma_1} q(x)\, u^*(\xi, x)\, d\Gamma(x) - \int_{\Gamma_2} \bar{q}(x)\, u^*(\xi, x)\, d\Gamma(x) + \int_{\Gamma_2} u(x)\, q^*(\xi, x)\, d\Gamma(x) + \int_{\Gamma_1} \bar{u}(x)\, q^*(\xi, x)\, d\Gamma(x) \tag{2.63}$$

or, generally,

$$\int_\Omega \nabla^2 u^*(\xi, x)\, u(x)\, d\Omega(x) = -\int_\Gamma q(x)\, u^*(\xi, x)\, d\Gamma(x) + \int_\Gamma u(x)\, q^*(\xi, x)\, d\Gamma(x). \tag{2.64}$$

Let us now remember that the Dirac delta function $\Delta(\xi, x)$ has the following properties:

$$\begin{aligned} \Delta(\xi, x) &= 0 \quad \text{for } \xi \neq x, \\ \Delta(\xi, x) &= \infty \quad \text{for } \xi = x. \end{aligned} \tag{2.65}$$

and

$$\int_\Omega u(x)\, \Delta(\xi, x)\, d\Omega(x) = u(\xi). \tag{2.66}$$

Assuming u^* to be the fundamental solution to Laplace's equation means that

$$\nabla^2 u^*(\xi, x) = -2\alpha\pi\, \Delta(\xi, x), \tag{2.67}$$

where $\alpha = 1$ for two-dimensional problems and $\alpha = 2$ for three-dimensional problems. Substituting Eq. (2.67) into Eq. (2.64) gives

$$2\alpha\pi u(\xi) + \int_\Gamma u(x)\, q^*(\xi, x)\, d\Gamma(x) = \int_\Gamma q(x)\, u^*(\xi, x)\, d\Gamma(x). \tag{2.68}$$

Equation (2.68) states that a harmonic function may be expressed as the superposition of a single-layer potential with density $q/2\alpha\pi$ and a double-layer potential with density $-u/2\alpha\pi$.

Another advantage of the direct formulation over the indirect one is that the restriction for the boundary surface to be a Liapunov (smooth) one can be relaxed. In fact, it can be applied to the more general Kellogg regular surfaces [3], thus allowing surfaces with corners or edges to be included. So, taking the point ξ to the boundary and accounting for the jump of the left-hand side integral in Eq. (2.68) yields the boundary integral equation

$$c(\xi)\, u(\xi) + \int_\Gamma u(x)\, q^*(\xi, x)\, d\Gamma(x) = \int_\Gamma q(x)\, u^*(\xi, x)\, d\Gamma(x). \tag{2.69}$$

This equation provides a functional constraint between u and q over Γ which ensures their compatibility as boundary data. If the solution of a Neumann problem is required, the right-hand side of Eq. (2.69) is known, and we have to solve a Fredholm equation of the second kind for the unknown boundary values of the function u. If the solution of a Dirichlet problem is required, values of u are prescribed throughout Γ and we obtain a Fredholm equation of the first kind for the unknown boundary values of the normal derivative q. Solution of Cauchy (mixed) boundary-value problems leads to a mixed integral equation for the unknown boundary data.

Two different procedures can be employed to calculate the value of the coefficient c: one is through the physical consideration that a constant potential applied over a closed body produces no flux, which is equivalent to the rigid-body translations of the theory of elasticity and will be discussed in detail in Chapter 3; the other is herein presented for two-dimensional problems, but a similar approach is also valid for three dimensions.

Assume that the body under consideration can be augmented by a small region Γ_ε which is part of a circle of radius ε centered at point ξ on the boundary Γ (Fig. 2.7). Proceeding as for evaluating the jump of the double-layer potential in Section 2.2 and assuming that the function $u(x)$ satisfies a Hölder condition at ξ, we have

$$c(\xi) = 2\pi + \lim_{\varepsilon \to 0} \int_{\Gamma_\varepsilon} \frac{\partial}{\partial n(x)} \left(\ln \frac{1}{r(\xi, x)} \right) d\Gamma(x) \tag{2.70}$$

which, referring to Fig. 2.7, reduces to

$$c(\xi) = 2\pi - \lim_{\varepsilon \to 0} \int_{\theta_1}^{\theta_2} \frac{1}{\varepsilon}\, \varepsilon\, d\theta = \pi + \alpha_1 - \alpha_2 \tag{2.71}$$

that is, $c(\xi)$ equals the internal angle of the boundary at ξ.

Note: In Eqns. (2.68, 2.69), $u^* = \ln(\frac{1}{r})$ for 2-D problems and $= \frac{1}{r}$ for 3-D cases.
∴ For a 2-D problem with constant boundary elements, (smooth!), $\alpha = 1$ and $c(\xi) = \pi$

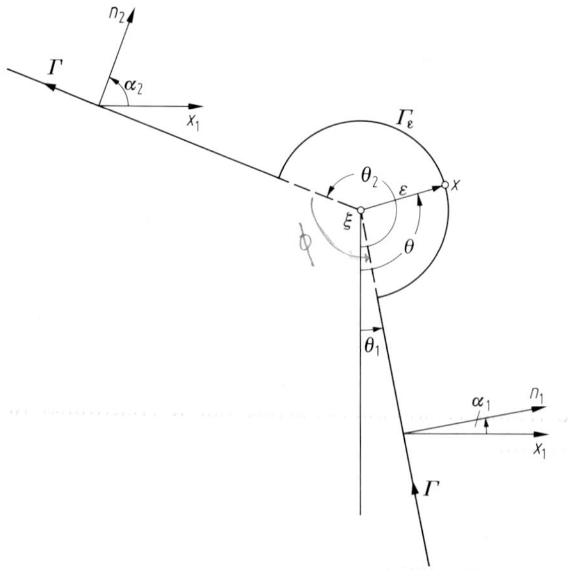

Fig. 2.7. Two-dimensional body augmented by region Γ_ε

Since in a well-posed boundary-value problem only half of the boundary variables in Eq. (2.69) is prescribed, this equation can be employed in order to obtain the unknown boundary data. In Section 2.6, a numerical scheme to solve this boundary integral equation will be presented. Then, values of the function u at any internal point ξ can be calculated by a numerical quadrature via Eq. (2.68). The derivatives of u at ξ (with Cartesian coordinates $x_i(\xi)$, $i = 1, 2, 3$), if required, can also be computed by a quadrature via the equation $\xi \in \Omega$

$$\frac{\partial u(\xi)}{\partial x_i(\xi)} = \frac{1}{2\alpha\pi}\left\{\int_\Gamma q(x)\frac{\partial u^*(\xi, x)}{\partial x_i(\xi)}\,d\Gamma(x) - \int_\Gamma u(x)\frac{\partial q^*(\xi, x)}{\partial x_i(\xi)}\,d\Gamma(x)\right\} \quad (2.72)$$

as we may generally differentiate beneath the integral signs in Eq. (2.68).

2.5. Boundary Element Method

Rather than attempting analytical solutions to Eq. (2.69) for particular geometries and boundary conditions, we seek a suitable reduction of the equation to an algebraic form that can be solved by a numerical approach. This approach generally consists of the following steps (see, for instance, [5, 15, 21]).

(a) The boundary Γ is discretized into a series of elements over which the potential and its normal derivative are assumed to vary according to interpolation functions. The geometry of these elements can be modelled using straight lines, circular arcs, parabolas, etc.;

(b) By using the method of collocation, the discretized equation is applied to a number of particular nodes within each element where values of the potential and its normal derivative are associated;

(c) The integrals over each element are carried out by using, in general, a numerical quadrature scheme;

(d) By imposing the prescribed boundary conditions of the problem, a system of linear algebraic equations is obtained. The solution of this system of equations, which can be effected using direct or iterative methods, produces the remaining boundary data.

Values of the function u at any internal point, if required, can then be calculated from the boundary data by a numerical quadrature via Eq. (2.68). Similarly, the derivatives of u at any internal point can also be computed by a quadrature via Eq. (2.72).

In the following section, the above-listed steps are examined in detail in connection with two-dimensional problems defined over finite regions of homogeneous isotropic media with Neumann, Dirichlet, Cauchy, or Robin boundary conditions. In subsequent sections, it is shown how the method can be extended to include internal sources. If the region is nonhomogeneous but is constituted of several homogeneous subregions with different physical properties, the method can be applied by first writing a system of equations for each subregion and then introducing compatibility (in terms of potentials) and equilibrium (in terms of normal derivatives) conditions between the subregions.

Fundamental solutions for orthotropic and anisotropic regions are derived and it is shown that all concepts presented in the previous sections are also valid for infinite regions fulfilling certain regularity conditions at infinity. By adopting a convenient fundamental solution which satisfies part of the boundary conditions of the problem under consideration, a reduction in the amount of numerical work can be achieved, as explained in Section 2.11. Finally, specific numerical procedures for three-dimensional and axisymmetric problems are derived.

2.6. Two-Dimensional Problems

$c(\xi)u(\xi) + \int_\Gamma u(x) q^*(\xi,x)\, d\Gamma(x) = \int_\Gamma q(x) u^*(\xi,x)\, d\Gamma(x), \quad \xi \in \Gamma$

The integral equation (2.69) can be discretized into a series of elements. Consider for simplicity that the body is two-dimensional and its boundary is divided into a series of "segments" or "boundary elements," as shown in Fig. 2.8. The points where the unknown values are considered are called "nodes" and taken to be in the middle of each segment for the so-called "constant" elements (Fig. 2.8a). These are going to be the only elements considered in this chapter, but in Chapter 3 we will also discuss the case of linear elements (i.e., elements for which the nodes are at the intersection between elements as shown in Fig. 2.8b) and curved elements such as the ones shown in Fig. 2.8c. For the latter case an extra midelement node is required and the elements are called quadratic.

For the constant element case the boundary is discretized into N elements, of which N_1 are assumed to belong to Γ_1 and N_2 to Γ_2. The values of u and q are taken to be constant on each element and equal to their values at the midnode of the element. Note that in each element the value of one of the two variables (u or q) is known.

Equation (2.69) can be written as

$$c_i u_i + \int_\Gamma u\, q^*\, d\Gamma = \int_\Gamma q\, u^*\, d\Gamma, \tag{2.73}$$

With the factor $2\alpha\pi$ absorbed in u^* →

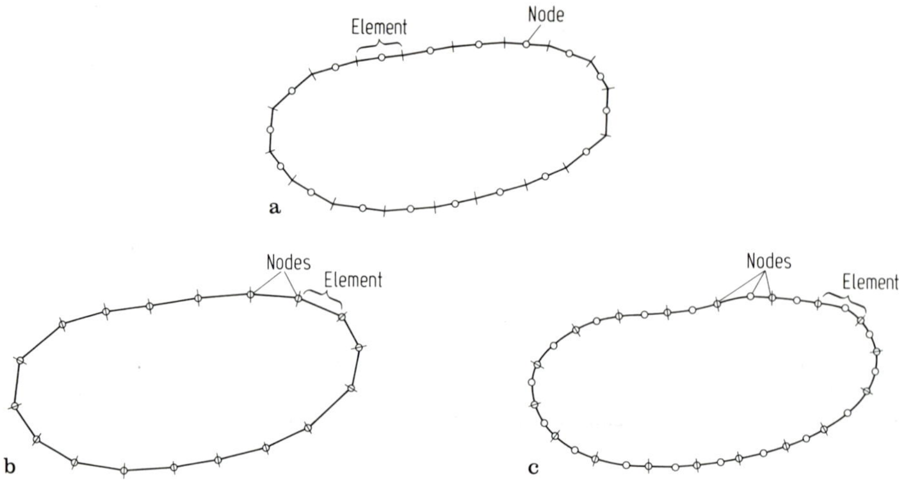

Fig. 2.8. Boundary elements: **a** constant; **b** linear, and **c** quadratic

where we have taken

$$u^* = \frac{1}{2\pi} \ln\left(\frac{1}{r}\right) \tag{2.74}$$

for two-dimensional problems and

$$u^* = \frac{1}{4\pi}\left(\frac{1}{r}\right) \tag{2.75}$$

for three-dimensional cases.

Notice that we have taken ξ as the point i on which the fundamental solution is applied, i.e., $u(\xi) = u_i$. Similar consideration applies to the coefficient $c(\xi)$. We have also dropped the ξ, x notation written between brackets for simplicity.

Equation (2.73) can be discretized as follows:

$$c_i u_i + \sum_{j=1}^{N} \int_{\Gamma_j} u\, q^*\, d\Gamma = \sum_{j=1}^{N} \int_{\Gamma_j} u^*\, q\, d\Gamma . \tag{2.76}$$

Notice that for constant elements the boundary is always "smooth," hence the c_i coefficient is identically equal to $\frac{1}{2}$. Γ_j is the length of element j. Equation (2.76) represents, in discrete form, the relationship between the node i at which the fundamental solution is applied and all the j elements (including the one $i = j$) on the boundary (Fig. 2.9).

$c_i = \frac{\pi}{2\pi} = \frac{1}{2}$

The u and q values inside the integrals in (2.76) are constant within each element and consequently can be taken out of the integrals. This gives

$$\tfrac{1}{2} u_i + \sum_{j=1}^{N} \left(\int_{\Gamma_j} q^*\, d\Gamma \right) u_j = \sum_{j=1}^{N} \left(\int_{\Gamma_j} u^*\, d\Gamma \right) q_j . \tag{2.77}$$

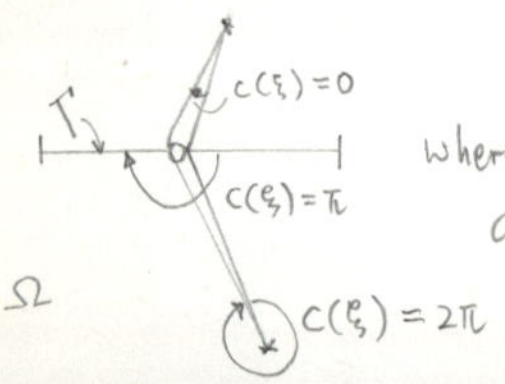

where $C(\xi)$ is defined by (2.69) and (2.71) in 2-D case $\Rightarrow C_i = \begin{cases} 0 & \text{for } i \text{ outside } \Omega \\ \frac{1}{2} & \text{for } i \text{ on } \Gamma \\ 1 & \text{for } i \text{ in } \Omega \end{cases}$ where C_i is defined by (2.76)/

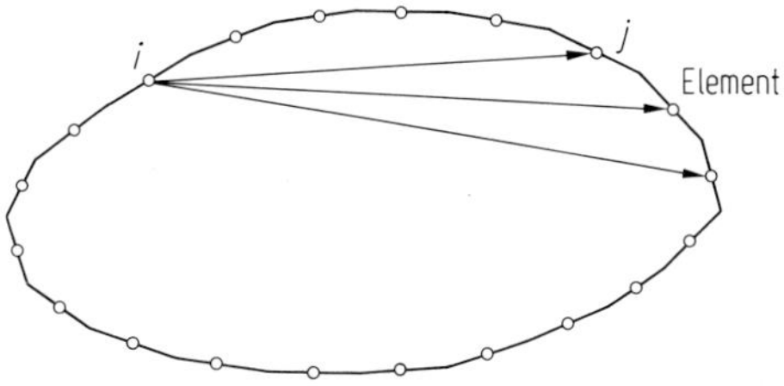

Fig. 2.9. Relationship between the fundamental solution at boundary node i and the boundary elements

The integrals $\int q^* \, d\Gamma$ relate the i node with the element j over which the integral is carried out. Hence these integrals will be called $\hat{H}_{ij}$. Similarly, the integrals of type $\int u^* \, d\Gamma$ will be called G_{ij}. Hence, Eq. (2.77) can be written

$$\tfrac{1}{2} u_i + \sum_{j=1}^{N} \hat{H}_{ij} u_j = \sum_{j=1}^{N} G_{ij} q_j . \tag{2.78}$$

The integrals in this case can be computed analytically, as the fundamental solution and element geometry are very simple. In general, it is necessary or more convenient to integrate them numerically.

We can rewrite Eq. (2.78) for each i node under consideration. Let us now define

$$H_{ij} = \begin{cases} \hat{H}_{ij} & \text{when } i \neq j \\ \hat{H}_{ij} + \frac{1}{2} & \text{when } i = j . \end{cases} \tag{2.79}$$

Hence Eq. (2.78) can be written

$$\sum_{j=1}^{N} H_{ij} u_j = \sum_{j=1}^{N} G_{ij} q_j . \tag{2.80}$$

The whole set of equations can also be expressed in matrix form as

$$\boldsymbol{H}\boldsymbol{U} = \boldsymbol{G}\boldsymbol{Q} \tag{2.81}$$

Note that N_1 values of u and N_2 values of q are known on Γ, hence one has a set of only N unknowns in formula (2.81) which now can be reordered in accordance with the unknown under consideration. We can reorder Eq. (2.81) with all the unknowns on the left-hand side and a vector on the right-hand side obtained by multiplying matrix elements by the known values of potential and flux. This gives

$$\boldsymbol{A}\,\boldsymbol{Y} = \boldsymbol{F} . \tag{2.82}$$

where $\boldsymbol{Y}$ is the vector of unknowns u's and q's.

Notice that $\boldsymbol{A}$ is a fully populated matrix of order N. In computational terms H_{ij} and G_{ij} can be assembled directly into $\boldsymbol{A}$ so that Eq. (2.81) does not need to be formed.

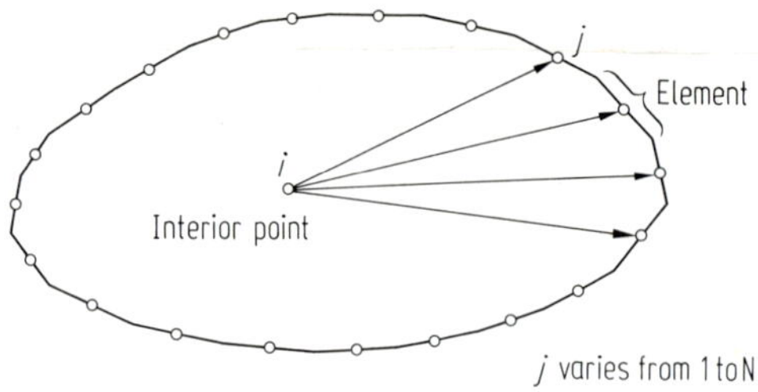

Fig. 2.10. Relationship between the fundamental solution at internal point i and the boundary elements

Once Eq. (2.82) is solved, all the values of potential and fluxes on the boundary are known and one could calculate the values of potentials and fluxes at any interior point using

$$u_i = \int_\Gamma q\, u^*\, d\Gamma - \int_\Gamma q^*\, u\, d\Gamma\,. \tag{2.83}$$

This equation represents the integral relationship obtained between an internal point i and the boundary values of u and q (Fig. 2.10) and can be written in discretized form as

$$u_i = \sum_{j=1}^{N} q_j\, G_{ij} - \sum_{j=1}^{N} u_j\, \hat{H}_{ij}\,. \tag{2.84}$$

The values of internal fluxes can be calculated by differentiating Eq. (2.83) as was done in Eq. (2.72). Hence,

$$\left(\frac{\partial u}{\partial x_l}\right) = \int_\Gamma q \frac{\partial u^*}{\partial x_l}\, d\Gamma - \int_\Gamma u \frac{\partial q^*}{\partial x_l}\, d\Gamma\,. \tag{2.85}$$

x_l are the coordinates, $l = 1, 2$ for two-dimensional problems, and $l = 1, 2, 3$ for three-dimensional cases.

The integrals $\hat{H}_{ij}$ and G_{ij} can be calculated using simple Gauss quadrature rules for all elements (except the one corresponding to the node under consideration) as follows:

$$\hat{H}_{ij} = \int_{\Gamma_j} q^*\, d\Gamma = \frac{l_j}{2} \sum_{k=1}^{K} (q^*)_k\, w_k \tag{2.86}$$

and

$$G_{ij} = \int_{\Gamma_j} u^*\, d\Gamma = \frac{l_j}{2} \sum_{k=1}^{K} (u^*)_k\, w_k\,, \tag{2.87}$$

where l_j is the element length and w_k is the weight associated with the numerical integration point k. The function u^* or q^* has to be evaluated at the same point. The length l_j is divided by 2 as numerical integration formulas are usually given from -1 to $+1$ with a total weight of 2. Usually, four integration points are sufficient to provide the required accuracy for two-dimensional problems.

For the integrals corresponding to the singular elements, we need to use higher order integration rules such as the ones given in Appendix A. For the particular case of constant elements the $\hat{H}_{ii}$ and G_{ii} integrals can be easily computed analytically. He $\hat{H}_{ii}$ term, for instance, is identically zero due to the orthogonality between the normal and the surface of the element, i.e.,

$$\hat{H}_{ii} = \int_{\Gamma_i} q^* \, d\Gamma = \int \frac{\partial u^*}{\partial r} \frac{\partial r}{\partial n} \, d\Gamma \equiv 0 \,. \tag{2.88}$$

The evaluation of the G_{ii} terms give

$$G_{ii} = \int_{\Gamma_i} u^* \, d\Gamma = \frac{1}{2\pi} \int_{\Gamma_i} \ln\left(\frac{1}{r}\right) d\Gamma \tag{2.89}$$

or using the homogeneous coordinate η over a segment (Fig. 2.11)

$$G_{ii} = \frac{1}{2\pi} \int_{(1)}^{(2)} \ln\left(\frac{1}{r}\right) d\Gamma = \frac{1}{\pi} \int_{(0)}^{(2)} \ln\left(\frac{1}{r}\right) d\Gamma \,. \tag{2.90}$$

Transforming coordinates, $r = \eta\,|r_1|$, where $|r_1| = |r_2|$, one has

$$G_{ii} = \frac{1}{\pi} \int_{(0)}^{(2)} \ln\left(\frac{1}{r}\right) d\Gamma = \frac{|r_1|}{\pi} \left[\ln\left(\frac{1}{|r_1|}\right) + \int_0^1 \ln\left(\frac{1}{\eta}\right) d\eta \right] . \tag{2.91}$$

Noting that the last integral is equal to 1, we have

$$G_{ii} = \frac{1}{\pi} |r_1| \left[\ln\left(\frac{1}{|r_1|}\right) + 1 \right] . \tag{2.92}$$

Before closing this section, it is interesting to point out that other types of boundary conditions that frequently appear on practical problems, such as Robin-type conditions, can easily be incorporated into the formulation.

The Robin condition prescribes a linear relationship between the potential and its normal derivative at points along the boundary Γ as follows:

$$e\,u + f\,q = d \,, \tag{2.93}$$

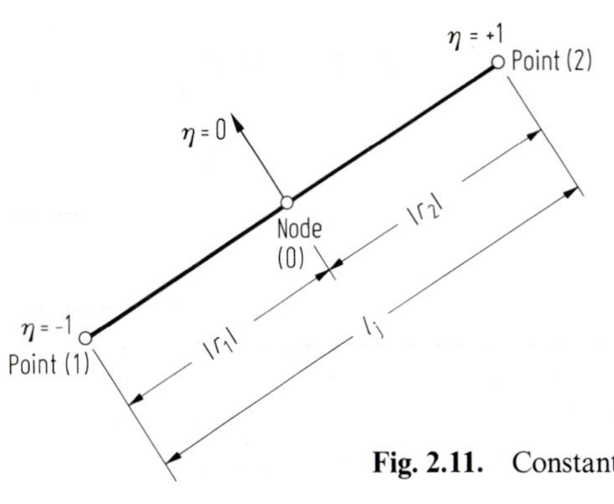

Fig. 2.11. Constant element

where d, e, and f are functions of the position. Notice that Eq. (2.93) includes all previous boundary conditions since for $f=0$ it becomes the Dirichlet condition, while for $e=0$ we have the Neumann condition. More generally, for $e, f \neq 0$ we can include the impedance boundary condition of electromagnetic problems, the convection boundary condition of heat conduction problems, etc.

If Eq. (2.93) is applied at all boundary nodes, we can write

$$\boldsymbol{Q} = \boldsymbol{D} - \boldsymbol{E}\boldsymbol{U} \,, \tag{2.94}$$

where the vector $\boldsymbol{D}$ and the diagonal matrix $\boldsymbol{E}$ contain the values of d/f and e/f, respectively, at each boundary node.

Substituting Eq. (2.94) into Eq. (2.81) yields the system of equations

$$(\boldsymbol{H} + \boldsymbol{G}\boldsymbol{E})\,\boldsymbol{U} = \boldsymbol{G}\boldsymbol{D} \tag{2.95}$$

or, more simply,

$$\boldsymbol{A}\;\boldsymbol{Y} = \boldsymbol{F} \,. \tag{2.96}$$

After solving the system of equations (2.96), the normal derivatives of potentials along the boundary can be evaluated pointwise by using condition (2.93).

2.6.1. Source Formulation

The source formulation can also be expressed in matrix from, starting with Eqs. (2.51) and (2.52), i.e.,

$$\begin{aligned} u_i &= \int_\Gamma \sigma\, u^*\, d\Gamma \,, \\ q_i &= -\tfrac{1}{2}\,\sigma_i + \int_\Gamma \sigma\, q^*\, d\Gamma \,. \end{aligned} \tag{2.97}$$

Once again we take the fundamental solution as Eqs. (2.74) and (2.75).

We can discretize the boundary into elements to obtain

$$\begin{aligned} u_i &= \sum_{j=1}^{N} \sigma_j\, G_{ij} \,, \\ q_i &= -\tfrac{1}{2}\,\sigma_i + \sum_{j=1}^{N} \sigma_j\, \hat{H}_{ij} \,. \end{aligned} \tag{2.98}$$

The difference between this formulation and the direct one is that the $\boldsymbol{G}$ and $\boldsymbol{H}$ contributions are uncoupled and that the diagonal terms in $\boldsymbol{H}$ are

$$H_{ii} = \hat{H}_{ii} - \tfrac{1}{2} \tag{2.99}$$

instead of $\hat{H}_{ii} + \frac{1}{2}$. Formula (2.98) can be written

$$\begin{aligned} u_i &= \sum_{j=1}^{N} \sigma_j\, G_{ij} \,, \\ q_i &= \sum_{j=1}^{N} \sigma_j\, H_{ij} \,. \end{aligned} \tag{2.100}$$

Equations (2.100) can now be applied at all boundary points with conditions

$$u_i = \bar{u}_i \qquad \text{at } N_1 \text{ points on } \Gamma_1$$

and (2.101)

$$q_i = \bar{q}_i \qquad \text{at } N_2 \text{ points on } \Gamma_2 .$$

The final system can be written

$$\boldsymbol{A}\,\boldsymbol{\sigma} = \boldsymbol{F}, \tag{2.102}$$

where the unknowns in the $\boldsymbol{\sigma}$ vector are the source intensities.

Example 2.1. This example assesses the accuracy of constant elements for the solution of a Dirichlet problem, that is, two confocal ellipses with temperatures u_I and u_E prescribed along the internal and external surfaces, respectively.

Taking the semiaxes of the ellipses to be [22]

$$\begin{aligned} a_E &= c \cosh \gamma_E , & b_E &= c \sinh \gamma_E , \\ a_I &= c \cosh \gamma_I , & b_I &= c \sinh \gamma_I , \end{aligned} \tag{a}$$

where c is a constant and $0 < \gamma_I < \gamma_E < \infty$, the exact solution of the problem is given by

$$u = u_E + \frac{\gamma_E - \gamma}{\gamma_E - \gamma_I}(u_I - u_E) , \qquad \gamma_I \leqq \gamma \leqq \gamma_E . \tag{b}$$

Figure 2.12 presents the *relative error* in the calculation of the temperature at the point $x_1 = 0$; $x_2 = c \sinh [(\gamma_I + \gamma_E)/2]$ obtained with several discretizations for two different aspect ratios a/b for the ellipses. Assuming a unit value for the constant c, the lower curves in the figure correspond to ellipses with aspect ratios $(a/b)_I = 1.313$ and $(a/b)_E = 1.037$, while the upper curves correspond to $(a/b)_I$

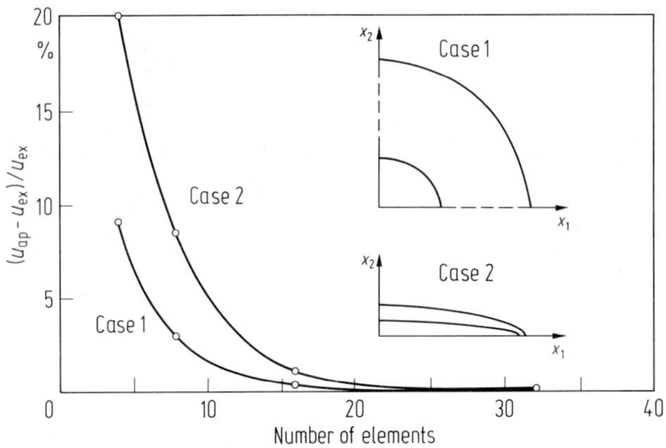

Fig. 2.12. Convergence of the solution. u_{ap} is the boundary element results, i.e., approximate; u_{ex} is the analytical solution, i.e., exact

$= 10.033$ and $(a/b)_E = 5.066$. In the second case, the inner ellipse is much more distorted than the outer one such that their longer axes almost touch each other (see Fig. 2.12). The convergence of the solutions is evident in the figure. Normal fluxes along the outer surface are plotted in Fig. 2.13 for the finest discretization employed. Due to the symmetry of the problem, only one-quarter of the cross section needed to be analyzed. Symmetry is taken into account through a direct condensation process with integration over reflected elements such that no discretization of the symmetry axes is necessary [41].

Example 2.2. The surface temperature distribution of an infinitely long circular cylinder of radius R is specified as shown in Fig. 2.14. One wants to find the steady temperature of the cylinder. The problem can be expressed by the equation

$$\frac{\partial^2 u}{\partial r^2} + \frac{1}{r}\frac{\partial u}{\partial r} + \frac{1}{r^2}\frac{\partial^2 u}{\partial \theta^2} = 0 \qquad \text{(a)}$$

which is the Laplace equation in polar coordinates. Note that there is a discontinuity on the temperature for the points $(R, 0)$ and (R, π). An analytical solution of the problem can be obtained as a Fourier series of the type [23]

$$\frac{u(r, \theta)}{u_0} = \frac{1}{2} + 2\sum_{n=1}^{\infty} \frac{1}{n\pi}\left(\frac{r}{R}\right)^n \sin n\theta\,, \qquad n = 1, 3, 5, \ldots\,. \qquad \text{(b)}$$

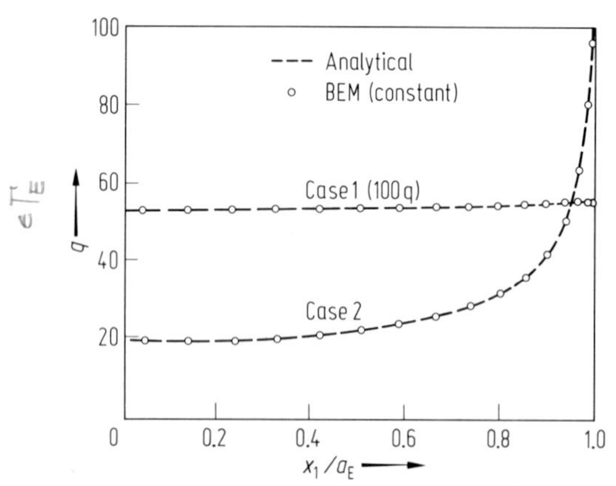

Fig. 2.13. Normal flux along outer surface

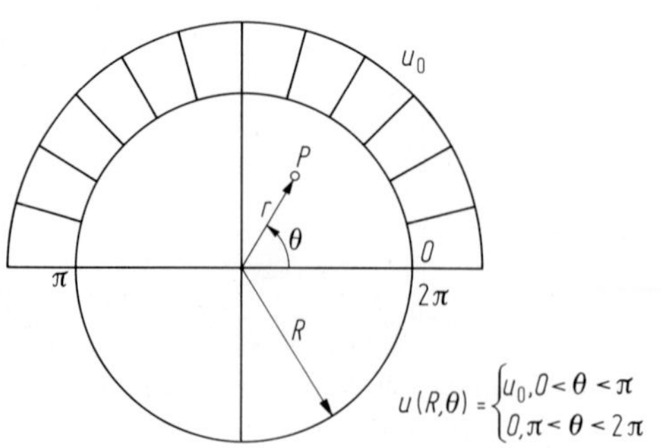

Fig. 2.14. Surface temperature of an infinitely long circular cylinder
(Given)

Table 2.1. Temperature at internal points.

Point	BEM (24 elements)	BEM (48 elements)	Analytical
1	0.500	0.500	0.500
2	0.799	0.796	0.795
3	0.776	0.774	0.773
4	0.689	0.688	0.687
5	0.500	0.500	0.500
6	0.311	0.312	0.313
7	0.224	0.226	0.227
8	0.201	0.204	0.205

Results for some internal points (Fig. 2.15) are compared in Table 2.1 for two discretizations using constant elements, assuming $u_0 = R = 1$. The agreement is very good even for the coarse discretization. Since the analytical solution is expanded as a sine series, it is not capable of predicting the distribution of fluxes. The radial fluxes present a singularity as shown in Fig. 2.16 for the boundary fluxes. The numerical results, for both discretizations, tend to represent this singularity as shown in Fig. 2.17.

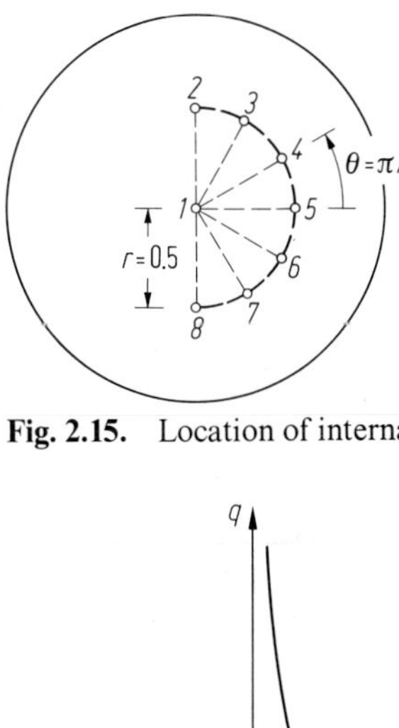

Fig. 2.15. Location of internal points

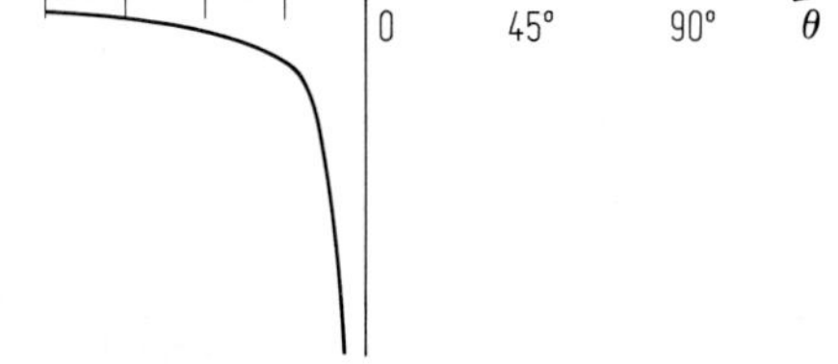

Fig. 2.16. Singularity of the radial fluxes

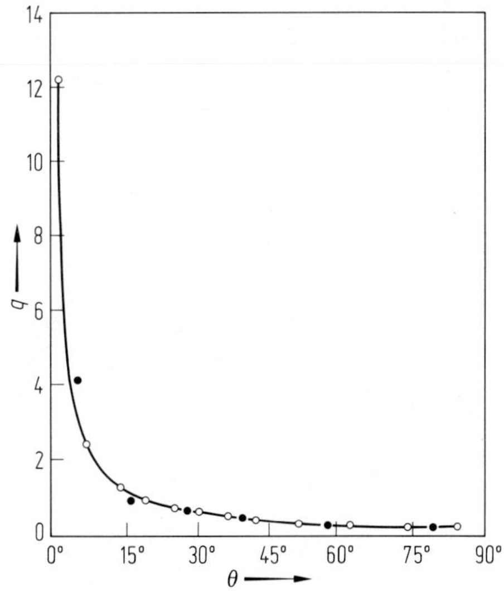

Fig. 2.17. Results for the radial fluxes. ○, Fine discretization; ●, coarse discretization

Example 2.3. This application deals with a concrete column of rectangular cross section where part of the boundary surface is subjected to an interior ambient condition, another part is subjected to outside weather conditions, and the remainder is in contact with an abutting wall which separates both. The boundary conditions of the problem are of the convective type,

$$q + h\,u = h\,u_s\,,$$

where h is the heat transfer coefficient and u_s is the temperature of the surrounding medium.

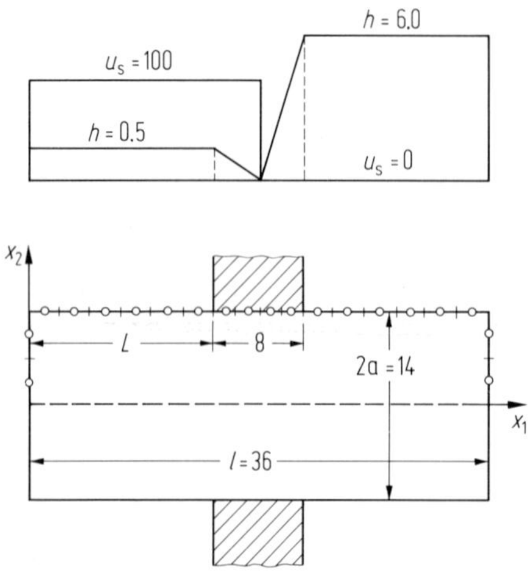

Fig. 2.18. Geometry, discretization, and boundary conditions for rectangular concrete column

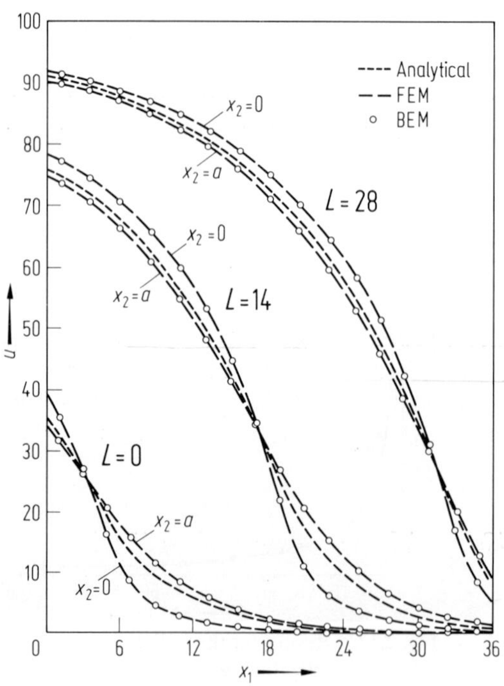

Fig. 2.19. Temperature distribution for three different positions of the abutting wall

The temperature and surface heat transfer coefficient on the interior face ($x_1 = 0$) are 100 °F and 0.5 Btu/(h ft^2 °F), respectively, and at the exterior face ($x_1 = l$) are 0 °F and 6.0 Btu/(h ft^2 °F). The variation of the temperature and surface heat transfer coefficient along the faces $x_2 = \pm a$ is indicated in Fig. 2.18. Note that the thermal conductivity was assumed to be 1.0 Btu/(h ft °F).

Results corresponding to three different positions for the abutting wall are presented in Fig. 2.19, compared with finite element results [24] and an analytical solution [25] (in terms of a mean temperature over the width of the cross section). The boundary element analyses were performed by discretizing one-half of the column into 20 constant elements (see Fig. 2.18) taking into account the symmetry with respect to the x_1 axis, while the finite element ones employed 252 quadrilateral elements.

2.7. Poisson Equation

Assuming that there exist sources inside the domain Ω, as for instance internal heat generation for heat conduction problems, the governing equation of the problem becomes the Poisson-type equation

$$\nabla^2 u = b \qquad \text{in } \Omega\,, \tag{2.103}$$

where b is a known function of position.

Boundary-value problems for Poisson's equation may be reduced to similar problems for Laplace's equation by subtracting a particular solution independent of the boundary conditions [15, 16].

For some practical problems, it may occur that the function b is only defined pointwise such that a particular solution of the problem is difficult to be found. For these cases, Eq. (2.69) can be generalized to include a domain integral involving function b of the form [5]

$$\int_{\Omega} b(x)\, u^*(\xi, x)\, d\Omega(x) \tag{2.104}$$

to be added to the left-hand side of Eq. (2.69). $c(\xi)u(\xi) + \int_{\Gamma} u(x)\, q^*(\xi,x)\, d\Gamma(x) = \int_{\Gamma} q(x)\, u^*(\xi,x)\, d\Gamma(x),\ \xi \in \Gamma$

The above integral can be performed by subdividing the domain Ω into a series of cells over which a numerical integration formula can be applied. The addition of the above integral can be justified by starting with Eq. (2.103) instead of the Laplace equation in formula (2.60). This gives the following weighted residual statement:

$$\int_{\Omega} (\nabla^2 u(x) - b(x))\, u^*(\xi, x)\, d\Omega(x) = \int_{\Gamma_2} [q(x) - \bar{q}(x)]\, u^*(\xi, x)\, d\Gamma(x) - \int_{\Gamma_1} [u(x) - \bar{u}(x)]\, q^*(\xi, x)\, d\Gamma(x)\,. \tag{2.105}$$

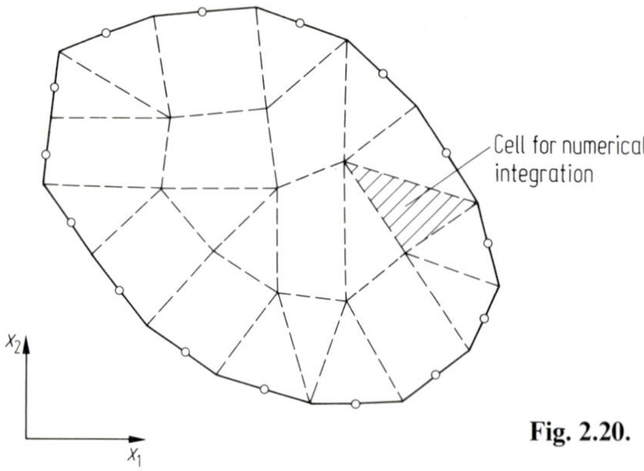

Fig. 2.20. Boundary elements and internal cells

Integrating this expression by parts twice and taking it to the boundary we find

$$c(\xi)\,u(\xi) + \int_\Gamma u(x)\,q^*(\xi, x)\,d\Gamma(x) + \int_\Omega b(x)\,u^*(\xi, x)\,d\Omega(x) = \int_\Gamma q(x)\,u^*(\xi, x)\,d\Gamma(x)\,. \tag{2.106}$$

Cells or regions of integration (Fig. 2.20) can now be employed to compute expression (2.104). The numerical integration can be written

$$B_i = \int_\Omega b\,u^*\,d\Omega = \sum_{N_e} \left(\sum_{k=1}^{K} w_k (b\,u^*)_k \right) A_e\,, \tag{2.107}$$

where w_k are the integration weights and the function $(b u^*)$ needs to be evaluated at the k integration points. N_e is the number of cells into which Ω has been divided and A_e is the area of each of them. B_i is valid for each position of the fundamental solution, assumed to be at i.

The whole set of equations for the N nodes can be expressed in matrix form as

$$\boldsymbol{B} + \boldsymbol{H}\boldsymbol{U} = \boldsymbol{G}\boldsymbol{Q}\,. \tag{2.108}$$

Note that N_1 values of u and N_2 values of q are known on the boundary. Hence Eq. (2.108) can be reordered in such a way that all the unknowns are on the left-hand side, i.e.,

$$\boldsymbol{A}\,\boldsymbol{Y} = \boldsymbol{F}\,, \tag{2.109}$$

where $\boldsymbol{Y}$ is the vector of unknown values of u's and q's.

Once the values of u and q on the whole boundary are known one can calculate the u at any interior point, i.e.,

$$u_i = \sum_{j=1}^{N} G_{ij}\,q_j - \sum_{j=1}^{N} \hat{H}_{ij}\,u_j - B_i\,. \tag{2.110}$$

An alternative way of including internal sources into the formulation is by transforming the corresponding domain integral (2.104) into equivalent boundary integrals for the cases when the function b is harmonic in Ω, i.e., $\nabla^2 b = 0$. If we compute a Galerkin-type function v^* such that $\nabla^2 v^* = u^*$, we can write Green's second identity [3] in the form

$$\int_\Omega (b\,\nabla^2 v^* - v^*\,\nabla^2 b)\, d\Omega = \int_\Gamma \left(b\frac{\partial v^*}{\partial n} - v^*\frac{\partial b}{\partial n} \right) d\Gamma \tag{2.111}$$

which reduces to

$$\int_\Omega b\, u^*\, d\Omega = \int_\Gamma \left(b\,\frac{\partial v^*}{\partial n} - v^*\,\frac{\partial b}{\partial n} \right) d\Gamma\,. \tag{2.112}$$

One such function v^* is given in [26] as

$$v^* = \frac{r^2}{8\pi}\left[\ln\left(\frac{1}{r}\right) + 1\right] \tag{2.113}$$

and it can easily be seen that

$$\nabla^2 v^* = \frac{1}{r}\frac{\partial}{\partial r}\left(r\,\frac{\partial v^*}{\partial r}\right) = \frac{1}{2\pi}\ln\frac{1}{r} = u^*\,. \tag{2.114}$$

Notice that as a special case we could have concentrated sources at an interior point l. For this case, b becomes

$$b = Q_l \Delta_l\,, \tag{2.115}$$

where Δ_l is a Dirac delta function. For the case of several such sources Eq. (2.106) can be generalized to give

$$c_i\, u_i + \int_\Gamma u\, q^*\, d\Gamma + \int_\Omega b\, u^*\, d\Omega + \sum (Q_l\, u_l^*) = \int_\Gamma q\, u^*\, d\Gamma\,. \tag{2.116}$$

These concentrated sources are very simple to handle as they do not require any special integration.

Example 2.4. The equation of motion of a uniform incompressible viscous fluid in steady unidirectional flow (in the x_3 direction) is [27]

$$-\frac{\partial p}{\partial x_3} + \mu\left(\frac{\partial^2 u}{\partial x_1^2} + \frac{\partial^2 u}{\partial x_2^2}\right) = 0\,, \tag{a}$$

where μ is the viscosity of the fluid, $\partial p/\partial x_3 = -G$ is a constant pressure gradient, and u is the velocity component in the x_3 direction. This equation can be rewritten as

$$\nabla^2 u = -\frac{G}{\mu}\,. \tag{b}$$

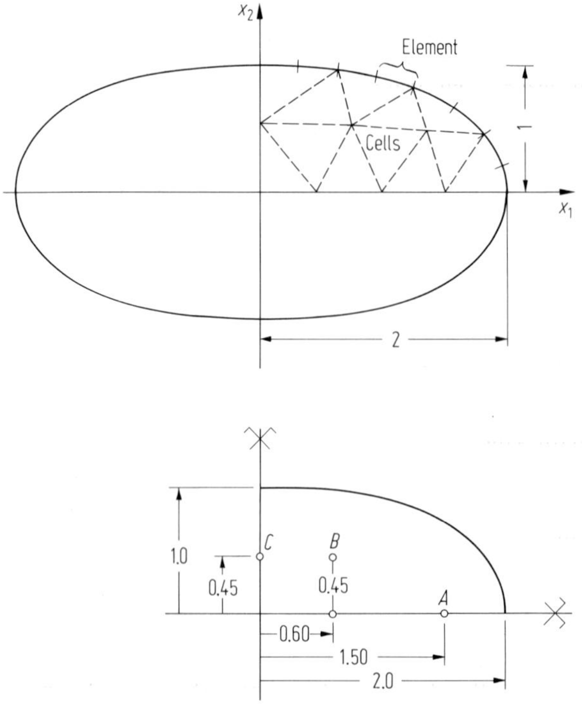

Fig. 2.21. Elliptical cross-section

For a pipe of elliptical cross section the velocity distribution is of the form (Fig. 2.21)

$$u = \frac{G}{2\mu(a^{-2}+b^{-2})}\left(1 - \frac{x_1^2}{a^2} - \frac{x_2^2}{b^2}\right), \tag{c}$$

where a and b are the semiaxes of the ellipse. Taking the value of the constant $G/\mu = 2$ and the semiaxes $a = 2$ and $b = 1$, the problem to be solved is

$$\nabla^2 u = -2 \tag{d}$$

with boundary conditions

$$u = 0 \qquad \text{on } \Gamma . \tag{e}$$

The solution of the above Poisson equation can be divided into two parts,

$$u = u_1 + u_2 , \tag{f}$$

where $u_1 = -(x_1^2 + x_2^2)/2$ is a particular solution and u_2 a complementary one, which satisfies $\nabla^2 u_2 = 0$ with boundary condition $u_2 = -u_1$ on Γ.

Table 2.2

Point	Variable	Poisson Eq. anal. [a]	Laplace Eq. anal. [a]	Exact
A	u	0.345	0.351	0.350
(1.50, 0)	$\partial u/\partial x_1$	−0.597	−0.604	−0.600
	$\partial u/\partial x_2$	0.0	0.0	0.0
B	u	0.563	0.569	0.566
(0.60, 0.45)	$\partial u/\partial x_1$	−0.253	−0.240	−0.240
	$\partial u/\partial x_2$	0.718	0.720	0.720
C	u	0.634	0.641	0.638
(0, 0.45)	$\partial u/\partial x_1$	0.0	0.0	0.0
	$\partial u/\partial x_2$	0.722	0.720	0.720

[a] Allowing eight elements to discretize a quarter of the boundary.

Results for the velocity u and for the derivatives $\partial u/\partial x_1$ and $\partial u/\partial x_2$ (necessary for the evaluation of the tangential stresses $\tau_{x_1 x_3}$ and $\tau_{x_2 x_3}$) are presented in Table 2.2, compared to the analytical solution.

First, the solution using the transformation (f) was used and the Laplace equation problem solved. Second, the problem was solved as a Poisson equation (d) and the domain divided into 12 cells using Hammer's (seven points) numerical integration scheme [34]. Subdividing the domain into more cells or employing a more refined numerical integration scheme resulted in no significant improvement on the solution. Due to the double symmetry of the problem only one-quarter of the cross section needed to be analyzed.

2.8. Subregions

If the problem under consideration is defined over a region which is only piecewise homogeneous, the numerical procedures described can be applied to each homogeneous subregion as they were separated from the others. The final system of equations for the whole region is obtained by adding the set of equations (2.81) for each subregion together with compatibility and equilibrium conditions between their interfaces [21].

To illustrate these ideas in more detail, consider for simplicity a region Ω consisting of two subregions Ω^1 and Ω^2 (Fig. 2.22). Over subregion Ω^1, we define the following:

$\boldsymbol{U}^1$, $\boldsymbol{Q}^1$ – nodal potentials and fluxes ($q_1 = k_1\,\partial u/\partial n$) at the external boundary Γ^1;
$\boldsymbol{U}_I^1$, $\boldsymbol{Q}_I^1$ – nodal potentials and fluxes at the interface Γ_I, considering it belongs to Ω^1;

Similarly, we define over subregion Ω^2:

$\boldsymbol{U}^2$, $\boldsymbol{Q}^2$ – nodal potentials and fluxes ($q_2 = k_2\,\partial u/\partial n$) at the external boundary Γ^2;
$\boldsymbol{U}_I^2$, $\boldsymbol{Q}_I^2$ – nodal potentials and fluxes at the interface Γ_I, considering it belongs to Ω^2;

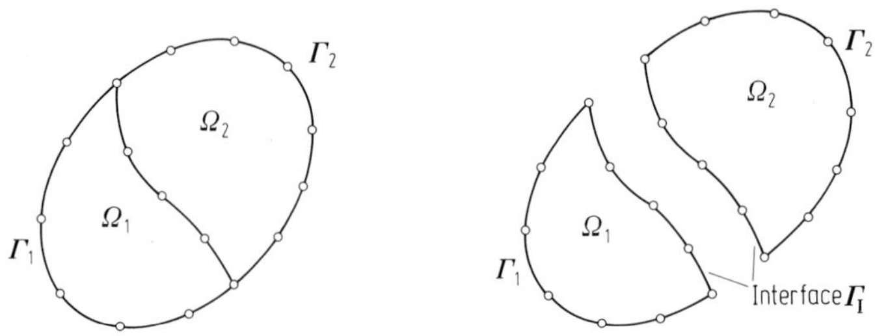

Fig. 2.22. Domain divided into two subregions

The system of equations (2.81) corresponding to subregion Ω^1 can be written

$$[\boldsymbol{H}^1 \ \boldsymbol{H}_I^1] \begin{Bmatrix} \boldsymbol{U}^1 \\ \boldsymbol{U}_I^1 \end{Bmatrix} = [\boldsymbol{G}^1 \ \boldsymbol{G}_I^1] \begin{Bmatrix} \boldsymbol{Q}^1 \\ \boldsymbol{Q}_I^1 \end{Bmatrix} . \tag{2.117}$$

For subregion Ω^2, we have

$$[\boldsymbol{H}^2 \ \boldsymbol{H}_I^2] \begin{Bmatrix} \boldsymbol{U}^2 \\ \boldsymbol{U}_I^2 \end{Bmatrix} = [\boldsymbol{G}^2 \ \boldsymbol{G}_I^2] \begin{Bmatrix} \boldsymbol{Q}^2 \\ \boldsymbol{Q}_I^2 \end{Bmatrix} . \tag{2.118}$$

The compatibility and equilibrium conditions to be applied at the interface Γ_I between Ω^1 and Ω^2 are, respectively,

$$\begin{aligned} \boldsymbol{U}_I^1 &= \ \ \boldsymbol{U}_I^2 = \boldsymbol{U}_I , \\ \boldsymbol{Q}_I^1 &= -\boldsymbol{Q}_I^2 = \boldsymbol{Q}_I . \end{aligned} \tag{2.119}$$

Equations (2.117)–(2.119) can be combined to form the system

$$\begin{vmatrix} \boldsymbol{H}^1 & \boldsymbol{H}_I^1 & \boldsymbol{0} \\ \boldsymbol{0} & \boldsymbol{H}_I^2 & \boldsymbol{H}^2 \end{vmatrix} \begin{Bmatrix} \boldsymbol{U}^1 \\ \boldsymbol{U}_I \\ \boldsymbol{U}^2 \end{Bmatrix} = \begin{vmatrix} \boldsymbol{G}^1 & \boldsymbol{G}_I^1 & \boldsymbol{0} \\ \boldsymbol{0} & -\boldsymbol{G}_I^2 & \boldsymbol{G}^2 \end{vmatrix} \begin{Bmatrix} \boldsymbol{Q}^1 \\ \boldsymbol{Q}_I \\ \boldsymbol{Q}^2 \end{Bmatrix} \tag{2.120}$$

or, more simply,

$$\boldsymbol{HU} = \boldsymbol{GQ} . \tag{2.121}$$

This system of equations is formally similar to Eq. (2.81) except that the matrices $\boldsymbol{H}$ and $\boldsymbol{G}$ are now banded. By imposing the boundary conditions of the problem and remembering that both the potentials and fluxes at the interface are considered as unknowns, the system (2.120) can be reordered as

$$\begin{bmatrix} \boldsymbol{H}^1 & \boldsymbol{H}_I^1 & -\boldsymbol{G}_I^1 & \boldsymbol{0} \\ \boldsymbol{0} & \boldsymbol{H}_I^2 & \boldsymbol{G}_I^2 & \boldsymbol{H}^2 \end{bmatrix} \begin{Bmatrix} \boldsymbol{U}^1 \\ \boldsymbol{U}_I \\ \boldsymbol{Q}_I \\ \boldsymbol{U}^2 \end{Bmatrix} = \begin{bmatrix} \boldsymbol{G}^1 & \boldsymbol{0} \\ \boldsymbol{0} & \boldsymbol{G}^2 \end{bmatrix} \begin{Bmatrix} \boldsymbol{Q}^1 \\ \boldsymbol{Q}^2 \end{Bmatrix} . \tag{2.122}$$

According to the prescribed boundary conditions, the submatrices corresponding to Γ^1 (and Γ^2) may interchange their positions. Notice that the final system matrix in Eq. (2.122) is also banded.

Detailed explanations of the computer implementation of the above procedures, including numerical results, can be found in [21, 28, 29].

Example 2.5. This example studies an impermeable dam with sheet piles resting on a permeable isotropic soil with an impermeable stratum underneath. The right- and left-hand side boundaries are also assumed to be impervious and potential boundary conditions are applied at the nodes in contact with the water. The coefficient of permeability of the soil is 0.03048 m/min and the base width of the dam is 79 m.

This problem was solved analytically by Lambe and Whitman [30] and numerically using the BEM and the FEM by Chang [29]. The discretizations employed are shown in Fig. 2.23 and it can be noted that for the BEM analysis, the

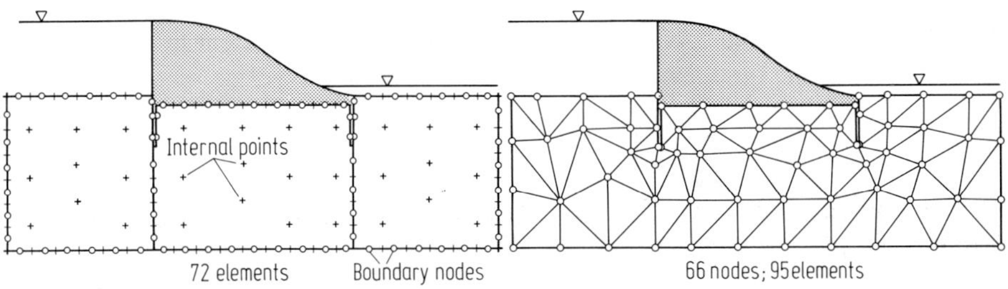

Fig. 2.23. Constant boundary element versus finite element grid

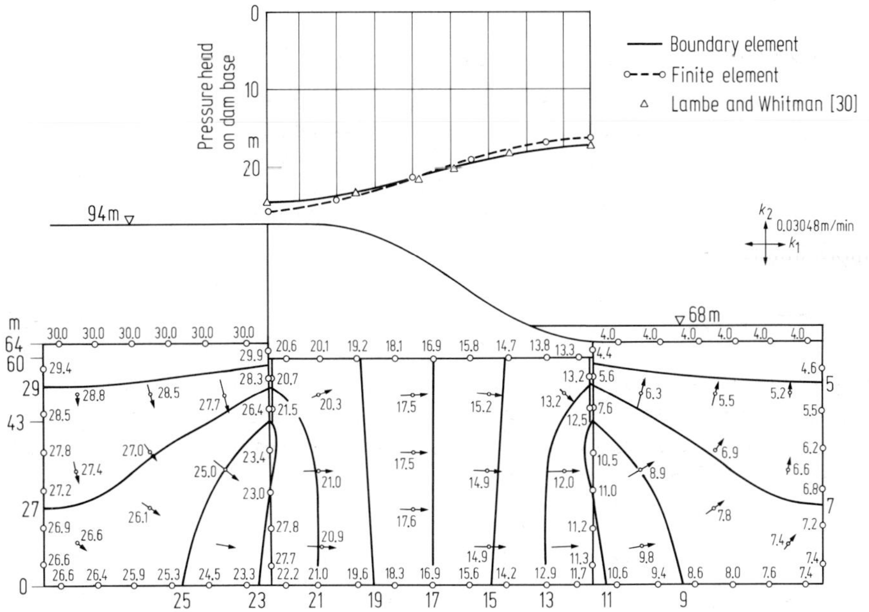

Fig. 2.24. Pressure distribution and flow under two-piled dam

region was subdivided into three different subregions due to the presence of the sheet piles.

The pressure head distribution on the base of the dam is plotted in Fig. 2.24 for the three methods. The equipotentials and the velocities at some internal points obtained with the BEM are also depicted in the figure.

2.9. Orthotropy and Anisotropy

Let us now assume that the medium over which the problem is defined is orthotropic (see Fig. 2.25). The governing equation in the directions of orthotropy can be written

$$k_1 \frac{\partial^2 u}{\partial y_1^2} + k_2 \frac{\partial^2 u}{\partial y_2^2} = 0 \tag{2.123}$$

for the two-dimensional case; k_i is the medium property coefficient in the direction of orthotropy i. The fundamental solution to this equation is [21]

$$u^*(\xi, x) = \frac{1}{(k_1 k_2)^{1/2}} \ln \frac{1}{r(\xi, x)}, \tag{2.124}$$

where

$$r(\xi, x) = \left\{ \frac{1}{k_1} [y_1(\xi) - y_1(x)]^2 + \frac{1}{k_2} [y_2(\xi) - y_2(x)]^2 \right\}^{1/2}. \tag{2.125}$$

Applying the divergence theorem to the terms of Eq. (2.123) yields

$$\int_\Omega \left(k_1 \frac{\partial^2 u}{\partial y_1^2} + k_2 \frac{\partial^2 u}{\partial y_2^2} \right) d\Omega = \int_\Gamma \left(k_1 \frac{\partial u}{\partial y_1} n_{y_1} + k_2 \frac{\partial u}{\partial y_2} n_{y_2} \right) d\Gamma, \tag{2.126}$$

where n_{y1} and n_{y2} are the direction cosines of the outward normal n to surface Γ (Fig. 2.25). The term enclosed in brackets in the right-hand side integral is the

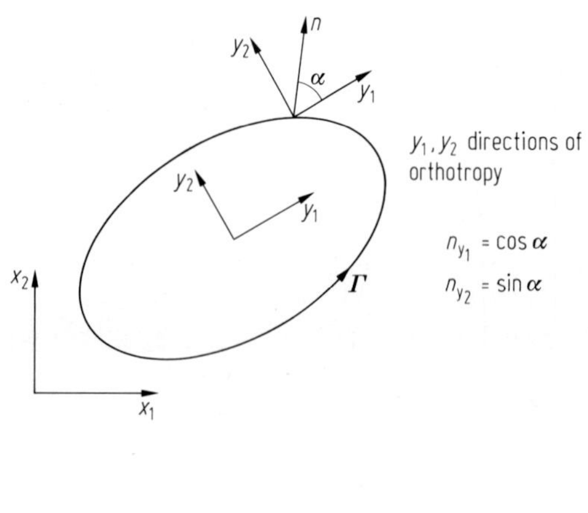

Fig. 2.25. Orthotropic medium

normal boundary flux q. Analogously, we can define

$$q^* (\xi, x) = k_1 \frac{\partial u^* (\xi, x)}{\partial y_1 (x)} n_{y_1} + k_2 \frac{\partial u^* (\xi, x)}{\partial y_2 (x)} n_{y_2} . \tag{2.127}$$

The problem can then be solved in the same manner as for isotropic problems, i.e., by transforming the governing equation (2.123) plus boundary conditions into a boundary integral equation similar to Eq. (2.69).

For fully anisotropic media, the governing equation becomes

$$k_{11} \frac{\partial^2 u}{\partial x_1^2} + 2k_{12} \frac{\partial^2 u}{\partial x_1 \partial x_2} + k_{22} \frac{\partial^2 u}{\partial x_2^2} = 0 \tag{2.128}$$

for two-dimensional problems, the coefficients k_{ij} defining the medium properties. This equation has the following fundamental solution [31]:

$$u^* (\xi, x) = \frac{1}{|k_{ij}|^{1/2}} \ln \frac{1}{r(\xi, x)}, \tag{2.129}$$

where $|k_{ij}|$ is the determinant of the medium property coefficients matrix and

$$r(\xi, x) = \left\{ \frac{1}{k_{11}} [x_1(\xi) - x_1(x)]^2 + \frac{2}{k_{12}} [x_1(\xi) - x_1(x)] [x_2(\xi) - x_2(x)] + \frac{1}{k_{22}} [x_2(\xi) - x_2(x)]^2 \right\}^{1/2} . \tag{2.130}$$

The normal boundary flux q is now given by

$$q = \left(k_{11} \frac{\partial u}{\partial x_1} + k_{12} \frac{\partial u}{\partial x_2} \right) n_{x_1} + \left(k_{12} \frac{\partial u}{\partial x_1} + k_{22} \frac{\partial u}{\partial x_2} \right) n_{x_2} . \tag{2.131}$$

Analogously, we have

$$q^* (\xi, x) = \left(k_{11} \frac{\partial u^* (\xi, x)}{\partial x_1 (x)} + k_{12} \frac{\partial u^* (\xi, x)}{\partial x_2 (x)} \right) n_{x_1} + \left(k_{12} \frac{\partial u^* (\xi, x)}{\partial x_1 (x)} + k_{22} \frac{\partial u^* (\xi, x)}{\partial x_2 (x)} \right) n_{x_2} \tag{2.132}$$

and the problem can now be solved as before.

Example 2.6. The case of groundwater flow under a dam with two different orthotropic strata of soil is considered. The coefficients of permeability of the lower layer are $k_1 = 0.25 \times 10^{-5}$ m/s and $k_2 = 0.075 \times 10^{-5}$ m/s; the principal permeability of the upper stratum makes an angle of 45° with the horizontal and the value of its coefficients is $k_1 = 4.0 \times 10^{-5}$ m/s and $k_2 = 1.0 \times 10^{-5}$ m/s. The dam retains 20 m of water upstream and has 5 m of tailwater downstream.

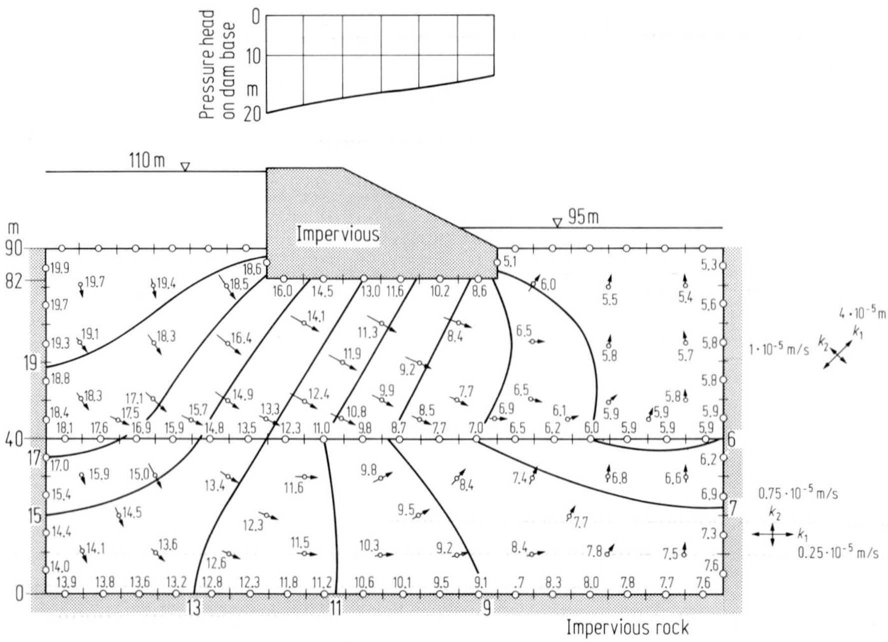

Fig. 2.26. Seepage flow in two orthotropic soils

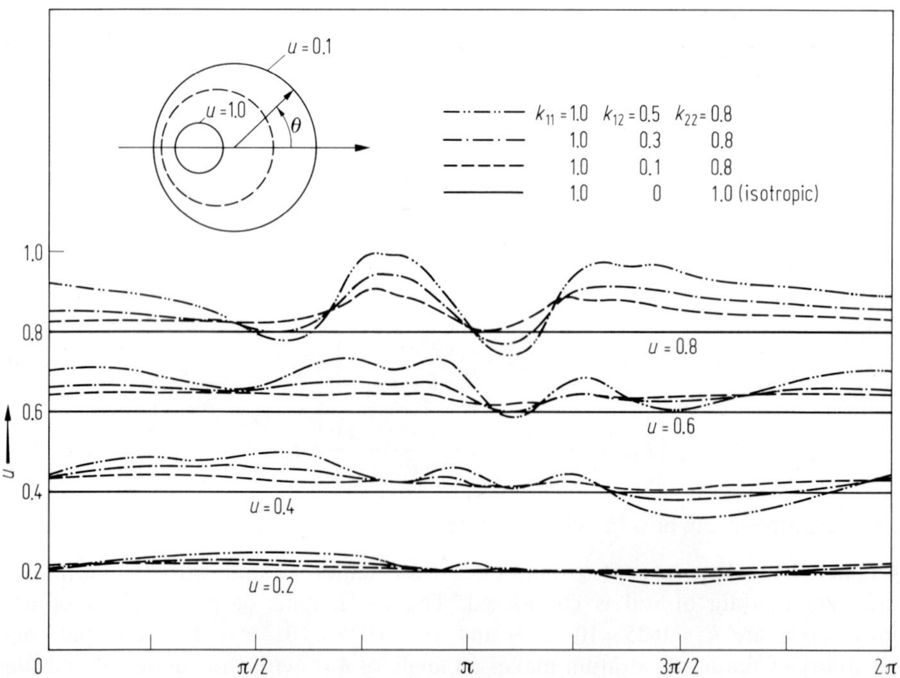

Fig. 2.27. Isotherms in anisotropic and isotropic media for various values of k_{12}

The pressure distribution on the dam base, the equipotential lines, and the flow velocities at various points under the dam are plotted in Fig. 2.26. These results were obtained by using 74 constant boundary elements [29].

Example 2.7. In order to verify the effects of anisotropy in temperature distributions, Chang *et al.* [31] studied an eccentric hollow cylinder subjected to uniform and constant temperatures applied at the inner and outer boundaries.

Results obtained by using the BEM are plotted in Fig. 2.27. It is interesting to note that the most significant quantity to characterize the anisotropy is the determinant of the conductivity coefficients, i.e., $|k_{ij}| = k_{11} k_{22} - k_{12}^2$ and that the smaller the value of $|k_{ij}|$, the more asymmetric are the temperature fields.

2.10. Infinite Regions

Although the boundary integral equation (2.69) has been derived considering that the domain Ω is bounded, all concepts presented thus far are also valid for infinite regular regions in the sense defined by Kellogg [3], i.e., regions bounded by a regular surface (hence a bounded surface) and containing all sufficiently distant points. However, for this extension to be valid, certain regularity conditions concerning the behavior of the functions in Eq. (2.69) on a surface which is infinitely remote from the origin must be fulfilled.

Let $\bar{\Gamma}$ be the surface of a circle (or a sphere if the problem is three-dimensional) of radius R surrounding the surface Γ and centered at the point ξ (Fig. 2.28). A boundary integral equation similar to Eq. (2.69) for the finite domain Ω enclosed by the actual surface Γ and the fictitious surface $\bar{\Gamma}$ can be written

$$\begin{aligned} c(\xi)\, u(\xi) + \int_{\Gamma} u(x)\, q^*(\xi, x)\, d\Gamma(x) + \int_{\bar{\Gamma}} u(x)\, q^*(\xi, x)\, d\Gamma(x) \\ = \int_{\Gamma} q(x)\, u^*(\xi, x)\, d\Gamma(x) + \int_{\bar{\Gamma}} q(x)\, u^*(\xi, x)\, d\Gamma(x)\,. \end{aligned} \tag{2.133}$$

If we let the radius $R \to \infty$, Eq. (2.133) will only be valid for points in Γ (and Ω) if

$$\lim_{R \to \infty} \int_{\bar{\Gamma}} [q(x)\, u^*(\xi, x) - u(x)\, q^*(\xi, x)]\, d\Gamma(x) = 0\,. \tag{2.134}$$

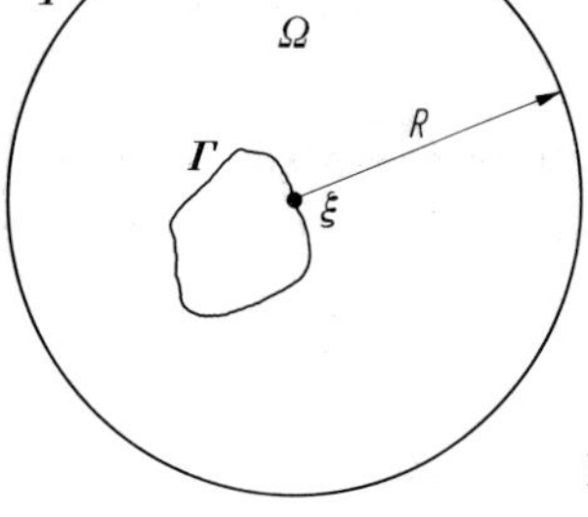

Fig. 2.28. Infinite region with cavity

For three-dimensional problems, since

$$\begin{aligned} d\Gamma(x) &= |\boldsymbol{J}|\, d\varphi\, d\Phi\,, & |\boldsymbol{J}| &= O(R^2)\,, \\ u^*(\xi, x) &= O(R^{-1})\,, & x &\in \bar{\Gamma}\,, \\ q^*(\xi, x) &= O(R^{-2})\,, \end{aligned} \tag{2.135}$$

where $|\boldsymbol{J}|$ is the Jacobian and $O(\,)$ represents the asymptotic behavior of the functions as $R \to \infty$; the condition of Eq. (2.134) is satisfied if the function $u(x)$ behaves at most as $O(R^{-1})$, such that its derivative $q(x) = O(R^{-2})$. These are the regularity conditions at infinity [3, 16] and they ensure that each term in the integral in Eq. (2.134) behaves at most as $O(R^{-1})$, i.e., they approach zero as $R \to \infty$.

For two-dimensional problems, we have that the function $u^*(\xi, x)$ behaves as the logarithm of R and its derivative $q^*(\xi, x) = O(R^{-1})$ as $R \to \infty$. The regularity conditions at infinity for this case imply that $u(x)$ behaves at most as $\ln R$ such that $q(x) = O(R^{-1})$. Note that now the terms in the integral in Eq. (2.134) do not approach zero separately as $R \to \infty$ since $d\xi(x) = |\boldsymbol{J}|\, d\xi$ and $|\boldsymbol{J}| = O(R)$, but they cancel each other, thus fulfilling condition (2.134).

Therefore, applying condition (2.134) into Eq. (2.133) yields

$$c(\xi)\, u(\xi) + \int_\Gamma u(x)\, q^*(\xi, x)\, d\Gamma(x) = \int_\Gamma q(x)\, u^*(\xi, x)\, d\Gamma(x)\,, \tag{2.136}$$

that is, the boundary integral equation obtained for points on the internal surface Γ of the infinite regular region is of the same form as Eq. (2.69) for finite regions. The same is also valid for the integral equation for points inside the infinite region.

Consider a three-dimensional Neumann problem defined over the infinite domain Ω. Unlike the case of finite regions (see Section 2.3), Eq. (2.136) has a unique solution for arbitrary continuous values of q prescribed over the internal boundary Γ. Moreover, the Gauss condition (2.53) need not be satisfied by q since the integral of q around Γ is balanced by a compensating flux at infinity. Since the region enclosed by Γ and $\bar{\Gamma}$ is solenoidal, we can write

$$\int_\Gamma q(x)\, d\Gamma(x) + \int_{\bar{\Gamma}} q(x)\, d\Gamma(x) = 0\,, \tag{2.137}$$

where

$$\int_{\bar{\Gamma}} q(x)\, d\Gamma(x) = O(1) \tag{2.138}$$

since $u(x) = O(R^{-1})$ as $R \to \infty$. If $u(x)$ behaves as $O(R^{-2})$, the flux over $\bar{\Gamma}$ vanishes and so Eq. (2.137) becomes the Gauss condition

$$\int_\Gamma q(x)\, d\Gamma(x) = 0\,. \tag{2.139}$$

Conversely, if condition (2.139) is fulfilled, it follows that $u(x)$ behaves $O(R^{-2})$ as $R \to \infty$.

By analogous considerations we can state that, for two-dimensional Neumann problems, satisfaction of the Gauss condition (2.139) ensures that $u(x)$ behaves at most as $O(R^{-1})$ as $R \to \infty$.

If the function q tends towards a nonzero limiting value at infinity, this value can be included in the analysis through a particular solution as will be shown in a following example.

Example 2.8. Let us consider the problem of a circular cavity of unit radius in an infinite two-dimensional region with Neumann boundary conditions, i.e., a constant radial influx of 31.21 J/(m^2 s) specified along the cavity surface (Fig. 2.29).

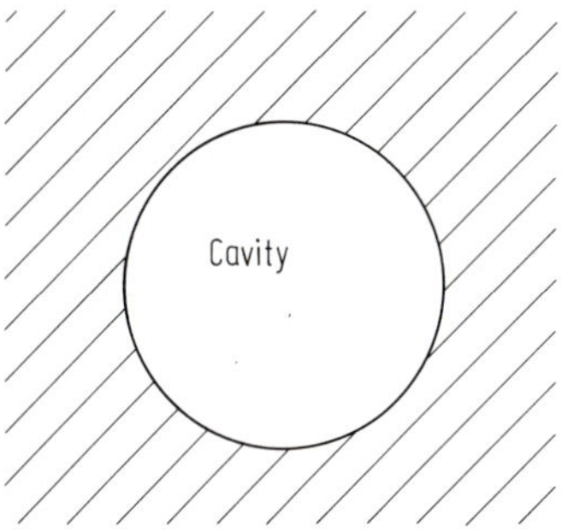

Fig. 2.29. Circular cavity in an infinite two-dimensional region

Since the Gauss condition (2.139) is not satisfied, the solution will have a logarithmic potential behavior at infinity. The exact solution of this problem is

$$u = -31.21 \ln R$$

which shows the expected behavior.

Results for the temperature at points on the boundary Γ and inside the domain Ω are given in Table 2.3 and for the radial flux at points in Ω in Table 2.4, compared to the exact solution. Taking symmetry into account, only one-quarter of the cavity surface was subdivided using constant elements.

Table 2.3. Temperature (−) at points on infinite region.

R	BEM ($N = 4$)	BEM ($N = 8$)	Exact
1.0	0.48	0.12	0.00
1.5	12.57	12.63	12.65
2.0	21.49	21.60	21.63
3.0	34.07	34.23	34.28
5.0	49.91	50.15	50.22
10.0	71.40	71.75	71.86
100.0	142.81	143.50	143.72
1000.0	214.21	215.24	215.58

Table 2.4. Radial flux at points in infinite region.

R	BEM ($N=4$)	BEM ($N=8$)	Exact
1.5	20.68	20.77	20.81
2.0	15.51	15.58	15.61
3.0	10.34	10.39	10.40
5.0	6.20	6.23	6.24
10.0	3.10	3.12	3.12
100.0	0.31	0.31	0.31
1000.0	0.03	0.03	0.03

Example 2.9. The previously discussed formulation for infinite regions can be readily applied to practical problems such as potential fluid flow past obstacles.

As an example, we study a two-dimensional potential flow with uniform onset velocity V in the x_1 direction around a NACA 0018 aerofoil, whose shape is shown in Fig. 2.30. For the solution of this problem, we employ a stream function ψ defined as

$$q_1 = \frac{\partial \psi}{\partial x_2}, \qquad q_2 = -\frac{\partial \psi}{\partial x_1}.$$

It is now convenient to separate the stream function ψ into two parts,

$$\psi = \psi_1 + \psi_2,$$

where $\psi_1 = V x_2$ defines the steady onset flow and ψ_2 is a perturbation stream function. Since the perturbation decays at infinity, we require that $\psi_2 = O(R^{-1})$ at most as $R \to \infty$. Furthermore, as $\nabla^2 \psi = 0$, we also have that $\nabla^2 \psi_2 = 0$ and the problem can now be solved in terms of the perturbation ψ_2.

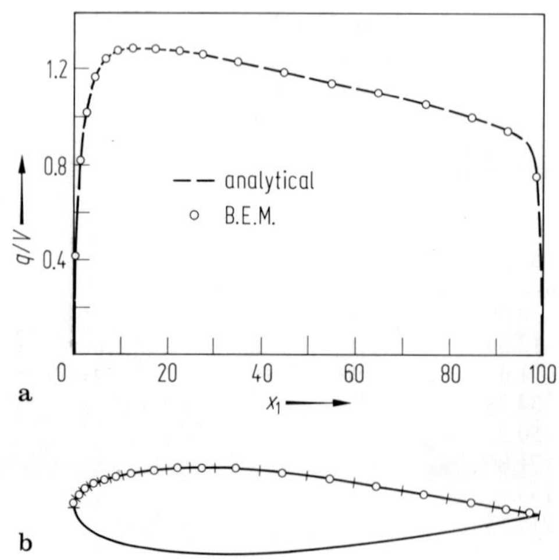

Fig. 2.30. NACA 0018 aerofoil: **a** results; **b** discretization

Considering the surface of the aerofoil as the streamline $\psi = 0$, the boundary conditions of the problem are

$$\psi_2 = -\psi_1 = -Vx_2 \qquad \text{on } \Gamma .$$

Since the problem is antisymmetric with respect to the x_1 axis, only one-half of the aerofoil needed to be analyzed. The constant boundary elements discretization employed for the solution is shown in Fig. 2.30. Results for the tangential velocity presented in the same figure are in good agreement with results given in [32].

2.11. Special Fundamental Solutions

The fundamental solutions that we have presented so far can be immediately recognized as Green's functions for an infinite region. Since they were derived without any proper attention to boundary conditions, the boundary conditions of the actual problem are introduced by requiring that the function or its normal derivative (or a linear relationship between both) satisfies prescribed values at points on the boundary, which was previously discretized. In some problems, the region may be confined in some regular way and it may be more convenient to find a fundamental solution specific to the region [21].

As an example, let us derive the fundamental solution for a semi-infinite region such as occurs in fluid mechanics or geotechnical problems (Fig. 2.31). In a problem of this nature, it is preferable to remove the infinite boundary $\bar{\Gamma}$. By choosing a fundamental solution which identically satisfies the boundary condition on the surface $\bar{\Gamma}$ we shall not need to discretize this surface, thus considerably reducing the amount of numerical work involved in the solution of the problem.

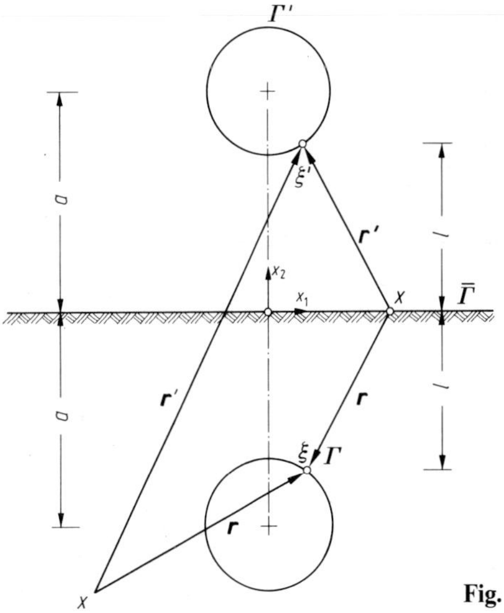

Fig. 2.31. Semi-infinite region

Consider a source of intensity $\sigma(\xi)$ at a point ξ belonging to Γ (Fig. 2.31). The potential generated by this source will somehow be reflected at the surface $\bar{\Gamma}$, depending on the boundary conditions applied there. In order to represent this reflection, we shall introduce an image source of intensity $\sigma(\xi')$ at a point ξ' symmetrically located with respect to $\bar{\Gamma}$. Thus, the potential at any field point x will be the superposition of the ones generated by both sources, i.e.,

$$u(x) = \sigma(\xi)\, u^*(\xi, x) + \sigma(\xi')\, u^*(\xi', x)\,, \tag{2.140}$$

where u^* is the infinite space fundamental solution.

Applying the boundary condition $u = 0$ at the surface $\bar{\Gamma}$, we obtain

Case 1: Dirichlet Problem

$$\sigma(\xi)\, u^*(\xi, x) + \sigma(\xi')\, u^*(\xi', x) = 0 \qquad \text{on } \bar{\Gamma}\,. \tag{2.141}$$

Taking for instance the two-dimensional Laplace equation, condition (2.141) implies that

$$\sigma(\xi) \ln \frac{1}{r} + \sigma(\xi') \ln \frac{1}{r'} = 0 \tag{2.142}$$

which gives

$$\sigma(\xi) = -\,\sigma(\xi')\,. \tag{2.143}$$

Note $r \equiv r'$ for x on $\bar{\Gamma}$ (Fig. 2.31).

Since, by definition, the fundamental solution is equivalent to the potential generated by unit sources, the fundamental solution for the semi-infinite region with zero potential at the interface is simply

$$u^*(\xi, x) = \frac{1}{2\pi} \ln \frac{r'}{r}\,. \tag{2.144}$$

If the boundary condition at the interface $\bar{\Gamma}$ is that of zero normal flux, i.e.,

Case 2: Neumann Problem

$$\sigma(\xi)\, q^*(\xi, x) + \sigma(\xi')\, q^*(\xi', x) = 0 \qquad \text{on } \bar{\Gamma} \tag{2.145}$$

implying the condition

$$-\,\sigma(\xi)\left(\frac{l}{r}\right) + \sigma(\xi')\left(\frac{l}{r'}\right) = 0 \tag{2.146}$$

which gives

$$\sigma(\xi) = \sigma(\xi')\,, \tag{2.147}$$

then the fundamental solution of the problem is

$$u^*(\xi, x) = \frac{1}{2\pi} \ln (r r')\,. \tag{2.148}$$

Fundamental solutions for other problems such as parallel layered regions can be constructed in the same way, as well as for three-dimensional problems.

Example 2.10. This example studies the steady-state heat conduction problem of a semi-infinite medium bounded internally by two parallel and equal cylinders, as shown in Fig. 2.32. The interface $\bar{\Gamma}$ is at zero temperature, the temperature at infinity is also zero, and the surfaces of the cylinders are isotherms.

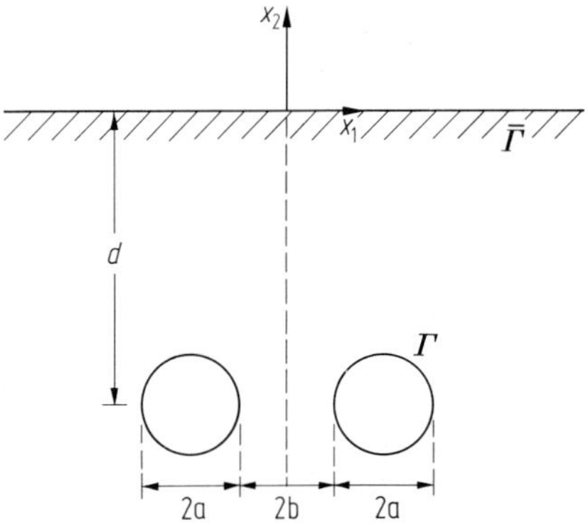

Fig. 2.32. Semi-infinite region bounded internally by two equal cylinders

If the depth d is much greater than the cylinder radius a, this problem can be seen as an approximation to the more practical one of two equally loaded electricity cables laid direct in the ground in horizontal formation. Of interest in this kind of problem is the determination of the external thermal resistance of each cable.

The thermal resistance G per unit length between the surface Γ at a temperature c and the surface $\bar{\Gamma}$ at zero temperature, through a medium with thermal conductivity k, is given by

$$G = -\frac{c}{k \int_{\Gamma} q \, d\Gamma}. \tag{a}$$

Results for the nondimensional ratio Gk are presented in Table 2.5 for several values of d/a, for a unit cylinder surface temperature c. Two different cases were

Table 2.5. Values of ratio Gk.

b	d/a	BEM	Analytical
$b=0$	10	0.810	0.810
	25	1.102	1.102
	50	1.322	1.322
	100	1.543	1.543
$b=a$	10	0.726	0.724
	25	1.016	1.014
	50	1.236	1.235
	100	1.457	1.456

considered, i.e., when the cables are touching ($b=0$) and when the cable spacing equals one diameter ($b=a$). These results were obtained by subdividing the surface of one cylinder into 32 elements and considering symmetry with respect to the x_2 axis. Also shown in the table are the results obtained through an approximate analytical solution [33]; the agreement between both solutions is very good.

2.12. Three-Dimensional Problems

The solution of Eq. (2.69) for three-dimensional problems can be attempted following basically the same steps as discussed in Section 2.4 for two dimensions. The boundary Γ, now a two-dimensional surface, can be modelled by using flat or curved (see Chapter 3) triangles or quadrilaterals and the potentials and normal derivatives over it assumed to be piecewise constant, linear, quadratic, etc. The interpolation functions adopted are generally the same as employed for two-dimensional finite element analysis.

In what follows, the numerical procedures necessary for the computer implementation of a simple element, namely, a flat triangle with constant potential and normal derivative will be described in detail. Numerical results are included at the end of the section in order to show the validity of these procedures. As for the two-dimensional case, extension to higher-order interpolation functions presents no further difficulties (see Chapter 3).

The functions u and q are assumed to be constant within each element and associated to their nodal values at the centroid of the element (Fig. 2.33). If the boundary Γ is discretized into j elements, Eq. (2.69) becomes

$$c_i\, u_i + \sum_{j=1}^{N} \left\{ \int_{\Gamma_j} q^*\, d\Gamma \right\} u_j = \sum_{j=1}^{N} \left\{ \int_{\Gamma_j} u^*\, d\Gamma \right\} q_j\,. \tag{2.149}$$

The transformation of the element of surface $d\Gamma$ from the global Cartesian system of coordinates to the intrinsic system of coordinates in a triangular element gives

$$d\Gamma = |\boldsymbol{J}|\, d\eta_1\, d\eta_2\,, \tag{2.150}$$

where the Jacobian $|\boldsymbol{J}|$ equals twice the area of the triangle – for further details see Chapter 3. The unit normal vector, necessary for the evaluation of the function q^*, can be calculated by considering the cross-product between the vectors $(2-1)$ and $(3-1)$ shown in Fig. 2.33.

The integrals to be calculated in Eq. (2.149) are of the type

$$\begin{aligned} \int_{\Gamma_j} q^*\, d\Gamma &= |\boldsymbol{J}| \int_0^1 \left[\int_0^{1-\eta_2} q^*(\eta)\, d\eta_1 \right] d\eta_2\,, \\ \int_{\Gamma_j} u^*\, d\Gamma &= |\boldsymbol{J}| \int_0^1 \left[\int_0^{1-\eta_2} u^*(\eta)\, d\eta_1 \right] d\eta_2\,. \end{aligned} \tag{2.151}$$

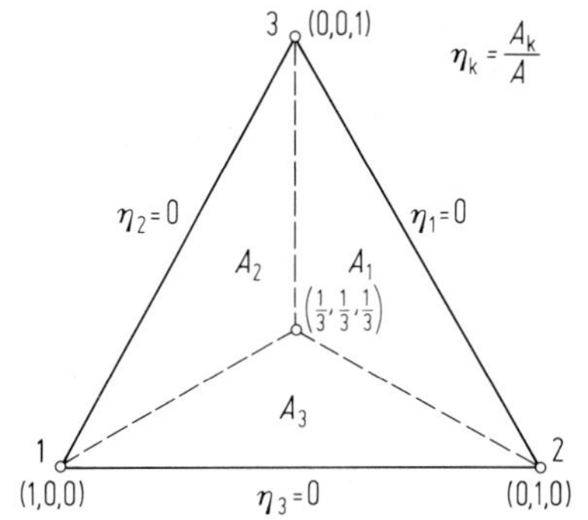

Fig. 2.33. Intrinsic triangular coordinates

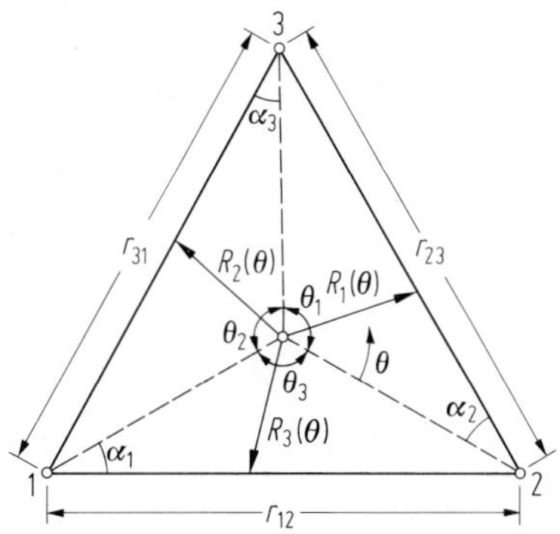

Fig. 2.34. Geometrical definitions for analytical integration

For the cases when the i node does not belong to the Γ_j element these integrals are computed numerically using Hammer's quadrature scheme [34]. Thus, the off-diagonal coefficients of matrices H and G in Eq. (2.81) are given by summations of the form

$$H_{ij} = 2A_j\, d_{ij} \sum_{k=1}^{K} \frac{1}{r_{ik}^3}\, w_k\,, \qquad G_{ij} = 2A_j \sum_{k=1}^{K} \frac{1}{r_{ik}}\, w_k \tag{2.152}$$

in which A_j and d_{ij} refer to the element j, A_j being its area and d_{ij} the perpendicular distance from the point i to the plane passing through the element. The H_{ii} coefficients are equal to $\frac{1}{2}$ for constant elements. The G_{ii} coefficients, which contain integrable singularities, can be evaluated analytically by employing polar coordinates (see Fig. 2.34),

$$G_{ii} = \int_0^{\theta_1} \int_0^{R_1(\theta)} dR\, d\theta + \int_{\theta_1}^{\theta_1+\theta_2} \int_0^{R_2(\theta)} dR\, d\theta + \int_{\theta_1+\theta_2}^{2\pi} \int_0^{R_3(\theta)} dR\, d\theta$$

which after evaluation yields

$$G_{ii} = \frac{2A_i}{3} \left\{ \frac{1}{r_{23}} \ln\left(\frac{\tan[(\theta_1+\alpha_2)/2]}{\tan[\alpha_2/2]}\right) + \frac{1}{r_{31}} \ln\left(\frac{\tan[(\theta_2+\alpha_3)/2]}{\tan[\alpha_3/2]}\right) + \frac{1}{r_{12}} \ln\left(\frac{\tan[(\theta_3+\alpha_1)/2]}{\tan[\alpha_1/2]}\right) \right\}. \tag{2.153}$$

Example 2.11. The first three-dimensional example studies the temperature distribution over a unit cube with Dirichlet boundary conditions as follows (see Fig. 2.35):

$$u = 1 \quad \text{at } x_1 = +0.5\,, \qquad u = 0 \quad \text{at } x_2 = \pm 0.5\,,$$
$$u = 2 \quad \text{at } x_1 = -0.5\,, \qquad u = 0 \quad \text{at } x_3 = \pm 0.5\,.$$

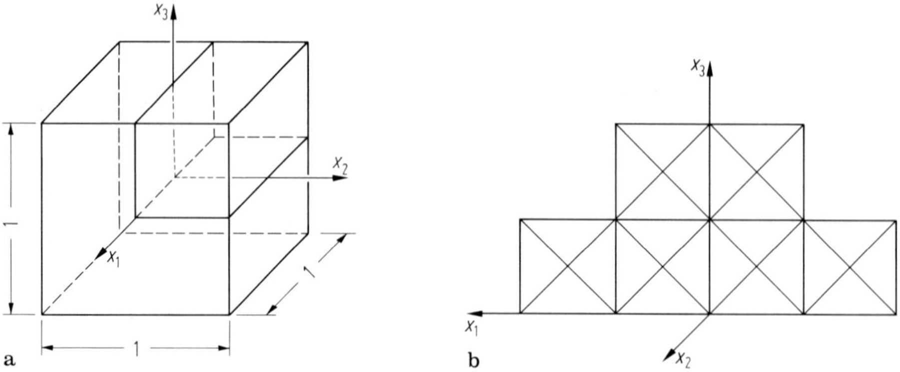

Fig. 2.35. Unit cube: **a** geometry; **b** discretization

Table 2.6. Temperature along the x_1 axis.

x_1	BEM ($N = 12$)	BEM ($N = 24$)	Analytical
− 0.375	1.637	1.472	1.430
− 0.250	1.044	0.979	0.967
− 0.125	0.678	0.661	0.659
0.	0.500	0.500	0.500
0.125	0.478	0.472	0.472
0.250	0.597	0.566	0.560
0.375	0.855	0.770	0.748

Due to symmetry with respect to the planes $x_1 - x_2$ and $x_1 - x_3$, only one-quarter of the actual cube needed to be analyzed. Two different meshes were employed, the finer of which is shown in Fig. 2.35. Results for the temperature at some internal points are presented in Table 2.6 and compared with an available analytical solution [7].

Example 2.12. In this example, we seek the temperature distribution over a rectangular parallelepiped with the following mixed boundary conditions (Fig. 2.36):

$$u = 10 \qquad \text{at } x_1 = -0.5\,,$$

$$\frac{\partial u}{n} + 5u = 0 \qquad \text{at } x_1 = +0.5\,,$$

$$\frac{\partial u}{\partial n} + 5u = 0 \qquad \text{at } x_2 = \pm 1\,,$$

$$\frac{\partial u}{\partial n} + 5u = 0 \qquad \text{at } x_3 = \pm 1\,.$$

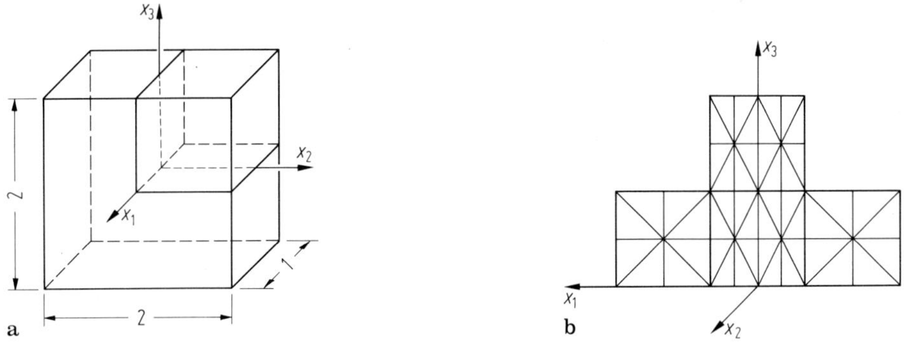

Fig. 2.36. Rectangular parallelepiped: **a** geometry; **b** discretization

Table 2.7. Temperature at internal points.

x_1	x_2	x_3	BEM ($N=24$)	BEM ($N=48$)	Analytical
− 0.25	0.	0.	7.387	7.282	7.259
0.	0.	0.	4.827	4.840	4.837
0.	0.50	0.50	3.745	3.843	3.843
0.25	0.	0.	2.816	2.843	2.844
0.25	0.25	0.25	2.612	2.658	2.658
0.25	0.50	0.50	2.000	2.073	2.089
0.25	0.75	0.75	1.050	1.144	1.180

As in the previous example, we take advantage of the symmetry of the problem with respect to the planes $x_1 - x_2$ and $x_1 - x_3$. The finer mesh employed in this analysis (48 elements) is shown in Fig. 2.36 and the numerical results at some internal points are compared with the ones obtained through an analytical solution [7] in Table 2.7.

Example 2.13. Let us now study the problem of a spherical cavity of unit radius in an infinite region with a constant radial influx of 10 J/(m^2 s) prescribed along the cavity surface. The exact solution of this problem is simply,

$$u = \frac{10}{R}$$

which shows that the expected solution behaves as $O(R^{-1})$ as $R \to \infty$ since the Gauss condition (2.139) is not satisfied.

By taking symmetry into account, only one-eighth of the cavity surface needed to be analyzed. Results for the averaged surface temperature and for the temperature at some points inside the domain Ω are shown in Table 2.8, compared to the exact solution. The slow convergence of the numerical solution on and near the cavity surface are attributed to the geometrical approximation of the sphere using flat elements.

Table 2.8. Temperature at points on infinite region.

R	BEM ($N = 7$)	BEM ($N = 16$)	Exact
1.0	9.676	9.727	10.000
1.5	6.505	6.569	6.667
2.0	4.899	4.922	5.000
3.0	3.274	3.281	3.333
6.0	1.639	1.640	1.667
10.0	0.983	0.984	1.000
100.0	0.098	0.098	0.100
1000.0	0.010	0.010	0.010

2.13. Axisymmetric Problems

In Section 2.2, it was pointed out that the fundamental solution to the two-dimensional Laplace equation (the logarithmic potential) can be derived by integrating the three-dimensional one (the Newtonian potential) for a line source at a point ξ. The same idea can be applied in order to derive the fundamental solution for Laplace's equation over an axisymmetric domain, which is equivalent to a ring source.

Assuming that all boundary values have axial symmetry (and consequently all domain values are also axisymmetric), Eq. (2.69) can be written in cylindrical polar coordinates (R, θ, Z) as

$$c(\xi)\, u(\xi) + \int_{\bar{\Gamma}} u(x) \int_0^{2\pi} q^*(\xi, x)\, d\theta(x)\, R(x)\, d\bar{\Gamma}(x) = \int_{\bar{\Gamma}} q(x) \int_0^{2\pi} u^*(\xi, x)\, d\theta(x)\, R(x)\, d\bar{\Gamma}(x) \tag{2.154}$$

since

$$d\Gamma(x_1, x_2, x_3) = R\, d\theta\, d\bar{\Gamma}(R, Z)\,. \tag{2.155}$$

Note that $\bar{\Gamma}$ is the generating boundary contour which is the intersection of Γ with the R^+-Z semiplane (Fig. 2.37).

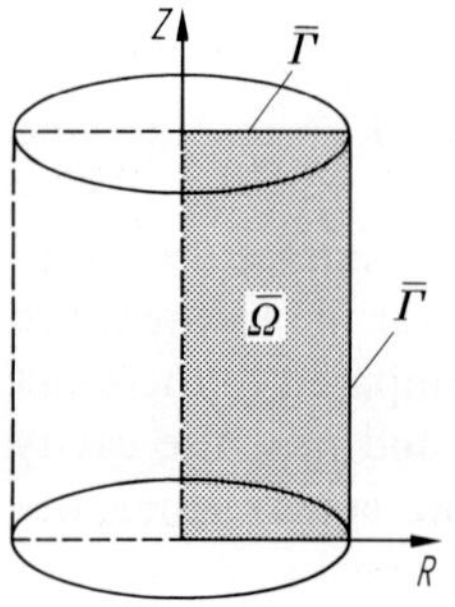

Fig. 2.37. Generating area and boundary contour of solid of revolution

Writing the three-dimensional fundamental solution in cylindrical polar coordinates,

$$u^*(\xi, x) = \frac{1}{r(\xi, x)} \tag{2.156}$$

$$= \frac{1}{\{R^2(\xi) + R^2(x) - 2R(\xi)\,R(x)\cos[\theta(\xi) - \theta(x)] + [Z(\xi) - Z(x)]^2\}^{1/2}},$$

the axisymmetric one can be calculated explicitly in terms of the complete elliptic integral of the first kind $K(m)$ as

$$\bar{u}^*(\xi, x) = \int_0^{2\pi} u^*(\xi, x)\,d\theta(x) = \frac{4K(m)}{(a+b)^{1/2}}, \tag{2.157}$$

where

$$\begin{aligned} m &= \frac{2b}{a+b}, \\ a &= R^2(\xi) + R^2(x) + [Z(\xi) - Z(x)]^2, \\ b &= 2R(\xi)\,R(x). \end{aligned} \tag{2.158}$$

The range of variation of the parameter m is $0 \leqq m \leqq 1$. Unlike the two- and three-dimensional cases, the axisymmetric fundamental solution cannot be written as simply a function of the distance between two points, but it also depends on the distance of the points to the axis of revolution.

The normal derivative of the fundamental solution along the boundary contour $\bar{\Gamma}$ is given by

$$\begin{aligned} \bar{q}^*(\xi, x) = {} & \frac{4}{(a+b)^{1/2}} \\ & \cdot \left\{ \frac{1}{2R(x)} \left[\frac{R^2(\xi) - R^2(x) + [Z(\xi) - Z(x)]^2}{a-b} E(m) - K(m) \right] n_R(x) \right. \\ & \left. + \frac{Z(\xi) - Z(x)}{a-b} E(m)\, n_Z(x) \right\}, \end{aligned} \tag{2.159}$$

where $E(m)$ is the complete elliptic integral of the second kind.

From expressions (2.157)–(2.159), it can be seen that as $R(\xi) \to 0$ we have that $m \to 0$, $K(m) \to \pi/2$, $E(m) \to \pi/2$, so that the ring source tends to a point source with intensity 2π over the axis of revolution.

Substituting Eqs. (2.157) and (2.159) into Eq. (2.154) yields the following boundary integral equation:

$$c(\xi)\,u(\xi) + \int_{\bar{\Gamma}} u(x)\,\bar{q}^*(\xi, x)\,R(x)\,d\bar{\Gamma}(x) = \int_{\bar{\Gamma}} q(x)\,\bar{u}^*(\xi, x)\,R(x)\,d\bar{\Gamma}(x). \tag{2.160}$$

The solution of Eq. (2.160) can be attempted by using the same basic calculation procedures as discussed in Section 2.6 for two-dimensional problems.

For convenience of numerical computation, the complete elliptic integrals can be approximated by polynomial expressions [35].

After discretizing Eq. (2.160) and summing the contributions from all boundary elements, a system of equations of the form (2.81) is obtained. The terms H_{ij} and $G_{ij}\,(i \neq j)$ of this system are evaluated numerically using a standard Gaussian quadrature with four integration points. The diagonal terms H_{ii} and G_{ii}, however, are the result of evaluating singular integrals for which standard quadratures cannot be applied.

In order to facilitate the evaluation of these integrals, the fundamental solution and its normal derivative can be written in terms of Legendre functions of the second kind as

$$\bar{u}^*(\xi, x) = \frac{8^{1/2} Q_{-1/2}(\gamma)}{b^{1/2}}, \tag{2.161}$$

$$\begin{aligned} \bar{q}^*(\xi, x) = -\frac{8^{1/2}}{R(x)\, b^{1/2}} \Bigg\{ \Bigg[\frac{Q_{-1/2}(\gamma)}{2} &+ \frac{R^2(\xi) - R^2(x) + [Z(\xi) - Z(x)]^2}{b} \frac{dQ_{-1/2}(\gamma)}{d\gamma} \Bigg] n_R(x) \\ &+ \frac{Z(\xi) - Z(x)}{R(\xi)} \frac{dQ_{-1/2}(\gamma)}{d\gamma} n_Z(x) \Bigg\}, \end{aligned} \tag{2.162}$$

where

$$\gamma = 1 + \frac{a-b}{b}, \qquad 1 \leqq \gamma \leqq \infty. \tag{2.163}$$

This form of the fundamental solution is the same as given by Snow [36].

This Legendre function can be expanded, for small values of γ, as [37]

$$Q_{-1/2}(\gamma) = -\frac{1}{2} \ln\left(\frac{\gamma - 1}{32}\right), \tag{2.164}$$

$$\frac{dQ_{-1/2}(\gamma)}{d\gamma} = -\frac{1}{2(\gamma - 1)}. \tag{2.165}$$

The substitution of expressions (2.164) and (2.165) into Eqs. (2.161) and (2.162) permits the explicit evaluation of the singular integrals. Formulas obtained through analytical integration are given in [38].

Notice that for elements located near the axis of revolution (so with small $R(\xi)$) it is not always possible to integrate the whole element analytically in the manner described above since the value of the parameter γ will be large for points far from the singularity and, therefore, approximations (2.164) and (2.165) are no longer valid for these points. Thus, the scheme adopted for these cases was to integrate analytically over a short segment near the singularity and numerically integrate the rest of the element using a standard Gaussian quadrature, as if these parts were separate elements. For computational purposes, the length L of the analytically integrated part of the element was assumed to be [39]

$$\frac{L}{2} \leqq \left[\frac{R(\xi)\,R(x)}{50}\right]^{1/2} \leqq \frac{l}{2}, \tag{2.166}$$

where l is the total length of the element and $R(x)$ is the distance from the nearest point of that part to the axis of revolution.

Example 2.14. The problem of a spherical cavity of unit radius in an infinite medium already studied with three-dimensional elements in Example 2.13 is now restudied with axisymmetric constant elements in order to assess a comparison between both types of approximations.

Results are presented in Table 2.9 for two different discretizations of one-half of the generating contour of the sphere, taking symmetry into account. This provides a better geometrical representation of the cavity surface and the improvement of the results reflects this fact.

Table 2.9. Temperature at points on infinite region.

R	BEM ($N = 4$)	BEM ($N = 8$)	Exact
1.0	9.961	9.991	10.000
1.5	6.539	6.634	6.667
2.0	4.904	4.976	5.000
3.0	3.269	3.317	3.333
6.0	1.635	1.659	1.667
10.0	0.981	0.995	1.000
100.0	0.098	0.100	0.100
1000.0	0.010	0.010	0.010

Example 2.15. Finally, a more practical application is the analysis of a prototype nuclear reactor pressure vessel subjected to an increase of temperature applied on the inside. This problem was studied using 96 triangular finite elements in [40] and the results, as well as the mesh employed, are reproduced in Fig. 2.38.

Results for a constant boundary element analysis employing 31 elements and taking into account the symmetry with respect to the R axis are plotted in Fig. 2.39, and compare well with the finite element solution.

2.14. Axisymmetric Problems with Arbitrary Boundary Conditions

If an axisymmetric body is subjected to arbitrary (nonaxisymmetric) boundary conditions, the problem cannot be treated as in Eq. (2.154) any longer due to the angular dependence of the variables u and q. Nevertheless, formulations to deal

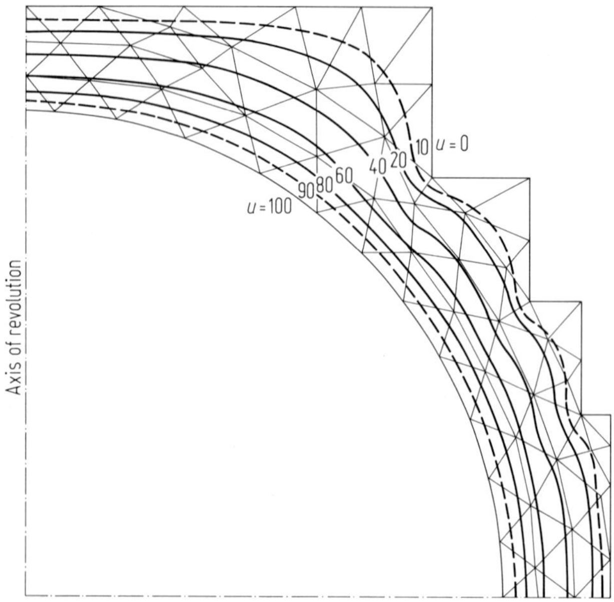

Fig. 2.38. FEM mesh and isotherms

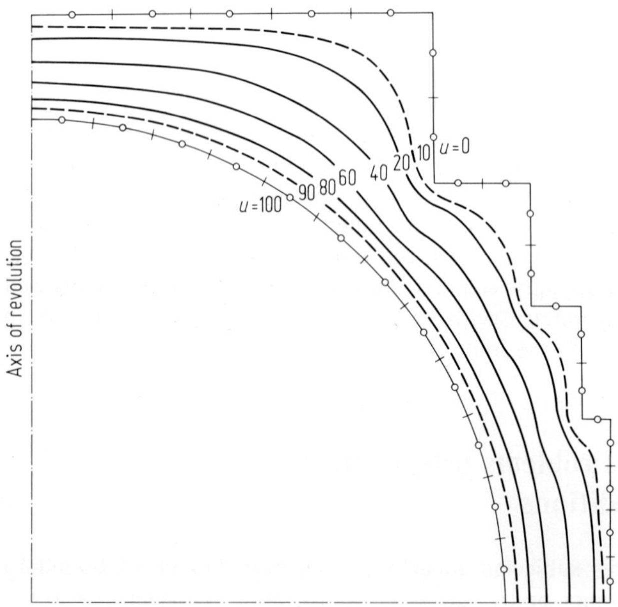

Fig. 2.39. BEM discretization and isotherms

with such problems can be developed by employing appropriate expansions of the variables in Fourier series, resulting in a sequence of uncoupled two-dimensional angularly independent problems [42].

In this section, we discuss (following [43]) the necessary procedures for the BEM solution of this type of problem. Note that the solution procedure for each element of the sequence of two-dimensional equations provides a further reduction in dimension such that boundary-value problems for axisymmetric bodies under arbitrary boundary conditions are reduced to a sequence of one-dimensional ones, involving only a line integration.

We start by reminding that the initial integral statement (2.68) for the BEM formulation of three-dimensional potential problems is of the form

$$u(\xi) = \frac{1}{4\pi} \int_{\Gamma} [q(x)\, u^*(\xi, x) - u(x)\, q^*(\xi, x)]\, d\Gamma(x)\,, \tag{2.167}$$

where $u^*(\xi, x) = 1/r(\xi, x)$.

Assuming that the boundary Γ is axisymmetric, the various functions appearing in Eq. (2.167) can be expanded in Fourier series as follows:

$$v(x) = \sum_{n=0}^{\infty} [v_n^c(\bar{x}) \cos n\theta(x) + v_n^s(\bar{x}) \sin n\theta(x)]\,, \tag{2.168}$$

$$v^*(\xi, x) = \sum_{n=0}^{\infty} [v_n^{*c}(\xi, \bar{x}) \cos n\theta(x) + v_n^{*s}(\xi, \bar{x}) \sin n\theta(x)] \tag{2.169}$$

in which v is symbolically either u or q and functions of $\bar{x}$ are regarded as functions of $R(x)$ and $Z(x)$ only, i.e.

$$v(x) = v[R(x), Z(x)]\,. \tag{2.170}$$

Coefficients of given v in expression (2.168) are obtained by standard Fourier analysis, while coefficients of unknown v are the solution of the sequence of one-dimensional boundary integral equations to be derived. Coefficients of the function v^* are transformed into expressions given by Eqs. (2.173) and (2.174).

Expressing $d\Gamma(x)$ as in Eq. (2.155) and taking into account the orthogonality of the trigonometric functions as discussed in [43], the following integral equations relating the coefficients of the Fourier expansions are obtained:

$$\begin{aligned} u_n^c(\bar{\xi}) = \frac{1}{4\pi} \int_{\bar{\Gamma}} [&q_n^c(\bar{x})\, u_n^{*c}(\bar{\xi}, \bar{x}) + q_n^s(\bar{x})\, u_n^{*s}(\bar{\xi}, \bar{x}) - u_n^c(\bar{x})\, q_n^{*c}(\bar{\xi}, \bar{x}) \\ &- u_n^s(\bar{x})\, q_n^{*s}(\bar{\xi}, \bar{x})]\, R(x)\, d\bar{\Gamma}(\bar{x})\,, \end{aligned} \tag{2.171}$$

$$\begin{aligned} u_n^s(\bar{\xi}) = \frac{1}{4\pi} \int_{\bar{\Gamma}} [&q_n^s(\bar{x})\, u_n^{*c}(\bar{\xi}, \bar{x}) - q_n^c(\bar{x})\, u_n^{*s}(\bar{\xi}, \bar{x}) - u_n^s(\bar{x})\, q_n^{*c}(\bar{\xi}, \bar{x}) \\ &+ u_n^c(\bar{x})\, q_n^{*s}(\bar{\xi}, \bar{x})]\, R(x)\, d\bar{\Gamma}(\bar{x})\,, \end{aligned} \tag{2.172}$$

where

$$v_n^{*c}(\bar{\xi}, \bar{x}) = \int_{-\pi}^{\pi} v^*(\bar{\xi}, \bar{x}, \theta) \cos n\theta \, d\theta \,, \tag{2.173}$$

$$v_n^{*s}(\bar{\xi}, \bar{x}) = \int_{-\pi}^{\pi} v^*(\bar{\xi}, \bar{x}, \theta) \sin n\theta \, d\theta \tag{2.174}$$

with $\theta = \theta(x) - \theta(\xi)$.

The angle integrals in Eqs. (2.173) and (2.174) can be evaluated explicitly and expressed in terms of the complete elliptic integrals of the first and second kinds. The resulting line integrals in Eqs. (2.171) and (2.172) can only be evaluated numerically.

The numerical solution of the sequence of boundary integral equations obtained by taking Eqs. (2.171) and (2.172) to the boundary is discussed in [44], where some results of analyses are presented. Note that the term $n = 0$ corresponds to the case of axisymmetric boundary conditions dealt with in the previous section.

Extension of the above ideas to the analysis of elasticity problems are briefly discussed in Section 5.15, where related references are given. Some more specific problems involving axisymmetric bodies under arbitrary boundary conditions are treated in Sections 9.3 and 12.4.

2.15. Nonlinear Materials and Boundary Conditions

In many practical applications in potential problems the conductivity is a function of the potential, i.e., $k = k(u)$. In this case the governing equation becomes

$$\frac{\partial}{\partial x_1}\left(k \frac{\partial u}{\partial x_1}\right) + \frac{\partial}{\partial x_2}\left(k \frac{\partial u}{\partial x_2}\right) + \frac{\partial}{\partial x_3}\left(k \frac{\partial u}{\partial x_3}\right) + b = 0 \,, \tag{2.175}$$

where b is the source term. The boundary conditions for this equation are

$$\begin{aligned} u &= \bar{u} \qquad && \text{on } \Gamma_1 \,, \\ q &= k \frac{\partial u}{\partial n} = \bar{q} \qquad && \text{on } \Gamma_2 \,. \end{aligned} \tag{2.176}$$

Part of the boundary can also have convective conditions of the type seen in Example 2.3, i.e.,

$$q = h(u_f - u) \qquad \text{on } \Gamma_3 \,, \tag{2.177}$$

where u_f is the potential or temperature in the surrounding medium and h is a heat transfer coefficient, for instance.

Equation (2.175) is nonlinear as $k = k(u)$, and in order to solve it using boundary elements we need to define a source term which will contain all nonlinearities and apply a recursive procedure to find the correct solution. This technique implies using domain cells as explained in Section 2.7. Another way of solving the problem which is much more elegant and efficient is using Kirchhoff's transformation to render the nonlinear problem into a linear one [45–47].

We need to start by defining a new variable $\psi(u)$ such that

$$\nabla\psi = \frac{d\psi}{du}\nabla u\,. \tag{2.178}$$

Comparing the right-hand side of (2.178) with Eq. (2.175) we now define ψ in such a way that

$$\frac{d\psi}{du} = k(u) \tag{2.179}$$

or, in integral form,

$$\psi = K(u) = \int_{u_0}^{u} k(u)\,du\,. \tag{2.180}$$

This equation is called Kirchhoff's transformation, where u_0 is an arbitrary reference value. It follows from Eqs. (2.178) and (2.179) that

$$\frac{\partial\psi}{\partial x_i} = k\frac{\partial u}{\partial x_i}\,, \qquad i = 1, 2, 3\,. \tag{2.181}$$

Equation (2.175) can now be written

$$\nabla^2\psi + b = 0\,. \tag{2.182}$$

Hence we have arrived at the form of a homogeneous isotropic Poisson equation but written in terms of a new variable ψ. Solutions for the cases of constant conductivity may now be used replacing u by ψ. The problems are still linear provided that the boundary conditions are of the essential (Γ_1) or natural (Γ_2) type. If the boundary conditions are of a mixed type, i.e., conditions on Γ_3, the Kirchhoff transformation process introduces nonlinearities.

The conditions on Γ_1 and Γ_2 corresponding to Eq. (2.182) are

$$\begin{aligned} \psi &= \bar{\psi} \qquad \text{on } \Gamma_1\,,\\ \frac{\partial\psi}{\partial n} &= \bar{q} \qquad \text{on } \Gamma_2\,. \end{aligned} \tag{2.183}$$

These equations are related to the original ones by

$$\begin{aligned} \psi &= \bar{\psi} = \int_{u_0}^{\bar{u}} k\,du \qquad \text{on } \Gamma_1\,,\\ \frac{\partial\psi}{\partial n} &= k\frac{\partial u}{\partial n} = \bar{q} \qquad \text{on } \Gamma_2\,. \end{aligned} \tag{2.184}$$

The third type of boundary condition gives

$$\frac{\partial \psi}{\partial n} + h\{K^{-1}(\psi) - u_f\} = 0 \tag{2.185}$$

which is a nonlinear equation. K^{-1} is the inverse Kirchhoff transform, i.e.,

$$K^{-1}(\psi) = u\,. \tag{2.186}$$

Transforming the problem into a boundary integral equation does not present now any special difficulties. The resulting equation will not present any domain integrals other than those due to the source term b and can be written as

$$\begin{aligned} c(\xi)\,\psi(\xi) + \int_{\Gamma_{1+2+3}} \psi(x)\,q^*(\xi,x)\,d\Gamma(x) + \int_{\Gamma_3} h\,K^{-1}[\psi(x)]\,u^*(\xi,x)\,d\Gamma(x) \\ = \int_{\Gamma_{1+2}} q(x)\,u^*(\xi,x)\,d\Gamma(x) + \int_{\Omega} b(x)\,u^*(\xi,x)\,d\Omega(x) + \int_{\Gamma_3} h\,u_f u^*(\xi,x)\,d\Gamma(x)\,. \end{aligned} \tag{2.187}$$

This equation can be discretized in the usual way. In the presence of a Γ_3-type boundary, the system of equations is nonlinear and the equations need to be solved iteratively.

The dependence of the conductivity on the potential can be of different types:

(i) *Exponential Law.* In this case,

$$k = k_0 \exp\left\{\beta\left(\frac{u-u_0}{u_0}\right)\right\}, \tag{2.188}$$

where u_0, β, and k_0 are material constants. For this case,

$$\psi = \bar{\psi} = \int_{u_0}^{\bar{u}} k_0 \exp\left\{\beta\left(\frac{u-u_0}{u_0}\right)\right\} du = k_0 \frac{u_0}{\beta} \exp\left\{\beta\left(\frac{\bar{u}-u_0}{u_0}\right)\right\} - k_0 \frac{u_0}{\beta} \tag{2.189}$$

and the potential can be written

$$u = u_0 + \frac{u_0}{\beta} \ln\left(\frac{\psi\beta}{k_0 u_0} + 1\right). \tag{2.190}$$

Notice that Eqs. (2.189) and (2.190) are valid for all cases, i.e., $\beta \gtrless 0$, $(u - u_0) \gtrless 0$.

(ii) *Power Law.* Another possibility is to define

$$k = k_0 \left\{1 + \beta\left(\frac{u-u_0}{u_0}\right)\right\}^n \tag{2.191}$$

where n is given. Applying the transformation (2.180) in this case gives

$$\psi = \bar{\psi} = \frac{k_0 u_0}{(n+1)\beta}\left\{1 + \beta\left(\frac{\bar{u}-u_0}{u_0}\right)\right\}^{n+1} - \frac{k_0 u_0}{(n+1)\beta} \tag{2.192}$$

with the following function for the potential at any point:

$$u = u_0 + \frac{u_0}{\beta}\left\{1 + \psi\frac{(n+1)\,\beta}{k_0\,u_0}\right\}^{1/(n+1)} - \frac{u_0}{\beta}. \tag{2.193}$$

For the special case of linear function, i.e., $n = 1$, we have

$$\bar{\psi} = \frac{k_0\,u_0}{2\beta}\left\{1 + \beta\left(\frac{\bar{u} - u_0}{u_0}\right)\right\}^2 - \frac{k_0\,u_0}{2\beta} \tag{2.194}$$

and

$$u = u_0 + \frac{u_0}{\beta}\left\{1 + \frac{2\beta\,\psi}{k_0\,u_0}\right\}^{1/2} - \frac{u_0}{\beta}. \tag{2.195}$$

In this case, Eq. (2.195) has two roots, a positive one and a negative one. From the physical point of view the positive root is the only valid one, as the other will imply negative conductivity which has no physical meaning. Equations (2.194) and (2.195) are valid for all cases, i.e., $\beta \gtrless 0$, $(u - u_0) \gtrless 0$ provided $k(u) > 0$.

Example 2.16. Bialecki and Nowak [46] analyzed a two-dimensional steady-state temperature field in the square shown in Fig. 2.40. Heat conductivity was assumed to vary with temperature in accordance with a linear law, i.e.,

$$k = k_0(1 + \beta\,u)\left[\frac{\mathrm{W}}{\mathrm{m\,K}}\right], \tag{a}$$

where $k_0 = 1$. The value of β was assumed to vary to study the influence of this coefficient. The boundaries $x_1 - 0$, $x_2 - 0$ were isolated; on $x_1 = 1$ the temperature was prescribed as $u = 300$ K; on $x_2 = 1$ heat convection with a fluid at a temperature $u_f = 500$ K was assumed to occur. The heat transfer coefficient was taken to be constant, $h = 10$ W/(m^2 K).

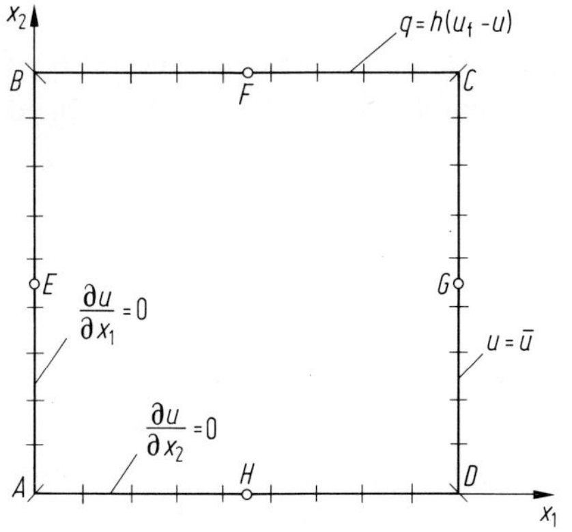

Fig. 2.40. Boundary conditions and discretization of square

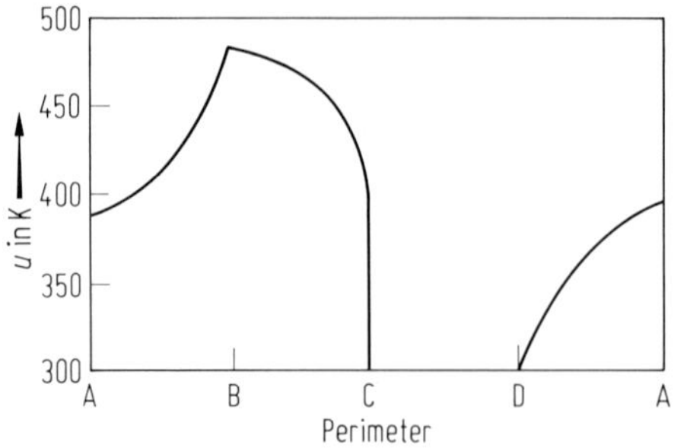

Fig. 2.41. Temperature along perimeter for linear case $\beta = 0$

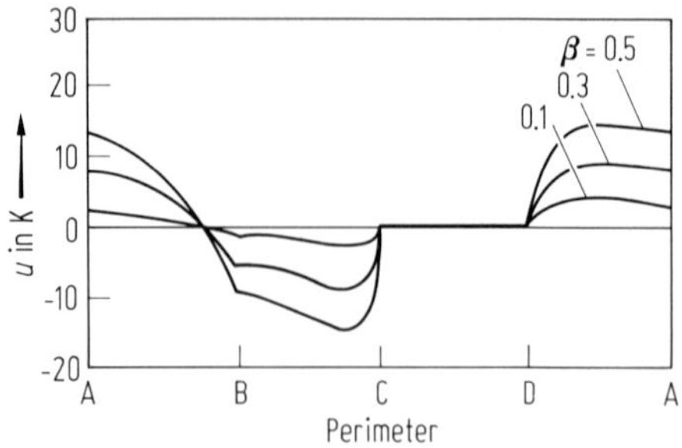

Fig. 2.42. Difference between temperatures for nonlinear and linear cases

The temperature distribution along the perimeter of the square for some values of β is shown in Figs. 2.41 and 2.42. An accuracy in the order of 10^{-4} (i.e., difference between the values at two different steps) was achieved in only 5 to 10 steps.

2.15.1. Nonlinear Boundary Conditions

Let us assume that the nonlinearity now occurs because of heat radiation on the Γ_3 part of the boundary. In this case condition (2.177) can be replaced by

$$q = h(u_f - u) - \sigma\,\varepsilon(u^4 - u_s^4) \qquad \text{on } \Gamma_3, \tag{2.196}$$

where σ is the Steffan – Boltzman constant, ε is the temperature-dependent emissivity between the surface Γ_3 and the radiating medium at a temperature u_s, and h is the temperature-dependent heat transfer coefficient.

The introduction of condition (2.196) into the governing equation in the usual weighted residual way gives rise to a further nonlinear term in Γ_3 corresponding to the radiation condition.

Example 2.17. The following example of nonlinear boundary conditions is also due to Bialecki and Nowak [46] who solved the problem using a special iterative procedure.

The square of Example 2.16 was analyzed and assumed to have a constant heat conductivity $k_0 = 1$ W/(m K) and $\beta = 0$. On the boundary $x_2 = 1$ heat is exchanged

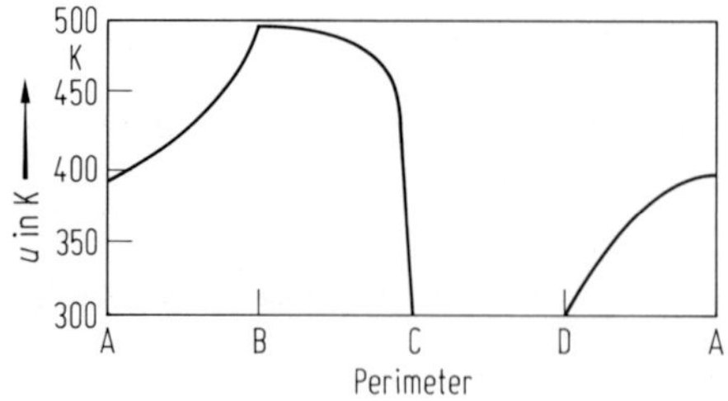

Fig. 2.43. Temperature along perimeter of square with heat radiation on surface BC

by both convection with a fluid having temperature $u_f = 500$ K and radiation with a surface having a temperature $u_s = 500$ K. The heat transfer coefficient is constant, $h = 10$ W/(m^2 K) and the emissivity $\varepsilon = 1$. The Steffan–Boltzman constant is $\sigma = 5.667 \times 10^{-8}$ W/(m^2 K^4). The other boundary conditions are the same as in the previous example. The results shown in Fig. 2.43 were obtained using a fourth-order Runge–Kutta–Gill procedure with an automatic step correction. Direct iterations did not converge when applied to this problem.

References

1. Fredholm, I., Sur une classe d'equations fonctionelles, Acta Math. **27**, 365–390 (1903).
2. Kupradze, V. D., *Potential Methods in the Theory of Elasticity*, Israel Program for Scientific Translations, Jerusalem, 1965.
3. Kellogg, O. D., *Foundations of Potential Theory*, Springer-Verlag, Berlin, 1929.
4. Somigliana, C., Sopra l'equilibrio di un corpo elastico isotropo, *Il Nuovo Ciemento* 17–19 (1886).
5. Brebbia, C. A., *The Boundary Element Method for Engineers*, Pentech Press, London; Halstead Press, New York, 1978 (Second edition, 1980).
6. Morse, P. M., and Feshbach, H., *Methods of Theoretical Physics*, McGraw-Hill, New York, 1953.
7. Carslaw, H. S., and Jaeger, J. C., *Conduction of Heat in Solids*, 2nd ed., Clarendon Press, Oxford, 1959.
8. Roach, G. F., *Green's Functions: Introductory Theory with Applications*, Van Nostrand Reinhold, London, 1970.
9. Jaswon, M. A., Integral equation methods in potential theory, I, Proc. Roy. Soc. Ser. A **275**, 23–32 (1963).
10. Symm, G. T., Integral equation methods in potential theory, II, Proc. Roy. Soc. Ser. A **275**, 33–46 (1963).
11. Jaswon, M. A., and Ponter, A. R., An integral equation solution of the torsion problem, Proc. Roy. Soc. Ser. A **273**, 237–246 (1963).
12. Hess, J. L., and Smith, A. M. O., Calculation of potential flow about arbitrary bodies, Progress in Aeronautical Sciences Vol. 8, (D. Küchemann, Ed.), Pergamon Press, London, 1967.
13. Harrington, R. F., Pontoppidan, K., Abrahamsen, P., and Albertsen, N. C., Computation of Laplacian potentials by an equivalent-source method, Proc. IEE **116**, 1715–1720 (1969).
14. Mautz, J. R., and Harrington, R. F., Computation of rotationally symmetric Laplacian potentials, Proc. IEE **117**, 850–852 (1970).
15. Jaswon, M. A., and Symm, G. T., *Integral Equation Methods in Potential Theory and Elastostatics*, Academic Press, London, 1977.
16. Cruse, T. A., Boundary integral equation methods in solid mechanics, Report SM-73-17 Dept. of Mechanical Engineering, Carnegie-Mellon University, Pittsburgh, 1973.
17. Courant, R., and Hilbert, D., *Methods of Mathematical Physics*, Vol. 2, Interscience Publishers, New York, 1962.
18. Sternberg, W. J., and Smith, T. L., *The Theory of Potential and Spherical Harmonics*, University of Toronto Press, Toronto, 1944.
19. Kantorovich, L. V., and Krylov, V. I., *Approximate Methods of Higher Analysis*, Noordhoff, Groningen, 1958.
20. Miller, G. F., Fredholm equations of the first kind, in *Numerical Solution of Integral Equations* (L. M. Delves and J. Walsh, Eds.), Clarendon Press, Oxford, 1974.
21. Brebbia, C. A., and Walker, S., *Boundary Element Techniques in Engineering*, Butterworths, London, 1980.
22. Moon, P., and Spencer, D. E., *Field Theory Handbook*, 2nd edn., Springer-Verlag, Berlin, 1971.
23. Arpaci, V. S., *Conduction Heat Transfer*, Addison – Wesley, Reading, Mass., 1966.

24. Wilson, E. L., and Nickell, R. E., Application of the finite element method to heat conduction analysis, Nuclear Engng. Design **4,** 276–286 (1966).
25. Peavy, B. A., Steady-state heat conduction in an exposed exterior column of rectangular cross-section, J. Res. Nat. Bur. Stand. **69 C,** 145–151 (1965).
26. Fairweather, G., Rizzo, F. J., Shippy, D. J., and Wu, Y. S., On the numerical solution of two-dimensional potential problems by an improved boundary integral equation method, J. Comput. Phys. **31,** 96–112 (1979).
27. Batchelor, G. K., *An Introduction to Fluid Dynamics,* Cambridge University Press, Cambridge, 1967.
28. Nakaguma, R. K., Three-dimensional elastostatics using the boundary element method, Ph. D. Thesis, Southampton University, 1979.
29. Chang, O. V., Boundary elements applied to seepage problems in zoned anisotropic soils, MSc. Thesis, Southampton University, 1979.
30. Lambe, T. W., and Whitman, R. V., *Soil Mechanics,* Wiley, New York, 1969.
31. Chang, Y. P., Kang, C. S., and Chen, D. J., The use of fundamental Green's functions for the solution of heat conduction in anisotropic media, Int. J. Heat Mass Transfer **16,** 1905–1918 (1973).
32. Abbott, I. H., and von Doenhoff, A. E., *Theory of Wing Sections,* Dover, New York, 1959.
33. Goldenberg, H., External thermal resistance of two buried cables, Proc. IEE **116,** 822–826 (1969).
34. Hammer, P. C., Marlowe, O. J., and Stroud, A. H., Numerical integration over simplexes and cones, Math. Comput. **10,** 130–137 (1956).
35. Abramowitz, M., and Stegun, I. A. (Eds.), *Handbook of Mathematical Functions,* Dover, New York, 1965.
36. Snow, C., *Hypergeometric and Legendre Functions with Application to Integral Equations of Potential Theory,* Applied Mathematical Series No. 19, National Bureau of Standards, Washington, D.C. 1952.
37. Erdelyi, A., *et al., Higher Transcendental Functions,* Vol. 1, Bateman Manuscript Project, McGraw–Hill, New York, 1953.
38. Wrobel, L. C., Potential and viscous flow problems using the boundary element method, Ph. D. Thesis, Southampton University, 1981.
39. Wrobel, L. C., and Brebbia, C. A., Axisymmetric potential problems, in *New Developments in Boundary Element Methods,* Butterworths, London, 1980. CML Publications, Southampton, 1980.
40. Zienkiewicz, O. C., and Cheung, Y. K., Finite elements in the solution of field problems, The Engineer **220,** 507–510 (1965).
41. Ricardella, P. C., An implementation of the boundary-integral technique for planar problems in elasticity and elasto-plasticity, Ph. D. Thesis, Carnegie-Mellon University, 1973.
42. Wilson, E. S., Structural analysis of axisymmetric solids, AIAA J. **3,** 2269–2274 (1965).
43. Rizzo, F. J., and Shippy, D. J., A boundary integral approach to potential and elasticity problems for axisymmetric bodies with arbitrary boundary conditions, Mech. Res. Commun. **6,** 99–103 (1979).
44. Shippy, D. J., Rizzo, F. J., and Gupta, A. K., Boundary integral solution of potential problems involving axisymmetric bodies and nonsymmetric boundary conditions, in *Proc. 10th South Eastern Conf. on Theoretical and Applied Mechanics,* 1980.
45. Akkuratov, Yu. N., and Mikhailov, V. N., The method of boundary integral equations for solving nonlinear heat transmission problems, USSR Comput. Maths. Math. Phys. **20,** 117–125 (1980).
46. Bialecki, R., and Nowak, A. J., Boundary value problems for nonlinear material and non-linear boundary conditions, Applied Mathematical Modelling, **5,** 417–421 (1981).
47. Svelek, P., and Brebbia, C., Nonlinear potential problems, in *Progress in Boundary Element Methods, Vol. 2,* Pentech Press, London, Springer-Verlag, N.Y., 1983.
48. Anza, J. J., Ahedo, E., Da Riva, I., and Alarcon, E., A new boundary condition solved with BEM, in *Boundary Element Methods in Engineering* (C. Brebbia, Ed.), Springer-Verlag, Berlin, 1982.

Chapter 3

Interpolation Functions

3.1. Introduction

It was assumed in Chapter 2 that u and q were constant over each element. In general, however, u and q can have a linear or higher-order variation. In addition, their functional behavior need not be of the same order and, for instance, it may be more consistent to take q of one order less than u since q is given by the derivative of the potential. In practice it is simpler to take both functions u and q of the same order as otherwise the computational procedures become more involved.

In this chapter we will start by discussing how elements with linear variations of u and q, i.e., linear elements, can be developed for two-dimensional problems. We then progress to quadratic and higher-order elements for two-dimensional problems.

If the body is three-dimensional, the boundary elements are part of the external surface of the body and they are usually of two types: quadrilateral and triangular elements. To describe these elements we need to pass from the global three-dimensional coordinate system to a two-dimensional system defined over the surface of the body. Different types of elements are presented in this section together with the necessary transformation laws.

Three-dimensional cell elements, resembling classical finite elements, are also discussed. These elements may be useful to represent part of the body where integration over the domain is required (e.g., some types of body force terms). Finally, the use of discontinuous elements is described. The boundary element method admits the use of elements for which compatibility ot the potential does not need to be enforced. The simplest of them is the constant element previously derived, but a whole family of discontinuous boundary elements is presented here.

3.2. Linear Elements for Two-Dimensional Problems

Let us consider here a linear variation for u and q for which case the nodes are now located at the intersection between two straight elements such as those shown in Fig. 2.8b.

Equation (2.69) can now be written

$$c_i u_i + \int_\Gamma u\, q^*\, d\Gamma = \int_\Gamma q\, u^*\, d\Gamma$$

or in discretized form

$$c_i u_i + \sum_{j=1}^{N} \int_{\Gamma_j} u\, q^*\, d\Gamma = \sum_{j=1}^{N} \int_{\Gamma_j} q\, u^*\, d\Gamma . \tag{3.1}$$

Note that the $\frac{1}{2}$ coefficient of u_i has been replaced by an unknown c_i value. This is because c_i is equal to $\frac{1}{2}$ only for a smooth boundary. The determination of c_i was already discussed in Section 2.4 but it will soon be shown that it does not need to be known explicitly and this simplifies the formulation.

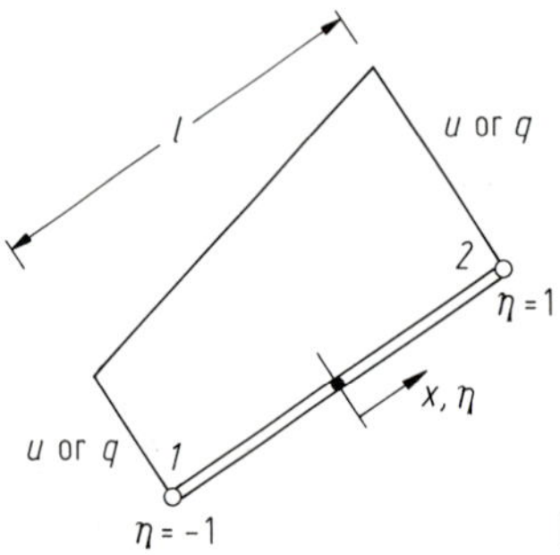

Fig. 3.1. Linear element

The integrals in Eq. (3.1) are now more difficult to evaluate than in Section 2.6 because the u's and q's vary linearly over the element, in the way shown in Fig. 3.1. The values of u and q at any point on the element can be defined in terms of their nodal values and two linear interpolation functions φ_1 and φ_2, functions of the homogeneous coordinate η such that

$$\begin{aligned} u(\eta) &= \varphi_1 u_1 + \varphi_2 u_2 = [\varphi_1\ \varphi_2] \begin{Bmatrix} u_1 \\ u_2 \end{Bmatrix} = \boldsymbol{\varphi}^T \boldsymbol{u}^n , \\ q(\eta) &= \varphi_1 q_1 + \varphi_2 q_2 = [\varphi_1\ \varphi_2] \begin{Bmatrix} q_1 \\ q_2 \end{Bmatrix} = \boldsymbol{\varphi}^T \boldsymbol{q}^n . \end{aligned} \tag{3.2}$$

The dimensionless coordinate η is equal to $x/(l/2)$ and the φ_1, φ_2 functions are given by

$$\varphi_1 = \tfrac{1}{2}(1-\eta), \qquad \varphi_2 = \tfrac{1}{2}(1+\eta). \tag{3.3}$$

The integrals along a segment j on the left-hand side of Eq. (3.1) can be written

$$\int_{\Gamma_j} u\, q^*\, d\Gamma = \int_{\Gamma_j} [\varphi_1\ \varphi_2]\, q^*\, d\Gamma \begin{Bmatrix} u_1 \\ u_2 \end{Bmatrix} = [h_{ij}^1\ h_{ij}^2] \begin{Bmatrix} u_1 \\ u_2 \end{Bmatrix} , \tag{3.4}$$

where

$$h_{ij}^1 = \int_{\Gamma_j} \varphi_1 q^*\, d\Gamma, \qquad h_{ij}^2 = \int_{\Gamma_j} \varphi_2 q^*\, d\Gamma .$$

The h_{ij}^k are influence coefficients defining the interaction between the point i under consideration and a particular k node on an element j.

For the integrals on the right-hand side of Eq. (3.1) we can write

$$\int_{\Gamma_j} q\, u^*\, d\Gamma = \int_{\Gamma_j} [\varphi_1\ \varphi_2]\, u^*\, d\Gamma \begin{Bmatrix} q_1 \\ q_2 \end{Bmatrix} = [g_{ij}^1\ g_{ij}^2] \begin{Bmatrix} q_1 \\ q_2 \end{Bmatrix} \tag{3.5}$$

with

$$g_{ij}^1 = \int_{\Gamma_j} \varphi_1\, u^*\, d\Gamma\,, \qquad g_{ij}^2 = \int_{\Gamma_j} \varphi_2\, u^*\, d\Gamma\,.$$

To write the equation corresponding to node i in discrete form we need to add the contribution from two adjoining elements, $(j-1)$ and (j), into one term, defining the nodal coefficient. This will give the following equation:

$$c_i\, u_i + [\hat{H}_{i1}\ \hat{H}_{i2} \ldots \hat{H}_{iN}] \begin{Bmatrix} u_1 \\ u_2 \\ \vdots \\ u_N \end{Bmatrix} = [G_{i1}\ G_{i2} \cdots G_{iN}] \begin{Bmatrix} q_1 \\ q_2 \\ \vdots \\ q_N \end{Bmatrix}, \tag{3.6}$$

where each $\hat{H}_{ij}$ term is equal to the h^2 term of element $(j-1)$ plus the h^1 term of element (j) for an anticlockwise numbering system. The same applies for G_{ij}. Hence, formula (3.6) represents the assembled equation for node i and it can be written as

$$c_i\, u_i + \sum_{j=1}^{N} \hat{H}_{ij}\, u_j = \sum_{j=1}^{N} G_{ij}\, q_j \tag{3.7}$$

or more simply

$$\sum_{j=1}^{N} H_{ij}\, u_j = \sum_{j=1}^{N} G_{ij}\, q_j\,, \tag{3.8}$$

where

$$H_{ij} = \hat{H}_{ij} \qquad \text{for } i \neq j,$$
$$H_{ij} = \hat{H}_{ij} + c_i \qquad \text{for } i = j.$$

When all the nodes are taken into consideration, Eq. (3.8) produces a $N \times N$ system of equations which can be represented in matrix form as

$$\boldsymbol{HU} = \boldsymbol{GQ}. \tag{3.9}$$

If the surface is not smooth at the point i, the $c_i = \frac{1}{2}$ is no longer valid, but one can calculate the diagonal terms of $\boldsymbol{H}$ by using the fact that when a uniform potential is applied on the whole body the values of q's must be zero. Under this condition Eq. (3.9) for closed domains produces

$$\boldsymbol{HI} = \boldsymbol{O}. \tag{3.10}$$

Equation (3.10) indicates that the sum of all the elements of $\boldsymbol{H}$ in a row should be zero, hence the values of the coefficients in the diagonal can be easily calculated once the off-diagonal coefficients are all known, i.e.,

$$H_{ii} = -\sum_{\substack{j=1 \\ (j \neq i)}}^{N} H_{ij}, \qquad i = 1, 2, \ldots, N. \tag{3.11}$$

For unbounded domains, it is also possible to evaluate the diagonal coefficients of matrix $\boldsymbol{H}$ through condition (3.10), but care must be taken because the regularity conditions at infinity (see Section 2.10) are violated as the function u is now assumed to be constant everywhere in Ω. Since it can easily be shown that

$$\lim_{R \to \infty} \int_{\bar{\Gamma}} q^* \, d\Gamma = -1 \tag{3.12}$$

for both two- and three-dimensional problems, the surface $\bar{\Gamma}$ being defined as in Section 2.10, we have instead of Eq. (3.11) the following relation:

$$H_{ii} = -\sum_{\substack{j=1 \\ (j \neq i)}}^{N} H_{ij} + 1, \qquad i = 1, 2, \ldots, N. \tag{3.13}$$

The G_{ii} terms for linear elements can be computed analytically and are

$$\begin{aligned} G_{ii} &= \frac{l_{i-1}}{4} \int_{-1}^{1} \ln \left[\frac{2}{l_{i-1}(1-\eta)} \right] (1+\eta) \, d\eta + \frac{l_i}{4} \int_{-1}^{1} \ln \left[\frac{2}{l_i(1+\eta)} \right] (1-\eta) \, d\eta \\ &= \frac{l_{i-1}}{2} \left[\frac{3}{2} - \ln (l_{i-1}) \right] + \frac{l_i}{2} \left[\frac{3}{2} - \ln (l_i) \right], \end{aligned} \tag{3.14}$$

where l_{i-1} and l_i are the length of the two elements contributing to the node i.

By applying the prescribed boundary conditions of the problem, Eq. (3.9) can be reordered in such a way that a final system of equations is obtained:

$$\boldsymbol{A}\,\boldsymbol{Y} = \boldsymbol{F}, \tag{3.15}$$

where $\boldsymbol{A}$ is a fully populated matrix of order N and $\boldsymbol{Y}$ is a vector containing all the boundary unknowns.

Notice that in computational terms $H_{ij}(i \neq j)$ and G_{ij} can be assembled directly into $\boldsymbol{A}$ so that Eq. (3.9) does not need to be formed.

Example 3.1. Flow in lakes and other water bodies can be approximated to provide an initial estimate of the circulation, which can then be checked against the full shallow water equations. This flow is governed by the following linearized equations, obtained by neglecting the inertia terms in the momentum equations [1]:

$$\begin{aligned} -h \varrho f q_2 + \varrho g H \frac{\partial \eta}{\partial x_1} + (\tau_1^s - \tau_1^b) &= 0, \\ h \varrho f q_1 + \varrho g H \frac{\partial \eta}{\partial x_2} + (\tau_2^s - \tau_2^b) &= 0 \end{aligned} \tag{a}$$

and the continuity formula

$$\frac{\partial q_1}{\partial x_1} + \frac{\partial q_2}{\partial x_2} = 0, \tag{b}$$

where

f = Coriolis parameter,
q_1, q_2 = vertically integrated velocity components in the x_1, x_2 directions,
ϱ = mass density,
g = acceleration of gravity,
H = $h + \eta$ = total depth of water,
h = depth with relation to the mean water level,
η = elevation of the free surface,
τ^s = wind stresses,
τ^b = bottom friction stresses.

If the η values are much smaller than the h, we can write $H \cong h$, hence

$$\begin{aligned} -h\varrho f q_2 + \varrho g h \frac{\partial \eta}{\partial x_1} + (\tau_1^s - \tau_1^b) = 0, \\ h\varrho f q_1 + \varrho g h \frac{\partial \eta}{\partial x_2} + (\tau_2^s - \tau_2^b) = 0. \end{aligned} \tag{c}$$

Assuming the τ^b terms to be linearly proportional to the mean velocity components

$$\tau_1^b = \gamma q_1, \qquad \tau_2^b = \gamma q_2, \tag{d}$$

we can cross-differentiate equations (c) and afterwards subtract both equations. Assuming that the derivatives of h are negligible (i.e., the bottom slope is small) this gives, taking continuity into consideration, the following equation:

$$\frac{\partial \tau_1^s}{\partial x_2} - \frac{\partial \tau_2^s}{\partial x_1} = \gamma \left(\frac{\partial q_1}{\partial x_2} - \frac{\partial q_2}{\partial x_1} \right). \tag{e}$$

We can propose a stream function ψ such that

$$q_1 = \frac{\partial \psi}{\partial x_2}, \qquad q_2 = -\frac{\partial \psi}{\partial x_1}$$

and formula (e) becomes

$$\nabla^2 \psi = \frac{1}{\gamma} w(x_1, x_2), \tag{f}$$

where

$$w(x_1, x_2) = \frac{\partial \tau_1^s}{\partial x_2} - \frac{\partial \tau_2^s}{\partial x_1}.$$

Note that we have included the Coriolis parameter but assumed it constant for all the lake, i.e., the lake is small enough to allow the neglect of local variations in the Coriolis forces. If we take

$$X_1 = \frac{x_1}{L}, \qquad X_2 = \frac{x_2}{L},$$

$$W(X_1, X_2) = \frac{w(x_1, x_2)}{T/L}, \tag{g}$$

$$\Psi = \frac{\psi}{(f\varepsilon/2H^2)^{1/2} L^2},$$

L being the lateral characteristic length of the lake, T the characteristic wind stress, and ε the eddy viscosity coefficient, equation (f) takes the nondimensional form

$$\nabla^2 \Psi = \frac{1}{d} W(X_1, X_2), \tag{h}$$

where

$$d = \frac{\gamma L (f\varepsilon/2)^{1/2}}{TH}.$$

The wind circulation in the dos Patos Lake, Brazil (Fig. 3.2a) has been analyzed using the above formulation. As a first numerical example, the streamlines for the flow in and out of the lake without wind effects were calculated, taking $\Psi = 0$ for the west shore and $\Psi = 1$ for the east shore. The results are shown in Fig. 3.2b. For this case the governing equation is simply a Laplace equation.

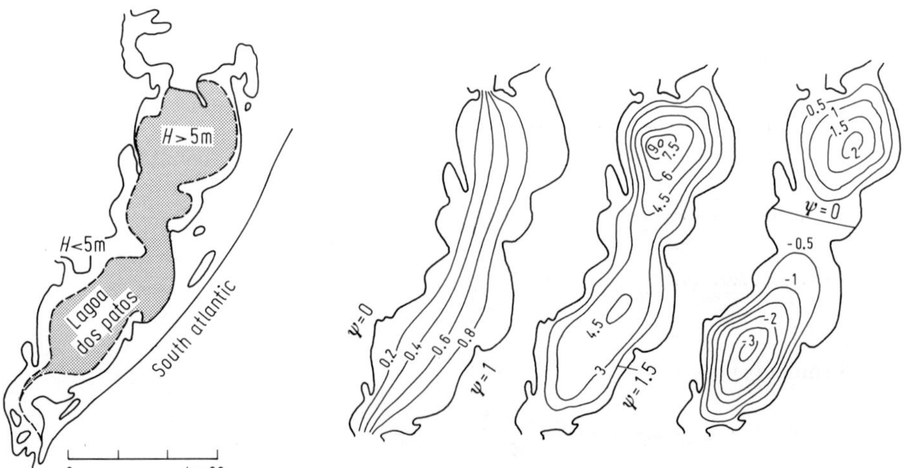

Fig. 3.2. dos Patos Lake: **a** geometry; **b** flow pattern for potential flow; **c** wind-driven mean circulation pattern due to a linear stress distribution; **d** wind-driven mean circulation pattern due to a quadratic stress distribution

If we consider the right-hand side of equation (h) equal to 1, X_1 and X_2, this allows for a superposition of three different sets of results in order of obtain any solution of the type

$$\nabla^2 \Psi = A + BX_1 + CX_2,$$

where the right-hand side represents a quadratic wind stress distribution. Results are shown in Fig. 3.2c for a linear wind stress distribution, $A = 1$, $B = C = 0$, and in Fig. 3.2d for a quadratic wind stress distribution, $A = 1$, $B = -3$, $C = 0$. All the previous results were obtained by discretizing the boundary of the lake using 93 linear elements and subtracting out a particular solution of equation (h).

Example 3.2. The problem represented in Fig. 3.3, that is, a two-dimensional problem of groundwater flow round a tunnel with permeable invert, was also studied with linear elements. If the medium is homogeneous and isotropic, the problem is reduced to that of Laplace's equation for the groundwater pressure u. The boundary conditions of the problem are

$$\begin{aligned} u &= d && \text{on } \bar{\Gamma}, \\ u &= 0 && \text{on } \Gamma_1, \\ q &= -\cos\theta && \text{on } \Gamma_2, \end{aligned} \qquad \text{(a)}$$

where d is the depth of the river and θ is the angle measured from the vertical (see Fig. 3.3). The surface Γ_1 is the permeable invert of the tunnel where we assume that water flows in freely and Γ_2 is the impermeable part of the tunnel lining, where the condition of no flow across the surface holds. Notice that for a point at infinity we have the condition $u = d - x_2$.

The problem can be reformulated by subtracting out the solution at infinity. The groundwater pressure u is divided into two parts,

$$u = u_1 + u_2, \qquad \text{(b)}$$

where $u_1 = d - x_2$ satisfies the infinity condition. Then we have that u_2 tends to zero at infinity and, furthermore, that $\nabla^2 u_2 = 0$ such that the problem can now be

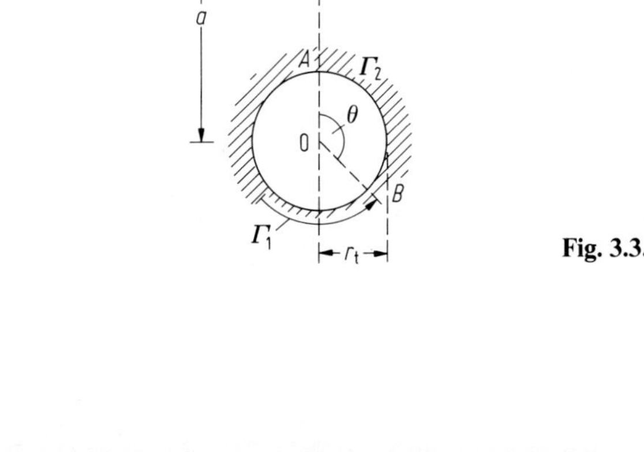

Fig. 3.3. Groundwater flow round a tunnel

solved in terms of u_2. The boundary conditions for u_2 become

$$\begin{aligned} u_2 &= 0 && \text{on } \bar{\Gamma}, \\ u_2 &= -(d + a - r_t \cos\theta) && \text{on } \Gamma_1, \\ q_2 &= 0 && \text{on } \Gamma_2, \end{aligned} \tag{c}$$

where r_t is the radius of the tunnel and a is the distance of the center of the tunnel below the bottom of the river. The numerical values adopted for the parameters are $d = 60$, $a = 30$, $r_t = 3.5$, and $A\hat{O}B = 3\pi/4$.

The boundary element analyses of the problem employed the fundamental solution for semi-infinite regions discussed in Section 2.11; thus, no discretization of the interface $\bar{\Gamma}$ was necessary. Furthermore, due to the symmetry of the problem with respect to the x_2 axis, only one-half of the tunnel surface needed to be modelled.

Results for the function u_2 at some boundary points are presented in Table 3.1 for three different discretizations using linear boundary elements and compared to

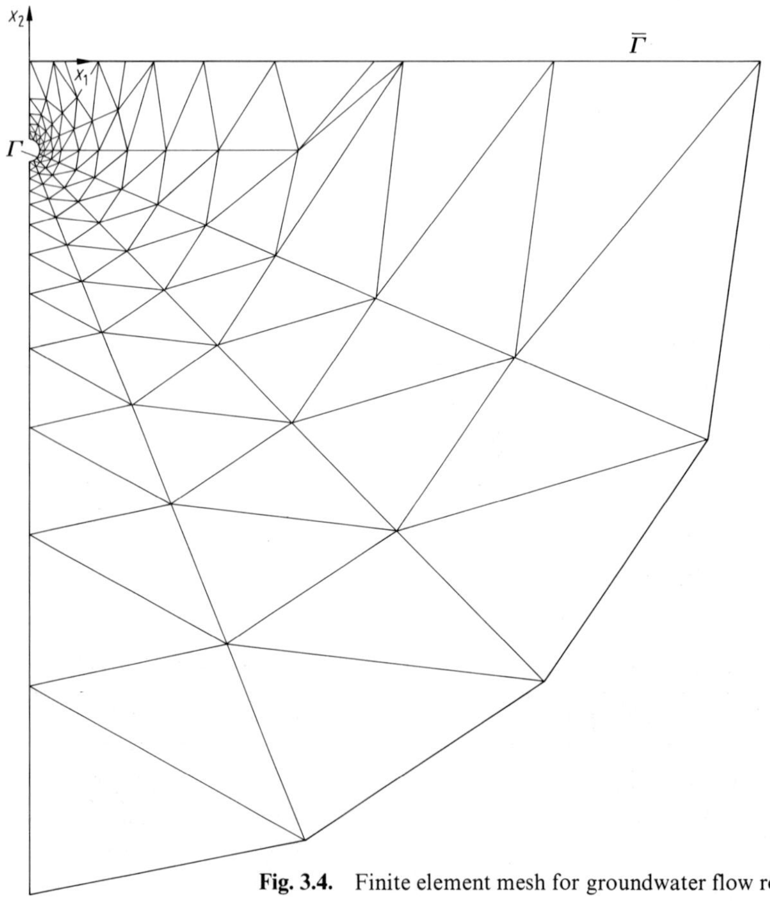

Fig. 3.4. Finite element mesh for groundwater flow round a tunnel

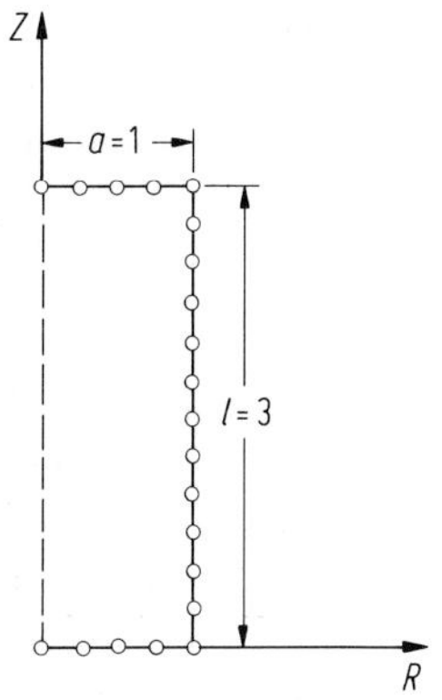

Fig. 3.5. Discretization of solid cylinder

a finite element solution [2] obtained by discretizing the whole semi-infinite region into 152 triangular finite elements plus some infinite elements (see Fig. 3.4). The discrepancy between both solutions is due to the coarseness of the finite element mesh around the tunnel (see Fig. 3.4) which does not properly take into account the discontinuity on the radial flux at the point B.

Example 3.3. This application considers an axisymmetric finite solid cylinder $0 \leqq R < a$, $0 < Z < l$, over one face of which the convection boundary condition (see Example 2.3) is prescribed. The total boundary conditions of the problem are

$$\begin{aligned} u &= 0 && \text{at } Z = l, \\ u &= V && \text{at } Z = 0, \\ q + h\,u &= 0 && \text{at } R = a. \end{aligned}$$

Table 3.1. Values of u_2 (–) on the tunnel surface.

θ	BEM (N=8)	BEM (N=16)	BEM (N=32)	FEM
0	30.51	29.08	28.87	33.44
$\pi/8$	31.52	30.03	29.83	34.51
$\pi/4$	34.60	32.87	32.73	37.76
$3\pi/8$	39.96	38.05	37.85	43.46
$\pi/2$	48.24	45.96	45.72	52.34
$5\pi/8$	61.43	58.42	58.10	66.46

Table 3.2. Temperature at $R = 0.25$.

Z	BEM	Analytical
0.5	0.781	0.781
1.0	0.585	0.585
1.5	0.416	0.416
2.0	0.267	0.267
2.5	0.130	0.130

Table 3.3. Temperature at $R = 1.00$.

Z	BEM	Analytical
0.5	0.751	0.751
1.0	0.560	0.560
1.5	0.397	0.397
2.0	0.254	0.254
2.5	0.124	0.124

Again linear boundary elements were employed and the cylinder surface discretized into 20 equal elements (Fig. 3.5). Notice that there is no need for elements over the axis of revolution, which is not part of the generating contour. The numerical values assumed for this analysis were $a = 1$, $l = 3$, $h = 0.1$, $V = 1$. Results are compared in Tables 3.2 and 3.3 with an available analytical solution [3].

3.3. Quadratic and Higher-Order Elements

Quadratic elements are also frequently used to represent better the geometry of the body. They do not present any special difficulties, but in order to apply them we need to transform from Cartesian to curvilinear coordinates.

Consider the curved boundary depicted in Fig. 3.6 and the element shown in Fig. 3.7. The u and q functions can be written in terms of the homogeneous η coordinate as follows:

$$u(\eta) = \varphi_1 u_1 + \varphi_2 u_2 + \varphi_3 u_3 = [\varphi_1\ \varphi_2\ \varphi_3] \begin{Bmatrix} u_1 \\ u_2 \\ u_3 \end{Bmatrix} = \boldsymbol{\varphi}^T \boldsymbol{u}^n,$$

$$q(\eta) = \varphi_1 q_1 + \varphi_2 q_2 + \varphi_3 q_3 = [\varphi_1\ \varphi_2\ \varphi_3] \begin{Bmatrix} q_1 \\ q_2 \\ q_3 \end{Bmatrix} = \boldsymbol{\varphi}^T \boldsymbol{q}^n, \tag{3.16}$$

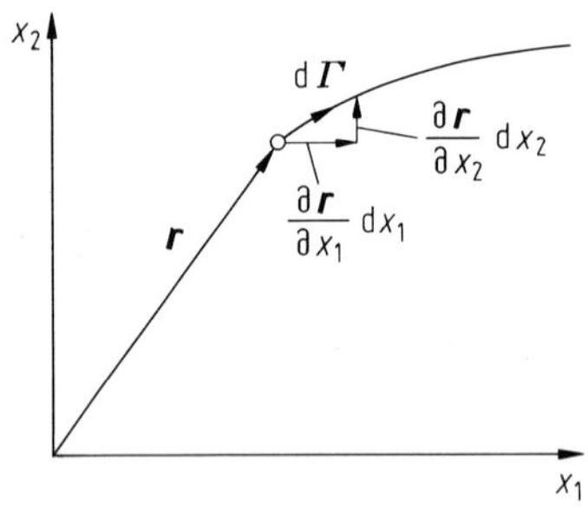

$$\boldsymbol{r} = \boldsymbol{r}(x_1\, x_2)$$

$$(\mathrm{d}\Gamma)^2 = \left[\left(\frac{\mathrm{d}x_1}{\mathrm{d}\eta}\right)^2 + \left(\frac{\mathrm{d}x_2}{\mathrm{d}\eta}\right)^2\right]\mathrm{d}\eta^2 \Rightarrow \frac{d\Gamma}{d\eta} = \sqrt{\left(\frac{dx_1}{d\eta}\right)^2+\left(\frac{dx_2}{d\eta}\right)^2}$$

Fig. 3.6. Curved boundary notation

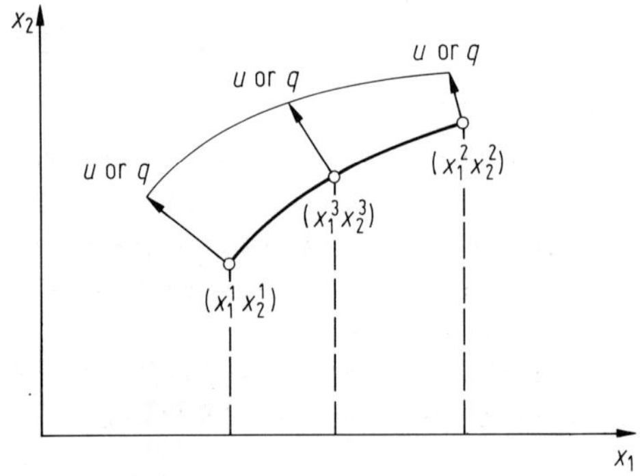

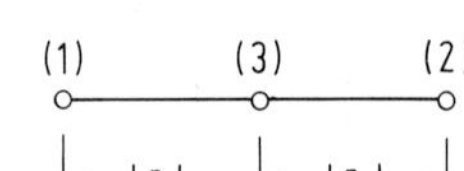

Node	η	φ_1 ϕ_1	φ_3 ϕ_3	φ_2 ϕ_2
1	-1	1	0	0
3	0	0	1	0
2	1	0	0	1

Fig. 3.7. Quadratic element

where

$$\varphi_1 = \tfrac{1}{2}\,\eta(\eta-1), \qquad \varphi_2 = \tfrac{1}{2}\,\eta(1+\eta), \qquad \varphi_3 = (1-\eta)(1+\eta).$$

Note that these functions are such that they give the nodal values of the variable under consideration when specialized for the nodes (see table in Fig. 3.7) and that they vary quadratically.

The integrals along a j element in Eq. (3.1) are similar to those for a linear element. The integral for u, for instance, is

$$\int_{\Gamma_j} u(\eta)\, q^*\, d\Gamma = \int_{\Gamma_j} [\varphi_1\, \varphi_2\, \varphi_3]\, q^*\, d\Gamma \begin{Bmatrix} u_1 \\ u_2 \\ u_3 \end{Bmatrix} = [h_{ij}^1\, h_{ij}^2\, h_{ij}^3] \begin{Bmatrix} u_1 \\ u_2 \\ u_3 \end{Bmatrix} \tag{3.17}$$

and similarly for q.

The evaluation of the integrals requires the use of a Jacobian as φ_i are functions of η, but the integrals are with respect to Γ. For a two-dimensional problem such as this, the transformation is simple as the Jacobian becomes

$$|\mathbf{G}| = \sqrt{\left(\frac{dx_1}{d\eta}\right)^2 + \left(\frac{dx_2}{d\eta}\right)^2} = \frac{d\Gamma}{d\eta}. \tag{3.18}$$

Thus,

$$d\Gamma = |\mathbf{G}|\, d\eta. \tag{3.19}$$

Substituting this relationship into Eq. (3.17) we obtain

$$\int_{\Gamma_j} u(\eta)\, q^*\, d\Gamma = \int_{(1)}^{(2)} u(\eta)\, q^*\, |\mathbf{G}|\, d\eta \tag{3.20}$$

which can now be integrated numerically. Similar considerations apply for the integrals for q.

Note that to calculate Eq. (3.18) we need to express the variations of the x_1 and x_2 coordinates on the boundary in terms of η. This can be done by writing them in the same way as the functions for u and q were expressed, i.e.,

$$\begin{aligned} x_1 &= \varphi_1\, x_1^1 + \varphi_2\, x_1^2 + \varphi_3\, x_1^3, \\ x_2 &= \varphi_1\, x_2^1 + \varphi_2\, x_2^2 + \varphi_3\, x_2^3, \end{aligned} \tag{3.21}$$

where x_1^i and x_2^i are the coordinates of a node referred to the global system (see Fig. 3.6).

It will be shown that in order to be able to reproduce the constant-potential-type conditions, the order of the u function has to be at least the same as the function for x_1 and x_2 used to describe the geometry of the body.

Higher-order elements can also be generated. For instance, a cubic variation for u or q can be obtained by taking four nodes over each element (Fig. 3.8). Here,

$$u = \varphi_1\, u_1 + \varphi_2\, u_2 + \varphi_3\, u_3 + \varphi_4\, u_4, \tag{3.22}$$

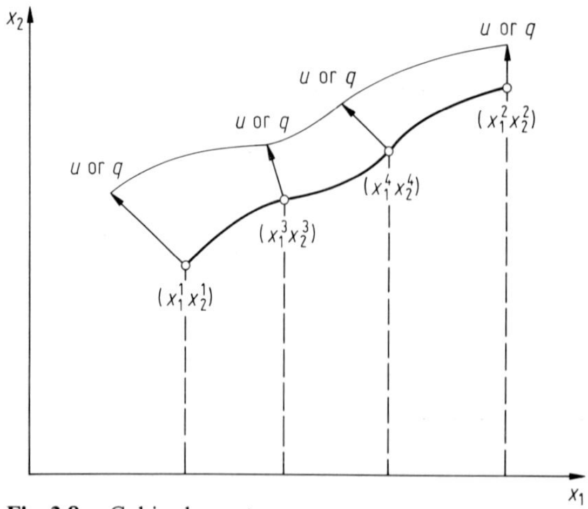

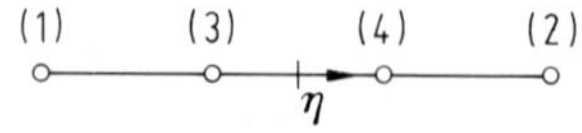

Node	η	ϕ_1	ϕ_3	ϕ_4	ϕ_2
1	-1	1	0	0	0
3	-1/3	0	1	0	0
4	1/3	0	0	1	0
2	1	0	0	0	1

Fig. 3.8. Cubic element

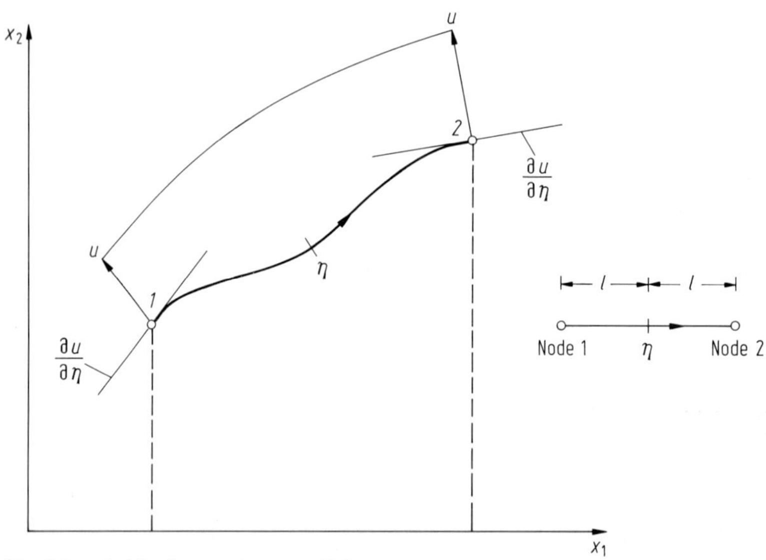

Fig. 3.9. Cubic element in term of slopes

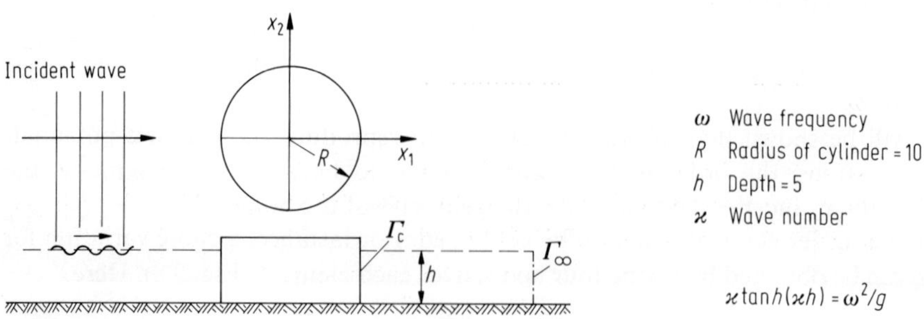

Fig. 3.10. Vertical circular cylinder under wave forces

where

$$\varphi_1 = \frac{1}{16}(1-\eta)[-10+9(\eta^2+1)], \qquad \varphi_2 = \frac{1}{16}(1+\eta)[-10+9(\eta^2+1)],$$

$$\varphi_3 = \frac{9}{16}(1-\eta^2)(1-3\eta), \qquad \varphi_4 = \frac{9}{16}(1-\eta^2)(1+3\eta).$$

We can also define the cubic variation of the function u (the same applies for q and the $x_1 x_2$ coordinates), taking as unknowns the function and its derivative at the two extreme nodes (Fig. 3.9). For this case,

$$u = \varphi_1 u_1 + \varphi_2 \left(\frac{\partial u}{\partial \eta}\right)_1 + \varphi_3 u_2 + \varphi_4 \left(\frac{\partial u}{\partial \eta}\right)_2 \tag{3.23}$$

and

$$\varphi_1 = \tfrac{1}{4}(\eta-1)^2(\eta+2), \qquad \varphi_2 = \tfrac{1}{4}(\eta-1)^2(\eta+1),$$
$$\varphi_3 = \tfrac{1}{4}(\eta+1)^2(\eta-2), \qquad \varphi_4 = \tfrac{1}{4}(\eta+1)^2(\eta-1).$$

Example 3.4. The importance of curved elements can be seen in the calculation of wave forces acting on offshore structures. Consider for simplicity the case of the vertical circular cylinder shown in Fig. 3.10, for which results can be compared with well-known analytical solutions.

The problem can be expressed in terms of the wave potential as shown in Ref. [4]. Using the linear wave theory for constant h, the problem can be reduced to the solution of a Helmholtz equation:

$$\frac{\partial^2 u}{\partial x_1^2} + \frac{\partial^2 u}{\partial x_2^2} + \varkappa^2 u = 0 \qquad \text{in } \Omega \tag{a}$$

with the boundary condition

$$\frac{\partial u}{\partial n} = -\frac{\partial u_I}{\partial n} = -\bar{q} \qquad \text{on the body surface } \Gamma_c \tag{b}$$

and the Sommerfeld radiation condition

$$\frac{\partial u}{\partial n} = i \varkappa u \qquad \text{at infinity } \Gamma_\infty . \tag{c}$$

The u_I incident wave potential is known from Airy's theory.

The above equations and boundary conditions can be put into BE form by using weighted residuals. One can start with

$$\int_\Omega (\nabla^2 u + \varkappa^2 u)\, u^*\, d\Omega = \int_{\Gamma_c} u^* \left(\frac{\partial u}{\partial n} + \bar{q}\right) d\Gamma + \int_{\Gamma_\infty} u^* \left(\frac{\partial u}{\partial n} - i \varkappa u\right) d\Gamma . \tag{d}$$

Integrating by parts twice, one obtains

$$- \int_{\Gamma} u \frac{\partial u^*}{\partial n} d\Gamma + \int_{\Omega} u (\nabla^2 u^* + \varkappa^2 u^*) \, d\Omega = \int_{\Gamma_c} u^* \bar{q} \, d\Gamma - \int_{\Gamma_\infty} i \varkappa u^* u \, d\Gamma . \qquad \text{(e)}$$

Notice that the fundamental solution of the Helmholtz equation is known, i.e.,

$$\nabla^2 u^* + \varkappa^2 u^* = - \Delta_i , \qquad \text{(f)}$$

where

$$u^* = \frac{i}{4} H_0^{(1)} (\varkappa r) \qquad \text{(g)}$$

with $r = \{[x_1(x) - x_1(\xi)]^2 + [x_2(x) - x_2(\xi)]^2\}^{1/2}$. $H_0^{(1)}(\,)$ is the Hankel function of the first kind and of zero order.

Hence, for a boundary point, equation (e) gives

$$c_i u_i + \int_{\Gamma_c} \left(\frac{\partial u^*}{\partial n} u + u^* \bar{q} \right) d\Gamma = 0 . \qquad \text{(h)}$$

It should be pointed out that the fundamental solution (g) satisfies the radiation condition, i.e., it makes the Γ_∞ integral tend to zero as r increases,

$$\int_{\Gamma_\infty} \left(\frac{\partial u^*}{\partial n} - i \varkappa u^* \right) u \, d\Gamma \to 0 \qquad \text{for } r \to \infty . \qquad \text{(i)}$$

The vertical column has been studied by Isaacson [4] using a source distribution on the cylinder surface (i.e., an indirect boundary element method) and by MacCamy and Fuchs [5] who found the exact solution for the forces,

$$F_x = H_1^{(1)} (\varkappa h) \varrho g a_0 Rh \frac{\tanh \varkappa h}{\varkappa h} ; \qquad \text{(j)}$$

a_0 is the wave amplitude. A convergence study was carried out using constant, linear, and quadratic elements [6]; some discretizations are shown in Fig. 3.11.

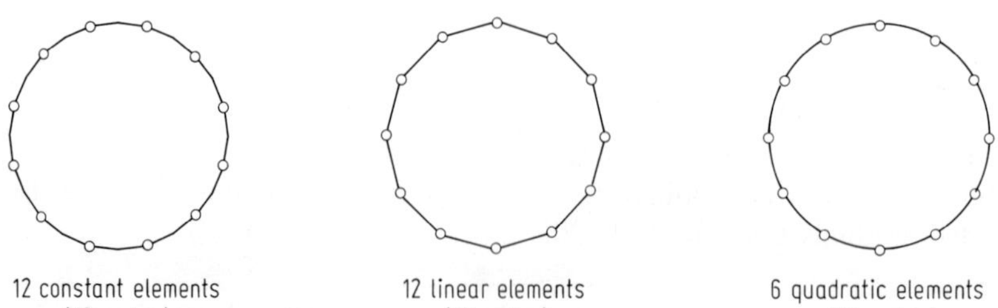

12 constant elements (12 nodes) — 12 linear elements (12 nodes) — 6 quadratic elements (12 nodes)

Fig. 3.11. Discretization

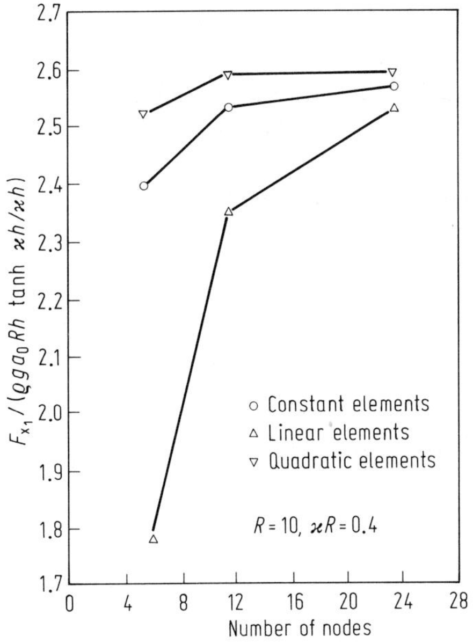

Fig. 3.12. Convergence of horizontal force on a vertical cylinder

Figure 3.12 shows the convergence of the results for the total horizontal force which can be compared with the exact solution given by formula (j). It is evident that the first convergence has been obtained using the quadratic elements which can more accurately represent the surface of the cylinder. More surprisingly, however, the convergence of the constant elements is superior to that of linear ones. This was thought to be due to the difficulty of representing properly the normal at corners in the case of linear elements.

Example 3.5. Another kind of problem for which the use of higher-order elements is generally recommended is the analysis of free-surface flows.

Free surfaces are generally related to problems of groundwater flow through saturated unconfined porous media governed by Darcy's law [1]. If the medium under consideration is homogeneous and isotropic, the problem is reduced to that of Laplace's equation for a velocity potential u with boundary conditions of the following types (Fig. 3.13): $q = 0$ at impervious boundaries, such as the surface of

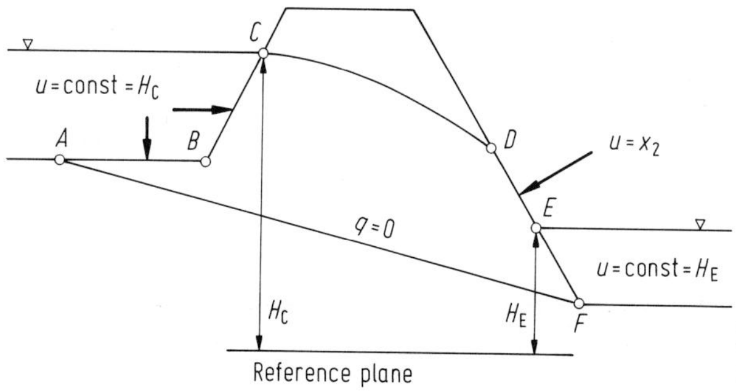

Fig. 3.13. Boundary conditions for free-surface problem

soil strata and rocks (surface AF in the figure); $u=$ constant at water boundaries (the upstream and downstream faces ABC and EF of the porous domain); $u=x_2$ at the seepage face DE where the water seeps out of the soil into the air; $u=x_2$ and $q=0$ at the free surface CD. In addition, the exact position of the free surface is not known a priori and its determination becomes part of the analysis of the problem.

For the numerical solution of these problems, an initial position of the free surface is arbitrarily assumed and the condition $q=0$ is applied at all points on it. The calculated potential at every nodal point at the free surface is then compared with its elevation; if the difference between these two values is greater than a maximum acceptable error, this difference is algebraically added to the elevation of the nodal point and a new iteration is carried out.

Notice that the coefficients of matrices $\boldsymbol{G}$ and $\boldsymbol{H}$ in Eq. (3.9) corresponding to the influence of fixed boundary nodes on other fixed boundary nodes remain constant during the analysis, hence they can be computed only once and stored. Potential values at internal points, if required, are calculated after the correct position of the free surface has been determined.

Figure 3.14 shows the linear boundary element discretization for a problem of free-surface flow through a block of porous media. The upstream and downstream water levels are maintained at 19.7 in. and 1.3 in., respectively, above the horizontal impervious base. The boundary conditions of the problem are $u=19.7$ on the upstream face (nodes 22 to 26), $q=0$ on the bottom (impervious) surface (nodes 1 to 8), $u=1.3$ on the downstream face (nodes 9 and 10) and $q=0$ on the free surface (nodes 11 to 21). Note that the initial shape of the free surface was arbitrarily assumed to be a straight line and its initial position was also guessed. The final position of the free surface was obtained by iteration as previously explained.

Results are presented in Fig. 3.15 together with a finite element solution and an experimental solution obtained from an analog model [7]. After the seventh iteration the maximum difference between the computed potential head and the elevation of each node along the free surface was less than 0.1% of the elevation and the solution was terminated.

Example 3.6. Gravity-driven free-surface flows can also be accurately solved using boundary elements [8].

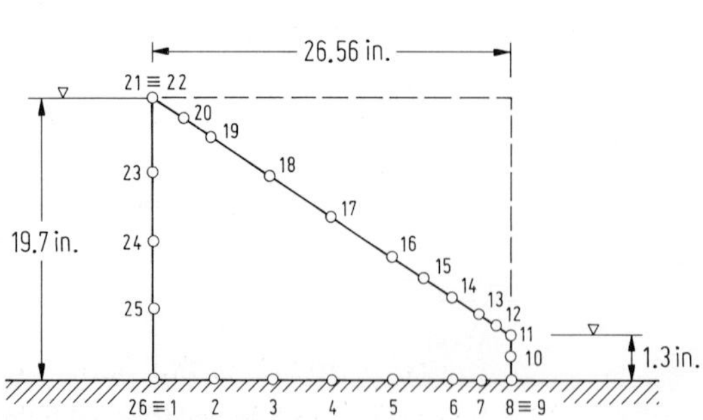

Fig. 3.14. Boundary element discretization of soil block with a free surface

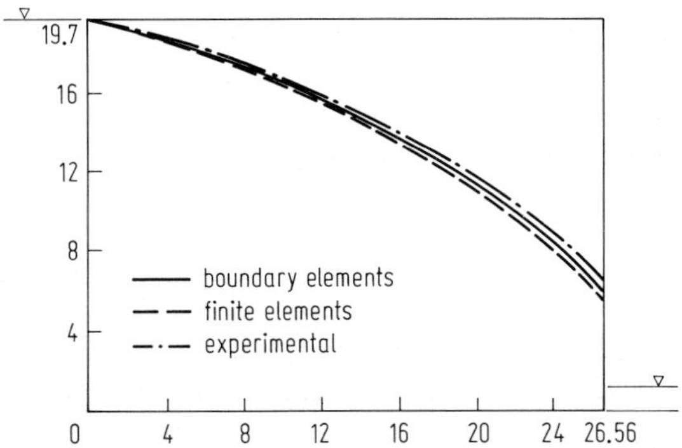

Fig. 3.15. Converged solution for the free surface

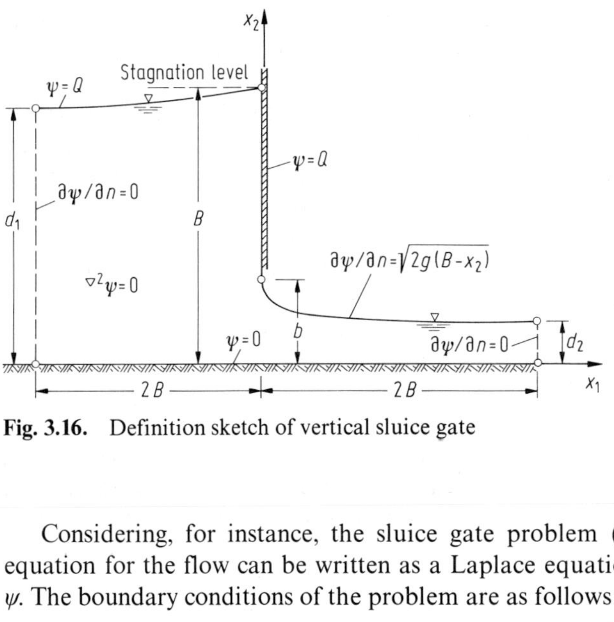

Fig. 3.16. Definition sketch of vertical sluice gate

Considering, for instance, the sluice gate problem (Fig. 3.16), the governing equation for the flow can be written as a Laplace equation for the stream function ψ. The boundary conditions of the problem are as follows:

$$\begin{aligned} \psi &= 0 \quad \text{on the lower boundary,} \\ \psi &= Q \quad \text{on the free surface and on the sluice gate} \end{aligned} \tag{a}$$

in which Q is the flow rate per unit width. On the free surface, the dynamic boundary condition requires that

$$\frac{1}{2g}\left(\frac{\partial \psi}{\partial n}\right)^2 + x_2 = B, \qquad p = 0, \tag{b}$$

where g is the acceleration of gravity, p is the pressure, and B is the Bernoulli constant. Either the flow rate Q or the Bernoulli constant B is known; the other is required as part of the solution.

The sluice gate flow generally extends to $\pm\infty$. For the purposes of the numerical solution the inflow and outflow streams are cut at right angles to the primary velocity. On the cut portions the boundary condition

$$\frac{\partial \psi}{\partial n} = 0 \tag{c}$$

is applied, which means that there is no velocity normal to the main flow.

The presence of the free-surface boundary condition (b) introduces an additional complication. This factor is treated in much the same way as it has been for most of the finite difference and finite element solutions. That is, for problems with a known flow rate Q, the position of the free-surface boundary is assumed and the problem is solved with boundary conditions (a) and (c). The "constant" B of equation (b) is then calculated on the free surface. If B is the same for all free surface points, the problem is solved. Otherwise, the assumed free surface is adjusted, iteratively, in order that B becomes constant at all points.

A similar procedure can be developed for problems with a given Bernoulli constant B. The problem is first solved by assuming the location of the free surface and applying the free-surface condition (b). The flow rate Q of equation (a) is part of the solution. If Q is a constant for all free surface points, the problem is solved; otherwise, an iteration scheme must be used to adjust the free surface elevation.

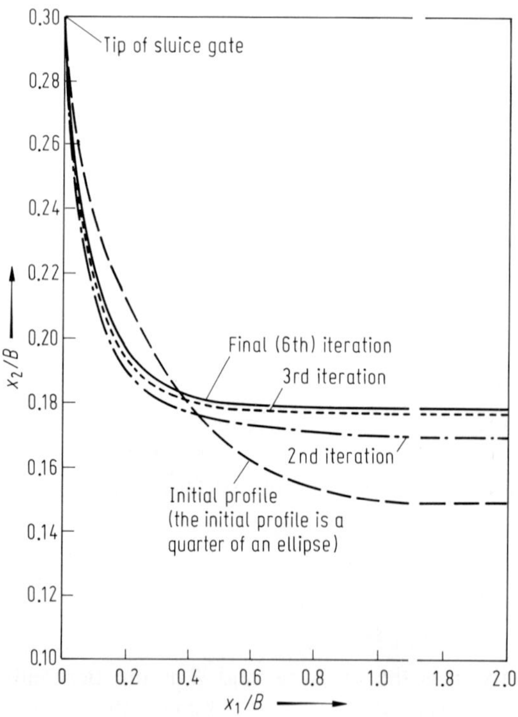

Fig. 3.17. Convergence of downstream profile for vertical sluice gate

This formulation was employed by Cheng *et al.* [8] to analyze the vertical sluice gate of Fig. 3.16. The numerical values adopted were $b = 0.3$ and $B = 1$. The initial trial upstream free-surface profile was considered to be flat ($x_2 = 1$) and the downstream profile was approximated by a quarter of an ellipse with an arbitrary downstream depth of $x_2 = 0.5\,b$. The flow boundary was discretized into 39 linear boundary elements and convergence achieved to $\pm 0.1\%$ after the sixth iteration (Fig. 3.17). An even more efficient solution to the problem can be obtained by using cubic spline boundary elements as described by Liggett and Salmon [9].

3.4. Boundary Elements for Three-Dimensional Problems

If the body is three-dimensional, the boundary elements are part of its external surface (Fig. 3.18). They are usually of two types: quadrilateral and triangular elements. The functions used to describe their geometry or the variation of u and q over them are similar to those used in finite elements.

In order to study these elements we need first to consider the way in which we can pass from the (x_1, x_2, x_3) global system to the system (η_1, η_2, ζ) defined over the surface of the body.

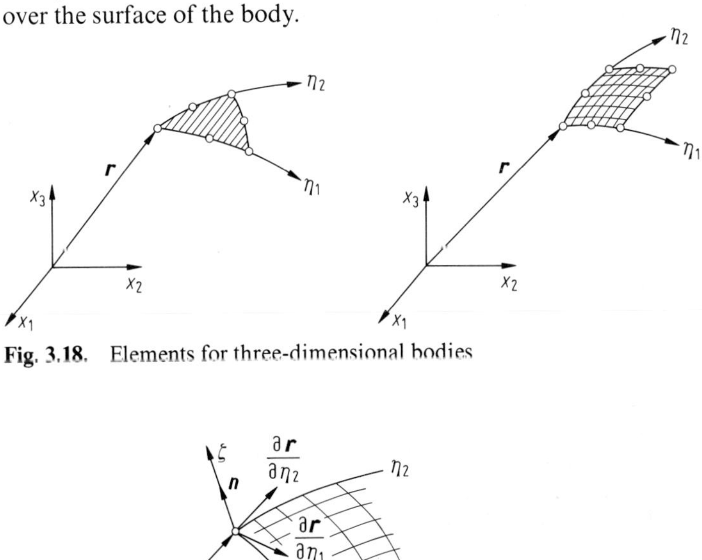

Fig. 3.18. Elements for three-dimensional bodies

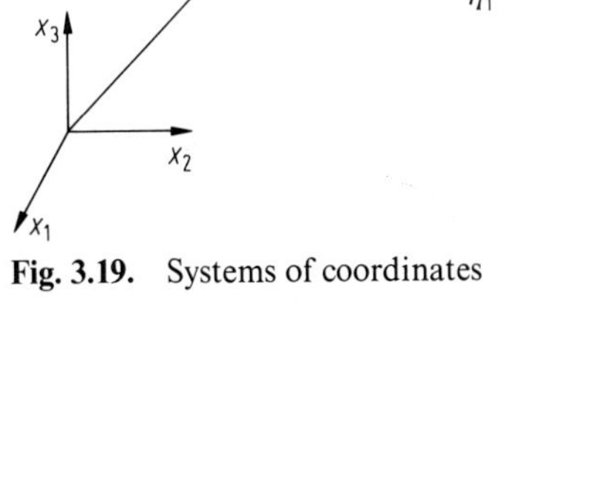

Fig. 3.19. Systems of coordinates

Consider the systems defined in Fig. 3.19. For a function u, the general transformation is given by

$$\begin{Bmatrix} \dfrac{\partial u}{\partial \eta_1} \\ \dfrac{\partial u}{\partial \eta_2} \\ \dfrac{\partial u}{\partial \zeta} \end{Bmatrix} = \begin{bmatrix} \dfrac{\partial x_1}{\partial \eta_1} & \dfrac{\partial x_2}{\partial \eta_1} & \dfrac{\partial x_3}{\partial \eta_1} \\ \dfrac{\partial x_1}{\partial \eta_2} & \dfrac{\partial x_2}{\partial \eta_2} & \dfrac{\partial x_3}{\partial \eta_2} \\ \dfrac{\partial x_1}{\partial \zeta} & \dfrac{\partial x_2}{\partial \zeta} & \dfrac{\partial x_3}{\partial \zeta} \end{bmatrix} \begin{Bmatrix} \dfrac{\partial u}{\partial x_1} \\ \dfrac{\partial u}{\partial x_2} \\ \dfrac{\partial u}{\partial x_3} \end{Bmatrix} \tag{3.24}$$

or

$$\begin{Bmatrix} \dfrac{\partial u}{\partial \eta_1} \\ \dfrac{\partial u}{\partial \eta_2} \\ \dfrac{\partial u}{\partial \zeta} \end{Bmatrix} = \boldsymbol{J} \begin{Bmatrix} \dfrac{\partial u}{\partial x_1} \\ \dfrac{\partial u}{\partial x_2} \\ \dfrac{\partial u}{\partial x_3} \end{Bmatrix}. \tag{3.25}$$

Hence we can find the inverse relationship, i.e.,

$$\begin{Bmatrix} \dfrac{\partial u}{\partial x_1} \\ \dfrac{\partial u}{\partial x_2} \\ \dfrac{\partial u}{\partial x_3} \end{Bmatrix} = \boldsymbol{J}^{-1} \begin{Bmatrix} \dfrac{\partial u}{\partial \eta_1} \\ \dfrac{\partial u}{\partial \eta_2} \\ \dfrac{\partial u}{\partial \zeta} \end{Bmatrix}. \tag{3.26}$$

Having defined these relationships we can take, for instance, $\mathbf{r}$ instead of u and define

$$\begin{aligned} d\Omega = d\,(\text{volume}) &= \text{magnitude of} \left(\frac{\partial \mathbf{r}}{\partial \eta_1} \times \frac{\partial \mathbf{r}}{\partial \eta_2} \cdot \frac{\partial \mathbf{r}}{\partial \zeta} \right) d\eta_1\, d\eta_2\, d\zeta \\ &= (\text{absolute value of } |\boldsymbol{J}|)\, d\eta_1\, d\eta_2\, d\zeta. \end{aligned} \tag{3.27}$$

A differential of area (such as $d\Gamma$) will be given by

$$d\,(\text{area}) = \left| \frac{\partial \mathbf{r}}{\partial \eta_1} \times \frac{\partial \mathbf{r}}{\partial \eta_2} \right| d\eta_1\, d\eta_2 = |\boldsymbol{G}|\, d\eta_1\, d\eta_2, \tag{3.28}$$

where

$$\begin{aligned} \frac{\partial \mathbf{r}}{\partial \eta_1} &= \left(\frac{\partial x_1}{\partial \eta_1}, \frac{\partial x_2}{\partial \eta_1}, \frac{\partial x_3}{\partial \eta_1} \right), \\ \frac{\partial \mathbf{r}}{\partial \eta_2} &= \left(\frac{\partial x_1}{\partial \eta_2}, \frac{\partial x_2}{\partial \eta_2}, \frac{\partial x_3}{\partial \eta_2} \right). \end{aligned} \tag{3.29}$$

Note that $|\boldsymbol{G}|$ is the magnitude of the normal vector,

$$\mathbf{n} = \frac{\partial \mathbf{r}}{\partial \eta_1} \times \frac{\partial \mathbf{r}}{\partial \eta_2} = \left(\frac{\partial x_1}{\partial \zeta}, \frac{\partial x_2}{\partial \zeta}, \frac{\partial x_3}{\partial \zeta} \right) = (g_1, g_2, g_3), \tag{3.30}$$

where

$$\begin{aligned} g_1 &= \frac{\partial x_2}{\partial \eta_1}\frac{\partial x_3}{\partial \eta_2} - \frac{\partial x_2}{\partial \eta_2}\frac{\partial x_3}{\partial \eta_1}, \\ g_2 &= \frac{\partial x_3}{\partial \eta_1}\frac{\partial x_1}{\partial \eta_2} - \frac{\partial x_1}{\partial \eta_1}\frac{\partial x_3}{\partial \eta_2}, \\ g_3 &= \frac{\partial x_1}{\partial \eta_1}\frac{\partial x_2}{\partial \eta_2} - \frac{\partial x_2}{\partial \eta_1}\frac{\partial x_1}{\partial \eta_2}, \end{aligned} \tag{3.31}$$

so the value of $|\boldsymbol{G}|$ is given by

$$|\boldsymbol{G}| = (g_1^2 + g_2^2 + g_3^2)^{1/2}. \tag{3.32}$$

These relationships will be used to integrate terms such as

$$\int u^* q \, d\Gamma \qquad \text{and} \qquad \int u \, q^* \, d\Gamma \tag{3.33}$$

which now become

$$\int u^* q \, |\boldsymbol{G}| \, d\eta_1 \, d\eta_2, \qquad \int u \, q^* \, |\boldsymbol{G}| \, d\eta_1 \, d\eta_2. \tag{3.34}$$

The domain integrals like

$$\int b \, u^* \, d\Omega \tag{3.35}$$

will now become

$$\int b \, u^* \, |\boldsymbol{J}| \, d\eta_1 \, d\eta_2 \, d\zeta; \tag{3.36}$$

u, b and q are assumed to be functions of η_1, η_2, and ζ where applicable.

3.4.1. Quadrilateral Elements

The simplest quadrilateral element is defined by its four corner points (Fig. 3.20). We can use a set of dimensionless coordinates (η_1, η_2) such that they vary from -1 to $+1$ and the interpolation functions φ_i. This allows us to write

$$u = \varphi_1 u_1 + \varphi_2 u_2 + \varphi_3 u_3 + \varphi_4 u_4, \qquad \varphi_i = \varphi_i(\eta_1, \eta_2) \tag{3.37}$$

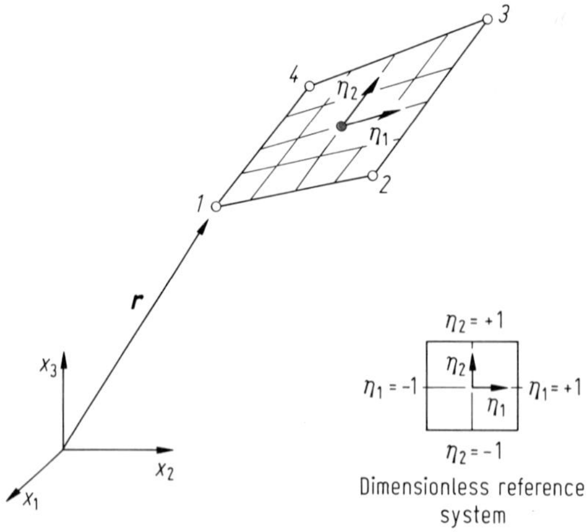

Fig. 3.20. Simplest quadrilateral element

where the φ_i are

$$\varphi_1 = \tfrac{1}{4}(1-\eta_1)(1-\eta_2), \qquad \varphi_2 = \tfrac{1}{4}(1+\eta_1)(1-\eta_2),$$
$$\varphi_3 = \tfrac{1}{4}(1+\eta_1)(1+\eta_2), \qquad \varphi_4 = \tfrac{1}{4}(1-\eta_1)(1+\eta_2),$$

and u_i are the potentials at the nodes 1 to 4. The same function can be used to describe the geometry of the element, i.e.,

$$x_1 = \varphi_1 x_1^1 + \varphi_2 x_1^2 + \varphi_3 x_1^3 + \varphi_4 x_1^4 = \sum_{i=1}^{4} \varphi_i x_1^i \tag{3.38}$$

and similarly for x_2 and x_3.

Note that once these functions are defined we can obtain the value of $|\boldsymbol{G}|$ and $|\boldsymbol{J}|$ using relations (3.32) and (3.25), respectively.

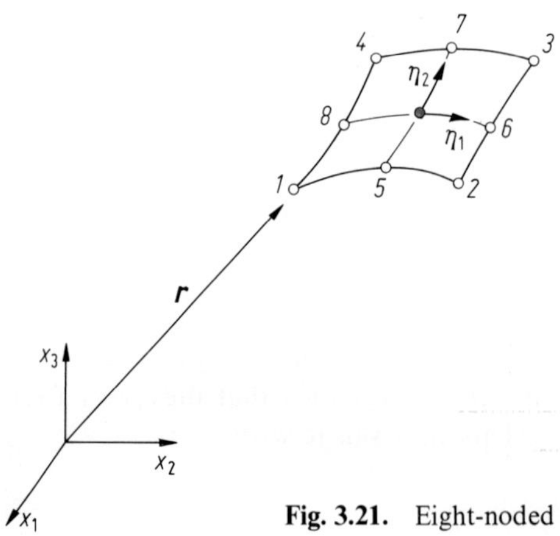

Fig. 3.21. Eight-noded quadrilateral element

3.4.2. Higher-Order Quadrilateral Elements

We can improve our approximation by taking a second-order quadrilateral surface element (Fig. 3.21), such that

$$x_1 = \sum_{i=1}^{8} \varphi_i x_1^i, \qquad x_2 = \sum_{i=1}^{8} \varphi_i x_2^i, \qquad x_3 = \sum_{i=1}^{8} \varphi_i x_3^i \tag{3.39}$$

with

$$\begin{aligned}
\varphi_1 &= \tfrac{1}{4}(1-\eta_1)(1-\eta_2)(-\eta_1-\eta_2-1),\\
\varphi_2 &= \tfrac{1}{4}(1+\eta_1)(1-\eta_2)(\eta_1-\eta_2-1),\\
\varphi_3 &= \tfrac{1}{4}(1+\eta_1)(1+\eta_2)(\eta_1+\eta_2-1),\\
\varphi_4 &= \tfrac{1}{4}(1-\eta_1)(1+\eta_2)(-\eta_1+\eta_2-1),\\
\varphi_5 &= \tfrac{1}{2}(1-\eta_1^2)(1-\eta_2),\\
\varphi_6 &= \tfrac{1}{2}(1-\eta_2^2)(1+\eta_1),\\
\varphi_7 &= \tfrac{1}{2}(1-\eta_1^2)(1+\eta_2),\\
\varphi_8 &= \tfrac{1}{2}(1-\eta_2^2)(1-\eta_1).
\end{aligned} \tag{3.40}$$

The same type of function could be used for u or q.

Higher-order quadrilateral elements can also be used as boundary elements, such as the cubic type (cubic variation of the function on the element side), shown in Fig. 3.22a. For this element, one has two additional nodes on each side, but one can also work with only corner nodes and include the derivatives as nodal quantities (Fig. 3.22b).

3.4.3. Lagrangian Quadrilateral Elements

A special type of quadrilateral elements are the Lagrangian ones shown in Fig. 3.23. The first element is a nine-noded quadrilateral, i.e., with one node in the

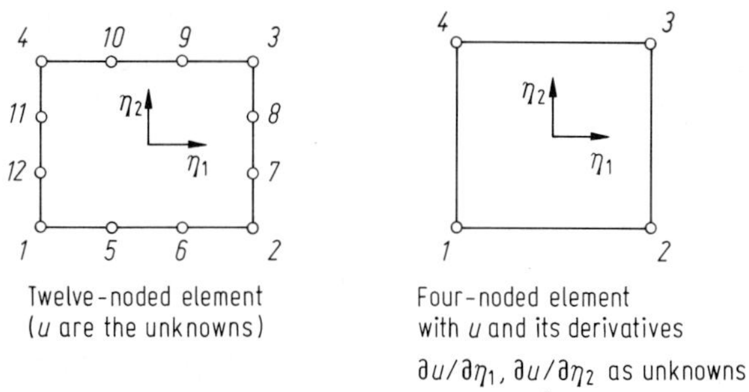

Fig. 3.22. Higher-order elements

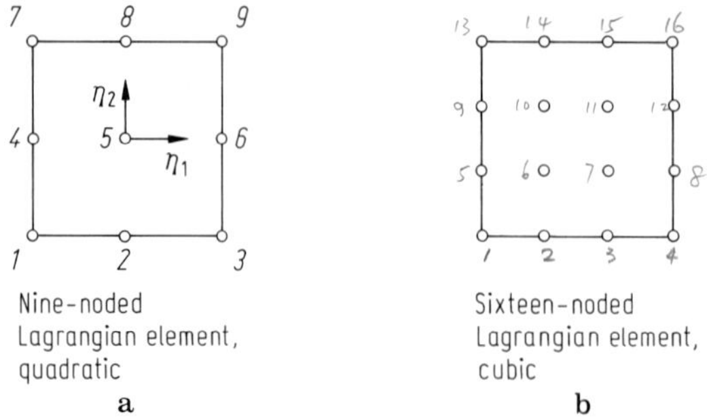

Fig. 3.23. Lagrangian quadrilateral elements. **a** Nine-noded Lagrangian element, quadratic. **b** Sixteen-noded Lagrangian element, cubic

middle and eight around the boundary; the second is a sixteen noded quadrilateral, which has four internal nodes.

These elements are generated by expanding one-dimensional interpolation functions in the η_1, η_2 directions. For instance, consider a quadratic one-dimensional element with the function expansion given by Eq. (3.16),

$$u(\eta_1) = \tfrac{1}{2}\,\eta_1(\eta_1 - 1)\,u_4 + (1-\eta_1^2)\,u_5 + \tfrac{1}{2}\,\eta_1(1+\eta_1)\,u_6. \tag{3.41}$$

If we now expand u_4, u_5, and u_6 in the η_2 direction, we have

$$\begin{aligned} u(\eta_1, \eta_2) = {} & \tfrac{1}{4}\,\eta_1(\eta_1-1)\,\eta_2(\eta_2-1)\,u_1 + \tfrac{1}{2}\,(1-\eta_1^2)\,\eta_2(\eta_2-1)\,u_2 \\ & + \tfrac{1}{4}\,\eta_1(1+\eta_1)\,\eta_2(\eta_2-1)\,u_3 + \tfrac{1}{2}\,\eta_1(\eta_1-1)(1-\eta_2^2)\,u_4 \\ & + (1-\eta_1^2)(1-\eta_2^2)\,u_5 + \tfrac{1}{2}\,\eta_1(1+\eta_1)(1-\eta_2^2)\,u_6 \\ & + \tfrac{1}{4}\,\eta_1(\eta_1-1)\,\eta_2(1+\eta_2)\,u_7 + \tfrac{1}{2}\,(1-\eta_1^2)\,\eta_2(1+\eta_2)\,u_8 \\ & + \tfrac{1}{4}\,\eta_1(1+\eta_1)\,\eta_2(1+\eta_2)\,u_9. \end{aligned} \tag{3.42}$$

The expression for the cubic element of Fig. 3.23b can also be easily obtained by expanding the one-dimensional cubic expansion (3.22) in the η_2 as well as the η_1 direction.

The Lagrangian elements are seldom used in finite element analysis, where generally the internal nodes are condensed and one effectively uses the quadrilateral elements previously discussed. Instead, in boundary elements the use of Lagrangian elements may be necessary since the extra conditions imposed by the internal nodes may improve dramatically the results as it is the case with the discontinuous elements discussed in Section 3.6.

3.4.4. Triangular Elements

The simplest triangular element is one as shown in Fig. 3.24. There is a linear variation of the function over the element and one can work with oblique coordinates (η_1, η_2), which vary from 0 to 1 over the two sides shown in the figure.

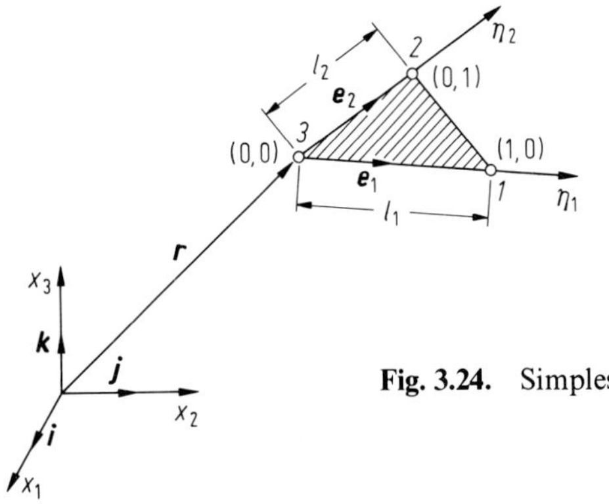

Fig. 3.24. Simplest triangular element

The position vector $\mathbf{r}$ of any point on the triangle is

$$\mathbf{r} = x_1\,\mathbf{i} + x_2\,\mathbf{j} + x_3\,\mathbf{k} = x_1^3\,\mathbf{i} + x_2^3\,\mathbf{j} + x_3^3\,\mathbf{k} + l_1\,\eta_1\,\mathbf{e}_1 + l_2\,\eta_2\,\mathbf{e}_2\,, \tag{3.43}$$

where

$$\begin{aligned} \mathbf{e}_1 &= \left(\frac{x_1^1 - x_1^3}{l_1}\right)\mathbf{i} + \left(\frac{x_2^1 - x_2^3}{l_1}\right)\mathbf{j} + \left(\frac{x_3^1 - x_3^3}{l_1}\right)\mathbf{k}\,, \\ \mathbf{e}_2 &= \left(\frac{x_1^2 - x_1^3}{l_2}\right)\mathbf{i} + \left(\frac{x_2^2 - x_2^3}{l_2}\right)\mathbf{j} + \left(\frac{x_3^2 - x_3^3}{l_2}\right)\mathbf{k}\,. \end{aligned} \tag{3.44}$$

Hence the position vector $\mathbf{r}$ can be defined as

$$\begin{aligned} \mathbf{r} = {} & [x_1^3 + (x_1^1 - x_1^3)\,\eta_1 + (x_1^2 - x_1^3)\,\eta_2]\,\mathbf{i} + [x_2^3 + (x_2^1 - x_2^3)\,\eta_1 + (x_2^2 - x_2^3)\,\eta_2]\,\mathbf{j} \\ & + [x_3^3 + (x_3^1 - x_3^3)\,\eta_1 + (x_3^2 - x_3^3)\,\eta_2]\,\mathbf{k} = x_1\,\mathbf{i} + x_2\,\mathbf{j} + x_3\,\mathbf{k}\,. \end{aligned} \tag{3.45}$$

Thus, we have that

$$\begin{aligned} x_1 &= \eta_1\,x_1^1 + \eta_2\,x_1^2 + (1 - \eta_1 - \eta_2)\,x_1^3\,, \\ x_2 &= \eta_1\,x_2^1 + \eta_2\,x_2^2 + (1 - \eta_1 - \eta_2)\,x_2^3\,, \\ x_3 &= \eta_1\,x_3^1 + \eta_2\,x_3^2 + (1 - \eta_1 - \eta_2)\,x_3^3\,. \end{aligned} \tag{3.46}$$

We can now define $\eta_3 = 1 - \eta_1 - \eta_2$, where η_3 is a new but not independent coordinate, such that

$$\eta_1 + \eta_2 + \eta_3 = 1\,.$$

Hence,

$$\begin{Bmatrix} x_1 \\ x_2 \\ x_3 \end{Bmatrix} = \begin{vmatrix} x_1^1 & x_1^2 & x_1^3 \\ x_2^1 & x_2^2 & x_2^3 \\ x_3^1 & x_3^2 & x_3^3 \end{vmatrix} \begin{Bmatrix} \eta_1 \\ \eta_2 \\ \eta_3 \end{Bmatrix}. \tag{3.47}$$

As η_3 is not independent, we can invert the first two equations to obtain

$$\eta_1 = \frac{1}{2A}(2A_1^0 + b_1 x_1 + a_1 x_2)\,,$$
$$\eta_2 = \frac{1}{2A}(2A_2^0 + b_2 x_1 + a_2 x_2) \tag{3.48}$$

and obtain η_3 as $1 - \eta_1 - \eta_2$, i.e.,

$$\eta_3 = \frac{1}{2A}(2A_3^0 + b_3 x_1 + a_3 x_2)\,. \tag{3.49}$$

The recurrence relationship for Eqs. (3.48) and (3.49) is

$$a_i = x_1^k - x_1^j\,, \qquad b_i = x_2^j - x_2^k\,, \qquad 2A_i^0 = x_1^j x_2^k - x_1^k x_2^j\,, \tag{3.50}$$

where $i = 1, 2, 3, j = 2, 3, 1$, and $k = 3, 1, 2$. The area A is

$$A = \tfrac{1}{2}(b_1 a_2 - b_2 a_1) \tag{3.51}$$

and represents the area of the projection of the element on the x_1, x_2 plane.

The transformation can now be carried out as previously. The reason for using η_3 as a dependent variable is that it renders a simple expression for the interpolation functions, but care should be taken not to confuse it with an independent variable.

Linear expressions for the coordinates or the u function are now

$$u = u_1 \eta_1 + u_2 \eta_2 + u_3 \eta_3 = \sum_{i=1}^{3} u_i \varphi_i\,,$$
$$x_1 = x_1^1 \eta_1 + x_1^2 \eta_2 + x_1^3 \eta_3 = \sum_{i=1}^{3} x_1^i \varphi_i\,. \tag{3.52}$$

Note that the interpolation functions are $\varphi_i = \eta_i$.

The homogeneous triangular coordinates have one more advantage. This is that the integrals for polynomial terms can be carried out using a simple rule, i.e.,

$$\iint \eta_1^i \eta_2^j \eta_3^k \, dA = \frac{i!\,j!\,k!}{(i+j+k+2)!}\, 2A\,. \tag{3.53}$$

This formula can sometimes be used and greatly simplifies the algebra when numerical integration is not required.

3.4.5. Higher-Order Triangular Elements

The next triangular model expresses the function as a complete second degree polynomial, which has a midside node (Fig. 3.25). The model satisfies interelement

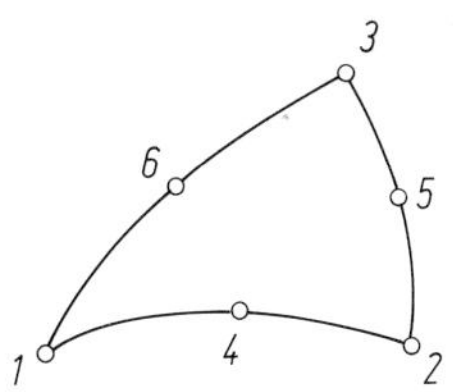

Fig. 3.25. Quadratic element

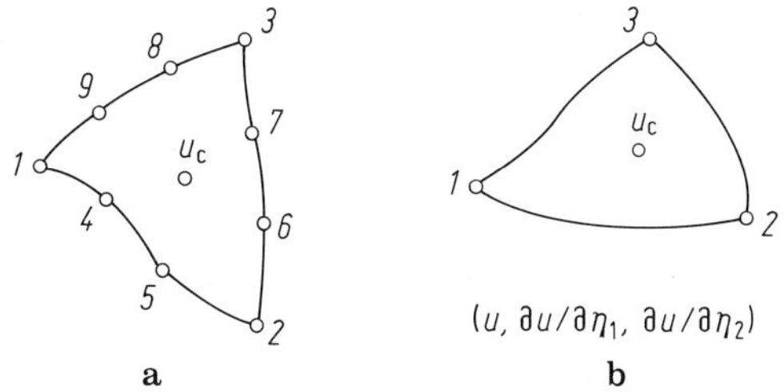

Fig. 3.26. Cubic elements

compatibility and gives

$$u = \sum_{i=1}^{6} \varphi_i u_i , \tag{3.54}$$

where

$$\begin{aligned} \varphi_1 &= \eta_1 (2\eta_1 - 1) , & \varphi_4 &= 4\eta_1 \eta_2 , \\ \varphi_2 &= \eta_2 (2\eta_2 - 1) , & \varphi_5 &= 4\eta_2 \eta_3 , \\ \varphi_3 &= \eta_3 (2\eta_3 - 1) , & \varphi_6 &= 4\eta_3 \eta_1 . \end{aligned} \tag{3.55}$$

The functions can also be expressed as complete cubic polynomials. Each component now involves 10 parameters. In order to satisfy interelement compatibility the function for a side (which is cubic) must be represented with four nodal quantities. One possibility is to work with corner point nodes and two interior nodes per side (Fig. 3.26a), taking as nodal quantities the u or q components. An additional interior node is needed to maintain completeness of the polynomial since if the polynomial is not complete, its properties will have preferred directions which is not desirable. It is convenient to take the interior node at the centroid. We can also attempt to work only with corner nodes. Since four quantities are necessary for interelement compatibility, one must include derivatives of the function as nodal quantities, i.e.,

$$\left(u , \frac{\partial u}{\partial \eta_1} , \frac{\partial u}{\partial \eta_2} \right)_{\text{node}\, i} . \tag{3.56}$$

As before, one additional quantity not associated with the boundaries is required, i.e., the value of u at the centroid, u_c.

3.5. Three-Dimensional Cell Elements

Three-dimensional cell elements may be useful to represent parts of the body where integration over the domain is required, namely, for terms of the type

$$\int_{\Omega} b\, u^* \, d\Omega . \tag{3.57}$$

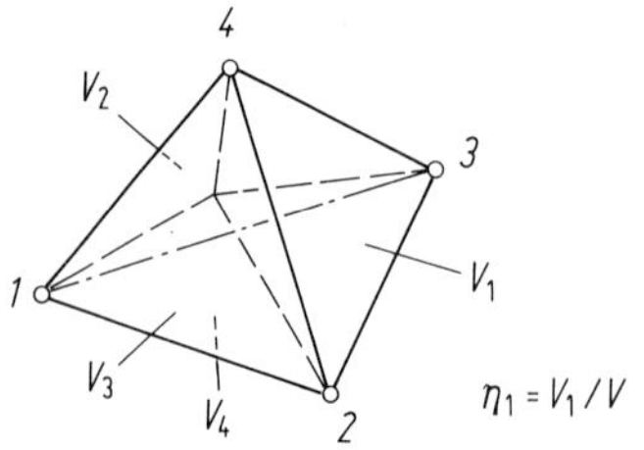

a

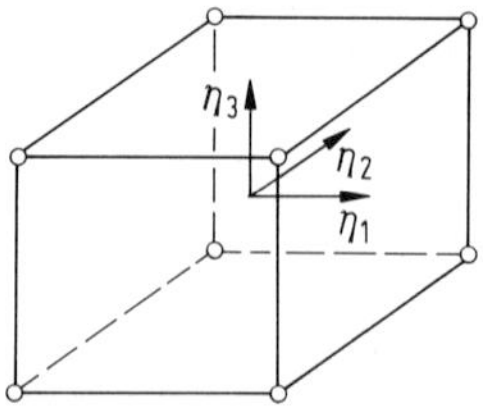

b

Fig. 3.27. First-order three-dimensional cell elements: **a** tetrahedron; **b** cube

These elements are similar to those used in finite elements and they are well known in the literature. The simplest are the tetrahedron and the cube (Fig. 3.27).

3.5.1. Tetrahedron

Dimensionless homogeneous coordinates can be defined similar to those used for the triangular elements previously seen. They are (Fig. 3.27)

$$\eta_1 = \frac{V_1}{V}, \qquad \eta_2 = \frac{V_2}{V}, \qquad \eta_3 = \frac{V_3}{V}, \qquad \eta_4 = \frac{V_4}{V}, \tag{3.58}$$

where the V_i volumes are defined in the figure and V is the volume of the tetrahedron. A function such as b can be written

$$b = b_1\,\eta_1 + b_2\,\eta_2 + b_3\,\eta_3 + b_4\,\eta_4\,, \tag{3.59}$$

where b_i are the nodal values of b and the interpolation functions are simply η_i.

Tetrahedrons with midside nodes can easily be defined (Fig. 3.28a) and their interpolation functions can be deduced from those of a triangle (Equation (3.55)).

3.5.2. Cube

The simplest cube-type element (they are undeformed cubes when plotted in dimensionless coordinates) is shown in Fig. 3.27b. We can also have a quadratic variation of the function on the sides (Fig. 3.28b).

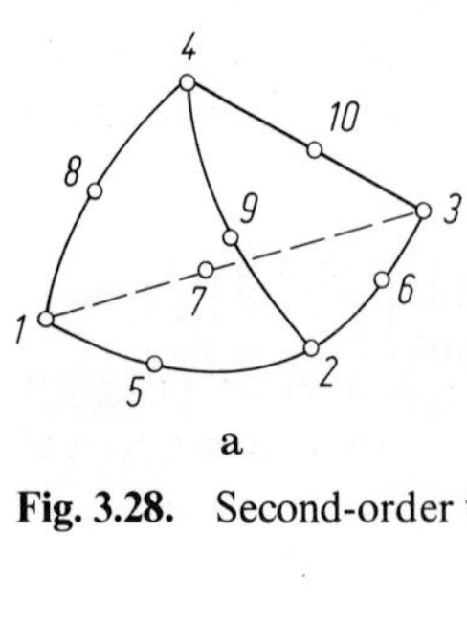

a

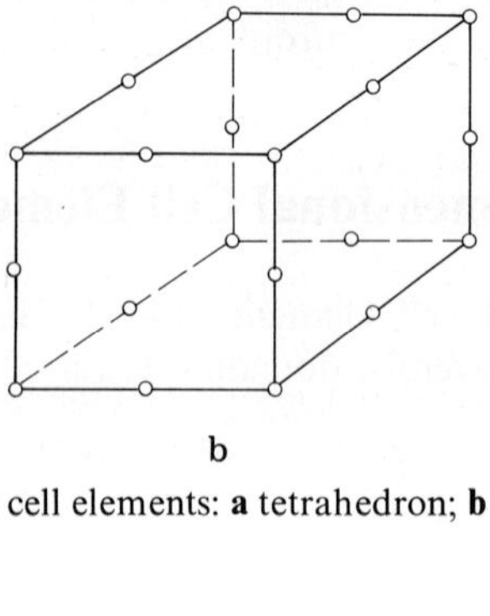

b

Fig. 3.28. Second-order three-dimensional cell elements: **a** tetrahedron; **b** cube

Interpolation functions for these and other three-dimensional cell elements can be seen in Connor and Brebbia [1].

3.6. Discontinuous Boundary Elements

One of the most interesting properties of boundary elements is that the basis functions for the approximation u can be constant within the element. This is a consequence of the inverse formulation (Section 1.7) which effectively transfers any derivatives from the approximate to the weighting function. This produces discontinuities in u between elements which do not invalidate the convergence of the technique. In fact, the constant element which can be identified as a subregion (i.e., element collocation technique) tends to give good results for many complex problems and its use is generally recommended before more refined elements are tried.

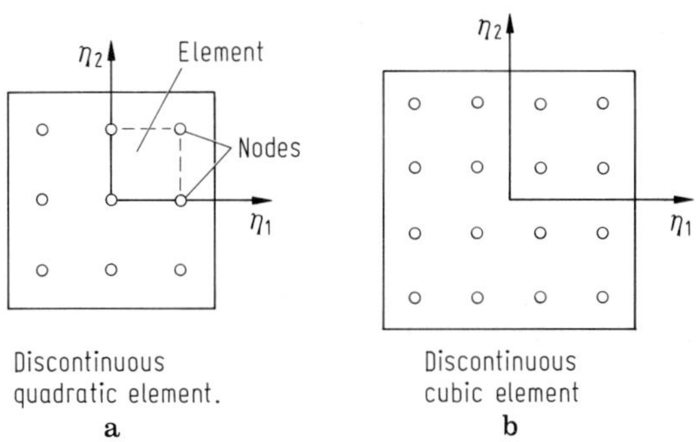

Fig. 3.29. Discontinuous elements. **a** Discontinuous quadratic element. **b** Discontinuous cubic element

This property of the boundary element method of accepting discontinuous functions motivated Danson [10] to develop a family of elements for which the approximate function is defined by Lagrangian polynomials but with all the nodes inside the elements, as shown in Fig. 3.29. The elements can be of the isoparametric type since their geometries are still defined by the coordinates of the corner points (i.e., $1, -1$).

These elements are of interest for problems in which the variable under consideration (u or q) is discontinuous between elements. The convergence of these elements is remarkably fast in many cases and they have the added advantage that different shape elements can be more easily combined since there are no continuity conditions between elements.

Example 3.7. In Fig. 3.30 we present the hollow cylinder under a certain temperature state studied by Danson [10] using discontinuous quadratic elements. The dimensions of the cylinder and the temperature at the inner and outer faces are given in the figure together with four discretizations used to study convergence of the results. The shape of the element and the variation of the potential were described using the quadratic Lagrangian polynomials defined in Section 3.4, but with all the nodal unknowns taken inside the element rather than on the boundaries.

Because of symmetry only one-eighth of the cylinder needs to be discretized and no elements are required on the planes of symmetry. The convergence of the solution can be seen in Fig. 3.30c where the results for the temperature at an internal point in the middle of the cylinder (i.e., at $R = 55$ mm) are plotted against the number of degrees of freedom. The result obtained with the most refined mesh is in excellent agreement with the exact solution (error is around 0.13%).

3.7. Order of Interpolation Functions

When working with curved elements we must always ensure that the expression for u in terms of the curvilinear coordinates contains a complete polynomial in $x_1 x_2 x_3$. That is, by suitably specializing the nodal potentials

$$u = \sum \varphi_i u_i , \tag{3.60}$$

we can reproduce

$$u = c_1 + c_2 x_1 + c_3 x_2 + c_4 x_3 . \tag{3.61}$$

If the nodal potentials are taken according to Eq. (3.61) and substituted into (3.60), we find

$$u = \sum \varphi_i (c_1 + c_2 x_1^i + c_3 x_2^i + c_4 x_3^i) . \tag{3.62}$$

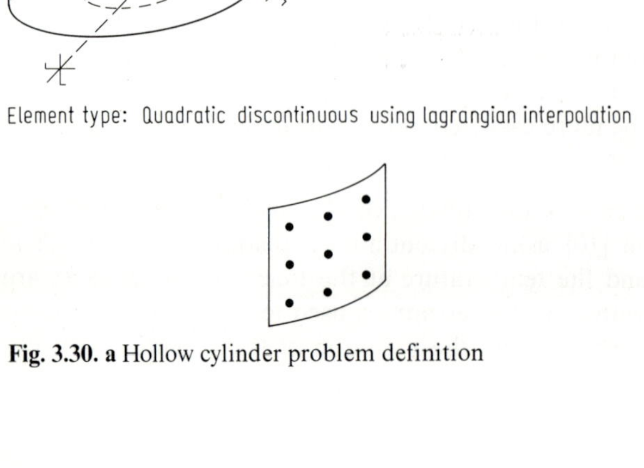

Fig. 3.30. a Hollow cylinder problem definition

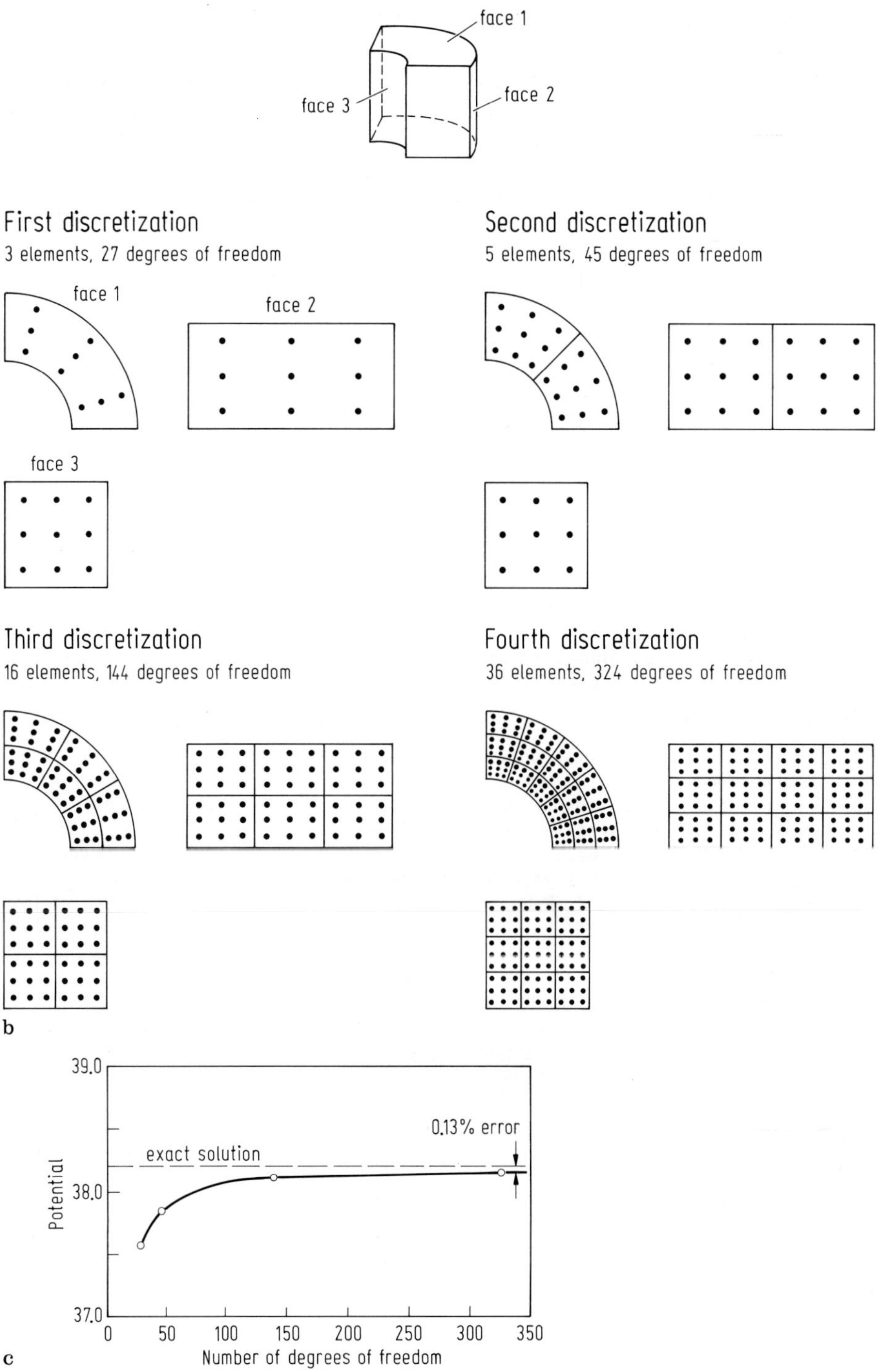

Fig. 3.30. **b** Hollow cylinder discretization using discontinuous elements. **c** Hollow cylinder; potential at $R = 55$ mm

Note that in order to be able to reproduce Eq. (3.61) everywhere in the element we have to have

$$\sum \varphi_i = 1 \tag{3.63}$$

and

$$\begin{aligned} \sum x_1^i \varphi_i &= x_1 = \boldsymbol{\varphi}^T \boldsymbol{x}_1^n, \\ \sum x_2^i \varphi_i &= x_2 = \boldsymbol{\varphi}^T \boldsymbol{x}_2^n, \\ \sum x_3^i \varphi_i &= x_3 = \boldsymbol{\varphi}^T \boldsymbol{x}_3^n. \end{aligned} \tag{3.64}$$

Relationship (3.63) is satisfied by any interpolation function, but Eqs. (3.64) are only satisfied if the order of u is at least the same as the order of the x_1, x_2, x_3 functions.

The same restriction applies for finite elements as well as boundary elements.

References

1. Connor, J. J., and Brebbia, C. A., *Finite Element Techniques for Fluid Flow,* Newnes-Butterworths, London, 1976.
2. Wood, W. L., On the finite element solution of an exterior boundary-value problem, Int. J. Numerical Methods Engng **10,** 885–891 (1976).
3. Carslaw, H. S., and Jaeger, J. C., *Conduction of Heat in Solids,* 2nd ed Clarendon Press, Oxford, 1959.
4. Isaacson, M. de St. Q., Vertical cylinder of arbitrary section in waves, J. Waterways, Port, Coastal and Ocean Div. ASCE, **104,** 309–324 (1978).
5. MacCamy, R. C., and Fuchs, R. A., Wave force on piles: A diffraction theory, Tech. Memo 69, US Army Corps of Engineers, Beach Erosion Board, Washington, D.C., 1954.
6. Au, M. C., and Brebbia, C. A., Diffraction of water waves for vertical cylinders using boundary elements, App. Math. Modelling **7,** 106–114 (1983).
7. France, P. W., Parekh, C. J., Peters, J. C., and Taylor, C., Numerical analysis of free surface seepage problems, J. Irrigation Drainage Div. ASCE **97,** 165–179 (1971).
8. Cheng, A. H.-D., Liggett, J. A., and Liu, P. L.-F., Boundary calculations of sluice and spillway flows, J. Hydraulics Div. ASCE **107,** 1163–1178 (1981).
9. Liggett, J. A., and Salmon, J. R., Cubic spline boundary elements, Int. J. Numerical Methods Engng **17,** 543–556 (1981).
10. Danson, D., BEASY Boundary Element Analysis System, Computational Mechanics Centre, MANUAL: Southampton, UK, 1982.

Chapter 4

Diffusion Problems

4.1. Introduction

In this chapter, we study boundary integral solutions to the diffusion equation

$$\nabla^2 u(x, t) - \frac{1}{k}\frac{\partial u(x, t)}{\partial t} = 0 \qquad x \in \Omega \tag{4.1}$$

with boundary conditions of the following types:

$$\begin{aligned} u(x, t) &= \bar{u}(x, t)\,, & x \in \Gamma_1\,, \\ q(x, t) &= \frac{\partial u(x, t)}{\partial n(x)} = \bar{q}(x, t) & x \in \Gamma_2\,. \end{aligned} \tag{4.2}$$

The coefficient k in Eq. (4.1) has different interpretations according to the physical problem concerned and is assumed to be constant both in space and time. Since the problem is time-dependent, some initial conditions at time $t = t_0$ must also be prescribed:

$$u(x, t) = u_0(x, t_0)\,, \qquad x \in \Omega\,. \tag{4.3}$$

The problem represented by Eq. (4.1) with boundary conditions (4.2) and initial conditions (4.3) can also be recast into an integral equation for the unknown function u, and different formulations can be employed in order to perform this transformation.

The first such formulation was proposed in 1970 by Rizzo and Shippy [1] who applied the direct formulation of the boundary element method in conjunction with Laplace transforms to solve problems of transient heat conduction. Assuming that all pertinent functions possess Laplace transforms, a boundary integral equation is derived and solved in the transform space for a sequence of real positive values of the transform parameter. A numerical transform inversion procedure is then employed to compute the physical variables in the real space. Using this approach, the time dependence of the problem is temporarily removed and a more tractable elliptic partial differential equation solved rather than the original parabolic one.

Butterfield and Tomlin [2, 3] employed the indirect (source) formulation for the analysis of zoned orthotropic media such as occurring in geotechnical engineering. Transient solutions were generated by distributing instantaneous sources over the problem region at zero time to reproduce the initial conditions and continuous sources over the region boundaries and interfaces, satisfying prescribed boundary and interface conditions.

Chang, Kang, and Chen [4] employed time-dependent fundamental solutions in the context of the direct method to solve two-dimensional problems of heat conduction in isotropic and anisotropic media. The discretization of the boundary integral equation was carried out using space and time piecewise constant values for the variables. A similar approach was discussed by Shaw [5] for the solution of three-dimensional problems, but emphasis was given to the analytical rather than numerical aspects of the method. This formulation was later extended by Wrobel and Brebbia [6] in order to allow higher-order space and time interpolation functions to be included, thus making possible the analysis of more practical problems. They also derived a numerical procedure to solve transient axisymmetric problems [7] where the complexity of the fundamental solution requires the introduction of series expansions to enable the time integrals in the boundary integral equation to be carried out analytically.

Another alternative integral formulation for the solution of transient problems is the coupled boundary element-finite difference method proposed by Brebbia and Walker [8]. In this approach, the time derivative is approximated in a finite difference form and a step-by-step finite difference-type procedure employed to advance the solution in time.

All previously discussed numerical schemes will be presented in the following sections, and the basic procedures for their numerical implementation in connection with two-dimensional problems will be derived. Although only problems defined over finite regions of homogeneous isotropic media are discussed, several other features such as internal sources, piecewise homogeneous regions, orthotropy and anisotropy, infinite or semi-infinite regions can be included in a manner similar to the way it has been done in Chapter 2 for potential problems.

Next, three-dimensional problems are briefly discussed and a fuller account of axisymmetric applications is included. The time-dependent axisymmetric fundamental solution is directly obtained from the three-dimensional one and a numerical formulation for the solution of Eq. (4.1) defined over an axisymmetric region is presented.

4.2. Laplace Transforms

Let us denote the Laplace transform of a function $u(x, t)$ when it exists (see, for instance, [9]) by

$$L[u(x, t)] = U(x, \lambda) = \int_0^\infty u(x, t)\, e^{-\lambda t}\, dt \tag{4.4}$$

and assume that the transform parameter λ is real and positive.

By integration by parts we can show that

$$L\left[\frac{\partial u(x, t)}{\partial t}\right] = \lambda\, U(x, \lambda) - u_0(x, t_0)\,. \tag{4.5}$$

Equation (4.1) in the transform space becomes

$$\nabla^2 U(x,\lambda) - \frac{\lambda}{k} U(x,\lambda) + \frac{1}{k} u_0(x,t_0) = 0 . \tag{4.6}$$

The boundary conditions (4.2) must also be transformed and we assume, for simplicity, that they are constant in time. This gives

$$\begin{aligned} U(x,\lambda) &= \bar{U}(x,\lambda) = \frac{\bar{u}(x,t)}{\lambda}, & x \in \Gamma_1 , \\ Q(x,\lambda) &= \bar{Q}(x,\lambda) = \frac{\bar{q}(x,t)}{\lambda}, & x \in \Gamma_2 . \end{aligned} \tag{4.7}$$

Proceeding as for Laplace's equation, we can write the following weighted residual statement:

$$\int_\Omega \left[\nabla^2 U(x,\lambda) - \frac{\lambda}{k} U(x,\lambda) + \frac{1}{k} u_0(x,t_0) \right] U^*(\xi,x,\lambda)\, d\Omega(x) \tag{4.8}$$

$$= \int_{\Gamma_2} [Q(x,\lambda) - \bar{Q}(x,\lambda)]\, U^*(\xi,x,\lambda)\, d\Gamma(x) - \int_{\Gamma_1} [U(x,\lambda) - \bar{U}(x,\lambda)]\, Q^*(\xi,x,\lambda)\, d\Gamma(x) ,$$

where $Q^*(\xi,x,\lambda) = \partial U^*(\xi,x,\lambda)/\partial n(x)$. Integrating by parts twice the Laplacian in the above equation gives

$$\int_\Omega \left[\nabla^2 U^*(\xi,x,\lambda) - \frac{\lambda}{k} U^*(\xi,x,\lambda) \right] U(x,\lambda)\, d\Omega(x) + \frac{1}{k} \int_\Omega u_0(x,t_0)\, U^*(\xi,x,\lambda)\, d\Omega(x)$$

$$= - \int_\Gamma Q(x,\lambda)\, U^*(\xi,x,\lambda)\, d\Gamma(x) + \int_\Gamma U(x,\lambda)\, Q^*(\xi,x,\lambda)\, d\Gamma(x) . \tag{4.9}$$

Assuming U^* to be the fundamental solution to Eq. (4.6), which satisfies the relation

$$k\nabla^2 U^*(\xi,x,\lambda) - \lambda U^*(\xi,x,\lambda) = - \Delta(\xi,x) , \tag{4.10}$$

Eq. (4.9) becomes

$$\begin{aligned} &U(\xi,\lambda) + k \int_\Gamma U(x,\lambda)\, Q^*(\xi,x,\lambda)\, d\Gamma(x) \\ &= k \int_\Gamma Q(x,\lambda)\, U^*(\xi,x,\lambda)\, d\Gamma(x) + \int_\Omega u_0(x,t_0)\, U^*(\xi,x,\lambda)\, d\Omega(x) . \end{aligned} \tag{4.11}$$

The fundamental solution U^* for three-dimensional problems is of the form

$$U^* = \frac{(k\lambda)^{1/4}}{r^{1/2}(2\pi k)^{3/2}} K_{1/2} \left[\left(\frac{\lambda}{k} \right)^{1/2} r \right] \tag{4.12}$$

and for two-dimensional problems

$$U^* = \frac{1}{2\pi k} K_0 \left[\left(\frac{\lambda}{k} \right)^{1/2} r \right], \tag{4.13}$$

where K_ν is the modified Bessel function of the second kind of order ν.

Let us now investigate the singularity of the above fundamental solutions. As $r \to 0$, so does the argument of the modified Bessel functions. The limiting form of $K_{1/2}(z)$ as $z \to 0$ is [10]

$$K_{1/2}(z) = \left(\frac{\pi}{2z} \right)^{1/2} \tag{4.14}$$

so that

$$U^* = \frac{(k\,\lambda)^{1/4}}{r^{1/2}(2\,\pi\,k)^{3/2}} \left(\frac{\pi}{2r} \right)^{1/2} \left(\frac{k}{\lambda} \right)^{1/4} = \frac{1}{4\,\pi\,k\,r} \quad \text{for } r \to 0 \tag{4.15}$$

which means that the singularity of the fundamental solution is of the same type as that of Laplace's equation.

Analogously, the limiting form of $K_0(z)$ as $z \to 0$ is [10]

$$K_0(z) = -\ln(z)\,, \tag{4.16}$$

thus

$$U^* = \frac{1}{2\,\pi\,k} \ln \frac{1}{r} - \frac{1}{4\,\pi\,k} \ln \frac{\lambda}{k} \quad \text{for } r \to 0\,. \tag{4.17}$$

The first term is the fundamental solution to the two-dimensional Laplace equation, while the second term is a nonsingular constant which adds nothing to the solution.

Taking the point ξ in Eq. (4.11) to the boundary and noting that the integral in Q^* presents a discontinuity as ξ approaches Γ produces the equation

$$c(\xi)\,U(\xi,\lambda) + k \int_\Gamma U(x,\lambda)\,Q^*(\xi,x,\lambda)\,d\Gamma(x)$$

$$= k \int_\Gamma Q(x,\lambda)\,U^*(\xi,x,\lambda)\,d\Gamma(x) + \int_\Omega u_0(x,t_0)\,U^*(\xi,x,\lambda)\,d\Omega(x)\,, \tag{4.18}$$

where the coefficient c has the same value as previously (see Chapter 2).

This equation is discretized and solved numerically for a sequence of N selected values of the transform parameter λ, chosen somewhat arbitrarily [1]. Notice that the presence of specified initial conditions gives rise to an integral over the domain Ω. One way of evaluating this integral is to divide the whole domain into cells and numerically integrate over each cell. However, if u_0 satisfies Laplace's equation, the domain integral in Eq. (4.18) can be transformed into equivalent boundary integrals [11]. Whatever the method of evaluating the domain integral may be, this integral introduces no further unknown since u_0 is prescribed, and Eq. (4.18) is still a boundary integral equation.

The remaining step is the transform inversion of the solution which is carried out numerically. Following, for instance, the method of Schapery [12] (cf. [1]), we assume that the value of u at any point can be represented as a finite series by

$$u(\xi, t) = u(\xi, \infty) + \sum_{n=1}^{N} a_n(\xi) \exp[-b_n(\xi)\, t], \tag{4.19}$$

where $u(\xi, \infty)$ is the steady-state solution and a_n and b_n are functions of the position. Transforming Eq. (4.19) gives

$$U(\xi, \lambda) = \frac{u(\xi, \infty)}{\lambda} + \sum_{n=1}^{N} \frac{a_n(\xi)}{\lambda + b_n(\xi)}. \tag{4.20}$$

The values of the coefficients b_n are now assumed to be equal to the previously selected λ. Thus, there remain the N values of the coefficient a to be computed at each boundary point (plus the internal points where the solution is required). The N solutions of Eq. (4.18) provide N values of U at each point, which allow the evaluation of the coefficients a_n using Eq. (4.20) and, consequently, the evaluation of the physical variable u using Eq. (4.19). A similar calculation is also required in order to obtain the real boundary (and internal) fluxes.

Notice that the transform inversion is essentially a curve-fitting process and as such, it is important for the analyst to have an idea of the expected behavior of the solution in order to select values of the transform parameter λ, since choosing too many values would quickly make Eq. (4.20) unstable, while choosing too few values would not represent the curve adequately [13]. Furthermore, as pointed out in [11], the formulation is not efficient when the time history of the boundary conditions is complex and in this case, step-by-step methods of the type subsequently discussed should be preferred.

Example 4.1. As a simple example to show the applicability of the numerical procedures just described, we present the problem of a circular region of unit radius with $u_0 = 0$ subject to the boundary condition $q = 2(1 - u)$ along Γ studied in Ref. [1]. The value of the parameter k was assumed to be unity, and due to symmetry only one-quarter of the boundary was subdivided into six constant boundary elements (Fig. 4.1).

In order to select the sequence of transform parameter values, it was pointed out by Schapery [12] that if a plot of λU versus $\log \lambda$ can be obtained the significant range of λ needed in the inversion scheme can be chosen by inspection. For example, the variation of λU with $\log \lambda$ for two locations in the circular region is shown in Fig. 4.2. Clearly, the significant range of λ referred to is between a minimum of about 0.1 and a maximum of about 100 for one curve, and 0.1 to 1000 for the other curve.

For the problem under consideration, a sequence of 14 terms was chosen beginning with $\lambda_1 = 0.1$ such that $\lambda_{i+1}/\lambda_i = 2$. A Gauss-Jordan matrix inversion scheme was used to solve for the coefficients a_n in Eq. (4.20). Numerical results are compared in Table 4.1 with the corresponding analytical values [14].

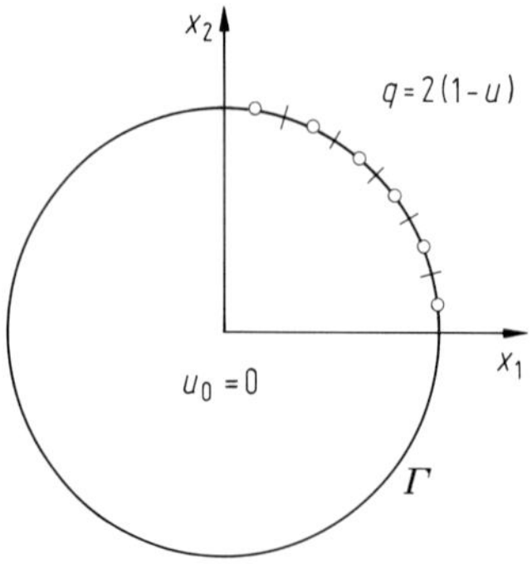

Fig. 4.1. Problem definition

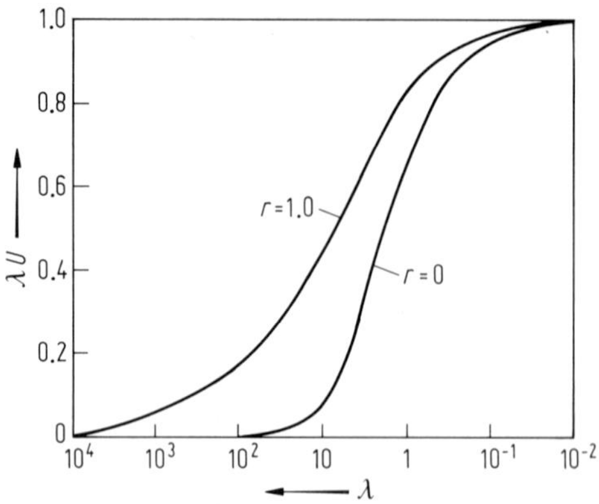

Fig. 4.2. Variation of U with λ

Table 4.1. Temperature variation in circular region.

Time ($\log_{10} t$)	$r = 0$		$r = 0.5$		$r = 1.0$	
	Analytical	Numerical	Analytical	Numerical	Analytical	Numerical
− 4.0	0	− 0.002	0	− 0.001	0.028	0.026
− 3.6	0	− 0.001	0	− 0.001	0.037	0.034
− 3.2	0	0.001	0	0	0.055	0.052
− 2.8	0	0.001	0	0.001	0.085	0.086
− 2.4	0	− 0.001	0	− 0.001	0.131	0.133
− 2.0	0	0.001	0	0.001	0.199	0.199
− 1.6	0	− 0.001	0.005	0.005	0.295	0.295
− 1.2	0.007	0.007	0.061	0.061	0.423	0.423
− 0.8	0.134	0.135	0.249	0.248	0.584	0.584
− 0.4	0.517	0.516	0.591	0.590	0.780	0.780
0	0.896	0.897	0.912	0.912	0.953	0.953
0.4	0.998	0.998	0.998	0.998	0.999	0.999
0.8	1.000	1.000	1.000	1.000	1.000	1.000

4.3. Coupled Boundary Element – Finite Difference Methods

Let us now assume that the time derivative in Eq. (4.1) can be approximated in a finite difference form, for a sufficiently small time step Δt, as

$$\frac{\partial u(x,t)}{\partial t} = \frac{u(x, t+\Delta t) - u(x,t)}{\Delta t} \tag{4.21}$$

Equation (4.1) can then be rewritten

$$\nabla^2 u(x, t+\Delta t) - \frac{1}{k\,\Delta t} u(x, t+\Delta t) + \frac{1}{k\,\Delta t} u(x,t) = 0\,. \tag{4.22}$$

This equation is similar in form to Eq. (4.6) and so its fundamental solutions are of the same type as Eqs. (4.12) and (4.13), with λ replaced by $1/\Delta t$.

The boundary integral equation for this formulation can be obtained through weighted residual considerations, in the same way as it was done in the previous section. By an analogy with Eq. (4.18), we can write

$$c(\xi)\,u(\xi,t+\Delta t)+k\int_{\Gamma} U(x,t+\Delta t)\,q^*(\xi,x,\Delta t)\,d\Gamma(x) \tag{4.23}$$
$$=k\int_{\Gamma} Q(x,t+\Delta t)\,u^*(\xi,x,\Delta t)\,d\Gamma(x)+\frac{1}{\Delta t}\int_{\Omega} u(x,t)\,u^*(\xi,x,\Delta t)\,d\Omega(x)\,.$$

Starting from known initial values of u at $t=t_0$, we can advance the process in time by solving Eq. (4.23) numerically. Values of u at time $t=t_0+\Delta t$ are then computed, at a sufficient number of internal points, in order to be used as pseudo-initial values for the next time step.

Numerical results using this formulation are presented in [15]. Notice that very small time steps have to be adopted if approximation (4.21) is to produce good results. As discussed in [15], the accuracy of this formulation can be significantly improved by employing second-order finite difference schemes, although convergence problems become more severe.

4.4. Time-Dependent Fundamental Solutions

Considering the time dependence of the problem directly in the integration by parts process, we can write the following weighted residual statement for the governing equation (4.1) with boundary conditions (4.2):

$$\int_{t_0}^{t_F}\int_{\Omega}\left[\nabla^2 u(x,t)-\frac{1}{k}\frac{\partial u(x,t)}{\partial t}\right]u^*(\xi,x,t_F,t)\,d\Omega(x)\,dt$$
$$=\int_{t_0}^{t_F}\int_{\Gamma_2}[q(x,t)-\bar{q}(x,t)]\,u^*(\xi,x,t_F,t)\,d\Gamma(x)\,dt \tag{4.24}$$
$$-\int_{t_0}^{t_F}\int_{\Gamma_1}[u(x,t)-\bar{u}(x,t)]\,q^*(\xi,x,t_F,t)\,d\Gamma(x)\,dt,$$

where $q^*(\xi,x,t_F,t)=\partial u^*(\xi,x,t_F,t)/\partial n(x)$.

Integrating by parts twice the Laplacian and once the time derivative in the above equation gives

$$\int_{t_0}^{t_F}\int_{\Omega}\left[\nabla^2 u^*(\xi,x,t_F,t)+\frac{1}{k}\frac{\partial u^*(\xi,x,t_F,t)}{\partial t}\right]u(x,t)\,d\Omega(x)\,dt$$
$$-\frac{1}{k}\left[\int_{\Omega} u(x,t)\,u^*(\xi,x,t_F,t)\,d\Omega(x)\right]_{t=t_0}^{t=t_F} \tag{4.25}$$
$$=-\int_{t_0}^{t_F}\int_{\Gamma} q(x,t)\,u^*(\xi,x,t_F,t)\,d\Gamma(x)\,dt+\int_{t_0}^{t_F}\int_{\Gamma} u(x,t)\,q^*(\xi,x,t_F,t)\,d\Gamma(x)\,dt\,.$$

The time-dependent fundamental solution u^* is of the form [14, 16]

$$u^* = \frac{1}{(4\pi k \tau)^{d/2}} \exp\left[-\frac{r^2}{4k\tau}\right] H(\tau), \tag{4.26}$$

where $\tau = t_F - t$ and d is the number of spatial dimensions of the problem, e.g., $d = 3$ for three-dimensional problems, etc. Note that Eqs. (4.12) and (4.13) are the Laplace transforms of Eq. (4.26) for $d = 3$ and $d = 2$, respectively. The Heaviside function $H(\tau)$ is included to emphasize the fact that the solution is identically zero for $t > t_F$. This condition is known as the causality condition [16].

The fundamental solution possesses the following properties:

$$k\nabla^2 u^*(\xi, x, t_F, t) + \frac{\partial u^*(\xi, x, t_F, t)}{\partial t} = -\Delta(\xi, x)\,\Delta(t_F, t), \tag{4.27}$$

$$\lim_{t \to t_F} u^*(\xi, x, t_F, t) = \Delta(\xi, x). \tag{4.28}$$

Let us now investigate the singularity that occurs in the integrals in Eq. (4.25) at time $t = t_F$. In order to avoid ending the integrations exactly at the peak of a Dirac delta function, we may subtract or add to the upper limit of the integrals an arbitrarily small quantity ε. In the former case, the first integral on the left-hand side is identically zero for t in the range 0, $t_F - \varepsilon$ and so, taking the limit as $\varepsilon \to 0$ and accounting for condition (4.28), Eq. (4.25) yields [14, 17]

$$\begin{aligned} &u(\xi, t_F) + k\int_{t_0}^{t_F}\int_{\Gamma} u(x, t)\, q^*(\xi, x, t_F, t)\, d\Gamma(x)\, dt \\ &= k\int_{t_0}^{t_F}\int_{\Gamma} q(x, t)\, u^*(\xi, x, t_F, t)\, d\Gamma(x)\, dt + \int_{\Omega} u_0(x, t_0)\, u^*(\xi, x, t_F, t_0)\, d\Omega(x). \end{aligned} \tag{4.29}$$

The same relation can be obtained by adding ε to the upper limit of the integrals in Eq. (4.25). In this case, $u^*(\xi, x, t_F, t_F + \varepsilon)$ equals zero due to the causality condition. Thus, taking the limit of Eq. (4.25) as $\varepsilon \to 0$, the inclusion of condition (4.27) into the first integral on the left-hand side produces the expected result [16].

Another property of the time-dependent fundamental solution (4.26) is that, as a steady-state is reached, it reduces to the fundamental solution to Laplace's equation. That is,

$$\lim_{t_F \to \infty} \int_0^{t_F} u^*(\xi, x, t_F, t)\, dt = u^*(\xi, x). \tag{4.30}$$

We shall now prove this property for three-dimensional problems, but a similar approach can also be applied in two dimensions. So we have to integrate

$$\int_0^{t_F} u^*\, dt = \int_0^{t_F} \frac{1}{(4\pi k \tau)^{3/2}} \exp\left(-\frac{r^2}{4k\tau}\right) dt. \tag{4.31}$$

This integral can be evaluated analytically by introducing the variable $x = r^2/4k\tau$. An interchange of variables then gives

$$\int_0^{t_F} u^* \, dt = \frac{1}{4\pi^{3/2} k r} \int_a^{\infty} x^{-1/2} e^{-x} \, dx = \frac{1}{4\pi^{3/2} k r} \Gamma(\tfrac{1}{2}, a), \tag{4.32}$$

where $a = r^2/4k\,t_F$ and Γ is the incomplete gamma function. Taking the limit of Eq. (4.32) as $t_F \to \infty$ [10],

$$\frac{1}{4\pi^{3/2} k r} \lim_{t_F \to \infty} \Gamma(\tfrac{1}{2}, a) = \frac{1}{4\pi k r} \tag{4.33}$$

which can be recognized as the fundamental solution to $k\nabla^2 u = 0$.

Note that the first two integrals in Eq. (4.29) represent the effects of boundary conditions, while the third term includes the effects of the initial value u_0 of the function u. As $t_F \to \infty$, the initial conditions effect vanishes, while the integrations over t for the boundary terms can be carried out assuming that u and q no longer depend on t (or at least that the contribution to the integral over t from 0 to ∞ from those values of t where u and q were still dependent on t is negligible compared to the total integral). Thus, by virtue of Eq. (4.30), the fundamental solution reduces to that of Laplace's equation and Eq. (4.29) becomes the integral equation (2.69) for steady potential problems.

Taking the point ξ in Eq. (4.29) to the boundary and accounting for the jump of the left-hand side integral yields the boundary integral equation

$$\begin{aligned} c(\xi)\, u(\xi, t_F) + k \int_{t_0}^{t_F} \int_{\Gamma} u(x, t)\, q^*(\xi, x, t_F, t)\, d\Gamma(x)\, dt \\ = k \int_{t_0}^{t_F} \int_{\Gamma} q(x, t)\, u^*(\xi, x, t_F, t)\, d\Gamma(x)\, dt + \int_{\Omega} u_0(x, t_0)\, u^*(\xi, x, t_F, t_0)\, d\Omega(x), \end{aligned} \tag{4.34}$$

where $c(\xi)$, as before, is a function of the solid angle of the boundary at point ξ (see Eq. (2.69)).

Since the time variation of functions u and q is not known a priori, a time-stepping technique (not to be confused with the previous finite difference one) has to be introduced for the numerical solution of Eq. (4.34). However, as the fundamental solution itself is time-dependent, large time steps can generally be adopted.

Two different time-marching schemes can be employed on this numerical solution: the first (hereafter referred to as *scheme* 1) treats each time step as a new problem and so, at the end of each step, computes values of the function u at a sufficient number of internal points in order to use them as pseudo-initial values for the next step; in the other (*scheme* 2), the time integration process always restarts from time t_0 and so, despite the increasing number of intermediate steps as the time progresses, values of u at internal points need not be recomputed. Furthermore, if u_0 satisfies Laplace's equation, the domain integral in Eq. (4.34) can be transformed into equivalent boundary integrals. The necessary procedures for numerical implementation of both time-marching schemes are discussed in the next section.

4.5. Two-Dimensional Problems

For the numerical solution of Eq. (4.34), the boundary Γ is discretized into a series of elements. The geometry of these elements can be modelled by straight lines, circular arcs, parabolas, etc., as discussed in Chapter 3. Furthermore, functions u and q are assumed to vary within each element and each time step according to space and time interpolation functions such that

$$\begin{aligned} u &= \boldsymbol{\varphi}^T \boldsymbol{\psi}\, \boldsymbol{u}^n, \\ q &= \boldsymbol{\varphi}^T \boldsymbol{\psi}\, \boldsymbol{q}^n, \end{aligned} \tag{4.35}$$

where $\boldsymbol{\varphi}$ and $\boldsymbol{\psi}$ are the space and the time interpolation functions, respectively.

For two-dimensional problems, the fundamental solution and its normal derivative are given by (see Eq. (4.26)),

$$u^* = \frac{1}{4\pi k \tau} \exp\left[-\frac{r^2}{4k\tau}\right], \tag{4.36}$$

$$q^* = \frac{d}{8\pi k^2 \tau^2} \exp\left[-\frac{r^2}{4k\tau}\right] \tag{4.37}$$

in which $d = [x_1(\xi) - x_1(x)]\, n_1(x) + [x_2(\xi) - x_2(x)]\, n_2(x)$.

If the boundary Γ is discretized into N elements, the domain Ω subdivided into L cells, and the time dimension subdivided into F time steps, the substitution of Eqs. (4.35) into Eq. (4.34) yields, for scheme 1, the equation

$$\begin{aligned} c_i u_F^i + k \sum_{j=1}^{N} \left(\int_{\Gamma_j} \boldsymbol{\varphi}^T \int_{t_{F-1}}^{t_F} q^* \boldsymbol{\psi}\, dt\, d\Gamma \right) \boldsymbol{u}^n \\ = k \sum_{j=1}^{N} \left(\int_{\Gamma_j} \boldsymbol{\varphi}^T \int_{t_{F-1}}^{t_F} u^* \boldsymbol{\psi}\, dt\, d\Gamma \right) \boldsymbol{q}^n + \sum_{l=1}^{L} \int_{\Omega_l} u^* \boldsymbol{u}_{F-1}\, d\Omega \end{aligned} \tag{4.38}$$

and for scheme 2 the equation

$$\begin{aligned} c_i u_F^i + k \sum_{j=1}^{N} \sum_{f=1}^{F} \left(\int_{\Gamma_j} \boldsymbol{\varphi}^T \int_{t_{f-1}}^{t_f} q^* \boldsymbol{\psi}\, dt\, d\Gamma \right) \boldsymbol{u}^n \\ = k \sum_{j=1}^{N} \sum_{f=1}^{F} \left(\int_{\Gamma_j} \boldsymbol{\varphi}^T \int_{t_{f-1}}^{t_f} u^* \boldsymbol{\psi}\, dt\, d\Gamma \right) \boldsymbol{q}^n + \sum_{l=1}^{L} \int_{\Omega_l} u^* \boldsymbol{u}_0\, d\Omega\,. \end{aligned} \tag{4.39}$$

4.5.1. Constant Time Interpolation

Assuming that functions u and q remain constant in time over each time step, i.e., the interpolation function ψ is unity, applying Eq. (4.39) to all boundary nodes yields the following system of equations (see Eq. (3.9)):

$$\sum_{f=1}^{F} \boldsymbol{H}_{fF} \boldsymbol{U}_f = \sum_{f=1}^{F} \boldsymbol{G}_{fF} \boldsymbol{Q}_f + \boldsymbol{B}_0 \bar{\boldsymbol{U}}_0. \tag{4.40}$$

The coefficients of matrices $\hat{\boldsymbol{H}}$ and $\boldsymbol{G}$ are constituted of terms, or combination of terms, of the form (see Eqs. (3.4) and (3.5))

$$\begin{aligned} h^m_{fFij} &= k \int_{\Gamma_j} \varphi_m \int_{t_{f-1}}^{t_f} q^* \, dt \, d\Gamma, \\ g^m_{fFij} &= k \int_{\Gamma_j} \varphi_m \int_{t_{f-1}}^{t_f} u^* \, dt \, d\Gamma \end{aligned} \tag{4.41}$$

where $H_{fFij} = \hat{H}_{fFij} + c_i \, \delta_{fF} \, \delta_{ij}$ and δ_{ij} is the Kronecker delta.

The computation of matrix $\boldsymbol{B}_0$, resulting from the domain integral, will be considered at a later stage. Note that, in Eq. (4.40), we are solving for the time $t = t_F$ and the values of $\boldsymbol{U}_f$ and $\boldsymbol{Q}_f$ for $f = 1, 2, \ldots, F-1$ are known from calculations at previous steps.

For time-marching scheme 1, the application of Eq. (4.38) to all boundary nodes gives

$$\boldsymbol{H} \boldsymbol{U}_F = \boldsymbol{G} \boldsymbol{Q}_F + \boldsymbol{B} \bar{\boldsymbol{U}}_{F-1}, \tag{4.42}$$

where the coefficients of matrices $\hat{\boldsymbol{H}}$ and $\boldsymbol{G}$ are also formed by terms of the type (4.41) with $f = F$.

The time integrals in Eqs. (4.41) can be carried out analytically. The integral in q^* gives

$$\int_{t_{f-1}}^{t_f} q^* \, dt = \frac{d}{2\pi k r^2} \int_{t_{f-1}}^{t_f} \frac{r^2}{4k\tau^2} \exp\left(-\frac{r^2}{4k\tau}\right) dt = \frac{d}{2\pi k r^2} \left[\exp(-a_{f-1}) - \exp(-a_f)\right], \tag{4.43}$$

where

$$a_f = \frac{r^2}{4k\,(t_F - t_f)}. \tag{4.44}$$

In order to perform the integral in u^*, we need to make an appropriate change of variables. Calling

$$x = \frac{r^2}{4k\tau}, \tag{4.45}$$

the integral becomes [18]

$$\begin{aligned} \int_{t_{f-1}}^{t_f} u^* \, dt &= \frac{1}{\pi r^2} \int_{t_{f-1}}^{t_f} \frac{r^2}{4k\tau} \exp\left(-\frac{r^2}{4k\tau}\right) dt \\ &= \frac{1}{4\pi k} \int_{a_{f-1}}^{a_f} \frac{e^{-x}}{x} \, dx = \frac{1}{4\pi k} \left[E_1(a_{f-1}) - E_1(a_f)\right], \end{aligned} \tag{4.46}$$

where E_1 is the exponential-integral function. From definition (4.44) we note that $\exp(-a_F) = 0$ in Eq. (4.43) and $E_1(a_F) = 0$ in Eq. (4.46).

4.5.2. Linear Time Interpolation

Let us now assume a linear variation in time for functions u and q within each time step according to the interpolation functions

$$\psi_1 = \frac{t_f - t}{\Delta t_f}, \qquad \psi_2 = \frac{t - t_{f-1}}{\Delta t_f}, \tag{4.47}$$

where $\Delta t_f = t_f - t_{f-1}$. The system of equations obtained from the application of Eq. (4.39) to all boundary nodes is now of the form

$$\sum_{f=1}^{F} (\boldsymbol{H}_{fF}^1 \boldsymbol{U}_{f-1} + \boldsymbol{H}_{fF}^2 \boldsymbol{U}_f) = \sum_{f=1}^{F} (\boldsymbol{G}_{fF}^1 \boldsymbol{Q}_{f-1} + \boldsymbol{G}_{fF}^2 \boldsymbol{Q}_f) + \boldsymbol{B}_0 \bar{\boldsymbol{U}}_0, \tag{4.48}$$

where the coefficients of the matrices involved are constituted of terms such as

$$\begin{aligned}
h_{fFij}^{1m} &= \frac{k}{\Delta t_f} \int_{\Gamma_j} \varphi_m \int_{t_{f-1}}^{t_f} (t_f - t)\, q^*\, dt\, d\Gamma, \\
h_{fFij}^{2m} &= \frac{k}{\Delta t_f} \int_{\Gamma_j} \varphi_m \int_{t_{f-1}}^{t_f} (t - t_{f-1})\, q^*\, dt\, d\Gamma, \\
g_{fFij}^{1m} &= \frac{k}{\Delta t_f} \int_{\Gamma_j} \varphi_m \int_{t_{f-1}}^{t_f} (t_f - t)\, u^*\, dt\, d\Gamma, \\
g_{fFij}^{2m} &= \frac{k}{\Delta t_f} \int_{\Gamma_j} \varphi_m \int_{t_{f-1}}^{t_f} (t - t_{f-1})\, u^*\, dt\, d\Gamma,
\end{aligned} \tag{4.49}$$

where $H_{fFij}^2 = \hat{H}_{fFij}^2 + c_i\, \delta_{fF}\, \delta_{ij}$.

Analogously, the application of Eq. (4.38) to all boundary nodes gives

$$\boldsymbol{H}^1 \boldsymbol{U}_{F-1} + \boldsymbol{H}^2 \boldsymbol{U}_F = \boldsymbol{G}^1 \boldsymbol{Q}_{F-1} + \boldsymbol{G}^2 \boldsymbol{Q}_F + \boldsymbol{B} \bar{\boldsymbol{U}}_{F-1} \tag{4.50}$$

in which, as previously, the coefficients of matrices $\hat{\boldsymbol{H}}$ and $\boldsymbol{G}$ are also formed by terms of the type (4.49) with $f = F$.

The time integrals can now be divided into integrals of the same form as previously (Eqs. (4.43) and (4.46)) plus integrals of the form

$$\begin{aligned}
\int_{t_{f-1}}^{t_f} t\, q^*\, dt &= \frac{d}{2\pi k} \int_{t_{f-1}}^{t_f} \frac{t}{4k\tau^2} \exp\left(-\frac{r^2}{4k\tau}\right) dt \\
&= \frac{d}{2\pi k} \left[\frac{t_F}{r^2} \int_{a_{f-1}}^{a_f} e^{-x}\, dx - \frac{1}{4k} \int_{a_{f-1}}^{a_f} \frac{e^{-x}}{x}\, dx \right],
\end{aligned} \tag{4.51}$$

$$\int_{t_{f-1}}^{t_f} t\, u^*\, dt = \frac{1}{4\pi k} \int_{t_{f-1}}^{t_f} \frac{t}{\tau} \exp\left(-\frac{r^2}{4k\tau}\right) dt$$

$$= \frac{1}{4\pi k}\left[t_F \int_{a_{f-1}}^{a_f} \frac{e^{-x}}{x}\, dx - \frac{r^2}{4k} \int_{a_{f-1}}^{a_f} \frac{e^{-x}}{x^2}\, dx\right]. \tag{4.52}$$

The integrals in the above equations are of the same type as the ones in Eqs. (4.43) and (4.46) apart from the last one in Eq. (4.52) which gives [18]

$$\int_{a_{f-1}}^{a_f} \frac{e^{-x}}{x^2}\, dx = \Gamma(-1, a_{f-1}) - \Gamma(-1, a_f), \tag{4.53}$$

where Γ is the incomplete gamma function.

Thus, adding up all terms and taking into account the relation between Γ and E_1, i.e. [10],

$$\Gamma(-n, a) = \frac{(-1)^n}{n!}\left[E_1(a) - e^{-a} \sum_{i=0}^{n-1} \frac{(-1)^i\, i!}{a^{i+1}}\right], \qquad n = 1, 2, \ldots, \tag{4.54}$$

the final expressions for the time integrals in Eq. (4.49) can be easily written by combining the appropriate terms explicitly calculated in Eqs. (4.43), (4.46), and (4.53) (see [19]).

4.5.3. Quadratic Time Interpolation

Consider now that functions u and q have a quadratic variation within each time step according to the interpolation functions

$$\psi_1 = 2\bar{t}^2 - 3\bar{t} + 1, \tag{4.55}$$

$$\psi_2 = 4\bar{t}(1 - \bar{t}), \qquad \psi_3 = \bar{t}(2\bar{t} - 1), \tag{4.56}$$

where $\bar{t} = (t - t_{f-1})/(t_f - t_{f-1})$ and the time stations are t_{f-1}, $t_{f-1/2}$, and t_f, where $t_{f-1/2} = (t_{f-1} + t_f)/2$.

The application of Eq. (4.38), for instance, to all boundary nodes now gives

$$\boldsymbol{H}^1\, \boldsymbol{U}_{F-1} + \boldsymbol{H}^2\, \boldsymbol{U}_{F-1/2} + \boldsymbol{H}^3\, \boldsymbol{U}_F$$

$$= \boldsymbol{G}^1\, \boldsymbol{Q}_{F-1} + \boldsymbol{G}^2\, \boldsymbol{Q}_{F-1/2} + \boldsymbol{G}^3\, \boldsymbol{Q}_F + \boldsymbol{B}\, \bar{\boldsymbol{U}}_{F-1}. \tag{4.57}$$

The coefficients of the above matrices are formed by terms obtained from space and time integrals similar to Eqs. (4.41) and (4.49).

After introducing the boundary conditions of the problem, Eq. (4.57) becomes a system of M equations with $2M$ unknowns (M being the number of boundary nodes) since all values of $\boldsymbol{U}_{F-1}$ and $\boldsymbol{Q}_{F-1}$ are prescribed (or have been previously calculated) but only half the boundary values at times $t_{F-1/2}$ and t_F are known.

This means that for the problem to be well-posed we need to double the total number of simultaneous equations involved in solving a single time step. This can be achieved by writing a boundary integral equation similar to Eq. (4.34) for the time $t = t_{F-1/2}$,

$$c(\xi)\, u(\xi, t_{F-1/2}) + k \int_{t_0}^{t_{F-1/2}} \int_{\Gamma} u(x, t)\, q^*(\xi, x, t_{F-1/2}, t)\, d\Gamma(x)\, dt$$

$$= k \int_{t_0}^{t_{F-1/2}} \int_{\Gamma} q(x, t)\, u^*(\xi, x, t_{F-1/2}, t)\, d\Gamma(x)\, dt \tag{4.58}$$

$$+ \int_{\Omega} u_0(x, t_0)\, u^*(\xi, x, t_{F-1/2}, t_0)\, d\Omega(x).$$

The upper limit of the time integrals was taken as $t_{F-1/2}$ because of the causality condition (see Section 4.4) which specifies that u^* and q^* are identically zero for $t > t_{F-1/2}$.

Discretizing the above equation and applying it to all boundary nodes yields the system of equations

$$\bar{\boldsymbol{H}}^1 \boldsymbol{U}_{F-1} + \bar{\boldsymbol{H}}^2 \boldsymbol{U}_{F-1/2} + \bar{\boldsymbol{H}}^3 \boldsymbol{U}_F = \bar{\boldsymbol{G}}^1 \boldsymbol{Q}_{F-1} + \bar{\boldsymbol{G}}^2 \boldsymbol{Q}_{F-1/2} + \bar{\boldsymbol{G}}^3 \boldsymbol{Q}_F + \bar{\boldsymbol{B}}\, \bar{\boldsymbol{U}}_{F-1} \tag{4.59}$$

in which the coefficients of the matrices can also be calculated by using expressions similar to Eqs. (4.41) and (4.49).

The simultaneous solution of Eqs. (4.57) and (4.59) now permits one to determine the unknown boundary values of $\boldsymbol{U}$ and $\boldsymbol{Q}$ at times $t_{F-1/2}$ and t_F from the knowledge of the initial conditions at t_{F-1} and prescribed boundary conditions at $t_{F-1/2}$ and t_F.

This procedure can be extended to time functions of higher orders, noting that the total number of simultaneous equations to be stepwise solved would be further increased.

4.5.4. Space Integration

The remaining step for the numerical solution of Eq. (4.34) is the computation of the space integrals. Although the space interpolation functions $\boldsymbol{\varphi}$ in Eq. (4.35) can be taken as constant, linear, quadratic, etc., only the constant case will be considered here, extension to higher-order elements being done by following the same procedures described in Chapter 3.

As for Laplace's equation, the off-diagonal coefficients of matrices $\boldsymbol{G}$ and $\boldsymbol{H}$ can be computed numerically, using Gauss quadrature. Usually, six points are sufficient to provide the required accuracy, although a more selective integration scheme, with fewer Gaussian points adopted for elements located far from the singularity, should be employed for computer efficiency.

The diagonal coefficients G_{FFii} in Eq. (4.40) contain integrals with a logarithmic (integrable) singularity. We can write such integrals, with reference to Fig. 4.3, as

$$G_{FFii} = \frac{1}{4\pi} \int_{-l/2}^{l/2} E_1\left[\frac{x^2}{4k\,\Delta t_F}\right] dx = \frac{l}{8\pi} \int_{-1}^{1} E_1(\alpha\, \eta^2)\, d\eta, \tag{4.60}$$

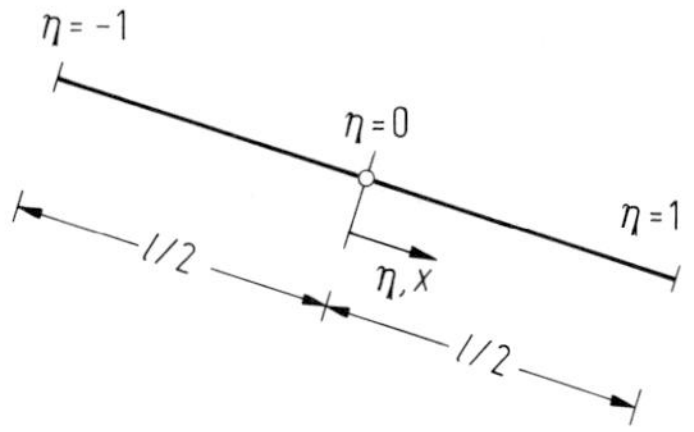

Fig. 4.3. Definitions for analytical integration

where

$$\alpha = \frac{l^2}{16 k \Delta t_F}. \tag{4.61}$$

Expanding the exponential-integral in series [10],

$$E_1(z) = -C - \ln z + \sum_{n=1}^{\infty} (-1)^{n-1} \frac{z^n}{n\, n!} \tag{4.62}$$

in which C is the Euler constant, the integral in Eq. (4.60) can be evaluated in closed form as

$$G_{FFii} = \frac{l}{4\pi} \left[2 - C - \ln(\alpha) + \sum_{n=1}^{\infty} (-1)^{n-1} \frac{\alpha^n}{n(2n+1)\, n!} \right]. \tag{4.63}$$

The series that appears in Eq. (4.63) converges very quickly for small values of α, but slowly as α increases. To overcome this problem, we can integrate analytically over a segment near the singularity thus ensuring that the coefficient α is always less or equal to unity (which guarantees convergence of the series within the required accuracy with a maximum of six terms) and numerically integrate the rest of the element using a standard Gaussian quadrature, as if these parts were separate elements. The length L of the analytically integrated part of the element is calculated by using Eq. (4.61), i.e.,

$$L \leqq 4 (k \Delta t_F)^{1/2} \leqq l. \tag{4.64}$$

This procedure can be extended to linear and higher-order functions. It should be pointed out that, for a certain space interpolation function, refining the order of the time approximation introduces only additional regular terms into the boundary integrals.

The diagonal coefficients H_{FFii} in Eq. (4.40) contain integrals with a stronger singularity which are only integrable in the Cauchy principal value sense. Although these coefficients can, similarly to Laplace's equation, be calculated through the application of a uniform potential over the whole body, the presence of the domain integral now makes this process ineffective. Thus, both terms c_i and $\hat{H}_{ii} (H_{ii} = c_i + \hat{H}_{ii})$ have to be computed directly.

The value of the free terms c_i are obtained by using the same limiting process as for Laplace's equation, which gives

$$c_i = 1 - \frac{1}{2\pi} \lim_{\varepsilon \to 0} \int_{\theta_1}^{\theta_2} \frac{1}{\varepsilon} \exp\left[-\frac{\varepsilon^2}{4k(t_F - t_0)} \right] \varepsilon \, d\theta = \frac{\pi + \alpha_1 - \alpha_2}{2\pi}. \tag{4.65}$$

For constant and linear elements, the terms $\hat{H}_{ii}$ are identically zero due to the orthogonality between r and n, which makes $d=0$ in Eq. (4.37). This is not so for higher-order elements, and the integrals must then be carried out in closed form (at least over a short straight-line segment around the singularity) in order to properly account for their principal values.

4.6. Time-Marching Schemes

The main difference between the two time-marching schemes presented here lies in the way values of the variables up to the actual instant of time are taken into account in order to solve for a new instant of time. In scheme 1, they are accounted for in the domain integral, as pseudo-initial values, while in scheme 2 their variation is considered through a summation of boundary integrals. Scheme 1 was employed in [2–6, 20, 21]; scheme 2 was suggested by Thaler and Mueller [22] in conjunction with the indirect formulation as early as 1970 and was subsequently used in [13, 19, 23].

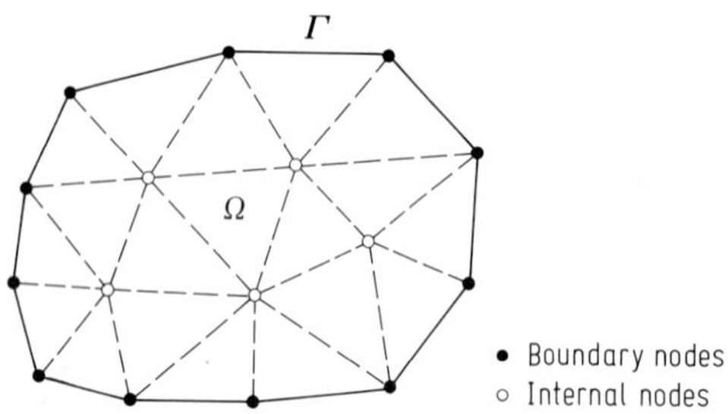

Fig. 4.4. Region $\Omega+\Gamma$ discretized into boundary elements and cells

Let us start by discussing scheme 1 in more detail. At the beginning of the process (time $t=t_0$), initial values u_0 of function u over Ω are specified. The domain is subdivided into L (triangular) cells (Fig. 4.4). The initial conditions are taken into account through a numerical integration over the domain and their values at a certain number of internal points considered.

Since half the boundary values of u and q are prescribed, Eq. (4.38) can then be employed to compute the remaining boundary data for the first time step ($F=1$). Note that if linear (or higher-order) time interpolation functions are adopted, initial values of function q along Γ must also be given (Eq. (4.50)).

At the end of the time step, the values of u at the previously selected internal points are recomputed to be used as initial values for the next step. This can be done by using Eq. (4.29) which, in matrix form, becomes (for instance, for the constant case)

$$\bar{\boldsymbol{U}}_F=\boldsymbol{G}'\,\boldsymbol{Q}_F-\boldsymbol{H}'\,\boldsymbol{U}_F+\boldsymbol{B}'\,\bar{\boldsymbol{U}}_{F-1}\,. \tag{4.66}$$

The coefficients of matrices $\boldsymbol{G}$ and $\boldsymbol{H}$ in Eq. (4.42), and $\boldsymbol{G}'$ and $\boldsymbol{H}'$ in Eq. (4.66) depend on geometrical data, properties of the medium, and the time step length. Thus, adopting a constant time step throughout the analysis, they can all be computed only once and stored.

The same also applies to the coefficients of matrices $\boldsymbol{B}$ and $\boldsymbol{B}'$, which result from integrals over the cells. We can employ Hammer's quadrature scheme [24] to numerically integrate over the cells and assume that the values of $\bar{\boldsymbol{U}}$ are calculated directly at each integration point. This gives, for the coefficients of matrix $\boldsymbol{B}$,

$$B_{ij} = \frac{1}{4\pi k \Delta t} \exp\left[-\frac{r_{ij}^2}{4k\Delta t}\right] |\boldsymbol{J}|_j w_j, \tag{4.67}$$

where $|\boldsymbol{J}|$ is the Jacobian and w the weighting factor of Hammer's quadrature.

Alternatively, we can assume a variation for u inside each cell according to a certain interpolation function and compute the coefficients B_{ij} using, for instance, a semianalytical integration scheme [25].

Although excellent results using this scheme have been reported in the literature, care must be taken in the choice of the time step value. As $\Delta t \to 0$, the integrand in the domain integral (the fundamental solution) becomes less and less smooth, its limit being a Dirac delta function (Fig. 4.5). The difficulty of numerically integrating a function with such behavior may introduce numerical problems into the solution, as reported in [21, 26].

For scheme 2, values of u at internal points need not be recomputed at the end of each time step. A domain integral, accounting for the initial conditions at $t = t_0$, is required only if $u_0 \neq 0$. Furthermore, if $\nabla^2 u_0 = 0$, the domain integral can be transformed into an equivalent boundary integral. As this is the case in most practical problems a reduction in the dimensionality of the problem is effectively achieved. But since the number of boundary integrals increases as time progresses, a selective numerical integration scheme has to be employed for computer efficiency.

In order to clarify the ideas, let us return to Eq. (4.40). From this equation, we note that computing the unknown boundary data at time $t = t_F$ requires the evaluation of matrices $\boldsymbol{G}_{fF}$ and $\boldsymbol{H}_{fF}$ for $f = 1, 2, \ldots, F$. The matrices $\boldsymbol{G}_{1F}$ to $\boldsymbol{G}_{(F-1)F}$, $\boldsymbol{H}_{1F}$

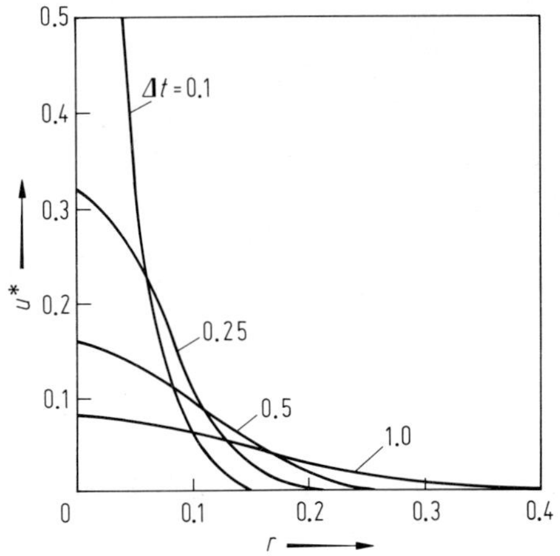

Fig. 4.5. Variation of u^* with r for several values of time steps

to $\boldsymbol{H}_{(F-1)F}$ will accordingly multiply the prescribed or calculated values of u and q at previous time steps to form the vector of independent coefficients. Because of the nature of the variation of the integrands in Eq. (4.41) with time, it is reasonable to use fewer Gaussian points to compute the contribution of the matrices corresponding to the initial steps.

Note that if a constant time step is adopted throughout the analysis, only two new matrices need to be evaluated for each step, making it possible for all the others to be kept in disc storage.

With regard to stability considerations, we note that the BEM formulation is implicit in character and thus relatively free from stability problems. In fact, a mathematical proof of uniform convergence and stability of the boundary element method as applied to linear two-dimensional transient heat conduction problems was recently reported in [27].

The transformation of the domain integral into equivalent boundary integrals for the case when u_0 is harmonic can be carried out by applying Green's second identity as follows:

$$\int_\Omega u_0 \nabla^2 U \, d\Omega = \int_\Gamma \left(u_0 \frac{\partial U}{\partial n} - U \frac{\partial u_0}{\partial n} \right) \mathrm{d}\Gamma . \tag{4.68}$$

Since the domain integral to be evaluated is of the form (Eq. (4.34))

$$\int_\Omega u_0 \, u^* \, d\Omega , \tag{4.69}$$

we have to determine a function U such that $\nabla^2 U = u^*$. One such function can be easily found as

$$U = \int \frac{1}{r} \left(\int r \, u^* \, dr \right) dr = \frac{1}{4\pi} E_1 \left(\frac{r^2}{4 k \tau} \right) \tag{4.70}$$

and Eq. (4.68) becomes

$$\begin{aligned} \int_\Omega u_0 \, u^* \, d\Omega = & \frac{1}{2\pi} \int_\Gamma \left\{ \frac{d}{r^2} \exp \left[- \frac{r^2}{4k(t_F - t_0)} \right] u_0 \right. \\ & \left. - \frac{1}{2} E_1 \left[\frac{r^2}{4k(t_F - t_0)} \right] q_0 \right\} d\Gamma , \end{aligned} \tag{4.71}$$

where $q_0 = \partial u_0 / \partial n$ and d is defined in Eq. (4.37). The above integrals can be evaluated numerically apart from the singular terms which are computed analytically, as discussed in the previous section.

Example 4.2. The object of the present investigation is a 3×3-m square region with initial temperature $u_0 = 30$ °F and thermal diffusivity $k = 1.25$ Btu/(h m °F), subjected to the Dirichlet boundary condition $u = 0$ along Γ for any $t > t_0$. These numerical values were chosen as to allow the results to be compared with an available finite element solution [28].

Since the initial conditions satisfy Laplace's equation, we can apply Eq. (4.71) in order to transform the domain integral of Eq. (4.39) into equivalent boundary integrals, as discussed in Section 4.5. The results obtained for this analysis (labelled BEM2), together with the ones obtained with time-marching scheme 1 (labelled BEM1), the finite element solution [28], and an analytical solution [28] are presented in Tables 4.2 and 4.3 for two different values of time steps, with the discretizations shown in Fig. 4.6. It can be seen that the BEM solutions are of the same level of accuracy and that they are superior to the finite element one at all points for both time steps, despite employing coarser discretizations.

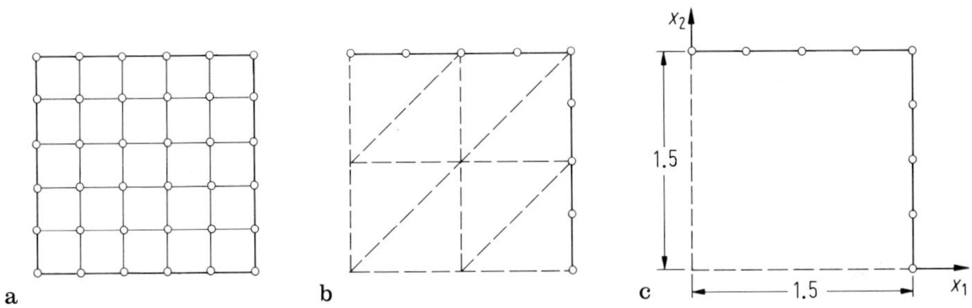

Fig. 4.6. Discretizations of one-quarter of square region: **a** FEM; **b** BEM1; **c** BEM 2

Table 4.2. Temperature values at $t = 1.2$ h for a time step $\Delta t = 0.1$ h.

x_1	x_2	BEM1	BEM2	FEM	Analytic
0.	0.	1.988	2.009	2.108	1.812
0.3	0.	1.893	1.913	2.005	1.723
0.6	0.	1.614	1.632	1.706	1.466
0.9	0.	1.180	1.194	1.239	1.065
1.2	0.	0.630	0.639	0.652	0.560
0.3	0.3	1.802	1.821	1.907	1.639
0.6	0.6	1.310	1.325	1.380	1.186
0.9	0.9	0.700	0.710	0.728	0.626
1.2	1.2	0.199	0.201	0.201	0.173

Table 4.3. Temperature values at $t = 1.2$ h for a time step $\Delta t = 0.5$ h.

x_1	x_2	BEM1	BEM2	FEM	Analytic
0.	0.	1.887	1.902	1.938	1.812
0.3	0.	1.798	1.809	1.843	1.723
0.6	0.	1.534	1.541	1.568	1.466
0.9	0.	1.114	1.122	1.139	1.065
1.2	0.	0.589	0.595	0.599	0.560
0.3	0.3	1.713	1.721	1.753	1.639
0.6	0.6	1.214	1.248	1.269	1.186
0.9	0.9	0.657	0.663	0.670	0.626
1.2	1.2	0.184	0.185	0.185	0.173

To verify if the use of Eq. (4.71) was introducing additional numerical errors, we restudied the problem by subtracting out a constant temperature of 30 °F so as to make the initial conditions equal to zero. This constant value was afterwards added to the new solution. Results obtained in this way agreed to the previous ones to the significant figures shown in Tables 4.2 and 4.3.

Example 4.3. The problem of a circular region of unit radius, with zero initial conditions, subjected to the time-dependent boundary conditions depicted in Fig. 4.7 was analyzed by discretizing one-quarter of the region into six linear boundary elements. The value of the parameter k was taken as 5.

Initially, a solution was attempted by adopting stepwise constant variations for functions u and q such that the prescribed values of u are equal to their average within each time step. The results obtained for both time-marching schemes were virtually coincident and are plotted in Fig. 4.8.

The problem was then restudied using stepwise linear variations for u and q and, in this way, the specified boundary conditions within each time step can be exactly accounted for. The results for both schemes agreed with the analytical solution given in [14] to three significant figures even for the first time step and are also plotted in Fig. 4.8. All numerical analyses adopted a time step value of $\Delta t = 0.02$.

Example 4.4. For problems involving regions extending to infinity, BEM solutions are much more economical than FEM ones (Fig. 4.9). In order to demonstrate this, we study in this example a circular opening in an infinite plane region with initial conditions $u_0 = 10$, subjected to the convection boundary condition of Example 2.3. The radius of the hole is unity, its ambient temperature equals zero, and the material properties of the medium are also assumed to be unity, for simplicity.

The variation of the surface temperature with time is presented in Fig. 4.10 for various values of the heat transfer coefficient, compared to an analytical solution given in [14]. The agreement between the two solutions is very good. A time step

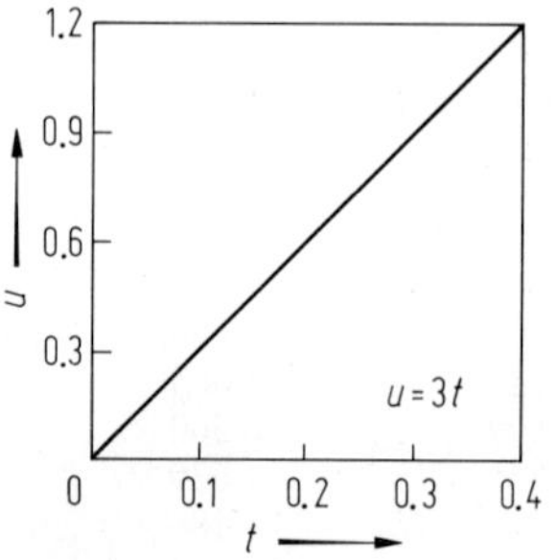

Fig. 4.7. Time variation of boundary conditions

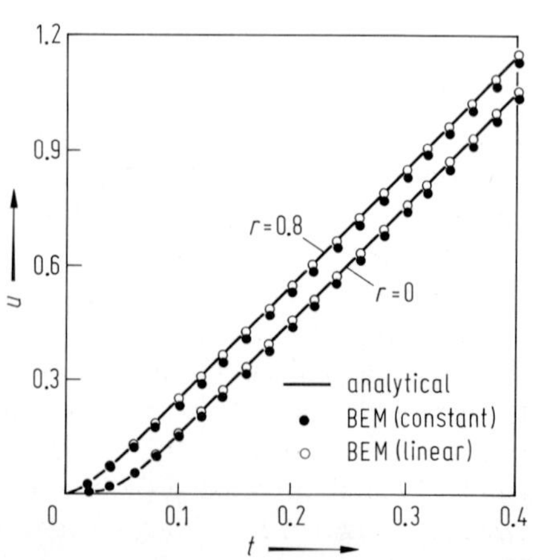

Fig. 4.8. Values of u at internal points

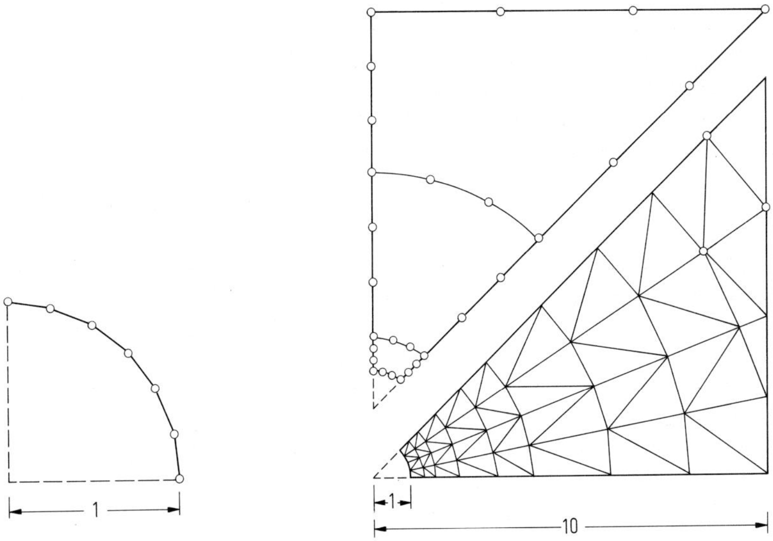

Fig. 4.9. Discretization of hole in infinite region: **a** BEM; **b** FEM

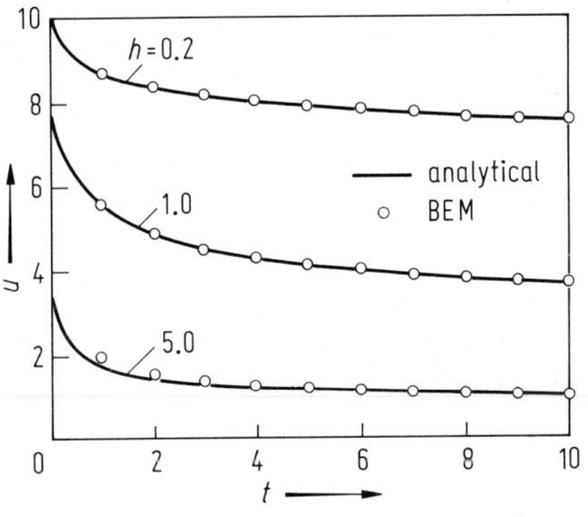

Fig. 4.10. Surface temperature of cooling hole in an infinite medium

value $\Delta t = 0.5$ was adopted and the analyses carried out until the surface temperature began to drop significantly. The BEM results were obtained with stepwise constant functions and, due to symmetry, only one-quarter of the interface between hole and medium was discretized into six linear boundary elements (Fig. 4.9).

This problem was also studied with the FEM in [29], but since the FEM is a domain-type technique, the infinite region has to be limited by a finite non-conducting boundary. In order to achieve the same level of accuracy, a time step ten times smaller ($\Delta t = 0.05$) was adopted and the domain discretized using 70 triangular elements or 3 cubic isoparametric elements (see Fig. 4.9). Note that a similar kind of approximation has to be introduced for the BEM if time-marching scheme 1 is utilized: although boundary elements are still restricted only to the hole–medium interface, cells have to be employed to integrate over the (infinite) medium.

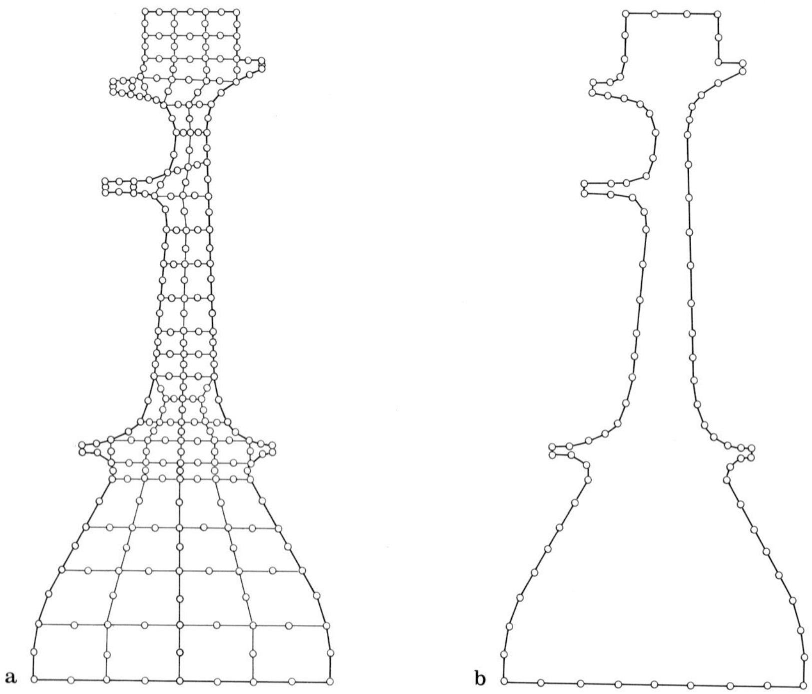

Fig. 4.11. Discretizations of turbine disc: **a** FEM; **b** BEM

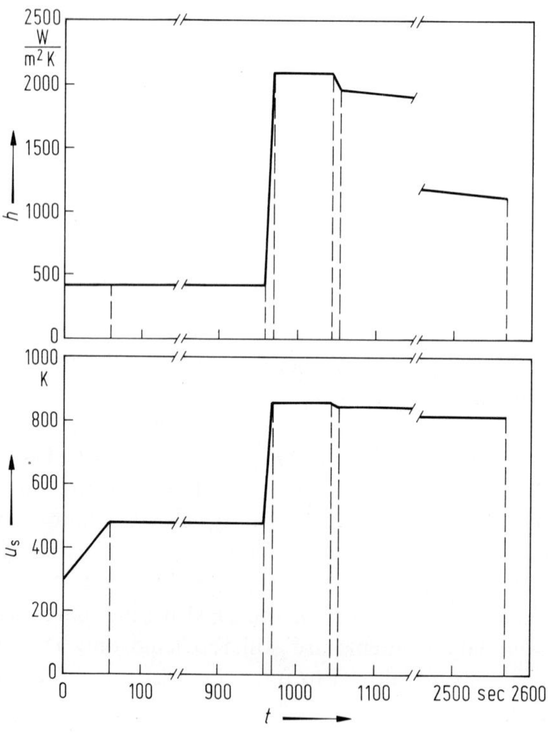

Fig. 4.12 Time variation of heat transfer coefficient and temperature of the surrounding gas for a typical boundary zone

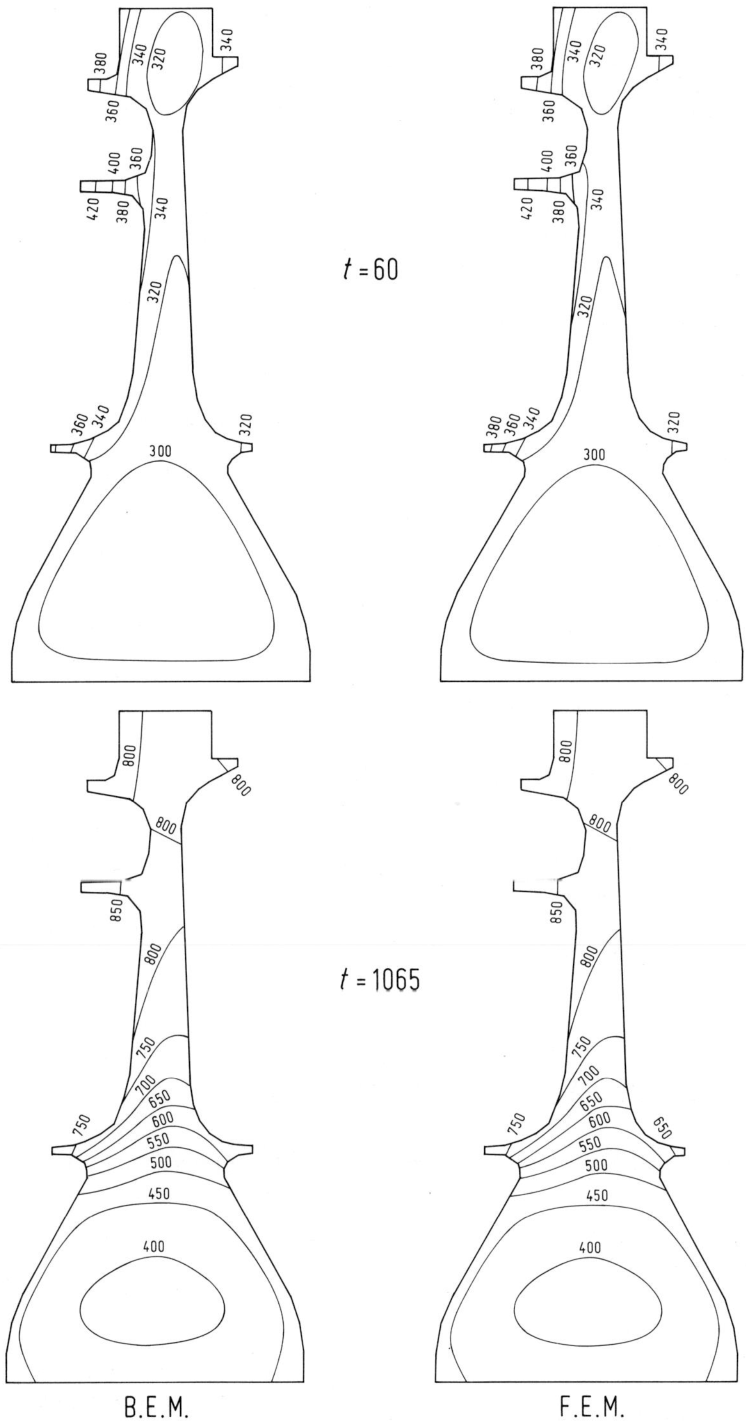

Fig. 4.13. Isotherms at some time levels

Example 4.5. A more practical problem with complex time-dependent boundary conditions is studied in this example, where the temperature distribution inside an actual turbine disc is sought. Although the real structure is axisymmetric, a two-dimensional FEM analysis was carried out for comparison purposes, employing 85 quadratic isoparametric elements and 348 nodes (Fig. 4.11).

The initial temperature of the turbine disc is 295.1 K and the values of the thermal conductivity, density, and specific heat of the material are 15 W/(m K), 8221 kg/m^3, and 550 J/(kg K), respectively. There are 18 different zones along the boundary, each with a different set of prescribed values for the heat transfer coefficient and the temperature of the surrounding gas. Their time variation at one such boundary zone is shown in Fig. 4.12.

The BEM discretization employed 90 linear elements and 106 nodes (there are 16 double nodes at the intersections of boundary zones). A stepwise linear variation was prescribed for the boundary temperature. For the boundary flux, it was assumed to be linear or quasi-quadratic, according to the variation of h and u_s within each time step [26]. In order to simplify the computation, a constant-temperature value was subtracted out so as to make the initial temperature equal to zero, thus avoiding the domain integration. This value was afterwards added to the solution.

Isotherms at some time levels are plotted in Fig. 4.13, compared to the FEM solutions. The agreement is, in general, excellent.

4.7. Three-Dimensional Problems

The three-dimensional fundamental solution to the diffusion equation and its normal derivative along Γ can be written (see Eq. (4.26))

$$u^* = \frac{1}{(4\pi k \tau)^{3/2}} \exp\left[-\frac{r^2}{4k\tau}\right], \tag{4.72}$$

$$q^* = \frac{d}{16\pi^{3/2}(k\tau)^{5/2}} \exp\left[-\frac{r^2}{4k\tau}\right], \tag{4.73}$$

where $d = [x_1(\xi) - x_1(x)]\, n_1(x) + [x_2(\xi) - x_2(x)]\, n_2(x) + [x_3(\xi) - x_3(x)]\, n_3(x)$.

The procedures for the numerical solution of Eq. (4.34) defined over a three-dimensional region follow basically the same ones as discussed for two dimensions. In this case, the time integrals (4.43) and (4.46) now give

$$\int_{t_{f-1}}^{t_f} q^*\, dt = \frac{d}{2\pi^{3/2} k r^3}\left[\Gamma\left(\frac{3}{2}, a_{f-1}\right) - \Gamma\left(\frac{3}{2}, a_f\right)\right], \tag{4.74}$$

$$\int_{t_{f-1}}^{t_f} u^*\, dt = \frac{1}{4\pi^{3/2} k r}\left[\Gamma\left(\frac{1}{2}, a_{f-1}\right) - \Gamma\left(\frac{1}{2}, a_f\right)\right]. \tag{4.75}$$

All surface boundary elements derived in Chapter 3 can be employed in the present case, including nonconforming elements. In particular, if a domain integration is required, the three-dimensional cell elements discussed in Section 3.5 should be employed.

4.8. Axisymmetric Problems

Assuming that all boundary and internal values have axial symmetry, Eq. (4.34) can be written in cylindrical polar coordinates (R, θ, Z) as

$$\begin{aligned} c(\xi)\, u(\xi, t_F) + k \int_{t_0}^{t_F} \int_{\bar{\Gamma}} u(x, t) \int_0^{2\pi} q^*(\xi, x, t_F, t)\, d\theta(x)\, R(x)\, d\bar{\Gamma}(x)\, dt \\ = k \int_{t_0}^{t_F} \int_{\bar{\Gamma}} q(x, t) \int_0^{2\pi} u^*(\xi, x, t_F, t)\, d\theta(x)\, R(x)\, d\bar{\Gamma}(x)\, dt \\ + \int_{\bar{\Omega}} u_0(x, t_0) \int_0^{2\pi} u^*(\xi, x, t_F, t_0)\, d\theta(x)\, R(x)\, d\bar{\Omega}(x), \end{aligned} \tag{4.76}$$

where $\bar{\Omega}$ and $\bar{\Gamma}$ are the generating area and boundary contour of the solid of revolution, i.e., the intersections of Ω and Γ, respectively, with the $R^+ - Z$ semiplane (see Fig. 2.37).

Writing the three-dimensional fundamental solution (4.72) in cylindrical polar coordinates and integrating over a ring of radius $R(x)$ at the plane $Z = Z(x)$ we have

$$\begin{aligned} \bar{u}^*(\xi, x, t_F, t) &= \int_0^{2\pi} u^*(\xi, x, t_F, t)\, d\theta(x) \\ &= \frac{1}{(4\pi k \tau)^{3/2}} \exp\left(-\frac{s}{4k\tau}\right) \int_0^{2\pi} \exp\left[\frac{R(\xi)\, R(x) \cos[\theta(\xi) - \theta(x)]}{2k\tau}\right] d\theta(x), \end{aligned} \tag{4.77}$$

where $s = R^2(\xi) + R^2(x) + [Z(\xi) - Z(x)]^2$.

The axisymmetric fundamental solution then becomes [30]

$$\bar{u}^* = \frac{2\pi}{(4\pi k \tau)^{3/2}} \exp\left(-\frac{s}{4k\tau}\right) I_0\left(\frac{l}{2k\tau}\right), \tag{4.78}$$

where $l = R(\xi)\, R(x)$ and I_0 is the modified Bessel function of the first kind of order zero. The normal derivative of the fundamental solution along the boundary contour can be obtained by differentiating expression (4.78):

$$\begin{aligned} \bar{q}^* = -\frac{1}{8\pi^{1/2}(k\tau)^{5/2}} \exp\left(-\frac{s}{4k\tau}\right) \Bigg\{ \Bigg[R(x)\, I_0\left(\frac{l}{2k\tau}\right) \\ - R(\xi)\, I_1\left(\frac{l}{2k\tau}\right)\Bigg] n_R(x) - [Z(\xi) - Z(x)]\, I_0\left(\frac{l}{2k\tau}\right) n_Z(x) \Bigg\}, \end{aligned} \tag{4.79}$$

where I_1 is the modified Bessel function of the first kind of order one.

From the above expressions, it can be seen that as $R(\xi) \to 0$ we have that $l \to 0$, $I_0(l/2k\tau) \to 1$, $I_1(l/2k\tau) \to 0$, so that the ring source tends to a point source with intensity 2π over the axis of revolution.

Substituting Eqs. (4.78) and (4.79) into Eq. (4.76) yields the following time-dependent boundary integral equation:

$$\begin{aligned} c(\xi)\, u(\xi, t_F) + k \int_{t_0}^{t_F} \int_{\bar{\Gamma}} u(x,t)\, \bar{q}^*(\xi, x, t_F, t)\, R(x)\, d\bar{\Gamma}(x)\, dt \\ = \int_{t_0}^{t_F} \int_{\bar{\Gamma}} q(x,t)\, \bar{u}^*(\xi, x, t_F, t)\, R(x)\, d\bar{\Gamma}(x)\, dt \\ + \int_{\bar{\Omega}} u_0(x, t_0)\, \bar{u}^*(\xi, x, t_F, t_0)\, R(x)\, d\bar{\Omega}(x)\,. \end{aligned} \tag{4.80}$$

The solution of Eq. (4.80) can be attempted by using the same calculation procedures as discussed in the previous sections. For simplicity, only the case of stepwise constant variation for the functions u and q will be considered in what follows.

After discretizing the surface and interior of the actual domain into boundary elements and cells, respectively, an equation similar to Eq. (4.39) is obtained, the coefficients of the matrices involved being computed as in Eq. (4.41). In order to perform the time integrals analytically, an appropriate change of variables is needed. Defining

$$c = \frac{l}{s}, \qquad y = \frac{s}{4k\tau}, \qquad a = \frac{s}{4k\Delta t_f}, \tag{4.81}$$

the integral in $\bar{u}^*$ becomes

$$\int_{t_{f-1}}^{t_f} \bar{u}^*\, dt = \frac{1}{2k(\pi s)^{1/2}} \int_{a_{f-1}}^{a_f} I_0(2cy)\, y^{-1/2} e^{-y}\, dy\,. \tag{4.82}$$

The Bessel function I_0 can be expanded in series as [10]

$$I_0(2cy) = \sum_{n=0}^{\infty} \frac{(cy)^{2n}}{n!^2} \tag{4.83}$$

and the integration in Eq. (4.82) is then performed term-by-term, giving

$$\int_{t_{f-1}}^{t_f} \bar{u}^*\, dt = \frac{1}{2k(\pi s)^{1/2}} \sum_{n=0}^{\infty} \frac{c^{2n}}{n!^2} \left[\Gamma\left(2n + \frac{1}{2}, a_{f-1}\right) - \Gamma\left(2n + \frac{1}{2}, a_f\right) \right]. \tag{4.84}$$

For the integral in $\bar{q}^*$, we have

$$\begin{aligned} \int_{t_{f-1}}^{t_f} \bar{q}^*\, dt = -\frac{1}{ks(\pi s)^{1/2}} \Big\{ \left[R(x)\, n_R(x) - [Z(\xi) - Z(x)]\, n_Z(x) \right] \\ \cdot \int_{a_{f-1}}^{a_f} I_0(2cy)\, y^{1/2} e^{-y}\, dy - R(\xi)\, n_R(x) \int_{a_{f-1}}^{a_f} I_1(2cy)\, y^{1/2} e^{-y}\, dy \Big\}\,. \end{aligned} \tag{4.85}$$

Expanding the Bessel function I_1 as [10]

$$I_1(2\,c\,y) = \sum_{n=0}^{\infty} \frac{(c\,y)^{2n+1}}{n!^2\,(n+1)}\,, \tag{4.86}$$

the integral becomes

$$\int_{t_{f-1}}^{t_f} \bar{q}^*\,dt = -\frac{1}{k\,s\,(\pi\,s)^{1/2}} \Bigg\{ \big[R(x)\,n_R(x) - [Z(\xi) - Z(x)]\,n_Z(x)\big]$$

$$\cdot \sum_{n=0}^{\infty} \frac{c^{2n}}{n!^2}\left[\Gamma\left(2n+\frac{3}{2},\,a_{f-1}\right) - \Gamma\left(2n+\frac{3}{2},\,a_f\right)\right] - R(\xi)\,n_R(x)$$

$$\cdot \sum_{n=0}^{\infty} \frac{c^{2n+1}}{n!^2\,(n+1)}\left[\Gamma\left(2n+\frac{5}{2},\,a_{f-1}\right) - \Gamma\left(2n+\frac{5}{2},\,a_f\right)\right]\Bigg\}\,. \tag{4.87}$$

All the incomplete gamma functions that appear in the above series can be evaluated in terms of $\Gamma(\frac{1}{2}, a)$ by using the following recurrence relation [18]:

$$\Gamma(n+1, a) = n\,\Gamma(n, a) + a^n\,e^{-a}\,. \tag{4.88}$$

From definition (4.81) we notice that the value of c varies between 0 (for $R(\xi)$ or $R(x) = 0$ or for $s \to \infty$) and 0.5 (for $\xi = x$). All the series that appear in expressions (4.84) and (4.87) converge very quickly for small values of c but slowly as $c \to 0.5$. In fact, they do not converge for $c = 0.5$ due to the singularity at $\xi = x$. So, from the computational point of view, it is not convenient to use expansions (4.83) and (4.86) for values of c in the vicinity of $c = 0.5$.

To overcome this problem, we can use asymptotic expansions of the Bessel functions that are valid for large values of their arguments. Thus, whenever y is large we can write [10]

$$I_0(2\,c\,y) = \frac{e^{2cy}}{2\,(\pi\,c\,y)^{1/2}}\left[1 + \sum_{n=1}^{\infty} \frac{f_1(n)}{n!\,(16\,c\,y)^n}\right], \tag{4.89}$$

$$I_1(2\,c\,y) = \frac{e^{2cy}}{2\,(\pi\,c\,y)^{1/2}}\left[1 + \sum_{n=1}^{\infty} \frac{f_2(n)}{n!\,(16\,c\,y)^n}\right], \tag{4.90}$$

$$\begin{aligned} f_1(n) &= (2n-1)^2\,(2n-3)^2 \cdots 1\,,\\ f_2(n) &= (-1)^n\,[4-(2n-1)^2][4-(2n-3)^2] \cdots [4-1]\,. \end{aligned} \tag{4.91}$$

The time integrals can then be carried out as follows:

$$\int_{t_{f-1}}^{t_f} \bar{u}^*\,dt = \frac{1}{4\pi\,k\,l^{1/2}} \Bigg\{ E_1(B_{f-1}) - E_1(B_f) \tag{4.92}$$

$$+ \sum_{n=1}^{\infty} \frac{f_1(n)\,b^n}{n!\,(16\,c)^n}\,[\Gamma(-n, B_{f-1}) - \Gamma(-n, B_f)]\Bigg\}\,,$$

$$\int_{t_{f-1}}^{t_f} \bar{q}^* \, dt = -\frac{1}{2\pi k s l^{1/2}} \left\{ -\frac{1}{b} [e^{-B_{f-1}} - e^{-B_f}] \right.$$
$$\cdot \left[[R(\xi) - R(x)] \, n_R(x) + [Z(\xi) - Z(x)] \, n_Z(x) \right]$$
$$+ \left[R(x) \, n_R(x) - [Z(\xi) - Z(x)] \, n_Z(x) \right]$$
$$\cdot \sum_{n=1}^{\infty} \frac{f_1(n) \, b^{n-1}}{n! \, (16 c)^n} [\Gamma(1-n, B_{f-1}) - \Gamma(1-n, B_f)] - R(\xi) \, n_R(x)$$
$$\left. \cdot \sum_{n=1}^{\infty} \frac{f_2(n) \, b^{n-1}}{n! \, (16 c)^n} [\Gamma(1-n, B_{f-1}) - \Gamma(1-n, B_f)] \right\}, \tag{4.93}$$

where $b = 1 - 2c$ and $B = a b$. The incomplete gamma functions can now be computed from $\Gamma(0, B)$ through the recurrence relation [10],

$$\Gamma(-n, B) = -\frac{1}{n} \left[\Gamma(1-n, B) - \frac{e^{-B}}{B^n} \right],$$
$$\Gamma(0, B) = E_1(B). \tag{4.94}$$

When the value of c tends to 0.5 but y is small over part of the integration interval (a_{f-1}, a_f), we cannot apply expansions (4.89) and (4.90) directly. Alternatively, expression (4.82) may be written

$$\int_{t_{f-1}}^{t_f} \bar{u}^* \, dt = \frac{1}{2k(\pi s)^{1/2}} \left[\int_{a_{f-1}}^{a'} I_0(2cy) \, y^{-1/2} e^{-y} \, dy + \int_{a'}^{a_f} I_0(2cy) \, y^{-1/2} e^{-y} \, dy \right], \tag{4.95}$$

where y is sufficiently large in the interval (a', a_f). Thus the first integral in the above equation can be computed as in Eq. (4.84), and expansion (4.89) is now used to evaluate the second integral. The same idea can be applied on calculating the time integral in $\bar{q}^*$.

The remaining step in the numerical solution of the boundary integral equation (4.80) is the computation of the space integrals. The terms H_{ij} and G_{ij} $(i \neq j)$ of the final system of equations (similar to Eq. (4.40)) can be calculated using a six-point Gauss quadrature rule. The diagonal terms H_{ii} and G_{ii}, however, need to be investigated more carefully since their calculation involves the evaluation of singular integrals.

The coefficients G_{ii} contain an integral with a logarithmic singularity. Expanding the exponential-integral in Eq. (4.92), we can isolate the logarithmic term and integrate it analytically (see [26]). The remainder is nonsingular and can be computed by using standard Gaussian quadrature.

The coefficients $\hat{H}_{ii}$ contain a logarithmic plus a $1/b$ singularity. The first one is directly integrable, but the second is only integrable in the Cauchy principal value sense. For the case of constant or linear elements, however, we can write with reference to Fig. 4.14,

$$n_R(x) = \cos\alpha, \qquad n_Z(x) = \sin\alpha,$$
$$R(\xi) - R(x) = -\eta \frac{l}{2} \sin\alpha, \tag{4.96}$$
$$Z(\xi) - Z(x) = \eta \frac{l}{2} \cos\alpha$$

such that the first term in the right-hand side of Eq. (4.93), which is the Cauchy singular one, becomes identically zero. Expanding the first term of each series in Eq. (4.93) in order to isolate the logarithmic singularity, we can then evaluate it analytically (see [26]) and the rest using a standard Gaussian quadrature.

The free coefficients c_i account for the jump that the integral in $\bar{q}^*$ experiences as it approaches the boundary Γ. For the present case, it can be shown that the values of $c(\xi)$ become the same as for two-dimensional problems, Eq. (4.65).

Example 4.6. The first example analyzed was that of a solid cylinder with unit initial conditions, subjected to the boundary conditions

$$u = 0 \qquad \text{at } R = a,$$
$$q = 2u \qquad \text{at } Z = \pm l.$$

The discretization adopted is shown in Fig. 4.14. Note that due to the symmetry with respect to the R axis, only one-half of the cross-section needed to be discretized. The numerical values assumed for the cross-section were $a = 1$, $l = 1$ and for simplicity, the material coefficient k was also taken to be unity.

Results are compared in Figs. 4.15 and 4.16 with an available analytical solution [14], showing good agreement. The analysis was performed with a time step value of $\Delta t = 0.025$.

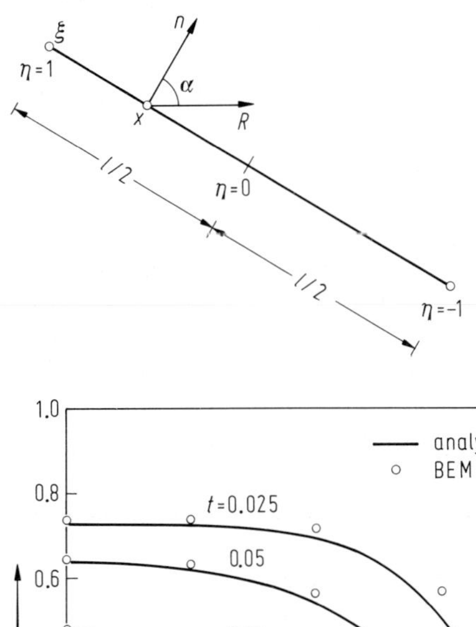

Fig. 4.14. Definitions for linear element

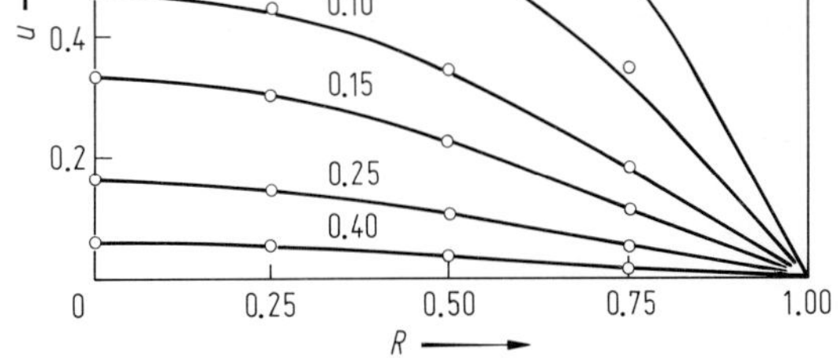

Fig. 4.15. Values of u along the faces $Z = \pm l$

Example 4.7. This example studies the heat conduction problem of a prolate spheroid initially at zero temperature and subjected to a unit surface temperature at $t = 0$. A parametric representation of points on its surface, in the $R-Z$ plane, may be written

$$R = L_1 \cos \varphi,$$
$$Z = L_2 \sin \varphi,$$

where the φ angle is indicated in Fig. 4.17.

The discretization adopted is also shown in Fig. 4.17 and the numerical values assumed for this analysis were $k = 1$, $L_1 = 1$, and $L_2 = 2$. Results for the temperature at the center point ($R = Z = 0$) are compared in Fig. 4.17 with an analytical solution [31] and a finite element solution [29] obtained with parabolic three-dimensional isoparametric elements. The finite element analysis was performed with a time step value $\Delta t = 0.025$, whereas the boundary element solution employed a $\Delta t = 0.050$.

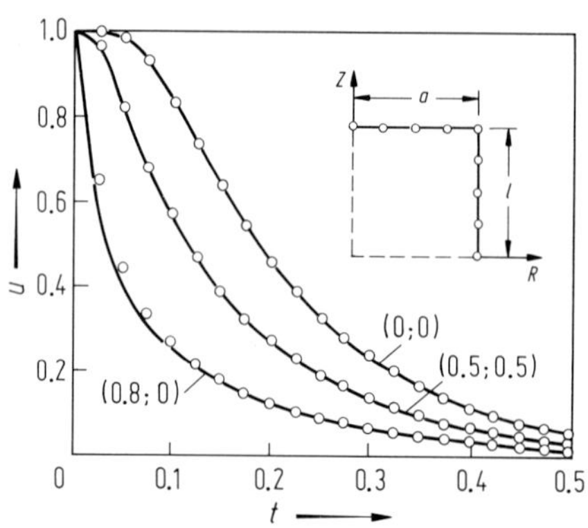

Fig. 4.16. Values of u at internal points

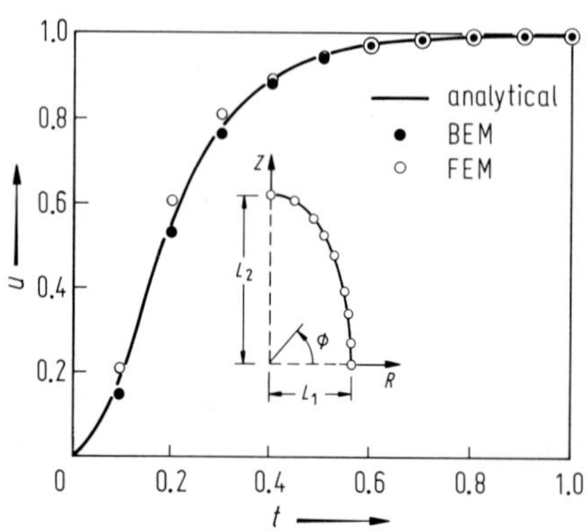

Fig. 4.17. Temperature at center of prolate spheroid

4.9. Nonlinear Diffusion

Up to now, this chapter has dealt only with linear diffusion problems. The linearity, however, is not preserved in many practical engineering applications. The most important nonlinear features which arise in connection with diffusion problems can be classified as follows:

(a) Nonlinear material, i.e., diffusivity coefficient dependent on the potential or its gradients;
(b) nonlinear boundary conditions, e.g., due to heat radiation;
(c) nonlinear sources inside the domain;
(d) moving interface problems, e.g., due to phase change.

The first and second types of nonlinearities can be treated in the same way as for steady potential problems, Section 2.15. The case of nonlinear sources can be dealt with by dividing the domain into cells and iterating over the cells (a case of practical interest is the problem of spontaneous ignition [32]).

In this section, we shall discuss a BEM formulation for the solution of moving interface problems due to phase change (solidification or melting), the so-called Stefan problem [33]. This formulation was developed by Chuang and others [34–37]. Another type of moving interface problem of interest in fluid mechanics is discussed in Section 12.3.

For simplicity, we shall restrict the following discussion to one-dimensional problems. Let us consider a solid slab, extending from $x = -l$ to $x = l$, at a given initial temperature distribution; at time $t = t_0$, the slab is immersed in its own melt and we are interested in the resulting melting or solidification process (see Fig. 4.18).

The boundary-value problem of heat conduction for the solid phase may be expressed by the linear diffusion equation

$$\frac{\partial^2 u(x,t)}{\partial x^2} - \frac{1}{k}\frac{\partial u(x,t)}{\partial t} = 0, \qquad -X(t) \leqq x \leqq X(t), \tag{4.97}$$

where $X(t)$ represents the instantaneous position of the moving boundary. The boundary conditions at $x = \pm X(t)$ are written

$$u(x,t) = u_m, \tag{4.98}$$

$$q(x,t) = K\frac{\partial u(x,t)}{\partial x} = f(t) - h\,u(x,t), \tag{4.99}$$

where the function $f(t)$ is related to the movement of the phase boundary, u_m is the melting temperature, K is the thermal conductivity of the solid, and h is the heat transfer coefficient (u_m, K, and h are all assumed to be constants). We also have the initial condition

$$u(x,t) = u_0(x,t_0), \qquad -l \leqq x \leqq l, \tag{4.100}$$

noting that $X(t) = l$ at $t = t_0$.

Equation (4.98) states that the solid is at melting temperature at the phase boundary, while Eq. (4.99) is a general expression for the conservation of heat at the moving boundary.

An integral statement equivalent to the above-described problem may be written [35]

$$
\begin{aligned}
u(\xi, t_F) &+ k\, u_m \int_{t_0}^{t_F} q^*(\xi, x, t_F, t)\Big|_{x=-X(t)}^{x=X(t)} dt \\
&= k \int_{t_0}^{t_F} [f(t) - h\, u(x,t)]\, u^*(\xi, x, t_F, t)\Big|_{x=-X(t)}^{x=X(t)} dt \\
&+ \int_{-X(t)}^{-l} u(x,t)\, u^*(\xi, x, t_F, t)\Big|_{x=-X(t)} dx \\
&+ \int_{l}^{X(t)} u(x,t)\, u^*(\xi, x, t_F, t)\Big|_{x=X(t)} dx \\
&+ \int_{-l}^{l} u_0(x, t_0)\, u^*(\xi, x, t_F, t_0)\, dx\,.
\end{aligned}
\tag{4.101}
$$

The one-dimensional fundamental solution u^* is given in expression (4.26).

On taking the point ξ in Eq. (4.101) to the boundary, we note that since the temperature at the interface is at melting point, the only unknown in the resulting boundary integral equation is the instantaneous moving boundary position $X(t)$. Thus, the problem can be solved numerically by using a time-stepping technique where an iterative procedure is adopted in order to adjust the position of the solidification front within each time step. Such technique is described in detail in [34–37].

Example 4.8. This example, taken from [35], considers a steel slab, initially at a uniform temperature, which at time $t=0$ is immersed into well-agitated pure molten iron, at some temperature above its melting point.

For such a problem (represented in Fig. 4.18), the boundary condition (4.99) takes the form

$$
K\frac{\partial u}{\partial x} = \varrho \Delta H \frac{dX(t)}{dt} + h\,(u_B - u)\,, \tag{a}
$$

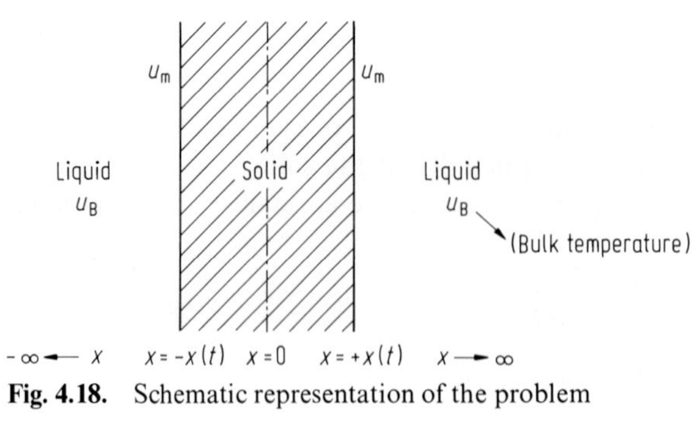

Fig. 4.18. Schematic representation of the problem

i.e.,

$$f(t) = h\, u_B + \varrho \Delta H \frac{dX(t)}{dt}. \tag{b}$$

Here ϱ is the density, ΔH is the latent heat of melting, and u_B is the bulk temperature of the melt.

The numerical property values adopted for the computation were:

$$\begin{aligned}
l &= 1 \text{ in.},\\
u_0 &= 70\,^\circ\text{F},\\
u_m &= 2800\,^\circ\text{F},\\
u_B &= 3000\,^\circ\text{F},\\
h &= 1000 \text{ Btu/(h ft}^2\,^\circ\text{F)},\\
\Delta H &= 110 \text{ Btu/lb.},\\
k &= 0.018 \text{ in.}^2/\text{s},\\
\varrho &= 445 \text{ lb./ft}^3,\\
K &= 20 \text{ Btu/(h ft}\,^\circ\text{F)}.
\end{aligned}$$

Some computed curves, giving the transient temperature profiles and the position of the melt line as a function of time are given in Figs. 4.19 and 4.20, respectively.

The position of the melt line, plotted in Fig. 4.20, appears to be a linear function of time after an initial time period; this is consistent with the fact that, by that time, all of the slab has been brought to the melting point so that melting

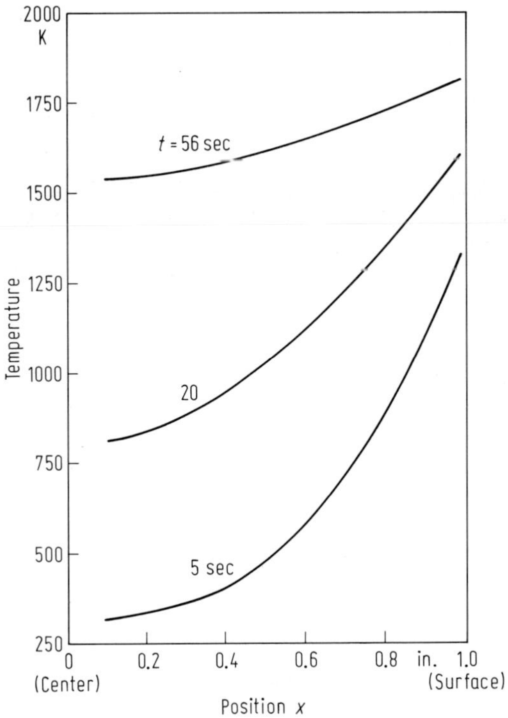

Fig. 4.19. Transient temperature profiles in the solid slab, prior to melting

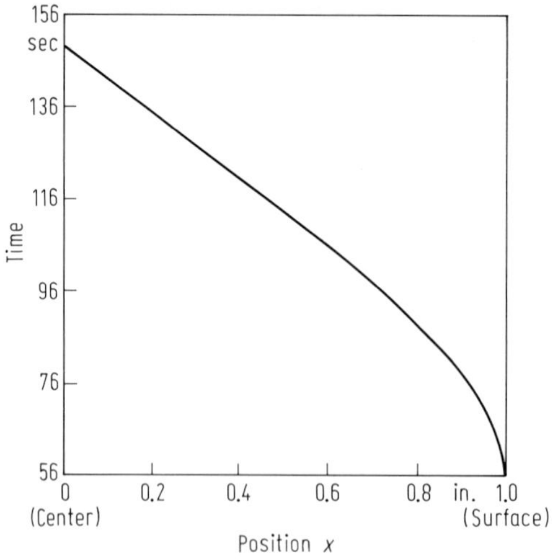

Fig. 4.20. Plot of the position of the melt line against time

would proceed at a uniform rate. This may be readily verified by comparing the numerical value of the melting rate that may be obtained from equation (a) upon setting $(\partial u/\partial x) = 0$ with that given in Fig. 4.20 for the final stages of the melting process. These numerical values of $|(dX/dt)|$ are 1.36×10^{-2} in./s and 1.34×10^{-2} in./s, respectively, which are indeed in very good agreement.

References

1. Rizzo, F. J., and Shippy, D. J., A method of solution for certain problems of transient heat conduction, AIAA J. **8,** 2004–2009 (1970).
2. Butterfield, R., and Tomlin, G. R., Integral techniques for solving zoned anisotropic continuum problems, in *Proc. Int. Conf. on Variational Methods in Engineering,* Vol. 2 (C. A. Brebbia and H. Tottenham, Eds.), Southampton University Press, Southampton 1972.
3. Tomlin, G. R., Numerical analysis of continuum problems in zoned anisotropic media, Ph.D. Thesis, Southampton University, 1972.
4. Chang, Y. P., Kang, C. S., and Chen, D. J., The use of fundamental Green's functions for the solution of problems of heat conduction in anisotropic media, Int. J. Heat Mass Transfer **16,** 1905–1918 (1973).
5. Shaw, R. P., An integral equation approach to diffusion, Int. J. Heat Mass Transfer **17,** 693–699 (1974).
6. Wrobel, L. C., and Brebbia, C. A., The boundary element method for steady-state and transient heat conduction, in *Numerical Methods in Thermal Problems* (R. W. Lewis and K. Morgan, Eds.), Pineridge Press, Swansea, Wales, 1979.
7. Wrobel, L. C., and Brebbia, C. A., A formulation of the boundary element method for axisymmetric transient heat conduction, Int. J. Heat Mass Transfer **24,** 843–850 (1981).
8. Brebbia, C. A., and Walker, S., *Boundary Element Techniques in Engineering,* Newnes–Butterworths, London, 1980.
9. Widder, V. D., *The Laplace Transform,* Princeton University Press, Princeton, 1946.
10. Abramowitz, M., and Stegun, I. A. (Eds.), *Handbook of Mathematical Functions,* Dover, New York, 1965.

11. Lachat, J. C., and Combescure, A., Laplace transform and boundary integral equation: application to transient heat conduction problems, in *Innovative Numerical Analysis in Applied Engineering Science* (T. A. Cruse *et al.*, Eds.), CETIM, Versailles, 1977.
12. Schapery, R. A., Approximate methods of transform inversion for visco-elastic stress analysis, in *Proc. Fourth U.S. National Congress on Applied Mechanics,* Vol. 2, 1962.
13. Liggett, J. A., and Liu, P. L. F., Unsteady flow in confined aquifers: A comparison of two boundary integral methods, Water Resources Res. **15,** 861 – 866 (1979).
14. Carslaw, H. S., and Jaeger, J. C., *Conduction of Heat in Solids,* 2nd ed., Clarendon Press, Oxford, 1959.
15. Curran, D. A. S., Cross, M., and Lewis, B. A., Solution of parabolic differential equations by the boundary element method using discretization in time, Appl. Math. Modelling **4,** 398 – 400 (1980).
16. Morse, P. M., and Feshbach, H., *Methods of Theoretical Physics,* Mc-Graw-Hill, New York, 1953.
17. Roach, G. F., *Green's Functions: Introductory Theory with Applications,* Van Nostrand Reinhold, London, 1970.
18. Gradshteyn, I. S., and Ryzhik, I. M., *Table of Integrals, Series and Products,* Academic Press, London, 1965.
19. Wrobel, L. C., and Brebbia, C. A., Time-dependent potential problems, in *Progress in Boundary Element Methods,* Vol. 1 (C. A. Brebbia, Ed.), Pentech Press, London, Halstead Press, NY, 1981.
20. Dubois, M., and Buysse, M., Transient heat transfer analysis by the boundary integral equation method, in *New Developments in Boundary Element Methods* (C. A. Brebbia, Ed.), CML Publications, Southampton, 1980.
21. Curran, D., Cross, M., and Lewis, B. A., A preliminary analysis of boundary element methods applied to parabolic partial differential equations, in *New Developments in Boundary Element Methods* (C. A. Brebbia, Ed.), CML Publications, Southampton, 1980.
22. Thaler, R. H., and Mueller, W. K., A new computational method for transient heat conduction in arbitrarily shaped regions, in *Fourth Int. Heat Transfer Conference* (U. Grigull and E. Hahne, Eds.), Elsevier Publishing Co., Amsterdam, 1970.
23. Onishi, K., and Kuroki, T., Boundary element method in transient heat transfer problems, Bull. Inst. Advan. Res. Fukuoka Univ. **52** (1981).
24. Hammer, P. C., Marlowe, O. J., and Stroud, A. H., Numerical integration over simplexes and cones, Math. Comput. **10,** 130 – 137 (1956).
25. Wrobel, L. C., and Brebbia, C. A., Boundary elements in thermal problems, in *Numerical Methods in Heat Transfer* (R. Lewis, K. Morgan, and O. C. Zienkiewicz, Eds.), Wiley, Chichester, 1981.
26. Wrobel, L. C., Potential and viscous flow problems using the boundary element method, Ph.D. Thesis, Southampton University, 1981.
27. Onishi, K., Convergence in the boundary element method for heat equation, TRU Math. **17,** 213 – 225 (1981).
28. Bruch, J. C., Jr., and Zyvoloski, G., Transient two-dimensional heat conduction solved by the finite element method, Int. J. Numerical Methods Engng. **8,** 481 – 494 (1974).
29. Zienkiewicz, O. C., and Parekh, C. J., Transient field problems: Two-dimensional and three-dimensional analysis by isoparametric finite elements, Int. J. Numerical Methods Engng. **2,** 61 – 71 (1970).
30. Watson, G. N., *A Treatise on the Theory of Bessel Functions,* 2nd ed., Cambridge University Press, Cambridge, 1944.
31. Haji-Sheik, A., and Sparrow, E. M., Transient heat conduction in a prolate spheroidal solid, J. Heat Transfer Trans. ASME **88 C,** 331 – 333 (1966).
32. Anderson, C. A., and Zienkiewicz, O. C., Spontaneous ignition: Finite element solutions for steady and transient conditions, J. Heat Transfer Trans. ASME **96 C,** 398 – 404 (1977).
33. Rubinstein, L. I., *The Stefan Problem,* AMS Translations of Mathematical Monographs, Vol. 27, Amer. Math. Soc., Providence, R. I., 1971.
34. Chuang, Y. K., The melting and dissolution of a solid in a liquid with a strong exothermic heat of solution, Ph.D. Thesis, State University of New York at Buffalo, 1971.

35. Chuang, Y. K., and Szekely, J., On the use of Green's functions for solving melting or solidification problems, Int. J. Heat Mass Transfer **14,** 1285–1294 (1971).
36. Chuang, Y. K., and Szekely, J., The use of Green's functions for solving melting or solidification problems in the cylindrical coordinate system, Int. J. Heat Mass Transfer **15,** 1171–1174 (1972).
37. Chuang, Y. K., and Ehrich, O., On the integral technique for spherical growth problems, Int. J. Heat Mass Transfer **17,** 945–953 (1974).

Chapter 5

Elastostatics

5.1. Introduction to the Theory of Elasticity

This chapter is partly devoted to introducing some basic concepts of the theory of elasticity needed for developing boundary element models. The chapter starts by reviewing the small strain theory of elasticity in accordance with standard texts on the subject [1 – 5].

In Chapters 6 to 8 plasticity, viscoplasticity, creep and no-tension materials, i.e., inelastic problems, are introduced but in the present chapter only elastic behavior is considered.

Throughout this work the so-called Cartesian tensor notation is used. This notation is not only a time-saver in writing long expressions, but is also extremely useful in derivation and in the proof of theorems. Such notation makes use of subscript indices (1, 2, 3) to represent (x, y, z) and also renders summation symbols unnecessary when the same letter subscript occurs twice in a term. Hence, in three dimensions,

$$a_i\, a_i = a_1^2 + a_2^2 + a_3^2$$

and (5.1)

$$a_{kk} = a_{11} + a_{22} + a_{33}.$$

In addition, the permutation symbol e_{ijk} and the Kronecker delta symbol δ_{ij} will be used, i.e.,

$e_{ijk} = 0$ when any two indices are equal

$= +1$ when i, j, k are 1, 2, 3 or an even permutation of 1, 2, 3 (in cyclic order)

$= -1$ when i, j, k are an odd permutation of 1, 2, 3 (in anti-cyclic order) (5.2)

and

$\delta_{ij} = 1$ if $i = j$

0 if $i \neq j$.

Herein, unless otherwise stated (locally), subscripts are assumed to have a range of three in three-dimensional problems, whereas for two dimensions (plane stress and plane strain) this range is only two. In this section, however, only three-dimensional expressions are considered.

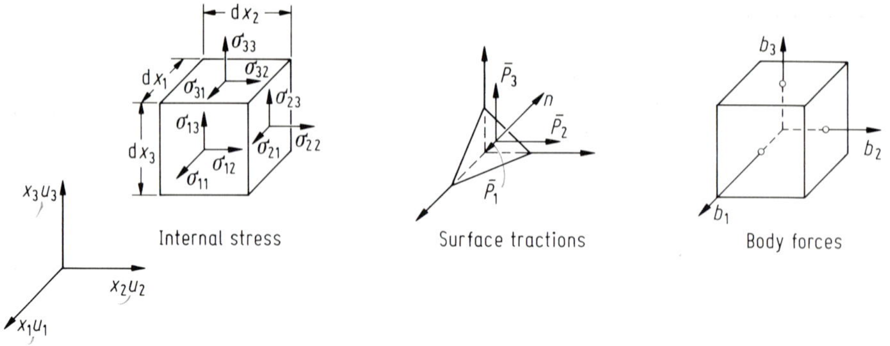

Fig. 5.1. Internal stress, body forces, and surface tractions

The external forces acting at any instant on a body are classified in two kinds: body forces and surface forces. Body forces act on the elements of volume or mass inside the body, e.g., gravity. These forces will be determined per unit volume. Surface forces act on the bounding surface of the body and will be determined per unit area of the surface across which they act. Such forces are designated tractions (Fig. 5.1).

If one considers an infinitesimal rectangular parallelepiped surrounding a given point within the body, it readily follows that static equilibrium of forces and moments requires satisfaction of

$$\sigma_{ij,i} + b_j = 0, \tag{5.3}$$

where the components of the stress tensor are represented by σ_{ij} and b_j represents the components of the body force. Space derivatives are indicated by a comma, i.e., $\partial\sigma_{ij}/\partial x_i = \sigma_{ij,i}$. Furthermore, if there are no body moments applied, equilibrium conditions also lead to

$$\sigma_{ij} = \sigma_{ji}. \tag{5.4}$$

If the six components of the stress tensor are assumed to be known at a certain point, the equivalent tractions (p_i) acting on any plane through this point can be computed by

$$p_i = \sigma_{ji} n_j, \tag{5.5}$$

where n_j represents the direction cosines of the normal to the plane.

Regardless to the state of stress at a given point, one can always choose a special set of axes through the point in such a way that the shear stress components vanish when the stresses are referred to it. The directions of these special axes are called principal directions and the normal stress acting on each plane perpendicular to the principal directions are called principal stresses.

The principal directions can be obtained by considering the relation

$$p_i = \lambda\, n_i \tag{5.6}$$

which indicates that the traction vector is parallel to the normal vector. Substitution of Eq. (5.6) into Eq. (5.5) leads to

$$(\sigma_{ij} - \lambda\, \delta_{ij})\, n_i = 0\,. \tag{5.7}$$

Equation (5.7) represents a system of three linear homogeneous equations which for $\sigma_{ij} \neq 0$ must admit a nontrivial solution ($n_i\, n_i = 1$). Consequently,

$$|\sigma_{ij} - \lambda\, \delta_{ij}| = 0 \tag{5.8}$$

or, in expanded form,

$$\lambda^3 - I_1\, \lambda^2 - I_2\, \lambda - I_3 = 0\,, \tag{5.9}$$

where

$$\begin{aligned} I_1 &= \sigma_{kk}\,, \\ I_2 &= \tfrac{1}{2}\,(\sigma_{ij}\, \sigma_{ij} - \sigma_{ii}\, \sigma_{jj}) = \tfrac{1}{2}\, \sigma_{ij}\, \sigma_{ij} - \tfrac{1}{2}\, I_1^2\,, \\ I_3 &= \tfrac{1}{6}\, e_{ijk}\, e_{pqr}\, \sigma_{ip}\, \sigma_{jq}\, \sigma_{kr}\,. \end{aligned} \tag{5.10}$$

The principal stresses (roots of the cubic equation (5.9)) are physical quantities whose values do not depend on the coordinate system in which the components of stress were initially given. They are, therefore, invariants of the stress state at the point. A direct consequence is that I_1, I_2, and I_3 are also scalar invariants with respect to any rotation of the Cartesian reference axis.

It can be demonstrated that the three principal directions are mutually perpendicular. Hence, if the reference axes are chosen to coincide with the principal axes of stress,

$$\begin{aligned} I_1 &= \sigma_1 + \sigma_2 + \sigma_3\,, \\ I_2 &= -\,(\sigma_1\, \sigma_2 + \sigma_2\, \sigma_3 + \sigma_3\, \sigma_1)\,, \\ I_3 &= \sigma_1\, \sigma_2\, \sigma_3\,, \end{aligned} \tag{5.11}$$

where σ_1, σ_2, σ_3 are the principal stresses.

It is convenient in many cases to split the stress tensor into two parts, one called the spherical stress tensor and the other the stress deviator or deviatoric stress tensor. The spherical stress tensor ($\bar{\sigma}_{ij}$) is related to the mean stress by

$$\bar{\sigma}_{ij} = \frac{1}{3}\, \sigma_{kk}\, \delta_{ij} = \frac{I_1}{3}\, \delta_{ij} \tag{5.12}$$

and the components of the deviatoric stress tensor are given by

$$S_{ij} = \sigma_{ij} - \bar{\sigma}_{ij} = \sigma_{ij} - \frac{I_1}{3}\, \delta_{ij}\,. \tag{5.13}$$

The principal directions of the stress deviator tensor are the same as those of the stress tensor and it is usually easier to compute the principal deviator stresses (S_k) than to calculate σ_k. If λ now denotes any one of the principal deviator stresses, a derivation similar to that of Eq. (5.9) yields

$$\lambda^3 - J_2 \lambda - J_3 = 0\,, \tag{5.14}$$

where J_1, J_2, and J_3 are the scalar invariants of the stress deviator analogous to those given in Eq. (5.10), but now calculated with S_{ij} instead of σ_{ij}. Hence,

$$\begin{aligned} J_1 &= 0\,, \\ J_2 &= \tfrac{1}{2} S_{ij} S_{ij}\,, \\ J_3 &= \tfrac{1}{3} S_{ij} S_{jk} S_{ki}\,. \end{aligned} \tag{5.15}$$

Equation (5.14) can be solved explicitly by the following substitution [3, 6]:

$$\lambda = 2\left(\frac{J_2}{3}\right)^{1/2} \sin\alpha \tag{5.16}$$

which leads to

$$2\left(\frac{J_2}{3}\right)^{3/2} [4\sin^3\alpha - 3\sin\alpha] = J_3\,. \tag{5.17}$$

The expression in square brackets is equal to $-\sin 3\alpha$; thus,

$$\sin 3\alpha = -\frac{J_3}{2}\left(\frac{3}{J_2}\right)^{3/2}. \tag{5.18}$$

Assuming that the first solution is obtained with 3α in the range $\pm\pi/2$ (i.e., $-\pi/6 \leqq \alpha \leqq \pi/6$), the other two solutions of Eq. (5.18) are found by the cyclic nature of $\sin(3\alpha + 2n\pi)$. This furnishes the three independent roots of Eq. (5.14), namely,

$$S_k = 2\left(\frac{J_2}{3}\right)^{1/2} \sin\alpha_k\,, \tag{5.19}$$

where for $S_1 > S_2 > S_3$ one has $\alpha_1 = \alpha + \frac{2}{3}\pi$, $\alpha_2 = \alpha$, and $\alpha_3 = \alpha + \frac{4}{3}\pi$. Note that the principal stresses can be calculated by the simple relation

$$\sigma_k = S_k + \frac{\sigma_{ll}}{3}\,. \tag{5.20}$$

In addition, $-\pi/6 \leqq \alpha \leqq \pi/6$ is also a stress invariant which can be used as an alternative to J_3 in representing the stress state at a point. These relationships will be of fundamental importance when dealing with inelastic problems.

Under the action of forces, a body is displaced from its original configuration. If x_i denotes the position of a point P of the body in its initial state and $x_i + u_i$ denotes the position of the same point when the body is deformed, u_i represents the displacement components and is a function of x_i. If the displacements are such that their first derivatives are so small that the square and product of the partial derivatives of u_i are negligible, then strains can be represented by the Cauchy infinitesimal strain tensor:

$$\varepsilon_{ij} = \tfrac{1}{2}\,(u_{i,j} + u_{j,i})\,. \tag{5.21}$$

In general, during the deformation process, any small element of the body is changed in shape, translated and rotated. Consider the point P' in the neighborhood of P with coordinates $x_i + dx_i$. Avoiding rigid-body translations, the relative displacement of P' with respect to P is given by

$$du_i = u_{i,j}\,dx_j \tag{5.22}$$

which can be further written

$$du_i = \tfrac{1}{2}\,(u_{i,j} + u_{j,i})\,dx_j - \tfrac{1}{2}\,(u_{j,i} - u_{i,j})\,dx_j \tag{5.23}$$

or

$$du_i = \varepsilon_{ij}\,dx_j - \omega_{ij}\,dx_j\,, \tag{5.24}$$

where ω_{ij} is the rotation tensor of the infinitesimal displacement field, i.e.,

$$\omega_{ij} = \tfrac{1}{2}\,(u_{j,i} - u_{i,j})\,. \tag{5.25}$$

From the above it is seen that although displacement uniquely defines the components of the strain tensor, the inverse problem is not so straightforward. In the first place, strains represent pure deformation, whereas displacements include rigid-body motion which has no effect on the strains. This problem however, can be made unique by specifying the rigid-body motion (i.e., displacement and rotation) at some point of the body. A more difficult problem is encountered in calculating the displacements from strains using Eq. (5.21). Here, a system of six differential equations for the three unknown functions u_i is obtained, and consequently one must expect as impossible solution unless some additional conditions are satisfied. These conditions are given by the compatibility equations and are found in standard texts on elasticity. They are as follows

$$\varepsilon_{ij,\,kl} + \varepsilon_{kl,\,ij} - \varepsilon_{ik,\,jl} - \varepsilon_{jl,\,ik} = 0\,. \tag{5.26}$$

Equation (5.26) is a necessary and sufficient condition that the strain components give single-valued displacements for simply connected regions. For multiply connected regions, however, this condition is necessary but generally not sufficient.

It should be emphasized that all the relations presented so far are independent of material properties; consequently, they hold for both elastic and inelastic material behavior (Chapter 6).

For an isotropic elastic material in which there is no change in temperature, Hooke's law relating stresses and strains can be stated in the form

$$\sigma_{ij} = 2\,G\,\varepsilon_{ij} + \frac{2\,G\,v}{1-2\,v}\,\varepsilon_{kk}\,\delta_{ij} \tag{5.27}$$

or inversely

$$\varepsilon_{ij} = \frac{1}{2\,G}\left(\sigma_{ij} - \frac{v}{1+v}\,\sigma_{kk}\,\delta_{ij}\right), \tag{5.28}$$

where v is Poisson's ratio and G is the shear modulus.

The shear modulus can be related to the Young's modulus and v as follows:

$$G = \frac{E}{2\,(1+v)}. \tag{5.29}$$

Alternatively, expression (5.27) can be written more concisely as

$$\sigma_{ij} = C_{ijkl}\,\varepsilon_{kl}, \tag{5.30}$$

in which C_{ijkl} is the fourth-order isotropic tensor of elastic constants given by

$$C_{ijkl} = \frac{2\,G\,v}{1-2\,v}\,\delta_{ij}\,\delta_{kl} + G\,(\delta_{ik}\,\delta_{jl} + \delta_{il}\,\delta_{jk}). \tag{5.31}$$

Equations (5.3), (5.21), and (5.27) represent a set of 15 equations for 6 stresses, 6 strains, and 3 displacements which can be further manipulated. A straightforward procedure is to substitute Eq. (5.21) into Eq. (5.27) to obtain stresses in terms of displacement gradients, and then substitute the result into Eq. (5.3) to obtain three second-order partial differential equations for the three displacement components. The result of these operations is the well-known Navier equation which may be written in the form

$$G\,u_{j,kk} + \frac{G}{1-2\,v}\,u_{k,kj} + b_j = 0. \tag{5.32}$$

This equation is particularly convenient when displacement boundary conditions are specified. By using Eqs. (5.21) and (5.27) as before, but now substituting into Eq. (5.5) for boundary points, the traction boundary conditions are obtained as

$$\frac{2\,G\,v}{1-2\,v}\,u_{k,k}\,n_i + G\,(u_{i,j} + u_{j,i})\,n_j = p_i, \tag{5.33}$$

where n_j stands for the direction cosines of the outward normal to the boundary of the body.

It is interesting to note that since the equilibrium condition is now expressed in terms of displacements in Eq. (5.32), the compatibility equations are no longer required. The displacement u_i is solved from the Navier equation to satisfy the boundary conditions. After u_i is known throughout the body, the strains are obtained by Eq. (5.21), and the stresses are calculated by Hooke's law.

5.1.1. Initial Stresses or Initial Strains

In certain problems, different effects such as temperature can be included in the above formulation as initial stresses or strains. If the thermal effect is considered in initial stress form one has

$$\sigma_{ij} = 2G\,\varepsilon_{ij} + \frac{2G\nu}{1-2\nu}\,\varepsilon_{kk}\,\delta_{ij} - \sigma_{ij}^T = \sigma_{ij}^e - \sigma_{ij}^T = C_{ijkl}\,\varepsilon_{kl} - \sigma_{ij}^T, \tag{5.34}$$

where the stresses σ_{ij} are now equal to the "elastic stresses" σ_{ij}^e and the thermal components σ_{ij}^T.

For a thermally isotropic material σ_{ij}^T is given by

$$\sigma_{ij}^T = 2G\left(\frac{1+\nu}{1-2\nu}\right)\varepsilon_{ij}^T \tag{5.35}$$

in which

$$\varepsilon_{ij}^T = \alpha\,T\,\delta_{ij}, \tag{5.36}$$

where α is the linear coefficient of thermal expansion and T is the difference in temperature.

5.2. Fundamental Integral Statement

In order to clarify the subsequent ideas, an initial remark is now due; throughout this book the concept of regular region is used in the sense defined by Kellogg [7]. More specifically, regular regions are always implied here, and these represent regions bounded by regular surfaces (not necessarily smooth everywhere) which may have corners or edges. The extension of this concept to infinite and semi-infinite regions will be discussed in another section.

Following the same ideas developed in the previous chapters one can write an extended weighted residual equation, which takes into consideration the equilibrium equation (5.3) and the boundary conditions. The traction or natural boundary conditions (5.5) are

$$p_i = \sigma_{ji}\,n_j = \bar{p}_i, \qquad \text{on } \Gamma_2 \tag{5.37}$$

where n_j is the outward normal and the prescribed tractions on the part of the boundary Γ_2 are denoted by $\bar{p}_i$.

The other type of boundary conditions are those given in terms of prescribed displacement components. Let Γ_1 denote the portion of the boundary on which displacements are prescribed. The displacements constraints are

$$u_i = \bar{u}_i \qquad \text{on } \Gamma_1 . \tag{5.38}$$

Note that the total external surface of the body is $\Gamma = \Gamma_1 + \Gamma_2$. The subdivision of the boundary Γ into two parts needs to be interpreted as a concept as one can have at the physical point the two types of boundary condition in different directions or even a combination of them, such as in the case of elastic supports.

The weighted residual statement can be written as

$$\int_\Omega (\sigma_{jk,j} + b_k)\, u_k^*\, d\Omega = \int_{\Gamma_2} (p_k - \bar{p}_k)\, u_k^*\, d\Gamma + \int_{\Gamma_1} (\bar{u}_k - u_k)\, p_k^*\, d\Gamma , \tag{5.39}$$

where u_k^* and p_k^* are the displacements and tractions corresponding to the weighting field, i.e.,

$$p_k^* = n_j\, \sigma_{jk}^* . \tag{5.40}$$

The strain-displacement relationship (5.21) and the constitutive equations (5.27) are assumed to apply for both the approximating and the weighting fields.

The first term in Eq. (5.39) can be integrated by parts, which gives

$$\begin{aligned} &-\int_\Omega \sigma_{jk}\, \varepsilon_{jk}^*\, d\Omega + \int_\Omega b_k\, u_k^*\, d\Omega \\ &\quad = -\int_{\Gamma_2} \bar{p}_k\, u_k^*\, d\Gamma - \int_{\Gamma_1} p_k\, u_k^*\, d\Gamma + \int_{\Gamma_1} (\bar{u}_k - u_k)\, p_k^*\, d\Gamma . \end{aligned} \tag{5.41}$$

The stresses can be written (Eq. (5.30)) as

$$\sigma_{jk} = C_{jkli}\, \varepsilon_{li} . \tag{5.42}$$

Taking this into consideration one can write Eq. (5.41) as

$$\begin{aligned} &-\int_\Omega C_{jkli}\, \varepsilon_{li}\, \varepsilon_{jk}^*\, d\Omega + \int_\Omega b_k\, u_k^*\, d\Omega \\ &\quad = -\int_{\Gamma_2} \bar{p}_k\, u_k^*\, d\Gamma - \int_{\Gamma_1} p_k\, u_k^*\, d\Gamma + \int_{\Gamma_1} (\bar{u}_k - u_k)\, p_k^*\, d\Gamma . \end{aligned} \tag{5.43}$$

Integrating by parts again the first term in Eq. (5.43) and taking into consideration the constitutive equation, i.e., the reciprocity principle due to the symmetry of C_{ijkl},

$$\int_\Omega \sigma_{jk}\, \varepsilon_{jk}^*\, d\Omega = \int_\Omega \varepsilon_{jk}\, \sigma_{jk}^*\, d\Omega , \tag{5.44}$$

one obtains

$$\begin{aligned} &\int_\Omega (\sigma_{jk,j}^*)\, u_k\, d\Omega + \int_\Omega b_k\, u_k^*\, d\Omega \\ &\quad = -\int_{\Gamma_2} \bar{p}_k\, u_k^*\, d\Gamma - \int_{\Gamma_1} p_k\, u_k^*\, d\Gamma + \int_{\Gamma_1} \bar{u}_k\, p_k^*\, d\Gamma + \int_{\Gamma_2} u_k\, p_k^*\, d\Gamma . \end{aligned} \tag{5.45}$$

Taking into consideration that the body forces are known functions, the second integral on the left-hand side of Eq. (5.45) does not introduce any unknowns. The first integral, however, presents unknown displacements in the Ω domain, while the boundary integrals on the right-hand side introduce unknown displacements and tractions only on the external surface of the body. Our objective in boundary elements is to eliminate the integral in the domain – first integral on the left-hand side – by proposing weighting field functions which satisfy the equilibrium equation in Ω.

5.2.1. Somigliana Identity

Throughout this book the concept of residuals and their minimization is applied to better understand the type of approximations involved and extend the boundary element method to study nonlinear problems. It is, however, usual in boundary integral equations in elastostatics to start by assuming Somigliana's identity. This identity can be deduced from the reciprocity relationship (5.44) as follows:

Consider a body defined by $\Omega + \Gamma$ (Γ is the boundary and Ω is the domain as shown in Fig. 5.2) which is in a state of equilibrium under some prescribed loads and displacements. This state is here represented by the set σ_{ij}, ε_{ij}, u_i, p_i, and b_i.

Let us assume a domain Ω^* with boundary Γ^* – which may be at infinity – that contains the body $\Omega + \Gamma$ under consideration (Fig. 5.3). As before, this new region is considered to be in an equilibrium state now denoted by σ^*_{ij}, ε^*_{ij}, u^*_i, p^*_i and a type of body force b^*_i which is discussed below. If elastic properties remain valid in both cases, the reciprocity integral (5.44) can be inferred by simple symmetry of the tensors involved, i.e.,

$$\int_{\Omega} \sigma_{jk}\, \varepsilon^*_{jk}\, d\Omega = \int_{\Omega} \varepsilon_{jk}\, \sigma^*_{jk}\, d\Omega\,. \tag{5.46}$$

Notice that we are assuming that a solution for σ^*_{jk} exists which satisfies the governing equations and, in addition, that the actual solution satisfies equilibrium. This assumption is valid for approximate solutions because the state of stress within the body will be given by a combination of the solution for the * field which is defined as being in equilibrium. Notice that this observation is not required in the weighted residual formulation.

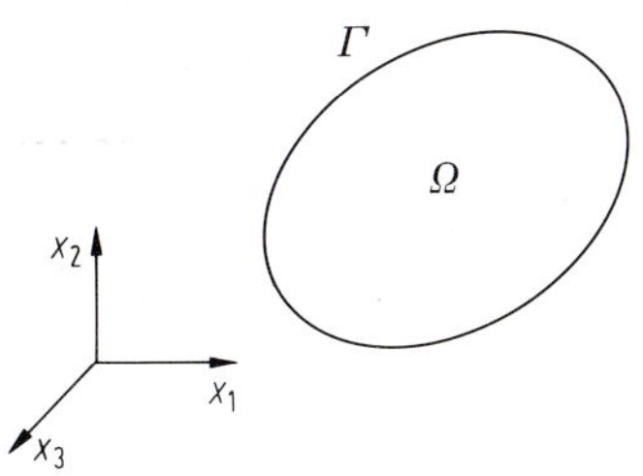

Fig. 5.2. Three-dimensional body with volume Ω and boundary Γ

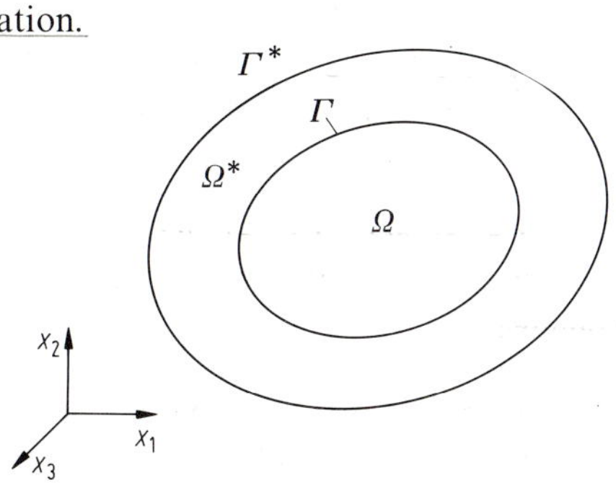

Fig. 5.3. General region $\Omega^* + \Gamma^*$ containing the body $\Omega + \Gamma$ with the same elastic properties

Integrating by parts both sides of Eq. (5.46) in order to obtain the equilibrium equations under the Ω integrals, we obtain

$$\int_{\Omega} b_k^* \, u_k \, d\Omega + \int_{\Gamma} p_k^* \, u_k \, d\Gamma = \int_{\Omega} b_k \, u_k^* \, d\Omega + \int_{\Omega} p_k \, u_k^* \, d\Gamma \tag{5.47}$$

which corresponds to Betti's second reciprocal work theorem.

Notice that the first integral in the above equation is equal (with sign changed) to the first integral in Eq. (5.45) i.e.,

$$\int_{\Omega} (\sigma_{jk,j}^*) \, u_k \, d\Omega = - \int_{\Omega} b_k^* \, u_k \, d\Omega \, . \tag{5.48}$$

In this way Eqs. (5.47) and (5.45) are identical once the boundary conditions have been applied.

Equation (5.47) can be further modified by assuming that the body force components b_k^* corresponds to positive *unit point* loads applied at a point $\xi \in \Omega^*$ in each of the three orthogonal directions given by the unit vectors e_i. This can be represented as

$$b_j^* = \Delta(\xi, x) \, e_j \, , \tag{5.49}$$

where $\Delta(\xi, x)$ represents the Dirac delta function, ξ is the singular – load – point, and $x \in \Omega^*$ is the field point.

The Dirac delta function has the following properties:

$$\begin{aligned} &\Delta(\xi, x) = 0 \qquad \text{if } \xi \neq x \, , \\ &\Delta(\xi, x) = \infty \qquad \text{if } \xi = x \, , \\ &\int_{\Omega^*} g(x) \, \Delta(\xi, x) \, d\Omega(x) = g(\xi) \, . \end{aligned} \tag{5.50}$$

Therefore, if $\xi \in \Omega$, the first integral in Eq. (5.47) can be represented as

$$\int_{\Omega} b_i^* \, u_i \, d\Omega = u_i(\xi) \, e_i \, . \tag{5.51}$$

Furthermore, if each point load is taken as independent, the starred displacements and tractions can be written in the form

$$\begin{aligned} u_j^* &= u_{ij}^*(\xi, x) \, e_i \, , \\ p_j^* &= p_{ij}^*(\xi, x) \, e_i \, , \end{aligned} \tag{5.52}$$

where $u_{ij}^*(\xi, x)$ and $p_{ij}^*(\xi, x)$ represent the displacements and tractions in the j direction at point x corresponding to a unit point force acting in the i direction (e_i) applied at point ξ.

From the above it is seen that Eq. (5.47) can be rewritten to represent the three separate components of the displacement at ξ in the form

$$u_i(\xi) = \int_{\Gamma} u_{ij}^*(\xi, x) \, p_j(x) \, d\Gamma(x) - \int_{\Gamma} p_{ij}^*(\xi, x) \, u_j(x) \, d\Gamma(x) + \int_{\Omega} u_{ij}^*(\xi, x) \, b_j(x) \, d\Omega(x) \, . \tag{5.53}$$

Equation (5.53) is known as Somigliana's identity for displacements [8] and was here obtained by reciprocity with a singular solution of the Navier equation satisfying

$$G u^*_{j,kk} + \frac{G}{1-2\nu} u^*_{k,kj} + \Delta(\xi, x)\, e_j = 0 . \tag{5.54}$$

Thus, solutions of the above equation are called *fundamental solutions*.

Equation (5.53) was alternatively justified through weighted residual considerations [9, 10]. Such procedure possesses the advantage of being more general and permits a straightforward extension to more complex differential equations.

5.3. Fundamental Solutions

Following the definition of fundamental solution introduced in the last section (see Eq. 5.54), the different singular solutions of the Navier equation considered here are now presented and classified according to the region $\Omega^* + \Gamma^*$ involved (see Fig. 5.3).

In the first class considered, Ω^* is assumed to be an infinite elastic medium and consequently Γ^* is taken to infinity. This case corresponds to the fundamental solution due to Kelvin [1], and the appropriate expressions for the fundamental displacements and tractions defined in Eq. (5.52) are given by [9, 10]

$$u^*_{ij}(\xi, x) = \frac{1}{16\pi(1-\nu)\, G r}\{(3-4\nu)\, \delta_{ij} + r_{,i} r_{,j}\} \tag{5.55}$$

for three-dimensional and

$$u^*_{ij}(\xi, x) = \frac{-1}{8\pi(1-\nu)\, G}\{(3-4\nu)\ln(r)\, \delta_{ij} - r_{,i} r_{,j}\}, \tag{5.56}$$

for two-dimensional plane strain problems

$$p^*_{ij}(\xi, x) = \frac{-1}{4\alpha\pi(1-\nu)\, r^{\alpha}}\left\{[(1-2\nu)\, \delta_{ij} + \beta\, r_{,i} r_{,j}]\frac{\partial r}{\partial n} - (1-2\nu)(r_{,i} n_j - r_{,j} n_i)\right\}, \tag{5.57}$$

where $\alpha = 2, 1$; $\beta = 3, 2$ for three- and two-dimensional plane strain, respectively. Also, $r = r(\xi, x)$ represents the distance between the load point ξ and the field point x and its derivatives are taken with reference to the coordinates of x, i.e.,

$$\begin{aligned} r &= (r_i r_i)^{1/2}, \\ r_i &= x_i(x) - x_i(\xi), \\ r_{,i} &= \frac{\partial r}{\partial x_i(x)} = \frac{r_i}{r}. \end{aligned} \tag{5.58}$$

In addition, the strains (ε^*_{jk}) at any point q due to a unit point load applied at ξ in i direction can be written

$$\varepsilon^*_{jki}(\xi,x) = \frac{-1}{8\alpha\pi(1-\nu)\,G r^{\alpha}}\{(1-2\nu)(r_{,k}\delta_{ij}+r_{,j}\delta_{ik})-r_{,i}\delta_{jk}+\beta r_{,i}\,r_{,j}r_{,k}\} \quad (5.59)$$

and the stresses can be written

	2 D	3 D
α	1	2
β	2	3

$$\sigma^*_{jki}(\xi,x) = \frac{-1}{4\alpha\pi(1-\nu)\,r^{\alpha}}\{(1-2\nu)(r_{,k}\delta_{ij}+r_{,j}\delta_{ki}-r_{,i}\delta_{jk})+\beta r_{,i}r_{,j}r_{,k}\}. \quad (5.60)$$

The plane strain expressions are valid for plane stress if ν is replaced by $\bar{\nu} = \nu/(1+\nu)$.

In order to illustrate some features of this fundamental solution, and also to clarify subsequent matters, the passage from three-dimensional to two-dimensional plane strain will be discussed. It is immediately apparent that the displacements u^*_{ij} for three-dimensional problems vanish as $r \to \infty$. This is not the same for two dimensions; in this case, $u^*_{ij} \to -\infty$ as $r \to \infty$ due to the logarithm of r included in expression (5.56). Such behavior of the two-dimensional case is entirely consistent and by all means expected. On physical grounds, for instance, one can consider the case of a semi-infinite bar extending from $x(A) = 0$ to $x(B) \to \infty$. If the extremity B is assumed to be fixed, a positive axial load applied at A would produce a state of constant strains. Consequently, by integration, $u(A) \to \infty$. Alternatively, if the displacement at a third point C is subtracted as a rigid-body motion (i.e., extremity B is not taken as a reference any more) and this point is at a finite distance from A, $u(A)$ would be finite and $u(C) = 0$, but $u(B) \to -\infty$.

This simple analysis clearly indicates the physical nature of expression (5.56) and can be used to justify the passage from three dimensions to two dimensions by integrating the former with respect to $x_3(\xi)$. Thus, consider the following alternative expression for displacements in the three-dimensional case:

$$\hat{u}^*_{ij}(\xi,x) = u^*_{ij} - \bar{u}^*_{ij} = \frac{1}{16\pi(1-\nu)\,G}\left\{(3-4\nu)\,\delta_{ij}\left(\frac{1}{r}-\frac{1}{\bar{r}}\right)+\frac{r_{,i}\,r_{,j}}{r}\right\}, \quad (5.61)$$

where u^*_{ij} was given by expression (5.55) and $\bar{u}^*_{ij} = u^*_{ij}(\xi,\bar{x})$ represents the displacements at a certain point $\bar{x}$, lying on a perpendicular to the load direction i passing through ξ. Also

$$x_3(\bar{x}) = x_3(x) \quad (5.62)$$

and

$$\bar{r} = r(\xi,\bar{x}) = (r_3^2+1)^{1/2}. \quad (5.63)$$

Expression (5.61) can be used to find the two-dimensional fundamental displacements as follows:

$$[u^*_{ij}(\xi,x)]_{2\text{-D}} = \int_{-\infty}^{\infty} \hat{u}^*_{ij}(\xi,x)\,dx_3(\xi) \qquad (i,j=1,2) \quad (5.64)$$

where the integrals involved are ($r=(r_1^2+r_2^2)^{1/2}$)

$$\int_{-\infty}^{\infty}\left\{\frac{1}{(r_3^2+r^2)^{1/2}}-\frac{1}{(r_3^2+1)^{1/2}}\right\}dr_3=-2\ln(r) \tag{5.65}$$

and

$$r_i r_j\int_{-\infty}^{\infty}\frac{1}{(r_3^2+r^2)^{3/2}}\,dr_3=\frac{2r_i r_j}{r^2}=2r_{,i}r_{,j}\,. \tag{5.66}$$

The above results can readily be seen to produce expression (5.56). Note that the displacements are zero at $\bar{x}$ (depending on the direction of the load).

The second class of fundamental solutions adopted corresponds to half-space problems. In this case the Kelvin region is subdivided by an infinite horizontal plane $\bar{\Gamma}$ and its lower part is considered as $\Omega^*+\Gamma^*$. Thus, the region of interest becomes a semi-infinite medium with the plane part of Γ^* being represented by the surface $\bar{\Gamma}$. This lower half-space is always assumed to contain the region $\Omega+\Gamma$ and the plane $x_1=0$ is taken to be the boundary surface $\bar{\Gamma}$ which is here considered free from tractions (see Fig. 5.4).

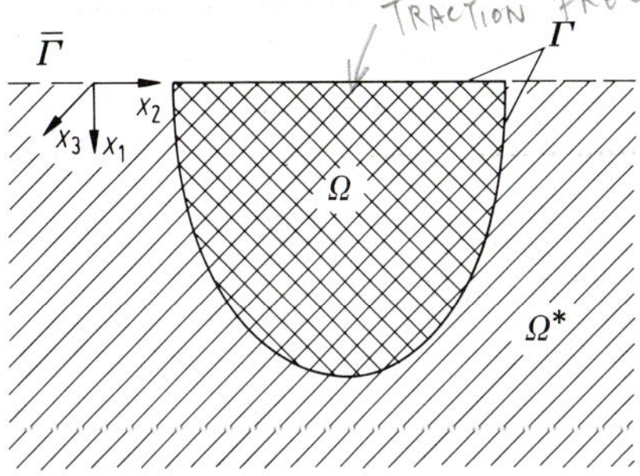

Fig. 5.4. Body $\Omega+\Gamma$ located within the semi-infinite space $x_1 \geqq 0$

The stress distribution due to point loads applied within the isotropic half plane was presented by Melan [11]. The solution to the equivalent three-dimensional problem (Fig. 5.5) was given by Mindlin [12] who produced not only the stresses, but also the corresponding displacements due to concentrated loads acting inside the half-space. The application of Mindlin's fundamental solution to boundary elements has been reported by Nakaguma [13] and the purpose of the present section is to present the complete solution to Melan's problem (including displacements), allowing for its general application to the boundary element technique.

Following Mindlin's procedure [12], it is seen that the complete half-space fundamental solution can be obtained by superposition of 18 nuclei of strain (see Love [1]) derived from Kelvin's solution (six for each of the three components of the force). Also, the first singular solution employed in each load direction is found to be the single Kelvin solution presented in expressions (5.55)–(5.60) for three dimensions. All the other nuclei involve the coordinates of the image of the load point with respect to the surface $\bar{\Gamma}$. This provides satisfaction of the traction-free

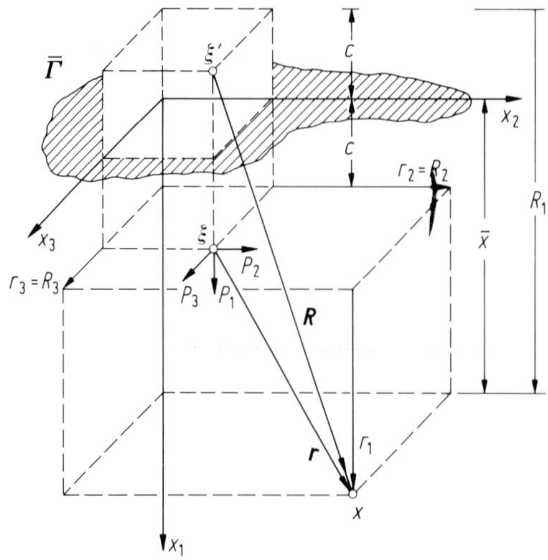

Fig. 5.5. Unit point loads applied within the half-space ($|P_1| = |P_2| = |P_3| = 1$)

condition over the surface of the semi-infinite space. Therefore, this class of solutions can be written as

$$()^* = ()^k + ()^c, \tag{5.67}$$

where $()^k$ and $()^c$ stand for the Kelvin part (expressions (5.55)–(5.60) for three dimensions and two dimensions) and complementary part, respectively.

In what follows, in order to avoid unnecessary repetition, only the complementary part of the solutions will be discussed. But it is always implied that the total expressions for the fundamental solutions are given by relation (5.67).

The complete set of expressions for the three-dimensional displacements and stresses are presented in Mindlin's paper. In order to obtain the plane strain fundamental displacements, the integration procedure already demonstrated for the Kelvin problem can be used and the complete half-plane fundamental solution will be obtained as shown by Telles and Brebbia [14]. The complementary solutions are given in Appendix B for reference purposes.

5.4. Stresses at Internal Points

Equation (5.53) is a continuous representation of displacements at any point $\xi \in \Omega$. Consequently, the stress state at this point can be obtained by combining the derivatives of Eq. (5.53) with respect to the coordinates of ξ to produce the strain tensor and then substituting the result into Hooke's law. The final expression is

$$\sigma_{ij}(\xi) = \int_\Gamma u^*_{ijk}(\xi, x)\, p_k(x)\, d\Gamma(x) - \int_\Gamma p^*_{ijk}(\xi, x)\, u_k(x)\, d\Gamma(x) + \int_\Omega u^*_{ijk}(\xi, x)\, b_k(x)\, d\Omega(x)\,. \tag{5.68}$$

$\xi \in \Omega$

Note that differentiation was carried out directly inside the integrals. This is obviously possible for the boundary integrals, but the body force term needs a proper proof. This matter will be taken up in Chapter 6 where the inelastic formulation is presented and a strong demonstration of the validity of this procedure will be discussed.

For the Kelvin fundamental solution, the new tensors are written as

$$u^*_{ijk} = -\sigma^*_{ijk}\,, \tag{5.69}$$

$$\begin{aligned} p^*_{ijk} = {} & \frac{G}{2\alpha\,\pi(1-\nu)\,r^\beta}\left\{\beta\frac{\partial r}{\partial n}\left[(1-2\nu)\,\delta_{ij}\,r_{,k} + \nu(\delta_{ik}\,r_{,j} + \delta_{jk}\,r_{,i}) - \gamma r_{,i}\,r_{,j}\,r_{,k}\right]\right. \\ & + \beta\,\nu(n_i\,r_{,j}\,r_{,k} + n_j\,r_{,i}\,r_{,k}) + (1-2\nu)\,(\beta\,n_k\,r_{,i}\,r_{,j} + n_j\,\delta_{ik} + n_i\,\delta_{jk}) \\ & \left. -(1-4\nu)\,n_k\,\delta_{ij}\right\}, \end{aligned} \tag{5.70}$$

where $\alpha = 2, 1$; $\beta = 3, 2$; $\gamma = 5, 4$ for three dimensions and two dimensions respectively. Note that the substitution

$$\frac{\partial r}{\partial x_i(\xi)} = -r_{,i} = -\frac{\partial r}{\partial x_i(x)} \tag{5.71}$$

was already effected.

The complementary expressions for the tensors corresponding to the half-plane fundamental solution have been presented by Telles and Brebbia [14]. The Mindlin solution follows the same approach.

5.5. Boundary Integral Equation

In the preceding sections the derivation of Somigliana's identity has been presented without need for distinction between the different fundamental solutions employed. In this section, however, it is instructive to consider first the Kelvin solution and then extend the complete formulation to half-space-type problems where full advantage of the traction-free condition can be taken.

Considering the Kelvin case, Somigliana's identity is not satisfactory for obtaining solutions unless the boundary displacements and tractions are known throughout the boundary Γ (body forces are always assumed to be prescribed). Therefore, it is interesting to examine the limiting form of Eq. (5.53) as ξ goes to the boundary. Assuming first that the body can be represented as shown in Fig. 5.6, i.e., with the point ξ as an internal point surrounded by part of a spherical surface of radius ε, Eq. (5.53) can be written

$$\begin{aligned} u_i(\xi) = {} & \int\limits_{\Gamma-\Gamma_\varepsilon+\bar{\Gamma}_\varepsilon} u^*_{ij}(\xi, x)\,p_j(x)\,d\Gamma(x) - \int\limits_{\Gamma-\Gamma_\varepsilon+\bar{\Gamma}_\varepsilon} p^*_{ij}(\xi, x)\,u_j(x)\,d\Gamma(x) \\ & + \int\limits_{\Omega'} u^*_{ij}(\xi, x)\,b_j(x)\,d\Omega(x)\,. \end{aligned} \tag{5.72}$$

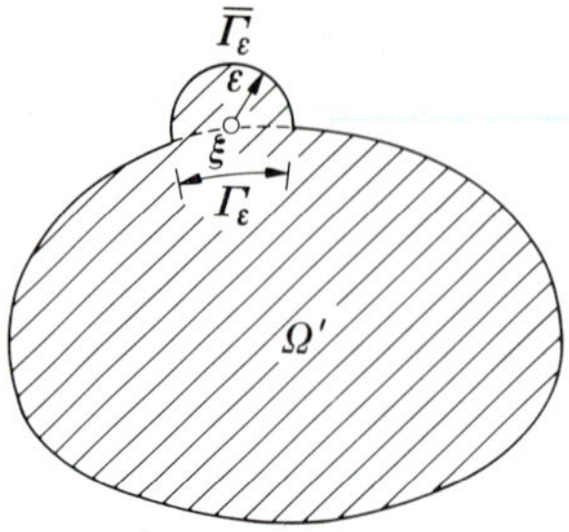

Fig. 5.6. Singular point ξ on the boundary surrounded by part of a spherical surface

Let us now study separately the limit of each integral in Eq. (5.72) as $\varepsilon \to 0$. The first integral can be written

$$\lim_{\varepsilon \to 0} \int_{\Gamma - \Gamma_\varepsilon + \bar{\Gamma}_\varepsilon} p_{ij}^*(\xi, x)\, u_j(x)\, d\Gamma(x) = \lim_{\varepsilon \to 0} \int_{\bar{\Gamma}_\varepsilon} p_{ij}^*(\xi, x)\, u_j(x)\, d\Gamma(x) + \lim_{\varepsilon \to 0} \int_{\Gamma - \Gamma_\varepsilon} p_{ij}^*(\xi, x)\, u_j(x)\, d\Gamma(x)\,, \tag{5.73}$$

where from the first integral on the right-hand side comes

$$\lim_{\varepsilon \to 0} \int_{\bar{\Gamma}_\varepsilon} p_{ij}^*(\xi, x)\, u_j(x)\, d\Gamma(x) = \lim_{\varepsilon \to 0} \int_{\bar{\Gamma}_\varepsilon} p_{ij}^*(\xi, x)\, [u_j(x) - u_j(\xi)]\, d\Gamma(x) + \lim_{\varepsilon \to 0} \left\{ u_j(\xi) \int_{\bar{\Gamma}_\varepsilon} p_{ij}^*(\xi, x) \right\} \tag{5.74}$$

Clearly, the first integral on the right of Eq. (5.74) vanishes from the condition of continuity of $u_j(x)$ and the second integral together with the left-hand side of Eq. (5.72) allow for the representation

$$c_{ij}(\xi) = \left\{ \delta_{ij} + \lim_{\varepsilon \to 0} \int_{\bar{\Gamma}_\varepsilon} p_{ij}^*(\xi, x)\, d\Gamma(x) \right\}. \tag{5.75}$$

Going back to expression (5.73), the second integral on the right-hand side is seen to be taken in the Cauchy principal value sense [15], the existence of which can be proved if $u_j(x)$ satisfies a Hölder condition [7] at ξ in the form

$$|u_j(x) - u_j(\xi)| \leqq B\, r^\alpha\,, \tag{5.76}$$

where B and α are positive constants.

The remaining integrals in expression (5.72) present no special singularities and can be interpreted in the normal sense of integration. Therefore, as $\varepsilon \to 0$, the following equation arises:

$$c_{ij}(\xi)\, u_j(\xi) + \int_\Gamma p_{ij}^*(\xi, x)\, u_j(x)\, d\Gamma(x) = \int_\Gamma u_{ij}^*(\xi, x)\, p_j(x)\, d\Gamma(x) + \int_\Omega u_{ij}^*(\xi, x)\, b_j(x)\, d\Omega(x)\,, \tag{5.77}$$

where the integral on the left-hand side is to be interpreted in the sense of Cauchy principal value.

Equation (5.77) is valid for three dimensions or two dimensions; it provides a relation that must be satisfied between surface displacements, surface tractions, and body forces. Taking into consideration that the body force term is always known, when boundary conditions are prescribed this equation becomes a boundary integral equation for the unknown boundary data. This important feature is the one that makes it most attractive and can be fully explored for numerical solutions.

The coefficient $c_{ij}(\xi)$ was defined in Eq. (5.75). If the tangent plane at ξ is continuous, $c_{ij}(\xi) = \delta_{ij}/2$, but if this is not the case, closed form expressions for this coefficient have been presented in Refs. [16, 17] for two-dimensions and three dimensions. For practical applications, however, it will be seen later that c_{ij} together with the corresponding principal value can be indirectly computed by applying Eq. (5.77) to represent rigid-body movements.

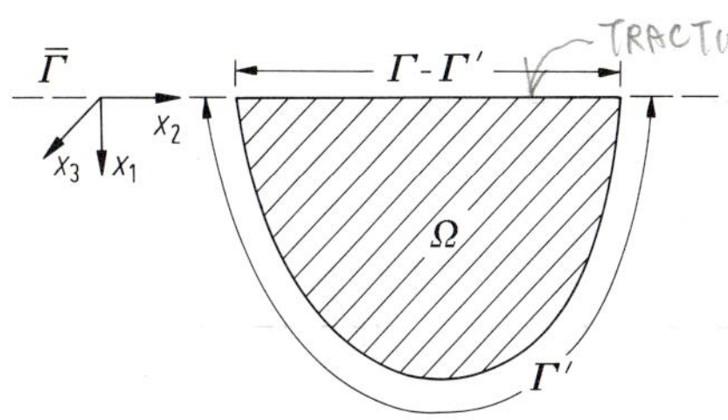

Fig. 5.7. Body with part of its boundary Γ coinciding with the surface of the semi-infinite space

Equation (5.77) is the starting equation for the boundary element technique using the Kelvin fundamental solution. For half-space-type solutions, it is interesting to start by reviewing Somigliana's identity to point out an important simplification in this equation which was not mentioned before. If the body that is being analyzed presents part of its boundary coinciding with the surface of the semi-infinite space $\bar{\Gamma}$ (see Fig. 5.7), the integral over this part which involves p_{ij}^* vanishes identically because of the traction-free condition included into the fundamental solution. Consequently, Somigliana's identity can be rewritten as follows (three dimensions or two dimensions):

$$u_i(\xi) = \int_{\Gamma} u_{ij}^*(\xi, x)\, p_j(x)\, d\Gamma(x) - \int_{\Gamma'} p_{ij}^*(\xi, x)\, u_j(x)\, d\Gamma(x) + \int_{\Omega} u_{ij}^*(\xi, x)\, b_j(x)\, d\Omega(x)\,, \tag{5.78}$$

where Γ' represents the part of Γ in which $x_1 > 0$.

In addition, Eq. (5.78) can also be specialized for load points located on the surface of the half-space $\Gamma - \Gamma'$ without any further modification. This is because when $c = 0$ (see Fig. 5.5) the singularity which occurs in the first integral on the right-hand side of Eq. (5.78) can be integrated in the usual sense. Furthermore, if the problem to be analyzed satisfies the traction-free condition ($p_j(x) = 0$) over some part of $\Gamma - \Gamma'$, this weak singularity is also removed, allowing for load points on such part of the boundary to be considered as internal points.

As will be seen later, the above-described characteristic of the half-space formulation plays an important role in its application to the boundary element method.

Due to the nonsingular nature of the complementary expressions (see formulas (B.1)−(B.12)), specialization of Eq. (5.78) for load points located along the boundary Γ' creates exactly the same singularities obtained for the single Kelvin formulation. Therefore, the following equation is obtained:

$$c_{ij}(\xi)\,u_j(\xi) + \int_{\Gamma'} p^*_{ij}(\xi, x)\,u_j(x)\,d\Gamma(x) = \int_{\Gamma} u^*_{ij}(\xi, x)\,p_j(x)\,d\Gamma(x) + \int_{\Omega} u^*_{ij}(\xi, x)\,b_j(x)\,d\Omega(x) \tag{5.79}$$

in which the integral on the left-hand side is to be interpreted in the Cauchy principal value sense and the expression of $c_{ij}(\xi)$ corresponds only to the Kelvin part of the fundamental solution. Hence, $c_{ij} = \delta_{ij}/2$ on smooth surfaces and can be taken from other references otherwise [16, 17].

The special case when the load is located at the intersection of Γ' and $\bar{\Gamma}$ can also be handled by Eq. (5.79). However, in this case limiting relations are involved resulting in a different expression for c_{ij}. This exception does not create any difficulty and the proper expression for c_{ij} can be obtained by applying Eq. (5.79) to represent rigid-body movements.

In conclusion, Eq. (5.79) can in general, be quoted as representing the specialization of Eq. (5.78) for any boundary load point if $c_{ij} = \delta_{ij}$ when referring to the boundary $\Gamma - \Gamma'$.

Following the procedure presented in Section 5.4, the derivatives of Eq. (5.78) with respect to the coordinates of the load point can be combined with Hooke's law to obtain a particular version of Eq. (5.68):

$$\sigma_{ij}(\xi) = \int_{\Gamma} u^*_{ijk}(\xi, x)\,p_k(x)\,d\Gamma(x) - \int_{\Gamma'} p^*_{ijk}(\xi, x)\,u_k(x)\,d\Gamma(x) + \int_{\Omega} u^*_{ijk}(\xi, x)\,b_k(x)\,d\Omega(x)\,. \tag{5.80}$$

Note that if the problem to be analyzed satisfies the traction-free condition $p_k(x) = 0$ over some part of the boundary $\Gamma - \Gamma'$, stresses at boundary points located on this part of the boundary can also be computed by Eq. (5.80).

Although body forces have been considered in the equations presented so far, in the following section this term will not be included for simplicity and their detailed treatment will be carried out in Section 5.14. However, it is worth mentioning that the domain integral involved can be avoided in many practical cases by using some alternative procedures, as for instance to find a particular solution for the body forces and as an *in situ* stress field, superimpose it to the boundary element solution. A more interesting procedure is presented by Rizzo and Shippy [18] where the possibility of transforming the body force integral into a surface integral is presented. This procedure is applicable to some of the most common types of body forces, such as constant gravitational load, centrifugal load due to a fixed axis of rotation, and the effect of a steady-state temperature distribution. A unified procedure for such cases (three-dimensional and two-dimensional) is to be presented in Section 5.14.

5.6. Infinite and Semi-Infinite Regions

Throughout the derivations presented in this chapter, bounded bodies were always considered. In this section the validity of the previous expressions (Kelvin fundamental solution) will be extended to external problems in infinite regular regions. As before, we borrow this concept from Kellogg [7]. Thus, infinite regular region means a region bounded by a regular surface (hence a bounded surface) and containing all sufficiently distant points. In addition, the same extension will be seen applicable to problems related to the semi-infinite space with or without cavities. Consequently, the ideas will be generalized to include such cases.

The extension of Eq. (5.77) to infinite regular regions is not valid without further hypothesis on the functions involved. Such hypotheses are concerned with the behavior of the functions on an infinitely distant surface and are referred to as regularity conditions.

Let ϱ be the radius of sphere of surface Γ_ϱ and centered at ξ, which encloses the cavity (or cavities) of the external problem depicted in Fig. 5.8.

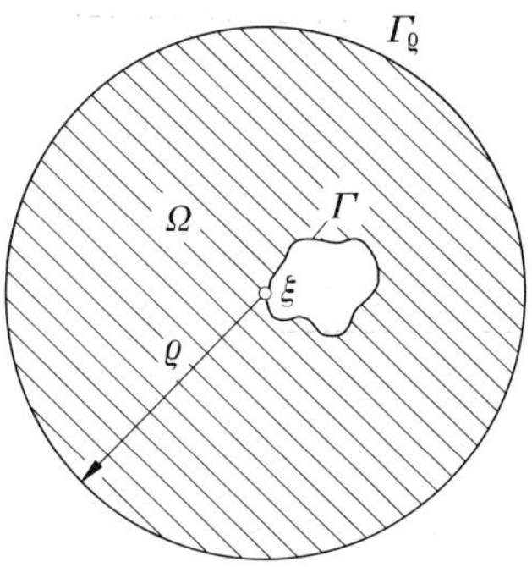

Fig. 5.8. Infinite region with cavity

Equation (5.77) can be written for the region within Γ and Γ_ϱ as follows

$$c_{ij}(\xi)\, u_j(\xi) + \int_\Gamma p^*_{ij}(\xi, x)\, u_j(x)\, d\Gamma(x) + \int_{\Gamma_\varrho} p^*_{ij}(\xi, x)\, u_j(x)\, d\Gamma(x)$$
$$= \int_\Gamma u^*_{ij}(\xi, x)\, p_j(x)\, d\Gamma(x) + \int_{\Gamma_\varrho} u^*_{ij}(\xi, x)\, p_j(x)\, d\Gamma(x)\,. \qquad (5.81)$$

Clearly, if the limiting case $\varrho \to \infty$ is considered, Eq. (5.81) can be expressed in terms of boundary integrals over Γ alone if

$$\lim_{\varrho\to\infty} \int_{\Gamma_\varrho} [p^*_{ij}(\xi, x)\, u_j(x) - u^*_{ij}(\xi, x)\, p_j(x)]\, d\Gamma(x) = 0\,. \qquad (5.82)$$

For three-dimensional problems, one has ($x \in \Gamma_\varrho$)

$$\begin{aligned} d\Gamma(x) &= |\boldsymbol{G}|\, d\varphi\, d\theta \qquad \text{with } |\boldsymbol{G}| = O(\varrho^2)\,,\\ u^*_{ij}(\xi, x) &= O(\varrho^{-1})\,,\\ p^*_{ij}(\xi, x) &= O(\varrho^{-2})\,, \end{aligned} \qquad (5.83)$$

where $O(\,)$ represents the asymptotic behavior as $\varrho \to \infty$.

Therefore, if at most $u_j(x)$ and $p_j(x)$ have the behavior ϱ^{-1} and ϱ^{-2} at infinity, the regularity conditions (5.82) are satisfied. Notice that if the total load applied over the boundary of the cavity is not self-equilibrated, Saint-Venant's principle [3] gives that $u_j(x)$ and $p_j(x)$ will behave like the fundamental solution corresponding to a concentrated force in the direction of the resultant load. Thus, $u_j = O(\varrho^{-1})$ and $p_j(x) = O(\varrho^{-2})$ are obtained and they ensure that each term of Eq. (5.82) vanishes separately from the other.

For two-dimensional problems, the equivalent of Eq. (5.83) is given by

$$\begin{aligned} d\Gamma(x) &= |\boldsymbol{G}|\, d\varphi \qquad \text{with } |\boldsymbol{G}| = O(\varrho), \\ u^*_{ij}(\xi, x) &= O(\ln(\varrho) + 1), \qquad i = j \\ &= O(1), \qquad i \neq j, \\ p^*_{ij}(\xi, x) &= O(\varrho^{-1}). \end{aligned} \tag{5.84}$$

From the above it is seen that in order to guarantee that each term of Eq. (5.82) vanishes separately, one needs $u_j(x) = O(\varrho^{-1})$ and $p_j(x) = O(\varrho^{-2})$ as before [19]. This case, however, does not correspond to the behavior of the fundamental solution at infinity. Based on the same argument as for three dimensions, one can substitute $u_j(x)$ and $p_j(x)$ by the tensors corresponding to the two-dimensional fundamental solution and indeed verify that Eq. (5.82) is satisfied. The only difference now is that the two terms do not approach zero separately, but they cancel each other as $\varrho \to \infty$.

The above discussion strongly suggests that the regularity conditions are always satisfied if $u_i(x)$ and $p_i(x)$ behave at most like the corresponding fundamental solution at infinity. This statement is also verified for semi-infinite problems where the half-space and half-plane fundamental solutions dictate the corresponding conditions. In these applications, the interesting case in which part of the boundary $\bar{\Gamma}$ is loaded – e.g., Fig. 5.9 – can also be included.

In conclusion, provided the regularity conditions are satisfied, problems of cavities in the external region can be represented by (see Fig. 5.8)

$$c_{ij}(\xi)\, u_j(\xi) + \int_{\Gamma} p^*_{ij}(\xi, u)\, u_j(x)\, d\Gamma(x) = \int_{\Gamma} u^*_{ij}(\xi, x)\, p_j(x)\, d\Gamma(x) \tag{5.85}$$

for infinite regions (Kelvin), and also

$$c_{ij}(\xi)\, u_j(\xi) + \int_{\Gamma'} p^*_{ij}(\xi, u)\, u_j(x)\, d\Gamma(x) = \int_{\Gamma} u^*_{ij}(\xi, x)\, p_j(x)\, d\Gamma(x) \tag{5.86}$$

for semi-infinite regions which may have a loaded boundary $\Gamma - \Gamma'$ (see Fig. 5.9).

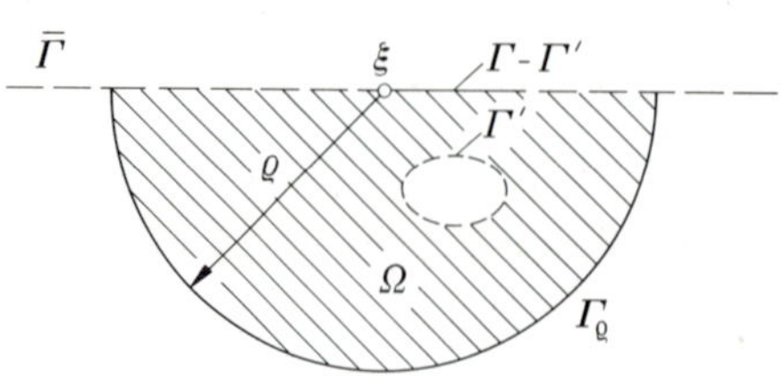

Fig. 5.9. Semi-infinite region with or without cavities

a b

Fig. 5.10. Definition of the normal: **a** External problems; **b** internal problems

The conclusions reached in this section are obviously valid for Somigliana's identity and also guarantee stronger regularity in the expressions for stresses at internal points. Note that for an external cavity problem the integral equations are of the same form as for the internal region counterpart, the difference being given by the normal n which is taken as pointing into the cavity (Fig. 5.10).

5.7. Numerical Implementation

In this section a general numerical procedure for the solution of boundary-value problems in solid mechanics will be described. In order to concentrate the attention on the main aspects of the process, the different forms of boundary integral equation introduced in the previous sections will be represented in a unified manner as follows (body forces are omitted here for simplicity):

$$c_{ij}(\xi)\, u_j(\xi) + \int_\Gamma p^*_{ij}(\xi, x)\, u_j(x)\, d\Gamma(x) = \int_\Gamma u^*_{ij}(\xi, x)\, p_j(x)\, d\Gamma(x)\,, \tag{5.87}$$

where depending on the fundamental solution employed (infinite or semi-infinite space) the appropriate expression for $c_{ij}(\xi)$ and the substitution of Γ by Γ' in the first integral are implied.

Instead of attempting closed form solutions to Eq. (5.87), which is a difficult task and only attainable for simple geometries and boundary conditions, the boundary element method employs a numerical approach. The basic steps involved in this approach constitute the numerical essence of the technique (see Chapter 2 and Refs. [19, 20]). They are summarized below:

(a) The boundary Γ is discretized into a series of elements over which displacements and tractions are chosen to be piecewise interpolated between the element nodal points;
(b) Equation (5.87) is applied in discretized form to each nodal point ξ of the boundary Γ and the integrals are computed (usually by a numerical quadrature scheme) over each boundary element. A system of N linear algebraic equations involving the set of N nodal tractions and N nodal displacements is therefore obtained;
(c) Boundary conditions are imposed and consequently N nodal values (traction or displacement in each direction per node) are prescribed. The system of N equations can therefore be solved by standard methods to obtain the remaining boundary data.

Values of displacements and stresses at any selected internal point can readily be computed by numerical quadrature using the appropriate equations (i.e., Eqs. (5.53) and (5.62) or Eqs. (5.78) and (5.80)) also in discretized fashion. Note that nonzero body forces can be included by a simple numerical integration scheme which leads to an additional contribution to the independent term of the system of equations, and a similar contribution to the internal displacements and stresses. For a full treatment of body forces see Section 5.14.

For the discretization of Eq. (5.87), the boundary Γ is approximated by using a series of elements. The Cartesian coordinates $\boldsymbol{x}$ of points located within each element Γ_j are expressed in terms of interpolation functions $\boldsymbol{\Psi}$ and the nodal coordinates $\boldsymbol{x}^m$ of the element by the matrix relation

$$\boldsymbol{x} = \boldsymbol{\Psi}^T \boldsymbol{x}^m, \tag{5.88}$$

where $\boldsymbol{x}$ represents the x_1, x_2, and x_3 coordinates in three-dimensional problems.

In a similar way, boundary displacements and tractions are approximated over each element through interpolation functions

$$\begin{aligned} \boldsymbol{u} &= \boldsymbol{\Phi}^T \boldsymbol{u}^n, \\ \boldsymbol{p} &= \boldsymbol{\Phi}^T \boldsymbol{p}^n, \end{aligned} \tag{5.89}$$

where $\boldsymbol{u}^n$ and $\boldsymbol{p}^n$ contain the nodal displacements and tractions, respectively. Note that the superscript m in Eq. (5.88) refers to the number of boundary points required to define the geometry of each boundary element, whereas the superscript n in Eq. (5.89) refers to the number of boundary nodes to which the nodal values of displacements and tractions are associated. These numbers may be different in general, provided that, as shown in Section 3.7, $m \leqq n$, to satisfy rigid-body-type conditions.

It is now more convenient to work with matrices instead of the indicial notation. In order to do so we can generalize $\boldsymbol{u}$ as the displacements and $\boldsymbol{p}$ as the traction vectors, such as

$$\boldsymbol{u} = \begin{Bmatrix} u_1 \\ u_2 \\ u_3 \end{Bmatrix}, \qquad \boldsymbol{p} = \begin{Bmatrix} p_1 \\ p_2 \\ p_3 \end{Bmatrix}. \tag{5.90}$$

In addition we can define the following two matrices:

$$\boldsymbol{u}^* = \begin{bmatrix} u_{11}^* & u_{12}^* & u_{13}^* \\ u_{21}^* & u_{22}^* & u_{23}^* \\ u_{31}^* & u_{32}^* & u_{32}^* \end{bmatrix}, \qquad \boldsymbol{p}^* = \begin{bmatrix} p_{11}^* & p_{12}^* & p_{13}^* \\ p_{21}^* & p_{22}^* & p_{23}^* \\ p_{31}^* & p_{32}^* & p_{33}^* \end{bmatrix}, \tag{5.91}$$

where the u_{ij}^* and p_{ij}^* are the displacements and tractions in the j direction due to a unit force at the point under consideration, acting in the i direction.

Let us first write Eq. (5.87) before any boundary conditions are applied, i.e.,

$$c_{ij}\, u_j + \int_\Gamma p_{ij}^*\, u_j \, d\Gamma = \int_\Gamma u_{ij}^*\, p_j \, d\Gamma. \tag{5.92}$$

This equation can now be expressed in matrix form as

$$\boldsymbol{c}\, \boldsymbol{u} + \int_\Gamma \boldsymbol{p}^*\, \boldsymbol{u}\, d\Gamma = \int_\Gamma \boldsymbol{u}^*\, \boldsymbol{p}\, d\Gamma. \tag{5.93}$$

This formulation is valid for a load point on the boundary at ξ_l. Note that $\boldsymbol{p}^*$ and $\boldsymbol{u}^*$ are known and $\boldsymbol{c}$ can be found analytically or from rigid-body conditions (see the next section). The unknowns are the values of $\boldsymbol{u}$ and $\boldsymbol{p}$ over the boundary.

5.8. Boundary Elements

We will now assume that the boundary is divided into elements. These can be constant, linear, quadratic, or higher order and they can be triangular or quadrilateral. Figure 5.11a shows the body divided into a series of constant elements. For these elements, the unknowns are assumed to be at the center and to have a constant value over the element. Those of Fig. 5.11b are such that the functions vary linearly over them. Finally, in Fig. 5.11c the variation of displacements and tractions is quadratic. In addition the geometry of the element is described by a quadratic function. Note that the constant element is such that the condition $m \leqq n$ (i.e., that the functions for the unknowns are of the same or higher order than the geometry) is violated. This will produce rigid-body-type straining and is evident in cases such as deflection of cantilever beams and other problems with considerable rigid-body movement.

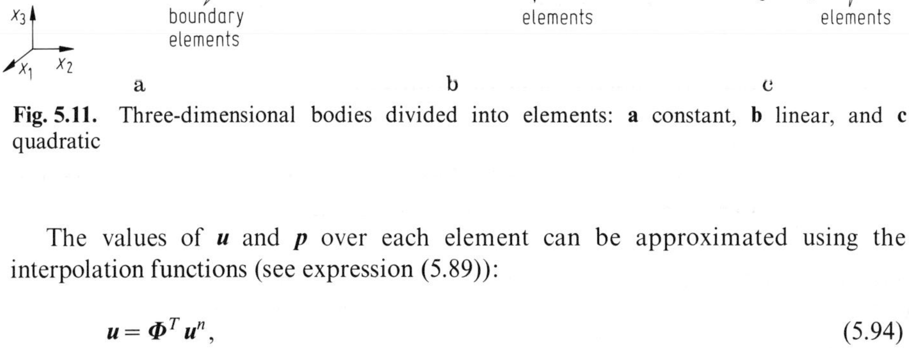

Fig. 5.11. Three-dimensional bodies divided into elements: **a** constant, **b** linear, and **c** quadratic

The values of $\boldsymbol{u}$ and $\boldsymbol{p}$ over each element can be approximated using the interpolation functions (see expression (5.89)):

$$\boldsymbol{u} = \boldsymbol{\Phi}^T \boldsymbol{u}^n, \tag{5.94}$$

$$\boldsymbol{p} = \boldsymbol{\Phi}^T \boldsymbol{p}^n. \tag{5.95}$$

Note that the interpolation functions for $\boldsymbol{u}$ and $\boldsymbol{p}$ may be generally different. For consistency it may be better to take the functions for $\boldsymbol{p}$ of one order less than those for $\boldsymbol{u}$. The interpolation functions are the ones discussed in Chapter 3. $\boldsymbol{u}^n$ and $\boldsymbol{p}^n$ are the nodal displacements and tractions.

We can substitute the above functions into Eq. (5.93) and obtain the following equation for the point ξ_l:

$$\boldsymbol{c}\,\boldsymbol{u} + \sum_{j=1}^{N} \left(\int_{\Gamma_j} \boldsymbol{p}^* \boldsymbol{\Phi}^T d\Gamma \right) \boldsymbol{u}^n = \sum_{j=1}^{N} \left(\int_{\Gamma_j} \boldsymbol{u}^* \boldsymbol{\Phi}^T d\Gamma \right) \boldsymbol{p}^n, \tag{5.96}$$

where the summation from $j=1$ to N indicates summation over the N elements on the surface and Γ_j is the surface of the j element.

The integrals in Eq. (5.96) are usually solved numerically and the functions $\boldsymbol{\Phi}$ are expressed in some of the homogeneous system of coordinates described in Chapter 3. The integrals need then to be written in terms of the homogeneous system, to be called η_1, η_2. Hence,

$$d\Gamma = (\text{absolute value of} |\boldsymbol{G}|)\, d\eta_1\, d\eta_2. \tag{5.97}$$

Due to the difficulties in integrating Eq. (5.96) it is usual to employ a numerical integration scheme. Hence formula (5.96) becomes

$$\begin{aligned} &\boldsymbol{c}\,\boldsymbol{u} + \sum_{j=1}^{N} \left(\sum_{l=1}^{L} |\boldsymbol{G}|_l\, w_l (\boldsymbol{p}^* \boldsymbol{\Phi}^T(\eta_1, \eta_2))_l \right) \boldsymbol{u}^n \\ &= \sum_{j=1}^{N} \left(\sum_{l=1}^{L} |\boldsymbol{G}|_l\, w_l (\boldsymbol{u}^* \boldsymbol{\Phi}^T(\eta_1, \eta_2))_l \right) \boldsymbol{p}^n. \end{aligned} \tag{5.98}$$

$|\boldsymbol{G}|$ is the Jacobian for the three-dimensional transformation and the w_l are the numerical integration weighting coefficients.

Note that Eq. (5.98) gives a set of equations for node i which can be written

$$\begin{aligned} &\boldsymbol{c}_i \boldsymbol{u}_i + [\hat{\boldsymbol{h}}_{i1}\, \hat{\boldsymbol{h}}_{i2} \cdots \hat{\boldsymbol{h}}_{ii} \cdots \hat{\boldsymbol{h}}_{ir}] \begin{Bmatrix} \boldsymbol{u}_1 \\ \boldsymbol{u}_2 \\ \vdots \\ \boldsymbol{u}_i \\ \vdots \\ \boldsymbol{u}_r \end{Bmatrix} \\ &= [\boldsymbol{g}_{i1}\, \boldsymbol{g}_{i2} \cdots \boldsymbol{g}_{ii} \cdots \boldsymbol{g}_{ir}] \begin{Bmatrix} \boldsymbol{p}_1 \\ \boldsymbol{p}_2 \\ \vdots \\ \boldsymbol{p}_i \\ \vdots \\ \boldsymbol{p}_r \end{Bmatrix}, \end{aligned} \tag{5.99}$$

where $\boldsymbol{u}_j$ and $\boldsymbol{p}_j$ are the unknowns at nodes j. $\hat{\boldsymbol{h}}_{ij}$ and $\boldsymbol{g}_{ij}$ are the interaction coefficients relating node i with all the nodes on the surface of the body. Note that with the exception of constant elements, more than one element will contribute to the $\hat{\boldsymbol{h}}_{ij}$ and $\boldsymbol{g}_{ij}$ submatrices. (They are 3×3 for the three-dimensional case and 2×2 for two dimensions.)

5.9. System of Equations

We can write a matrix equation such as Eq. (5.99) for each of the nodes under consideration. Writing them together we have

$$\begin{bmatrix} \boldsymbol{h}_{11} & \hat{\boldsymbol{h}}_{12}\ldots & \hat{\boldsymbol{h}}_{1i}\ldots & \hat{\boldsymbol{h}}_{1r} \\ \hat{\boldsymbol{h}}_{21} & \boldsymbol{h}_{22}\ldots & \hat{\boldsymbol{h}}_{2i}\ldots & \hat{\boldsymbol{h}}_{2r} \\ \vdots & \vdots & \vdots & \vdots \\ \hat{\boldsymbol{h}}_{i1} & \hat{\boldsymbol{h}}_{i2} & \boldsymbol{h}_{ii} & \hat{\boldsymbol{h}}_{ir} \\ \vdots & \vdots & \vdots & \vdots \\ \hat{\boldsymbol{h}}_{r1} & \hat{\boldsymbol{h}}_{r2} & \hat{\boldsymbol{h}}_{ri} & \boldsymbol{h}_{rr} \end{bmatrix} \begin{Bmatrix} \boldsymbol{u}_1 \\ \boldsymbol{u}_2 \\ \vdots \\ \boldsymbol{u}_i \\ \vdots \\ \boldsymbol{u}_r \end{Bmatrix} = \begin{bmatrix} \boldsymbol{g}_{11} & \boldsymbol{g}_{12}\cdots & \boldsymbol{g}_{1i}\cdots & \boldsymbol{g}_{1r} \\ \boldsymbol{g}_{21} & \boldsymbol{g}_{22}\cdots & \boldsymbol{g}_{2i}\cdots & \boldsymbol{g}_{2r} \\ \vdots & \vdots & \vdots & \vdots \\ \boldsymbol{g}_{i1} & \boldsymbol{g}_{i2}\cdots & \boldsymbol{g}_{ii}\cdots & \boldsymbol{g}_{ir} \\ \vdots & \vdots & \vdots & \vdots \\ \boldsymbol{g}_{r1} & \boldsymbol{g}_{r2}\cdots & \boldsymbol{g}_{ri}\cdots & \boldsymbol{g}_{rr} \end{bmatrix} \begin{Bmatrix} \boldsymbol{p}_1 \\ \boldsymbol{p}_2 \\ \vdots \\ \boldsymbol{p}_i \\ \vdots \\ \boldsymbol{p}_r \end{Bmatrix}, \tag{5.100}$$

where the submatrices $\boldsymbol{h}_{ii}$ on the diagonal are

$$\boldsymbol{h}_{ii} = \hat{\boldsymbol{h}}_{ii} + \boldsymbol{c}_i. \tag{5.101}$$

Equations (5.100) can be written as

$$\boldsymbol{H}\boldsymbol{U} = \boldsymbol{G}\boldsymbol{P}. \tag{5.102}$$

We now have to apply the boundary conditions in this system which are of two types: (i) $\boldsymbol{u}_i = \bar{\boldsymbol{u}}_i$ on Γ_1 and (ii) $\boldsymbol{p}_i = \bar{\boldsymbol{p}}_i$ on Γ_2. If the displacements are known, we can find the tractions and vice versa. This implies that the system of equations (5.102) can be reordered in such a way that all the unknowns are written on the left-hand side in an $\boldsymbol{Y}$ vector. The final result can be written

$$\boldsymbol{A}\,\boldsymbol{Y} - \boldsymbol{F}. \tag{5.103}$$

Note that we have not yet determined the $\boldsymbol{c}_i$ coefficients of Eq. (5.101). They can be obtained implicitly by consideration of rigid-body movements in a bounded body. Let us assume we have a unit rigid-body displacement in any one direction; Eq. (5.102) then becomes

$$\boldsymbol{H}\boldsymbol{I}_l = \boldsymbol{0}, \tag{5.104}$$

where $\boldsymbol{I}_l$ is a vector defining a unit rigid displacement in the direction l. Hence, the diagonal terms of $\boldsymbol{H}$ are simply

$$\boldsymbol{h}_{ii} = -\sum_{i \neq j} \boldsymbol{h}_{ij}, \tag{5.105}$$

which means that neither the $\boldsymbol{c}_i$ nor the $\hat{\boldsymbol{h}}_{ii}$ coefficients need be determined explicitly.

Once Eqs. (5.103) have been solved we know the tractions and displacements over the whole surface of the body. We can now compute the stresses and displacements at any internal point.

Expression (5.105) is valid for a finite or bounded body, for infinite or semi-infinite domains, however, a further term must be added. With reference to Section 5.6, one can readily verify that since a rigid-body movement is of order $O(1)$, the regularity conditions at infinity are no longer satisfied. Consequently, the following expression has to be considered in such cases:

$$c_{ij}(\xi)\, u_j + u_j \int_{\Gamma} p^*_{ij}(\xi, x)\, d\Gamma(x) + \lim_{\varrho \to \infty} u_j \int_{\Gamma_\varrho} p^*_{ij}(\xi, x)\, d\Gamma(x) = 0\,, \tag{5.106}$$

where u_j corresponds to any – constant – rigid-body translation and $\xi \in \Gamma$.

Since the tensor $p^*_{ij}(\xi, x)$ corresponds to a positive unit point load applied in the i direction, the condition of equilibrium within the fundamental solution region $\Omega^* + \Gamma^*$ leads to

$$\lim_{\varrho \to \infty} \int_{\Gamma_\varrho} p^*_{ij}(\xi, x)\, d\Gamma(x) = -\,\delta_{ij} \tag{5.107}$$

in which Γ_ϱ is the boundary of a spherical region for the infinite space or a hemispherical region for the semi-infinite space.

Substituting Eq. (5.107) into Eq. (5.106) and discretizing gives the following results for a rigid-body movement:

$$\boldsymbol{h}_{ii} = \boldsymbol{I} - \sum_{i \neq j} \boldsymbol{h}_{ij}\,, \tag{5.108}$$

where $\boldsymbol{I}$ is the identity matrix. This equation applies for the diagonal terms of unbounded bodies instead of Eq. (5.105).

5.10. Stresses and Displacements Inside the Body

Once that nodal values of boundary displacements and tractions are known, internal displacements and stresses can be computed by the corresponding equations also in discretized form. Here, since there are no singular integrals to be computed, simple standard numerical quadrature schemes can be used throughout.

It is interesting to note that the use of a semi-infinite fundamental solution renders the discretization of the traction-free part of the boundary unnecessary, the displacements and stresses on this part of the surface being computed as in the case of internal points. This feature, apart from generating a smaller system of equations, avoids any numerical approximation over the free surface. Therefore, semi-infinite or finite-sized problems can be handled with equal ease.

The displacements at internal points are given by Eq. (5.93) with $\boldsymbol{c} = \boldsymbol{I}$, i.e.,

$$\boldsymbol{u} = \int_{\Gamma} \boldsymbol{u}^* \boldsymbol{p}\, d\Gamma - \int_{\Gamma} \boldsymbol{p}^* \boldsymbol{u}\, d\Gamma\,. \tag{5.109}$$

Expressions for the internal stresses have been developed in Sections 5.4 and 5.5 for the infinite and semi-infinite domains, and including body force terms.

$\sigma_{ij}(\xi)=\int_{\Gamma} u^*_{ijk}(\xi,x)p_k(x)\,d\Gamma(x)-\int_{\Gamma'_x} p^*_{ijk}(\xi,x)u_k(x)\,d\Gamma(x)+\int_{\Omega} u^*_{ijk}(\xi,x)b_k(x)\,d\Omega(x)$

$\Gamma'_x=\begin{cases}\Gamma & \text{for } (5.68)\\ \Gamma' & \text{for } (5.80)\end{cases}$

5.11. Stresses on the Boundary

In most practical applications the designer is interested not only in stresses inside the body and tractions on the surface, but also in the complete stress tensor on the boundary. To this end, one can take the limit of the integral equation (5.68) or (5.80) when the load point goes to the boundary. Such procedure, however, does not seem to produce accurate results (except for the traction-free surface when using the half-space fundamental solution) and, furthermore, generates singular boundary integrals which can only be computed in the principal value sense [19, 21]. A less cumbersome alternative is to retrieve part of the stress tensor by expressing the tractions in a local coordinate system and employing the relationships between strain displacements on the tangent directions to the boundary [22, 23]. This procedure does not require any integration and is outlined in what follows.

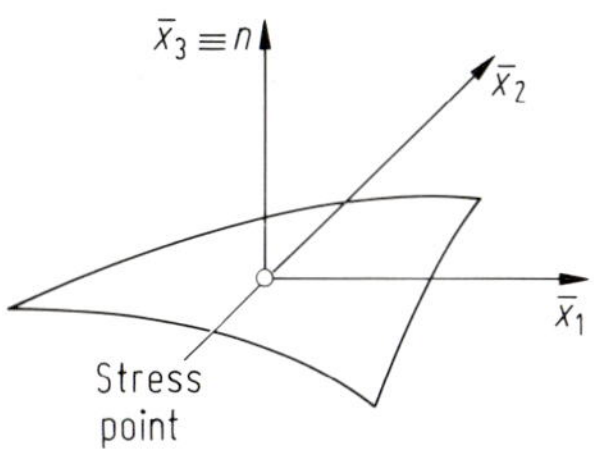

Fig. 5.12. Three-dimensional boundary element and local Cartesian coordinate system tangent to its surface at the stress point

Let us assume a general three-dimensional boundary element with a local Cartesian coordinate system tangent to its surface (Fig. 5.12). Recalling expression (5.94) with the nodal displacements expressed with reference to this local coordinate system, one obtains

$$\bar{\boldsymbol{u}} = \boldsymbol{\Phi}^T \bar{\boldsymbol{u}}^n . \tag{5.110}$$

The three components of the strain tensor ($\bar{\varepsilon}_{11}$, $\bar{\varepsilon}_{12}$, $\bar{\varepsilon}_{22}$) over the surface of the boundary element can, therefore, be obtained by

$$\bar{\varepsilon}_{ij} = \frac{1}{2}\left(\frac{\partial \bar{u}_j}{\partial \bar{x}_i} + \frac{\partial \bar{u}_i}{\partial \bar{x}_j}\right) \qquad (i, j = 1, 2), \tag{5.111}$$

where the appropriate coordinate transformations for higher-order elements are implied.

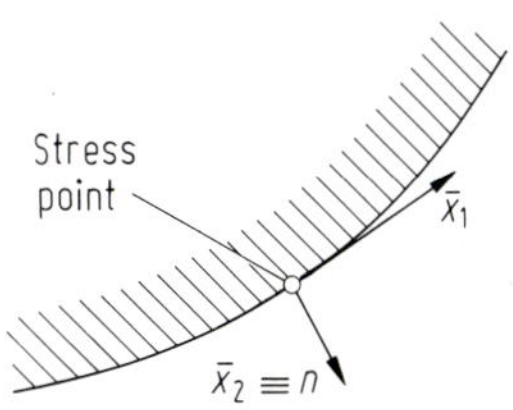

Fig. 5.13. General two-dimensional boundary element and local coordinate system defined at the stress point

The above equation together with Hooke's law provide three extra relations which can be used in addition to the equilibrium equations to determine the six different components of the stress tensor, i.e.,

$$\begin{aligned}
\bar{\sigma}_{11} &= \frac{1}{1-\nu}[\nu\,\bar{\sigma}_{33} + 2\,G\,(\bar{\varepsilon}_{11} + \nu\,\bar{\varepsilon}_{22})],\\
\bar{\sigma}_{12} &= 2\,G\,\bar{\varepsilon}_{12},\\
\bar{\sigma}_{22} &= \frac{1}{1-\nu}[\nu\,\bar{\sigma}_{33} + 2\,G\,(\bar{\varepsilon}_{22} + \nu\,\bar{\varepsilon}_{11})],\\
\bar{\sigma}_{13} &= \bar{p}_1,\\
\bar{\sigma}_{33} &= \bar{p}_3,\\
\bar{\sigma}_{23} &= \bar{p}_2.
\end{aligned} \tag{5.112}$$

For two-dimensional problems the procedure is analogous, the difference being that only one strain component is needed. With reference to Fig. 5.13 one can write

$$\bar{\varepsilon}_{11} = \frac{\partial \bar{u}_1}{\partial \bar{x}_1} \tag{5.113}$$

and the final expressions are given by (plane strain)

$$\begin{aligned}
\bar{\sigma}_{11} &= \frac{1}{1-\nu}(\nu\,\bar{\sigma}_{22} + 2\,G\,\bar{\varepsilon}_{11}),\\
\bar{\sigma}_{12} &= \bar{p}_1,\\
\bar{\sigma}_{22} &= \bar{p}_2,
\end{aligned} \tag{5.114}$$

where for plane stress ν is replaced by $\bar{\nu}$.

5.12. Surface Traction Discontinuities

In order to properly simulate traction discontinuities over the boundary, the concept of double nodes can be employed. In two-dimensional problems, for instance, this consists of having two boundary nodes with exactly the same coordinates without any boundary element in between [16, 24, 25] (see Fig. 5.14a).

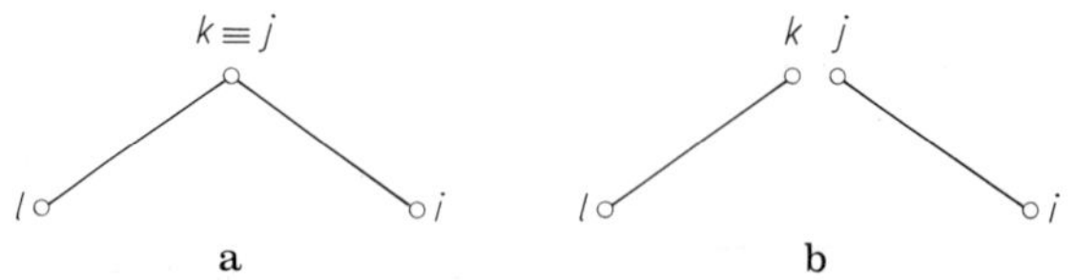

Fig. 5.14. The concept of double nodes illustrated for linear elements

In this procedure the connectivity of the elements is defined as shown in Fig. 5.14b. Therefore, traction discontinuities can be simulated by assigning different values for the tractions at nodes j and k. It should be noted that here, since the displacement at both nodes is the same, in each direction there are two acceptable combinations of prescribed boundary conditions:

(i) tractions at nodes j and k,
(ii) traction at node j and displacement at node k (or vice versa).

The only limitation of this concept is when both nodes have a displacement component in the same direction prescribed as a boundary condition. This generates a singular matrix $\boldsymbol{A}$ (such possibility would violate the displacement continuity condition). Nevertheless, this special case is likely to be associated with a corner problem as depicted in Fig. 5.15 (note that the other combinations may also occur over smooth parts of boundary). Consequently, the procedure proposed in Ref. [25] can be used in this situation; i.e., since the outward normal to the boundary presents a discontinuity, the assumption that the stress tensor is uniquely defined together with the invariance of trace of the strain tensor at the double node ($\varepsilon_{xx} + \varepsilon_{yy} + \varepsilon_{zz}$ invariant) provides two extra equations which can be used to remove the singularity of matrix $\boldsymbol{A}$.

Three-dimensional problems can be dealt with by following the same procedures. Here, since the idea of defining nodes with the same coordinates may require more than two nodes at the same point, the contribution of each element to the total matrix $\boldsymbol{G}$ can be left separate in such a way that there are only three unknowns per node [26]. This would be mathematically equivalent to applying multiple nodes [24] without actually increasing the size of the final matrix $\boldsymbol{A}$.

In many cases such as when different subregions are joined together (Fig. 5.16) any of the above treatment may be too cumbersome for computer implementation.

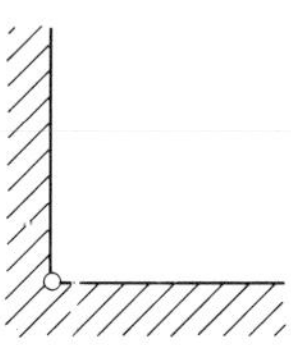

Fig. 5.15. Double node at a corner with displacement prescribed everywhere

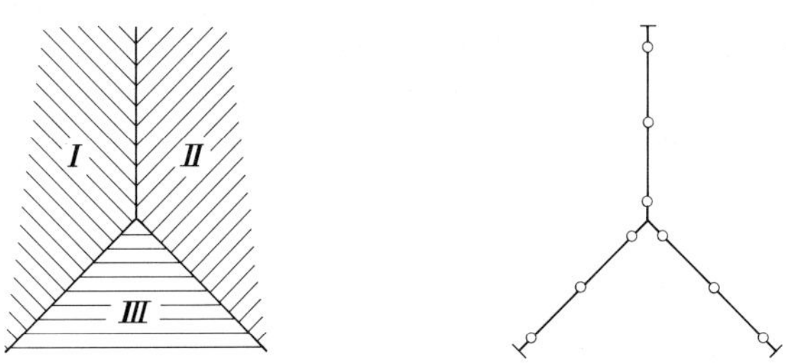

Fig. 5.16. Internal boundaries of subregions: **a** Three subregions (corner); **b** discontinuous elements

In this case the advantages of using discontinuous elements are more noticeable as the corner point problem has been completely removed. This treatment can be used for any other type of corners such as the one shown in Fig. 5.15 and generally gives accurate results. The technique is somewhat related to using two nodes at the corner, near to but not coincident, as proposed in Ref. [9]. This distance can of course be as small as one wishes – i.e., zero for the double node – the only restriction being the case for which a very small element is assumed to be spanning the two nodes, as shown in Fig. 5.17. This element, however, is not needed in practice as shown for double nodes, but some programs – such as those in Ref. [9] – still use this concept and in these cases care should be taken that the length ε does not become too small.

Example 5.1 [27]. As an example of the application of the above techniques, consider the problem of determining the stress concentration for a spherical cavity. Cruse used an equation similar to Eq. (5.53) as the starting point of his analysis. By discretizing the surface of the problem region (Fig. 5.18) he was able to represent the boundary using quadrilateral curved surface elements. (The triangular elements shown in Fig. 5.18 are just degenerate elements with essentially the same form). To represent the surface curvature Cruse used quadratic shape functions to map the elements and as he used quadratic interpolation functions on each element his formulation was of the isoparametric form (see Chapter 3). By exploiting the symmetry of the problem he was able to represent the surface using only six elements.

The results for the stress concentration depicted in Fig. 5.19 show good agreement with published ones; also included are the results obtained using the simpler linear boundary elements.

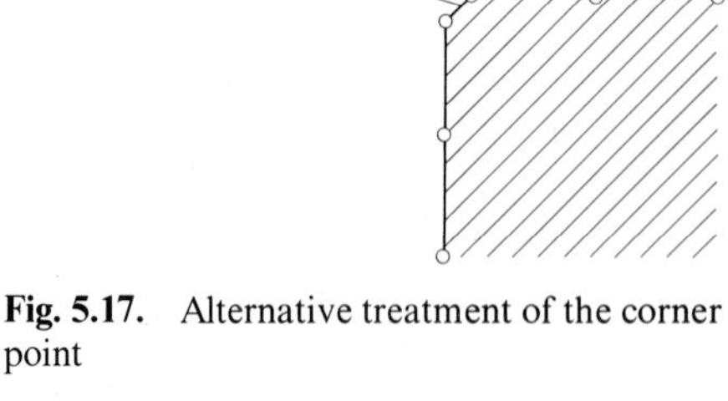

Fig. 5.17. Alternative treatment of the corner point

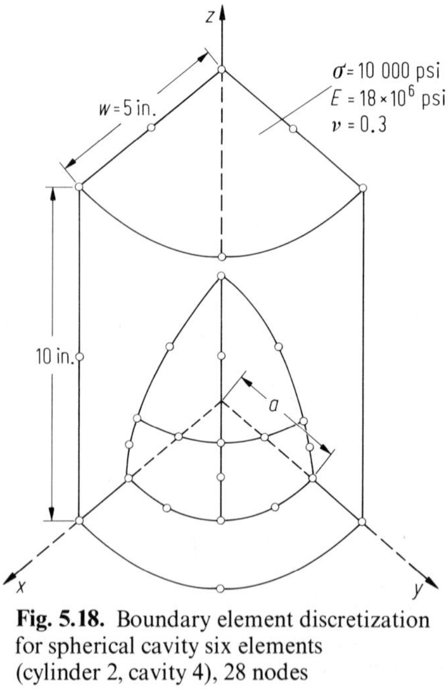

Fig. 5.18. Boundary element discretization for spherical cavity six elements (cylinder 2, cavity 4), 28 nodes

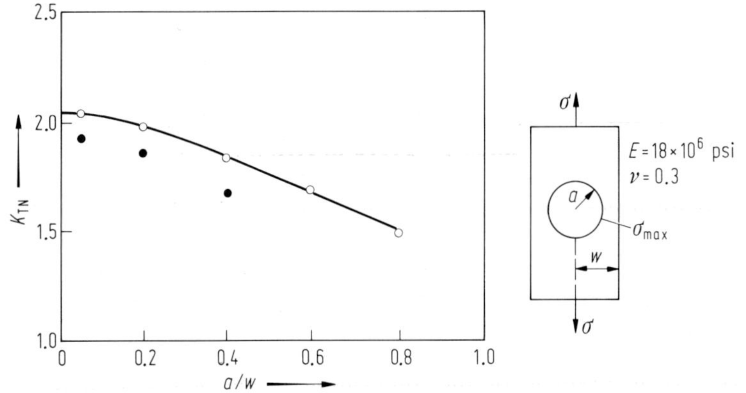

Fig. 5.19. Stress concentration for a spherical cavity. $K_{TN} = \sigma_{\max}/\bar{\sigma}$, where $\bar{\sigma} = \sigma/[1-(a/w)^2]$: the full curve is from Peterson and the data points are from boundary integral equations: ○, quadratic; ●, linear

Example 5.2 [27]. In the same paper referred to above, Cruse has studied the problem of crack propagation by looking at the stresses around a circular crack, again using quadratic boundary elements. Figure 5.20 shows the boundary element discretization used to model a quarter of a semicylinder with an interior or surface crack. Only one-quarter was considered because of the symmetry of the problem.

The numerical results for the circular crack are given in Fig. 5.21; for the buried crack, linear and quadratic variations are compared. The quadratic shape function representation more accurately reflects the displacements in the body. The

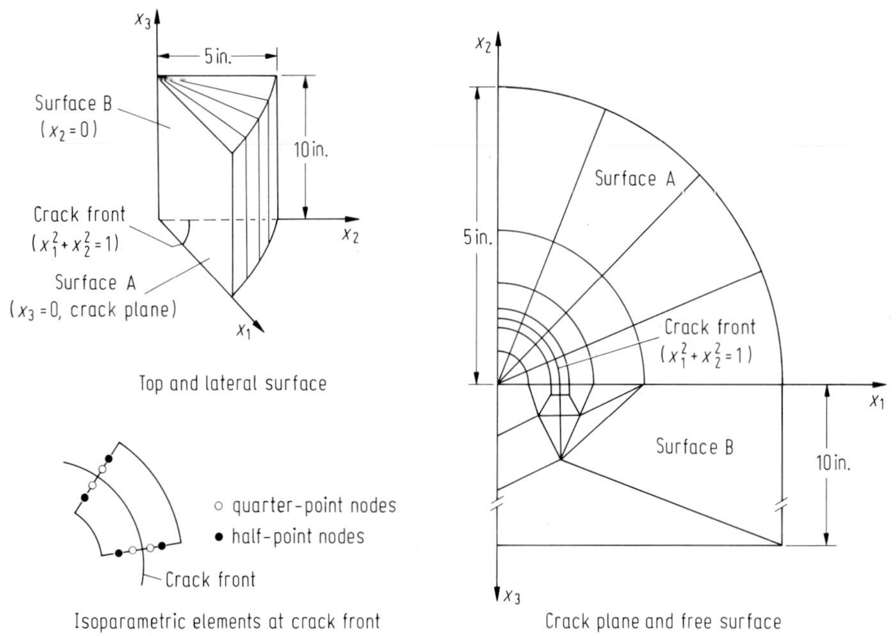

Fig. 5.20. Boundary element discretization for circular crack

quarter points referred to in the figure are midside nodes one-quarter of the element side length away from the crack for the two rows of elements adjacent to the crack tip.

Figure 5.22 shows the variation of the surface crack stress intensity factor using crack opening displacements as described in Cruse and Meyers [28]. The results are compared with those of Cruse and Meyers [28] using linear elements; it is clear that linear variation models are sufficiently accurate to determine the stress intensity.

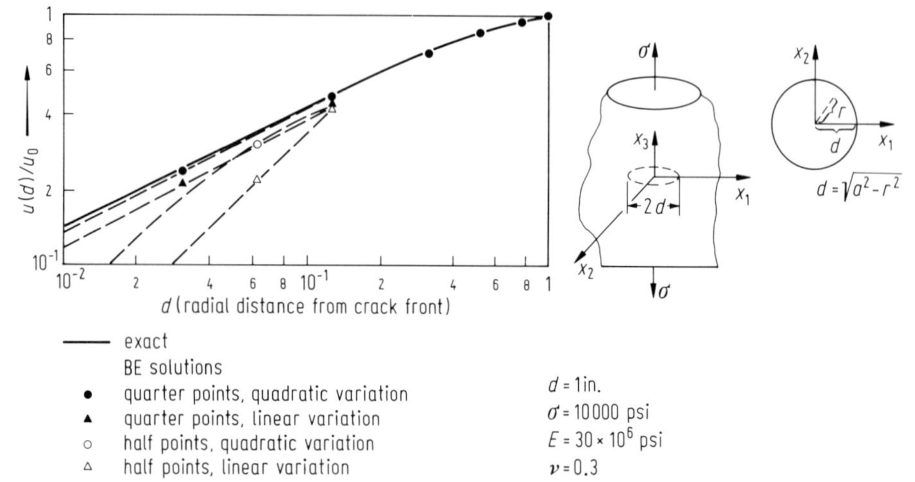

Fig. 5.21. Displacement variation for a buried circular crack

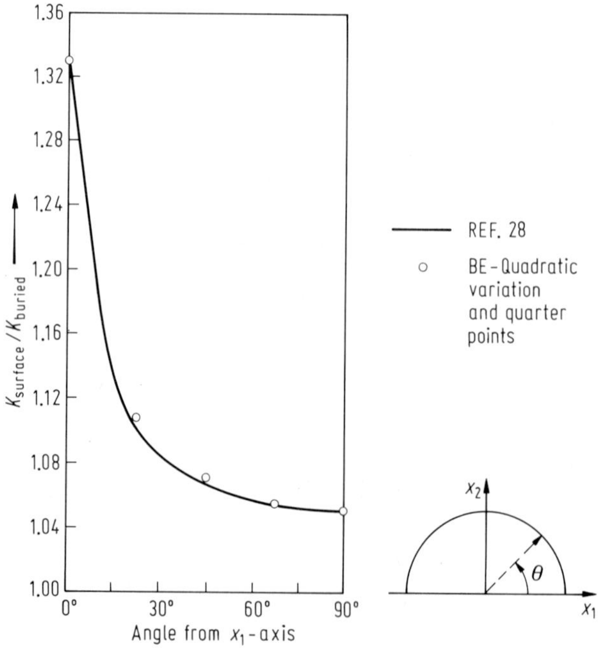

Fig. 5.22. Stress intensity factor variation for a semicircular crack

Example 5.3 [29]. At the intersection of two thick cylinders stress concentration problems may occur (Fig. 5.23). In Figs. 5.23 and 5.24 one particular type of joint is illustrated. Due to symmetry only one-eighth of the joint needs to be studied.

The problem was discretized using 76 constant triangular boundary elements as in Fig. 5.25 and stresses were calculated at five points at the intersection (Fig. 5.24). Results for normal stresses in the z direction are shown in Fig. 5.26. They indicate the distribution of stresses at these points.

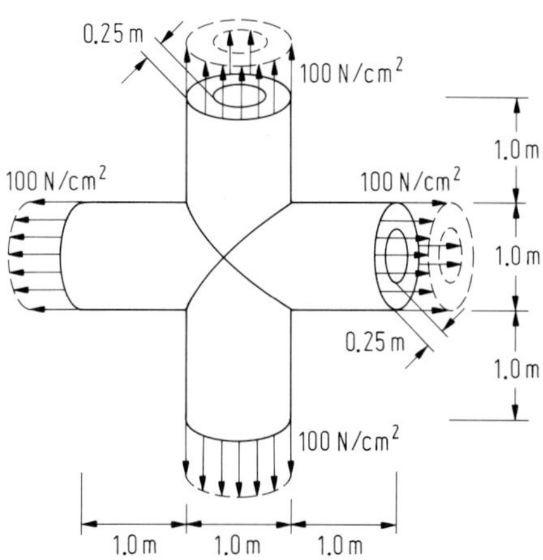

Fig. 5.23. Joint for two thick cylinders

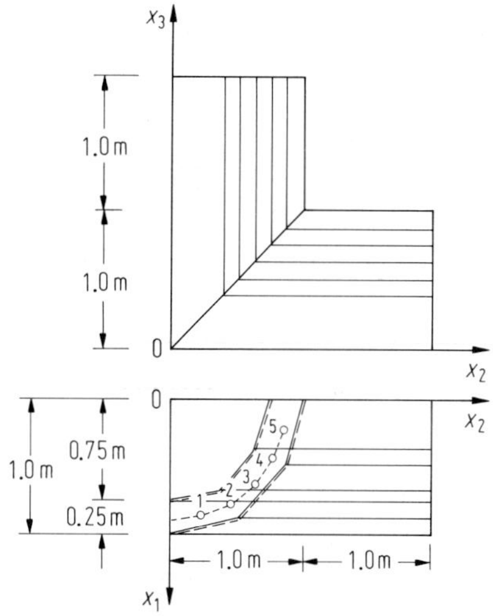

Fig. 5.24. Projection view of the joint

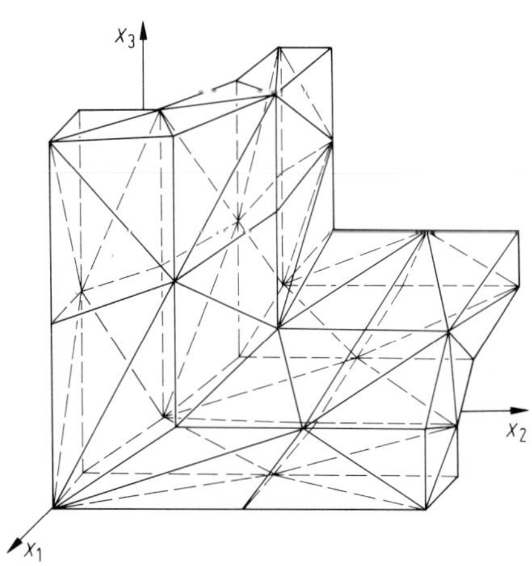

Fig. 5.25. Discretization of the joint

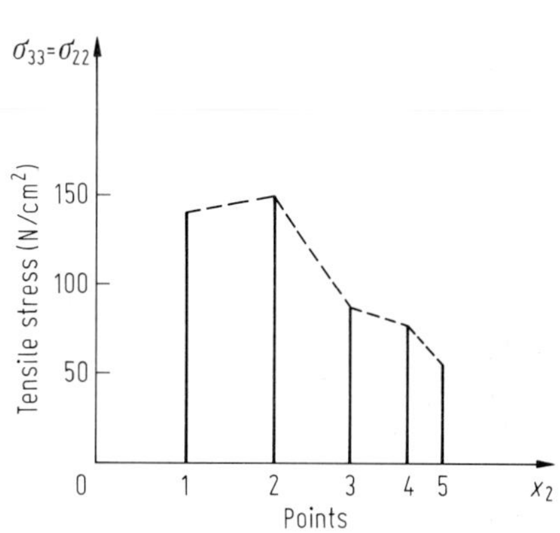

Fig. 5.26. Results for the joint at each point

5.13. Two-Dimensional Elasticity

Let us now particularize the elastostatics formulation for two-dimensional elasticity. The fundamental solution of the governing equation

$$\sigma^*_{jk,j} + \Delta(\xi, x)\, e_l = 0, \qquad j = 1, 2;\ k = 1, 2 \tag{5.115}$$

is as follows (plane strain) – Eqs. (5.56) and (5.57) –

$$u^*_{lk} = \frac{1}{8\pi G(1-\nu)}\left[(3-4\nu)\ln\left(\frac{1}{r}\right)\delta_{lk} + r_{,l}\, r_{,k}\right],$$

$$p^*_{lk} = \frac{-1}{4\pi(1-\nu)\, r}\left[\frac{\partial r}{\partial n}\left((1-2\nu)\,\delta_{lk} + 2\, r_{,k}\, r_{,l}\right) - (1-2\nu)\left(r_{,l}\, n_k - r_{,k}\, n_l\right)\right], \tag{5.116}$$

where p^*_{lk} and u^*_{lk} represent the tractions and displacements in the k direction due to a unit force in the l direction.

The elements of p^*_{lk} and u^*_{lk} can be written in matrix form as

$$\boldsymbol{p}^* = \begin{bmatrix} p^*_{11} & p^*_{12} \\ p^*_{21} & p^*_{22} \end{bmatrix}, \qquad \boldsymbol{u}^* = \begin{bmatrix} u^*_{11} & u^*_{12} \\ u^*_{21} & u^*_{22} \end{bmatrix}. \tag{5.117}$$

The displacements, tractions, and body force vectors are now

$$\boldsymbol{u} = \begin{Bmatrix} u_1 \\ u_2 \end{Bmatrix}, \qquad \boldsymbol{p} = \begin{Bmatrix} p_1 \\ p_2 \end{Bmatrix}, \qquad \boldsymbol{b} = \begin{Bmatrix} b_1 \\ b_2 \end{Bmatrix}. \tag{5.118}$$

The starting equation can be written in matrix form as for three-dimensional elasticity (Eq. (5.93)), i.e.,

$$\boldsymbol{c}\,\boldsymbol{u} + \int_\Gamma \boldsymbol{p}^*\boldsymbol{u}\, d\Gamma = \int_\Gamma \boldsymbol{u}^*\boldsymbol{p}\, d\Gamma + \int_\Omega \boldsymbol{u}^*\boldsymbol{b}\, d\Omega\,. \tag{5.119}$$

The rest of the matrix operations are similar to those described in Section 5.8 for the three-dimensional case, but here the boundary integrals are line integrals and the body force terms are obtained by integrating over area instead of volume of internal cells.

Once the displacements and tractions are known over all the surface, one can compute the internal displacements and stresses at any point using for the displacements

$$\boldsymbol{u} = \int_\Gamma \boldsymbol{u}^*\boldsymbol{p}\, d\Gamma - \int_\Gamma \boldsymbol{p}^*\boldsymbol{u}\, d\Gamma + \int_\Omega \boldsymbol{u}^*\boldsymbol{b}\, d\Omega\,,$$

or

$$u_i = \int_\Gamma u^*_{ij}\, p_j\, d\Gamma - \int_\Gamma p^*_{ij}\, u_j\, d\Gamma + \int_\Omega u^*_{ij}\, b_j\, d\Omega \tag{5.120}$$

and for the stresses (5.68)

$$\sigma_{ij}(\xi) = \int_\Gamma u^*_{ijk}\, p_k \, d\Gamma - \int_\Gamma p^*_{ijk}\, u_k \, d\Gamma + \int_\Omega u^*_{ijk}\, b_k \, d\Omega \,, \qquad (5.121)$$

where u^*_{ijk} and p^*_{ijk} are given by Eqs. (5.69) and (5.70).

As before, the integrals can be calculated using numerical integration formulas, or analytically for very simple cases.

Note that in order to pass from plane strain to two-dimensional plane stress analysis one has to replace the Poisson ratio by

$$\bar{\nu} = \nu/(1+\nu) \,. \qquad (5.122)$$

The above expressions are valid for the full space only; they can be extended to half-plane problems using the traction-free fundamental solution [14] indicated in Appendix B.

Among the different types of elements that can be employed in the numerical discretization of the integral equations [10], the linear element has been found to give acceptable accuracy without requiring much computer effort. Henceforth, the numerical implementation of this element will be discussed in more detail. The geometry of the element is represented by a straight line as indicated in Fig. 5.27.

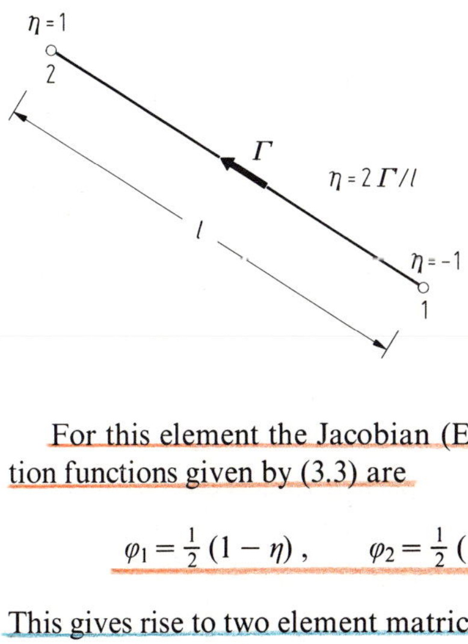

Fig. 5.27. Linear element and definition of intrinsic coordinate η

For this element the Jacobian (Eq. (3.24)) becomes $|G| = l/2$ and the interpolation functions given by (3.3) are

$$\varphi_1 = \tfrac{1}{2}(1-\eta) \,, \qquad \varphi_2 = \tfrac{1}{2}(1+\eta) \,. \qquad (5.123)$$

This gives rise to two element matrices $\boldsymbol{h}$ and $\boldsymbol{g}$ of order (2×4) such that

$$\begin{aligned} \boldsymbol{h} &= [\hat{\boldsymbol{h}}^1_{ij}\ \hat{\boldsymbol{h}}^2_{ij}] = \frac{l}{2} \int_{-1}^{1} [\varphi_1 \boldsymbol{p}^* \ \varphi_2 \boldsymbol{p}^*] \, d\eta \,, \\ \boldsymbol{g} &= [\boldsymbol{g}^1_{ij}\ \boldsymbol{g}^2_{ij}] = \frac{l}{2} \int_{-1}^{1} [\varphi_1 \boldsymbol{u}^* \ \varphi_2 \boldsymbol{u}^*] \, d\eta \,. \end{aligned} \qquad (5.124)$$

These integrals can generally be obtained numerically using Gauss integration formulas for all elements except those with a singular node. Four Gauss points have been found to produce acceptable results in most cases.

For the special case of the singular node being at the element, the difficult submatrix in $\boldsymbol{h}$ can be obtained by rigid-body-type considerations and the other part by using a numerical integration scheme. The $\boldsymbol{g}$ matrix presents integrable singularities which can be computed by using special numerical integration formulas (such as the logarithmic formula in Appendix A) or analytically for linear and constant elements.

Example 5.4. The problem of crack propagation in two dimensions has been examined by Lachat [30]. This situation is depicted in Fig. 5.28 and the discretizations used are shown in Figs. 5.29 and 5.30. Cubic boundary elements were used throughout and a traction was applied at node 12 in the direction indicated. The stress intensity factor was found for different specimens by evaluating Rice's integral along the contour J (see Fig. 5.29) and experimental results were compared with finite and boundary element solutions. The agreement between the three solutions was satisfactory (Table 5.1).

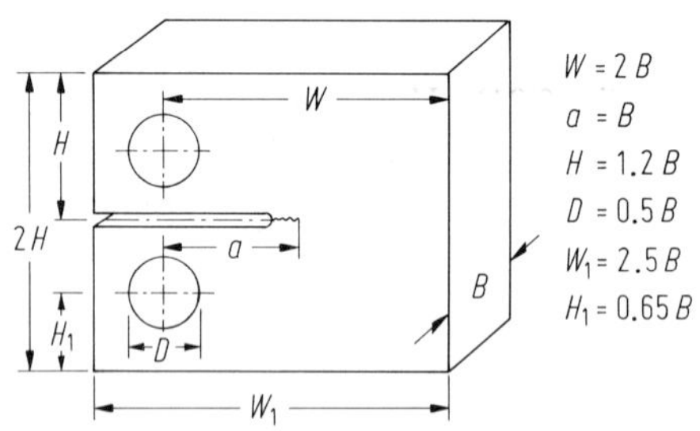

Fig. 5.28. Crack geometry

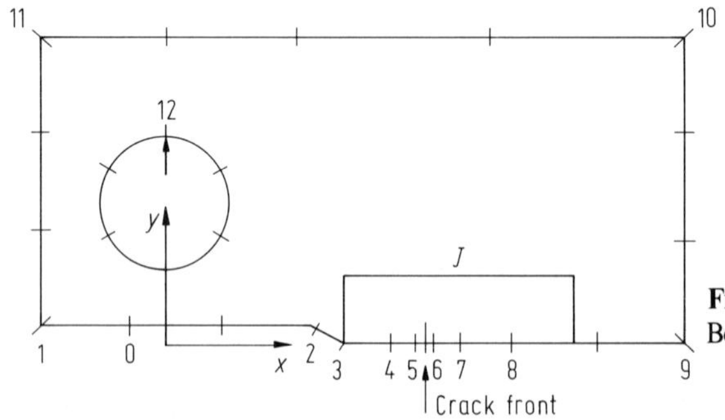

Fig. 5.29.
Boundary element discretization

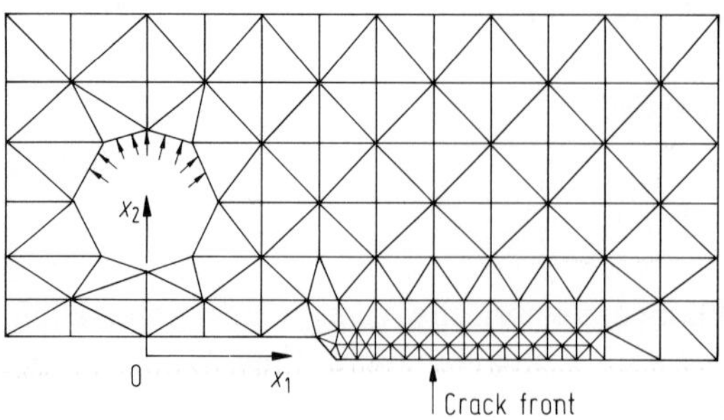

Fig. 5.30.
Finite element discretization

Table 5.1. Results for crack propagation problem.

Specimen number	Experimental	FEM	BEM
1	194	189	192
2	226	221	223
3	224	239	242
4	247	241	247
5	260	254	257

Cruse [27, 31] tackled a similar problem using a modified fundamental solution which contained the full elastic representation of a flat traction-free crack. This fundamental solution was obtained using the method of images and complex mapping. In this way the formulation automatically included the presence of the crack without having to model the crack surface geometry. This method, however, can only be applied in two-dimensional problems.

Example 5.5. Lachat [22] examined the case of a rubber sheet with embedded steel wires under plane longitudinal strain (Fig. 5.31). He used cubic elements and used two different boundary element discretizations: one with 9 segments and the other with 18 segments for comparison (Fig. 5.32). By the symmetry of the problem only the area ABCDE of Fig. 5.31 was considered.

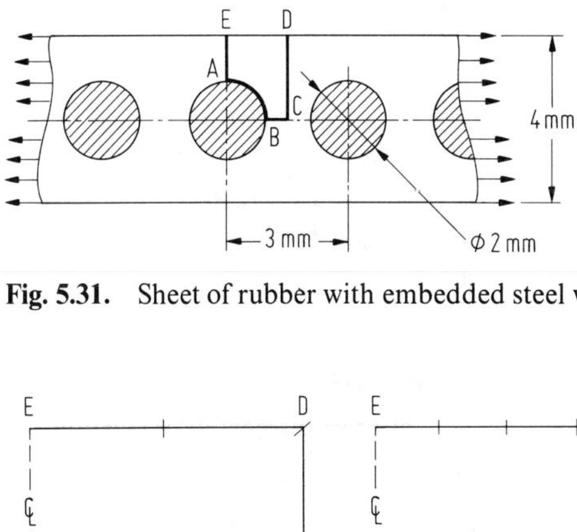

Fig. 5.31. Sheet of rubber with embedded steel wires

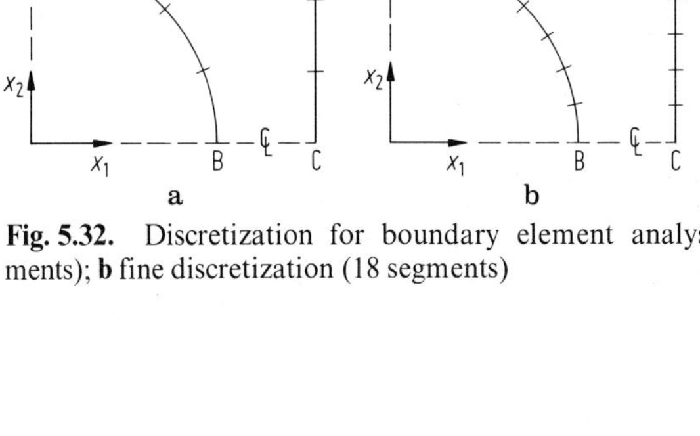

Fig. 5.32. Discretization for boundary element analysis: **a** Coarse discretization (9 segments); **b** fine discretization (18 segments)

For the purposes of the analysis the wires were taken to be perfectly rigid. The overall imposed strain was 1.667×10^{-2} in the x_1 direction. The modulus of elasticity was taken to be 20 N mm^{-2} and the two values of Poisson's ratio $\nu = 0.45$ and $\nu = 0.5$ were considered.

For $\nu = 0.45$ the stresses σ_{11} calculated using the two meshes were virtually indistinguishable (Fig. 5.33a). The circles represent the results for the coarse mesh and the solid line shows the values obtained using the finer 18-element mesh.

For $\nu = 0.5$ (Fig. 5.33b) this difference was greater, up to 5% of the maximum stress. However, there is very little difference between the displacements. (*Note:* The finite element displacement method is incapable of treating the case $\nu = 0.5$.)

Example 5.6 [14]. This example is a half-plane solution for an indentation produced by a rigid flat punch. The punch is considered to be perfectly smooth and indentation was carried out by prescribing the flat punch displacements. Boundary element analysis was performed by discretizing half of the contact surface into twelve unequally sized linear boundary elements. The thirteen boundary nodes were located along the discretized boundary according to the formula

$$y = \frac{6.5(1-\alpha)}{\alpha},$$

where α is the node number.

Additional results were obtained at four internal points (see Fig. 5.34a). Note that symmetry was considered by a direct condensation process which automatically integrates over reflected elements not requiring any boundary discretization of the symmetry axis.

Contact stresses along the discretized boundary are compared with analytical results [32] in Fig. 5.34b. Apart from some traction perturbation over the tiny element connected to node 13, the singularity at the edge of the punch does not seem to have disturbed the results. This is also confirmed by the accuracy of the computations at internal points presented in Table 5.2.

Table 5.2. Smooth punch problem. Results at internal points.

Point	$u - \bar{u}$	v	σ_x	σ_y	σ_{xy}	
14	− 0.724	0.	− 133.92	− 89.28	0.	BEM
15	− 0.826	− 1.210	− 330.64	− 113.16	107.93	BEM
16	− 1.824	− 2.038	0.	0.	0.	BEM
17	− 0.908	− 0.234	− 151.87	− 77.69	− 3.14	BEM
14	− 0.735	0.	− 133.91	− 89.27	0.	Exact
15	− 0.842	− 1.208	− 326.47	− 114.43	110.50	Exact
16	− 1.841	− 2.038	0.	0.	0.	Exact
17	− 0.917	− 0.234	− 152.00	− 77.67	− 3.35	Exact
	$m \times 10^2$	$m \times 10^2$	kN/m^2	kN/m^2	kN/m^2	

Note: $u - \bar{u}$ = obtained displacements minus the prescribed displacements for the punch.

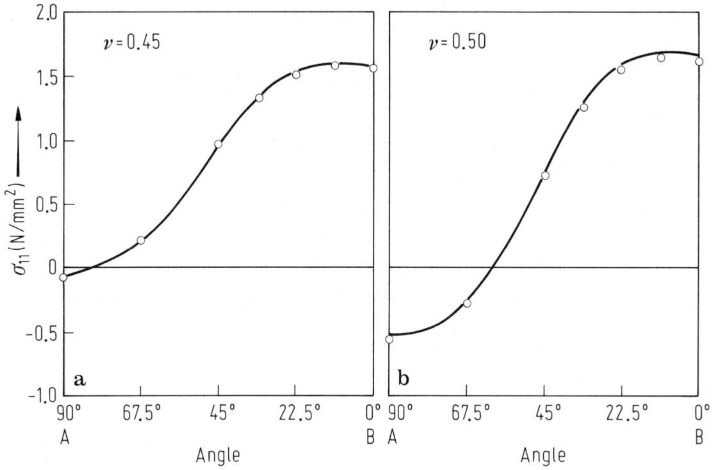

Fig. 5.33. Variation of σ_{11} along the curve AB; (**a**) $\nu = 0.45$, (**b**) $\nu = 0.5$

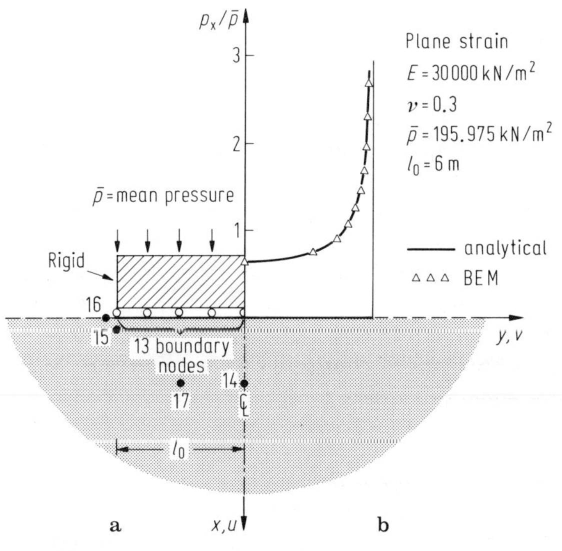

Fig. 5.34. Smooth punch problem: **a** Problem geometry; **b** contact stresses along the discretized boundary

Example 5.7 [14]. In this application a semi-infinite plate with a circular hole near the straight boundary is studied. The problem is here considered under two different loading cases, unit normal pressure applied over the surface of the hole and simple tension $\bar{\sigma}_y$ parallel to the straight edge. In both cases, the stress σ_y along the traction-free straight boundary is compared with analytical results presented by Jeffery [33] and Mindlin [34].

The relatively small distance between the center of the hole and the straight surface is 1.34 times the radius of the circle. For the boundary element analysis only the surface of the hole needs discretization and due to symmetry only half of this surface was considered.

Results for the first loading case (see Fig. 5.35) were computed at a series of points (considered as internal points) located along the straight boundary and 24 boundary elements of equal size were used to represent half the circle with the same area. The second loading case was analyzed by simple superposition; tractions p_y equal the scalar product of the simple tension and the unit normal to the surface of the hole were applied to the circular boundary and the corresponding results superimposed onto the constant stress field $\bar{\sigma}_y$.

To illustrate the convergence of the method the results for 6, 12, and 24 boundary element discretizations of the half-circle are compared with analytical results in Fig. 5.36.

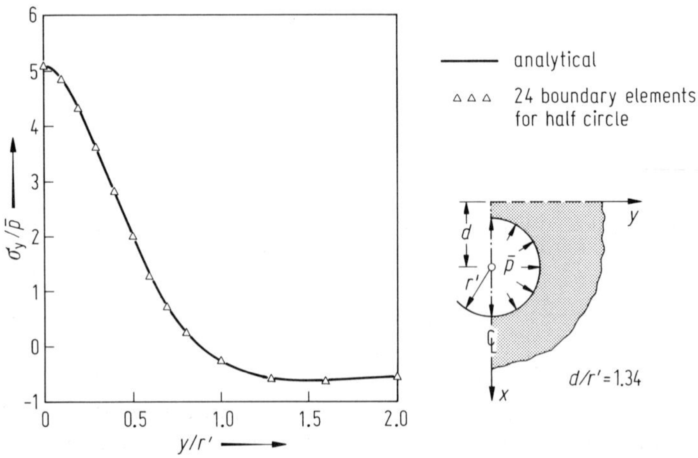

Fig. 5.35. Circular hole near straight boundary under uniform pressure. Stress σ_y along the traction-free straight boundary

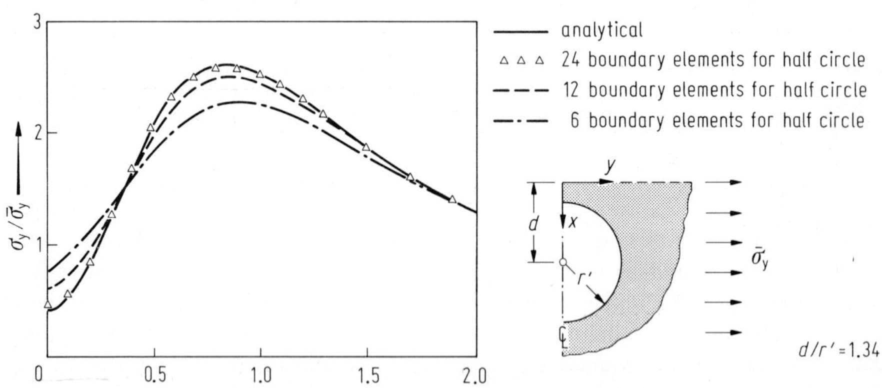

Fig. 5.36. Circular hole near the straight boundary under remote tension. Stress σ_y along the traction-free straight boundary

It is worth mentioning that if the distance between the hole and the straight edge were larger, fewer boundary elements would be required for the same accuracy of results. With reference to the first loading case, for $d/r' = 1.81$, 12 boundary elements produced an error at the peak stress of about -2.7%, whereas for $d/r' = 1.34$ this error was -6.5%.

The last two examples shown here illustrate some of the possible applications of the half-plane formulation. It is evident that such problems are more efficiently solved by this procedure than using the Kelvin fundamental solution which requires defining a series of elements on the traction-free surface. The number of these elements, in principle, needs to extend to infinity or at least should be large enough to produce accurate solutions. Special elements extending to infinity have been proposed [35] to reduce such large discretization of the free surface, but they require further tests to validate their application. The most accurate and computationally more efficient technique is to use the half-plane fundamental solution which eliminates the need for any numerical approximation over the free surface.

5.14. Body Forces

Since in many practical applications the problem to be studied presents nonzero body forces, attention will now be given to some alternative procedures for computing its influence into the analysis. It is immediately apparent that if body forces are considered in Eq. (5.87) of Section 5.7, domain integrals have to be computed (see Eqs. (5.77) and (5.79)). Unfortunately, this requires the domain of the problem to be divided into internal cells for integration. Though this is true for the general case, in many particular applications the domain integral can be suitably transformed into a surface integral [18, 36–38] which may be numerically evaluated at the same time as the previously defined boundary integrals (see Section 5.7). This procedure is applicable to some commonly encountered body forces such as constant gravitational load (i.e., self-weight), centrifugal load due to rotation about a fixed axis, and the effect of a steady-state thermal loading. In what follows this three particular cases are dealt with by employing the unified procedure presented by Danson [37]; i.e., the Galerkin vector corresponding to the Kelvin fundamental solution (three-dimensional and two-dimensional) is employed throughout. It should be noticed that for half-space-type fundamental solutions the procedure applies equally well.

Let us call G^*_{ij} the Galerkin tensor which is related to the fundamental solution u^*_{ij} (formulas (5.55) and (5.56) of Section 5.3) by the following expression [3, 4, 19]:

$$u^*_{ij} = G^*_{ij,kk} - \frac{1}{2(1-\nu)} G^*_{ik,kj}, \tag{5.125}$$

where in the same way as u^*_{ij} may be regarded as three (three-dimensional) or two (two-dimensional) displacement vectors each corresponding to the direction i of the unit load, G^*_{ij} may be regarded as three or two Galerkin vectors each corresponding to the direction i in which the unit load is applied.

$$u^*_{ij}(\xi, x) = \begin{cases} \dfrac{(3-4\nu)\delta_{ij} + r_{,i} r_{,j}}{16\pi(1-\nu)Gr} & \text{for 3D} \quad (5.55) \\ -\dfrac{(3-4\nu)\ln(r)\delta_{ij} - r_{,i} r_{,j}}{8\pi(1-\nu)G} & \text{for 2D} \quad (5.56) \end{cases}$$

In order to simplify this presentation, the body force integral of Eq. (5.77) will be represented as

$$B_i = \int_\Omega u_{ij}^* \, b_j \, d\Omega \, . \tag{5.126}$$

In addition, the effect of thermal strains can be introduced into the analysis by simply expressing σ_{jk} in expression (5.41) in the appropriate form, i.e. (see Eq. (5.34)),

$$\sigma_{jk} = C_{jkil} \, \varepsilon_{il} - 2G\left(\frac{1+v}{1-2v}\right) \alpha \, T \, \delta_{jk} \tag{5.127}$$

or

$$\sigma_{jk} = \sigma_{jk}^e - 2G\left(\frac{1+v}{1-2v}\right) \alpha \, T \, \delta_{jk} \, , \tag{5.128}$$

where α is the coefficient of linear thermal expansion and T is the difference in temperature.

The substitution of expression (5.128) in Eq. (5.41) generates two integrals of the form

$$-\int_\Omega \sigma_{jk} \, \varepsilon_{jk}^* \, d\Omega = -\int_\Omega \sigma_{jk}^e \, \varepsilon_{jk}^* \, d\Omega + 2G\left(\frac{1+v}{1-2v}\right) \alpha \int_\Omega T \, \delta_{ij} \, \varepsilon_{ij}^* \, d\Omega \, , \tag{5.129}$$

where the first integral on the right-hand side plays the role of the first integral on the left of expressions (5.43) and (5.44), and the last integral in (5.129) remains unchanged throughout. Therefore, after the consideration of each point load as acting independently, the following integral has to be added to the right-hand side of Eqs. (5.53) and (5.77):

$$\bar{B}_i = 2G\left(\frac{1+v}{1-2v}\right) \alpha \int_\Omega \varepsilon_{jki}^* \, T \, \delta_{jk} \, d\Omega \tag{5.130}$$

which can be further written

$$\bar{B}_i = 2G\left(\frac{1+v}{1-2v}\right) \alpha \int_\Omega u_{ik,k}^* \, T \, d\Omega \, . \tag{5.131}$$

Expression (5.131) is valid for three-dimensional and two-dimensional plane strain problems. For the plane stress case v is replaced by $\bar{v}$ as before and α is replaced by

$$\bar{\alpha} = \frac{\alpha}{1+\bar{v}} = \alpha\left(\frac{1+v}{1+2v}\right). \tag{5.132}$$

In order to transform the domain integrals of expressions (5.126) (self-weight and centrifugal load) and (5.131) into boundary integrals, the Galerkin tensor will be employed. Henceforth, for the infinite space one has

$$G_{ij}^* = \frac{1}{8\pi G} r\, \delta_{ij} \tag{5.133}$$

for three-dimensional problems and

$$G_{ij}^* = \frac{1}{8\pi G} r^2 \ln\left(\frac{1}{r}\right) \delta_{ij} \tag{5.134}$$

for two-dimensional plane strain.

With reference to the two-dimensional case, an important remark is now due. While for three-dimensional problems the substitution of expression (5.133) into (5.125) generates exactly the fundamental displacements presented in expression (5.55), this is not the same for two-dimensional problems. In this case the following expression is obtained:

$$u_{ij}^* = \frac{-1}{8\pi(1-\nu)\,G}\left\{(3-4\nu)\ln(r)\,\delta_{ij} - r_{,i}\,r_{,j} + \left(\frac{7-8\nu}{2}\right)\delta_{ij}\right\}, \tag{5.135}$$

where one can notice that the above expression differs from the fundamental displacements presented in expression (5.56) by a constant. This difference, however, is not of any theoretical importance since it simply corresponds to a rigid-body translation. Nevertheless, one has to be consistent in the choice of fundamental solution, so that when dealing with the body force problems discussed here, expression (5.135) must be used to substitute (5.56).

In what follows, B_i and $\bar{B}_i$ will be dealt with separately for each particular case. Throughout the manipulations the divergence theorem (or Gauss' theorem) will be employed in its various forms and for further details in the different passages the reader should refer to Ref. [37].

The substitution of expression (5.125) into Eq. (5.126) yields

$$B_i = \int_{\Omega}\left(G_{ij,kk}^* - \frac{1}{2(1-\nu)} G_{ik,kj}^*\right) b_j\, d\Omega\,, \tag{5.136}$$

where, as it will be seen, the transformation to a boundary integral can be easily accomplished.

5.14.1. Gravitational Loads

A body of constant mass density ϱ, in a constant gravitational field g_j, experiences a constant body force given by

$$b_j = \varrho\, g_j\,. \tag{5.137}$$

and applying the Divergence Theorem

The above expression when substituted into Eq. (5.136) leads to the following boundary integral:

$$B_i = b_j \int_\Gamma \left\{ G^*_{ij,k} - \frac{1}{2(1-\nu)} G^*_{ik,j} \right\} n_k \, d\Gamma \tag{5.138}$$

which can be concisely written

$$B_i = \int_\Gamma \hat{P}_i \, d\Gamma \, , \tag{5.139}$$

where

$$\hat{P}_i = \frac{1}{8\pi G} \left\{ b_i \, n_k \, r_{,k} - \frac{1}{2(1-\nu)} b_k \, r_{,k} \, n_i \right\} \quad \text{for three dimensions} \tag{5.140}$$

and

$$\hat{P}_i = \frac{r}{8\pi G} \left\{ \left[2 \ln\left(\frac{1}{r}\right) - 1 \right] \left(b_i \, n_k \, r_{,k} - \frac{1}{2(1-\nu)} b_k \, r_{,k} \, n_i \right) \right\} \tag{5.141}$$

for two-dimensional plane strain.

In addition, the equation for stresses at internal points becomes

$$\sigma_{ij} = \int_\Gamma u^*_{ijk} \, p_k \, d\Gamma - \int_\Gamma p^*_{ijk} \, u_k \, d\Gamma + \int_\Gamma \hat{S}_{ij} \, d\Gamma \tag{5.142}$$

in which

$$\hat{S}_{ij} = \frac{1}{8\pi r} \Big[n_m \, r_{,m} (b_i \, r_{,j} + b_j \, r_{,i}) + \frac{1}{1-\nu} \{ \nu \, \delta_{ij} (n_m \, r_{,m} \, b_s \, r_{,s} - b_m \, n_m) - \tfrac{1}{2} (b_m \, r_{,m} [n_i \, r_{,j} + n_j \, r_{,i}] + [1 - 2\nu] [b_i \, n_j + b_j \, n_i]) \} \Big] \tag{5.143}$$

for three dimensions and

$$\hat{S}_{ij} = \frac{1}{8\pi} \Bigg[2 n_m \, r_{,m} (b_i \, r_{,j} + b_j \, r_{,i}) + \frac{1}{1-\nu} \left\{ \nu \, \delta_{ij} \left(2 n_m \, r_{,m} \, b_s \, r_{,s} + \left[1 - 2 \ln \frac{1}{r} \right] b_m \, n_m \right) - b_m \, r_{,m} (n_i \, r_j + n_j \, r_i) + \frac{1-2\nu}{2} \left(1 - 2 \ln \frac{1}{r} \right) (b_i \, n_j + b_j \, n_i) \right\} \Bigg] . \tag{5.144}$$

for two-dimensional plane strain.

5.14.2. Centrifugal Load

Consider a body rotating with angular velocity ω_i. If the axis of rotation passes through the origin of the coordinate system, the problem is equivalent to a prescribed body force of the form

$$b_j = -\varrho\, e_{jik}\, \omega_i\, e_{kpq}\, \omega_p\, x_q\,, \tag{5.145}$$

where e_{ijk} is the permutation symbol.

Expression (5.145) can be concisely written as

$$b_j = g_{ij}\, x_i \tag{5.146}$$

in which g_{ij} can be represented in matrix form as

$$\boldsymbol{g} = \varrho \begin{bmatrix} \omega_2^2 + \omega_3^2 & -\omega_1\,\omega_2 & -\omega_3\,\omega_1 \\ & \omega_3^2 + \omega_1^2 & -\omega_2\,\omega_3 \\ \text{Sym.} & & \omega_1^2 + \omega_2^2 \end{bmatrix}. \tag{5.147}$$

or

$$g_{ij} = \varrho(\delta_{ij}\,\omega_m\,\omega_m - \omega_i\,\omega_j)$$

Substituting expression (5.146) into Eq. (5.136) given the following expression.

$$B_i = g_{jk} \int_\Omega x_j \left\{ G^*_{ik,mm} - \frac{1}{2(1-\nu)}\, G^*_{im,mk} \right\} d\Omega\,. \tag{5.148}$$

The above equation is equivalent to

$$B_i = g_{jk} \int_\Omega \left\{ \frac{\partial}{\partial x_m}(x_j\, G^*_{ik,m}) - G^*_{ik,j} + \frac{1}{2(1-\nu)} \left[G^*_{ij,k} - \frac{\partial}{\partial x_m}(x_j\, G^*_{im,k}) \right] \right\} d\Omega \tag{5.149}$$

which can be transformed into a boundary integral of the form

$$B_i = g_{jk} \int_\Gamma \left\{ x_j \left[G^*_{ik,m} - \frac{1}{2(1-\nu)}\, G^*_{im,k} \right] n_m - \frac{(1-2\nu)}{2(1-\nu)}\, G^*_{ik}\, n_j \right\} d\Gamma \tag{5.150}$$

or, with reference to Eq. (5.139), one has

$$\hat{P}_i = \frac{1}{8\pi G} \left\{ g_{ij}\, x_j\, r_{,m}\, n_m - \frac{1}{2(1-\nu)} \left[r_{,k}\, g_{jk}\, x_j\, n_i + (1-2\nu)\, r\, g_{ij}\, n_j \right] \right\}. \tag{5.151}$$ for 3-D

Similarly, with reference to Eq. (5.142)

$$\begin{aligned} \hat{S}_{ij} = \frac{1}{8\pi} \Bigg[& \left\{ n_s\, r_{,s}\, \frac{x_m}{r} + \frac{1-2\nu}{2(1-\nu)}\, n_m \right\} (g_{im}\, r_{,j} + g_{jm}\, r_{,i}) \\ & + \frac{1}{1-\nu} \Bigg\{ \nu\, \delta_{ij} (r_{,k}\, g_{km}\, \frac{x_m}{r}\, n_s\, r_{,s} - n_s\, g_{sm}\, \frac{x_m}{r} + r_{,s}\, g_{sm}\, n_m) \\ & - \frac{x_m}{2r}\, (r_{,s}\, g_{sm}\, [n_i\, r_{,j} + n_j\, r_{,i}] + [1-2\nu]\, [n_i\, g_{jm} + n_j\, g_{im}]) \Bigg\} \Bigg] \end{aligned} \tag{5.152}$$ for 3-D

Expressions (5.151) and (5.152) are valid for three-dimensional problems. For two dimensions in order to ensure that the problem remains plane, the axis of rotation must be either

(a) in the plane of the problem,
(b) at right angles to the plane of the problem.

Case (a) implies $\omega_3 = 0$, whence from Eq. (5.147) one has

$$\boldsymbol{g} = \varrho \begin{bmatrix} \omega_2^2 & -\omega_1\,\omega_2 \\ -\omega_1\,\omega_2 & \omega_1^2 \end{bmatrix}. \tag{5.153}$$

Case (b) implies $\omega_1 = \omega_2 = 0$. Thus,

$$g_{ij} = \varrho\,\omega_3^2\,\delta_{ij}\,. \tag{5.154}$$

For both cases one has

$$\hat{P}_i = \frac{r}{8\pi G}\left[\left(2\ln\frac{1}{r} - 1\right)\left\{n_m\,r_{,m}\,x_s\,g_{si} - \frac{r_{,m}\,g_{ms}\,x_s\,n_i}{2(1-\nu)}\right\} - \frac{1-2\nu}{2(1-\nu)}\,n_m\,g_{mi}\,r\ln\frac{1}{r}\right] \tag{5.155}$$

and

$$\hat{S}_{ij} = \frac{1}{8\pi}\left[2\,n_m\,r_{,m}(x_s\,g_{si}\,r_{,j} + x_s\,g_{sj}\,r_{,i}) + \frac{1}{1-\nu} \cdot \left\{\nu\,\delta_{ij}\left(2\,n_m\,r_{,m}\,r_{,s}\,g_{sq}\,x_q + \left[1 - 2\ln\frac{1}{r}\right][x_m\,g_{ms}\,n_s - n_m\,g_{ms}\,r_{,s}\,r]\right) + \frac{1-2\nu}{2}\left(1 - 2\ln\frac{1}{r}\right)(x_m\,g_{mi}\,n_j + x_m\,g_{mj}\,n_i - [n_m\,g_{mi}\,r_{,j} + n_m\,g_{mj}\,r_{,i}]\,r) - r_m\,g_{ms}\,x_s(n_i\,r_{,j} + n_j\,r_{,i})\right\}\right]. \tag{5.156}$$

5.14.3. Thermal Loading

The substitution of Eq. (5.125) into expression (5.131) yields

$$\bar{B}_i = G\left(\frac{1+\nu}{1-\nu}\right)\alpha\int_{\Omega} G^*_{ik,kjj}\,T\,d\Omega\,. \tag{5.157}$$

Taking into consideration that for steady-state heat conduction $T_{,jj} = 0$ one can write

$$\bar{B}_i = G\left(\frac{1+\nu}{1-\nu}\right)\alpha\int_{\Omega}\{G^*_{ik,kjj}\,T - G^*_{ik,k}\,T_{,jj}\}\,d\Omega \tag{5.158}$$

which is equivalent to

$$\bar{B}_i = G\left(\frac{1+\nu}{1-\nu}\right)\alpha \int\limits_{\Omega} \frac{\partial}{\partial x_j}\{G^*_{ik,kj}\,T - G^*_{ik,k}\,T_{,j}\}\,d\Omega\,. \tag{5.159}$$

Expression (5.159) can now be transformed into a boundary integral of the form

$$\bar{B}_i = G\left(\frac{1+\nu}{1-\nu}\right)\alpha \int\limits_{\Gamma} \{G^*_{ik,kj}\,T - G^*_{ik,k}\,T_{,j}\}\,n_j\,d\Gamma\,. \tag{5.160}$$

The above equation can be more conveniently written as

$$\bar{B}_i = \int\limits_{\Gamma} \bar{P}_i\,T\,d\Gamma - \int\limits_{\Gamma} \bar{Q}_i\,T_{,k}\,n_k\,d\Gamma\,, \tag{5.161}$$

where

$$\bar{P}_i = G\left(\frac{1+\nu}{1-\nu}\right)\alpha\,G^*_{ik,kj}\,n_j \tag{5.162}$$

and

$$\bar{Q}_i = G\left(\frac{1+\nu}{1-\nu}\right)\alpha\,G^*_{ik,k}\,. \tag{5.163}$$

For three-dimensional problems expressions (5.162) and (5.163) are given by

$$\bar{P}_i = \frac{(1+\nu)\,\alpha}{8\pi(1-\nu)\,r}\{n_i - r_{,i}\,r_{,k}\,n_k\} \tag{5.164}$$

and

$$\bar{Q}_i = \frac{(1+\nu)\,\alpha}{8\pi(1-\nu)}\,r_{,i}\,, \tag{5.165}$$

whereas for two-dimensional plane strain

$$\bar{P}_i = \frac{(1+\nu)\,\alpha}{4\pi(1-\nu)}\left\{\left[\ln\left(\frac{1}{r}\right) - \frac{1}{2}\right]n_i - r_{,i}\,r_{,k}\,n_k\right\} \tag{5.166}$$

and

$$\bar{Q}_i = \frac{(1+\nu)}{4\pi(1-\nu)}\,\alpha\,r\,r_{,i}\left[\ln\left(\frac{1}{r}\right) - \frac{1}{2}\right]. \tag{5.167}$$

In addition, the above expressions can be differentiated to obtain the stresses at internal points. In this case, Eq. (5.127) is used to substitute Hooke's law and the final expression is equal to Eq. (5.142) plus a term, $\bar{B}'_{ij}$, defined as

$$\bar{B}'_{ij} = \int\limits_{\Gamma} \bar{S}_{ij}\,T\,d\Gamma - \int\limits_{\Gamma} \bar{V}_{ij}\,T_{,k}\,n_k\,d\Gamma - 2G\left(\frac{1+\nu}{1-2\nu}\right)\alpha\,T\,\delta_{ij}\,, \tag{5.168}$$

where for three-dimensional problems

$$\bar{S}_{ij} = \frac{\alpha\, G(1+\nu)}{4\pi(1-\nu)\, r^2}\left\{n_m\, r_{,m}\left(\frac{\delta_{ij}}{1-2\nu} - 3\, r_{,i}\, r_{,j}\right) + n_i\, r_{,j} + r_{,i}\, n_j\right\},$$

$$\bar{V}_{ij} = \frac{\alpha\, G(1+\nu)}{4\pi(1-\nu)\, r}\left(r_{,i}\, r_{,j} - \frac{\delta_{ij}}{1-2\nu}\right) \qquad (5.169)$$

and for two-dimensional plane strain

$$\bar{S}_{ij} = \frac{\alpha\, G(1+\nu)}{2\pi(1-\nu)\, r}\left\{n_m\, r_{,m}\left(\frac{\delta_{ij}}{1-2\nu} - 2\, r_{,i}\, r_{,j}\right) + n_i\, r_{,j} + n_j\, r_{,i}\right\},$$

$$\bar{V}_{ij} = \frac{\alpha\, G(1+\nu)}{2\pi(1-\nu)}\left\{r_{,i}\, r_{,j} + \frac{\delta_{ij}}{1-2\nu}\left(\frac{1+2\nu}{2} - \ln\frac{1}{r}\right)\right\}. \qquad (5.170)$$

Note that all the plane strain expressions presented in this section are valid for plane stress if ν is replaced by $\bar{\nu}$ and α is replaced by $\bar{\alpha}$.

Finally, with reference to body force loads which are not included into the classes discussed in this section, one can always employ internal cells for integration. The geometry and interpolation functions associated with these cells have been presented in Chapter 3 and numerical integration formulas can be satisfactorily adopted. Here, the use of a semianalytical scheme which employs a polar coordinate system (r, φ, θ) for integration appears advantageous. In this procedure, due to the very nature of the Kelvin fundamental solution, integration with respect to r turns to be very simple and removes the singularities. Integration with respect to the angles φ and θ can then be performed numerically, but since there are no singularities involved, a reduced number of integration points can be used. An outline of this procedure is presented in Chapter 6 where transient thermal loading is also included.

5.15. Axisymmetric Problems

In many industrial applications the engineers are faced with the stress analysis of three-dimensional bodies in which the geometry and loading involved are axisymmetric. Problems such as pressure vessels, certain pipes, rotating discs, and many different types of containers are included into this category and plainly justify the use of more accurate and efficient solution procedures, which take advantage of the axial symmetry of the problem.

Starting with the three-dimensional fundamental solution, one can always make use of a cylindrical system (r, φ, z) to express the fundamental displacements; i.e., by a suitable coordinate transformation, expression (5.55) can be written in terms of unit point loads parallel to the base vectors (e_r, e_φ, e_z) at ξ and provide the displacements with reference to the cylindrical coordinate system at point x. This can be obtained as follows (see Fig. 5.37)

$$\hat{\boldsymbol{u}}^*(\xi, x) = \boldsymbol{T}^T(\xi)\, \boldsymbol{P}(\xi)\, \bar{\boldsymbol{u}}^*(\xi, x)\, \boldsymbol{T}(x), \qquad (5.171)$$

$$u^*_{ij}(\xi, x) = \frac{(3-4\nu)\delta_{ij} + r_{,i}\, r_{,j}}{16\pi(1-\nu)Gr} \quad (5.55)$$

where $\bar{\boldsymbol{u}}^*$ stands for expression (5.55) with $x_1 = r\cos\varphi$, $x_2 = r\sin\varphi$ and $x_3 = z$, $\boldsymbol{T}$ is the matrix that provides the coordinate transformation, and $\boldsymbol{P}$ is used to render $\boldsymbol{u}^*$ as corresponding to unit loads in the new directions. The above matrices can be written as

$$\boldsymbol{T} = \begin{bmatrix} \cos\varphi & -\sin\varphi & 0 \\ \sin\varphi & \cos\varphi & 0 \\ 0 & 0 & 1 \end{bmatrix}, \tag{5.172}$$

$$\boldsymbol{P} = \begin{bmatrix} 1/(\cos\varphi - \sin\varphi) & 0 & 0 \\ 0 & 1/(\cos\varphi + \sin\varphi) & 0 \\ 0 & 0 & 1 \end{bmatrix}. \tag{5.173}$$

Equation (5.171) can now be integrated to generate the required axisymmetric fundamental solution (ring loads) in the form

$$u_{ij}^*(\xi, x) = \frac{1}{2\pi}\int_0^{2\pi} \hat{u}_{ij}^*(\xi, x)\, d\varphi(\xi), \qquad i, j = r, \varphi, z, \tag{5.174}$$

where, as depicted in Fig. 5.38, u_{ij}^* is now independent of φ.

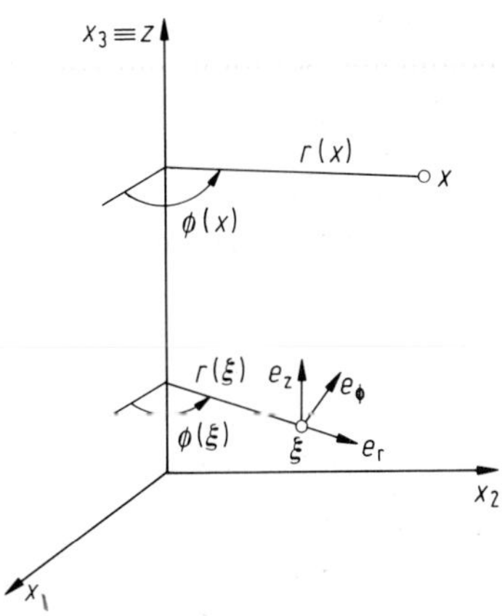

Fig. 5.37. Definition of cylindrical coordinate system

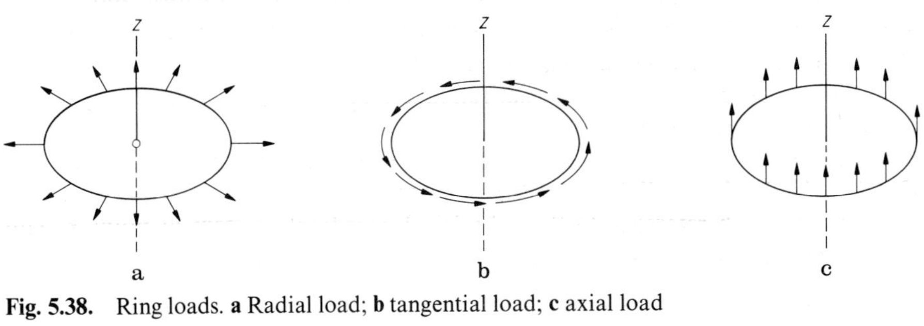

Fig. 5.38. Ring loads. **a** Radial load; **b** tangential load; **c** axial load

The displacements along any circular ring centered at $z(x)$ can be expressed in terms of Legendre functions of order zero and their first derivatives as indicated below (see Kermanidis [39] and Cruse *et al.* [40]).

$$\begin{aligned}
u_{rr}^* &= \frac{1}{16\pi^2(1-\nu)\,G\sqrt{R\,r}}\left\{(3-4\nu)\,Q_{+1/2}(\gamma)+\frac{\bar{Z}^2}{R\,r}\frac{dQ_{+1/2}}{d\gamma}\right\},\\
u_{r\varphi}^* &= 0,\\
u_{rz}^* &= \frac{\bar{Z}}{16\pi^2(1-\nu)\,G\sqrt{R\,r}\,r}\left\{\frac{Q_{+1/2}(\gamma)}{2}-\left(\gamma-\frac{r}{R}\right)\frac{dQ_{+1/2}}{d\gamma}\right\},\\
u_{\varphi r}^* &= u_{\varphi z}^* = 0,\\
u_{\varphi\varphi}^* &= \frac{1}{4\pi^2 G\sqrt{R\,r}}\,Q_{+1/2}(\gamma),\\
u_{zr}^* &= \frac{-\bar{Z}}{16\pi^2(1-\nu)\,G\sqrt{R\,r}\,r}\left\{\frac{Q_{-1/2}(\gamma)}{2}+\left(\gamma-\frac{r}{R}\right)\frac{dQ_{-1/2}}{d\gamma}\right\},\\
u_{z\varphi}^* &= 0,\\
u_{zz}^* &= \frac{1}{16\pi^2(1-\nu)\,G\sqrt{R\,r}}\left\{(3-4\nu)\,Q_{-1/2}(\gamma)-\frac{\bar{Z}^2}{R\,r}\frac{dQ_{-1/2}}{d\gamma}\right\},
\end{aligned}\tag{5.175}$$

where the following notation was used:

$$\begin{aligned}
R &= r(\xi), & z &= z(x),\\
Z &= z(\xi), & \bar{Z} &= Z - z,\\
r &= r(x), & \gamma &= 1+\frac{\bar{Z}^2+(R-r)^2}{2\,R\,r}.
\end{aligned}\tag{5.176}$$

In addition, the Legendre functions, $Q_{+1/2}$ and $Q_{-1/2}$, and their derivatives in Eqs. (5.175), can be written in terms of complete elliptic integrals of the first and second kind in the form

$$\begin{aligned}
Q_{+1/2}(\gamma) &= \gamma\,k\,K(m)-\frac{2}{k}\,E(m),\\
\frac{dQ_{+1/2}(\gamma)}{d\gamma} &= \frac{k}{2}\left\{K(m)-\frac{\gamma}{\gamma-1}\,E(m)\right\},\\
Q_{-1/2}(\gamma) &= k\,K(m),\\
\frac{dQ_{-1/2}(\gamma)}{d\gamma} &= -\frac{k}{2}\,\frac{1}{\gamma-1}\,E(m)
\end{aligned}\tag{5.177}$$

in which $K(m)$ and $E(m)$ are the complete elliptic integrals of the first and second kind, respectively. Also, m and k are called parameter and modulus of the respective elliptic integral; they are given by

$$m = \frac{2}{1+\gamma}, \qquad k = \sqrt{m}\,. \tag{5.178}$$

The expressions for the corresponding stresses can be obtained from the stress-displacement relations in cylindrical polar coordinates. Such relations are given below for completeness:

$$\begin{aligned}
\sigma_{rr} &= \frac{2G\nu}{1-2\nu}\left[\frac{u_r}{r} + \frac{\partial u_z}{\partial z} + \left(\frac{1-\nu}{\nu}\right)\frac{\partial u_r}{\partial r}\right],\\
\sigma_{r\varphi} &= G\left(\frac{\partial u_\varphi}{\partial r} - \frac{u_\varphi}{r}\right),\\
\sigma_{rz} &= G\left(\frac{\partial u_z}{\partial r} + \frac{\partial u_r}{\partial z}\right),\\
\sigma_{\varphi\varphi} &= \frac{2G\nu}{1-2\nu}\left[\frac{\partial u_r}{\partial r} + \frac{\partial u_z}{\partial z} + \left(\frac{1-\nu}{\nu}\right)\frac{u_r}{r}\right],\\
\sigma_{\varphi z} &= G\frac{\partial u_\varphi}{\partial z},\\
\sigma_{zz} &= \frac{2G\nu}{1-2\nu}\left[\frac{u_r}{r} + \frac{\partial u_r}{\partial r} + \left(\frac{1-\nu}{\nu}\right)\frac{\partial u_z}{\partial z}\right].
\end{aligned} \tag{5.179}$$

Substitution of expressions (5.175) into Eqs. (5.179) leads to the fundamental stresses; these stresses can then be converted to tractions on the boundary $\Gamma(x)$ by using the unit outward normal vector. The resulting expressions are

$$\begin{aligned}
p^*_{rr} &= \frac{2G}{1-2\nu}\left\{(1-\nu)\frac{\partial u^*_{rr}}{\partial r} + \nu\left(\frac{u^*_{rr}}{r} + \frac{\partial u^*_{rz}}{\partial z}\right)\right\} n_r + G\left(\frac{\partial u^*_{rr}}{\partial z} + \frac{\partial u^*_{rz}}{\partial r}\right) n_z,\\
p^*_{r\varphi} &= 0,\\
p^*_{rz} &= \frac{2G}{1-2\nu}\left\{(1-\nu)\frac{\partial u^*_{rz}}{\partial z} + \nu\left(\frac{u^*_{rr}}{r} + \frac{\partial u^*_{rr}}{\partial r}\right)\right\} n_z + G\left(\frac{\partial u^*_{rr}}{\partial z} + \frac{\partial u^*_{rz}}{\partial r}\right) n_r,\\
p^*_{\varphi r} &= p^*_{\varphi z} = 0,\\
p^*_{\varphi\varphi} &= G\left\{\left(\frac{\partial u^*_{\varphi\varphi}}{\partial r} - \frac{u^*_{\varphi\varphi}}{r}\right) n_r + \frac{\partial u^*_{\varphi\varphi}}{\partial z} n_z\right\},
\end{aligned} \tag{5.180}$$

$$p^*_{zr} = \frac{2G}{1-2\nu}\left\{(1-\nu)\frac{\partial u^*_{zr}}{\partial r} + \nu\left(\frac{u^*_{zr}}{r} + \frac{\partial u^*_{zz}}{\partial z}\right)\right\} n_r + G\left(\frac{\partial u^*_{zr}}{\partial z} + \frac{\partial u^*_{zz}}{\partial r}\right) n_z,$$

$$p^*_{z\varphi} = 0, \tag{5.180}$$

$$p^*_{zz} = \frac{2G}{1-2\nu}\left\{(1-\nu)\frac{\partial u^*_{zz}}{\partial z} + \nu\left(\frac{u^*_{zr}}{r} + \frac{\partial u^*_{zr}}{\partial r}\right)\right\} n_z + G\left(\frac{\partial u^*_{zr}}{\partial z} + \frac{\partial u^*_{zz}}{\partial r}\right) n_r,$$

where in order to carry out the differentiation of the fundamental displacements, one can make use of the following relations [41]:

$$\frac{dK(m)}{dm} = \frac{E(m) - (1-m)\,K(m)}{2m\,(1-m)},$$

$$\frac{dE(m)}{dm} = \frac{E(m) - K(m)}{2m}. \tag{5.181}$$

As can be seen from expressions (5.175) – (5.180), not only the resulting equations are independent of φ, but also the physical problem is now uncoupled, with the r and z displacement and traction components on one side and the φ components ($u^*_{\varphi\varphi}$ and $p^*_{\varphi\varphi}$) on the other. Herein, attention will be given to the (r, z) axes only; the interested reader can refer to Mayr [42] and Rizzo *et al.* [46], where applications of the isolated transverse part of the fundamental solution are described for pure torsion problems.

With reference to the boundary integral equation, the implementation of the (r, z) part of the fundamental solution leads to

$$\begin{aligned} c_{ij}(\xi)\, u_j(\xi) &+ 2\pi \int_\Gamma p^*_{ij}(\xi, x)\, u_j(x)\, r(x)\, d\Gamma(x) \\ &= 2\pi \int_\Gamma u^*_{ij}(\xi, x)\, p_j(x)\, r(x)\, d\Gamma(x) \\ &\quad + 2\pi \int_\Omega u^*_{ij}(\xi, x)\, b_j(x)\, r(x)\, d\Omega(x), \qquad i, j = r, z \end{aligned} \tag{5.182}$$

where the factor $2\pi r(x)$ appears into the integrals because integration with respect to φ was already accomplished and, consequently, Γ and Ω correspond to a two-dimensional region only. Notice that the above equation is also valid for interior points ($\xi \in \Omega$) if $c_{ij}(\xi) = \delta_{ij}$.

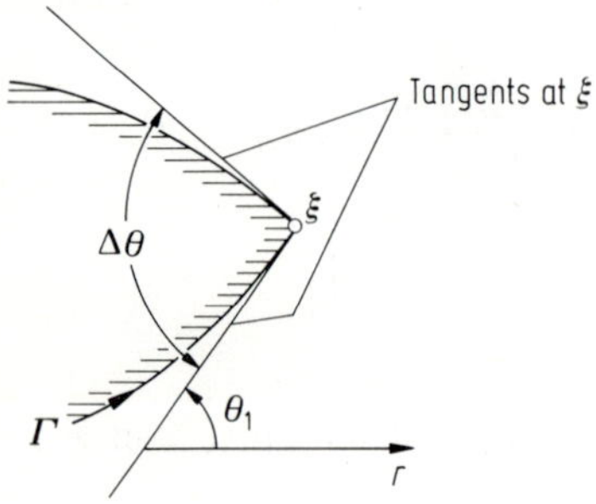

Fig. 5.39. Definition of angles for computation of c_{ij}

$C_{ij}(\xi) = \{\delta_{ij} + \lim_{\varepsilon \to 0} \int_{\Gamma_\varepsilon} p^*_{ij}(\xi, x)\, d\Gamma(x)\}$ (5.75)

Another feature of equation (5.182) is that, in contrast with the previous three- and two-dimensional cases, computation of the coefficient $c_{ij}(\xi)$ together with the associated principal value integral cannot be carried out by rigid-body movements in the direction of r. There are, however, some simple analytical solutions (inflating modes, for instance) which can be used instead. Moreover, a close examination of the definition of c_{ij} (see Eq. (5.75)) indicates that it is equivalent to the plane strain case, i.e. [17],

$$c = \frac{1}{8\pi(1-\nu)} \tag{5.183}$$

$$\times \begin{bmatrix} 4(1-\nu)(\theta_1-\theta_2)+\sin 2\theta_1 - \sin 2\theta_2 & \sin^2\theta_1 - \sin^2\theta_2 \\ \sin^2\theta_1 - \sin^2\theta_2 & 4(1-\nu)(\theta_1-\theta_2) - \sin 2\theta_1 + \sin 2\theta_2 \end{bmatrix}$$

in which θ_1 is defined in Fig. 5.39 and

$$\theta_2 = \theta_1 - \Delta\theta, \tag{5.184}$$

where $\Delta\theta$ is the absolute value of the internal angle at ξ.

Following the general boundary element procedure, stresses at internal points can be computed by substituting Eq. (5.182) ($c_{ij} = \delta_{ij}$) into Eqs. (5.179) with the derivatives being computed with reference to the cylindrical coordinates of ξ. Such derivatives can be obtained by using expressions (5.181).

It should be noted that, for some particular problems, the body force integral in Eq. (5.182) can also be transformed into boundary integrals. Here, the procedure outlined in Section 5.14 can be applied as indicated in Ref. [40].

In order to illustrate the efficiency of the axisymmetric formulation, the case of a cylinder with a spherical cavity subjected to a uniaxial tension is presented. This example was solved by Mayr [44] using constant boundary elements. A comparison between the relative error in the stress concentration factor against computer time is presented in Fig. 5.40. Also included are two finite element results for the same

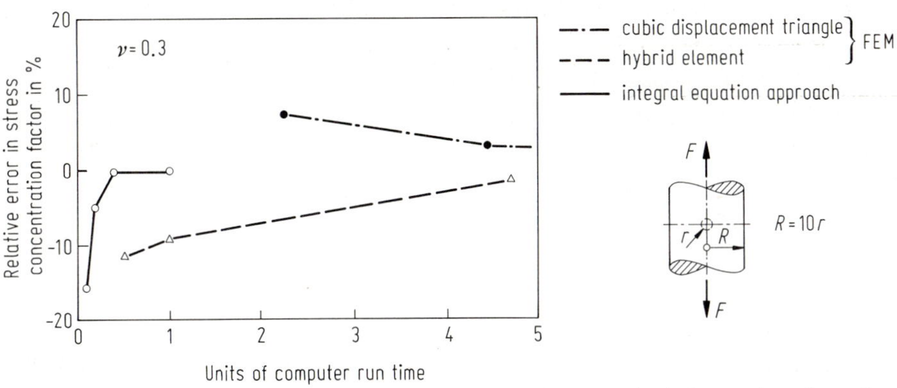

Fig. 5.40. Cylinder with spherical cavity, computer time required for various boundary element and finite element solutions

problem. Clearly, the boundary element solution is seen to be very efficient in this case.

5.15.1. Extension to Nonaxisymmetric Boundary Values

If a body with axisymmetric geometry experiences nonaxisymmetric boundary conditions, the problem becomes dependent not only on the radial (r) and axial (z) directions, but also on the angular direction (φ). Such cases can still be handled by a somewhat sophisticated boundary element formulation which employs Fourier series expansions of the variables involved (see Section 2.14). The key to this type of problem is a sufficiently accurate representation of the displacements and tractions (together with the fundamental solution tensors) in a series of orthogonal functions of the variable φ. This allows analytical integration with respect to φ and, consequently, despite the increase in complexity, the three-dimensional problem is still reduced to a two-dimensional one. An outline of the technique is included in Ref. [44], for the detailed treatment of the analytical formulation, including the fundamental solutions, the reader should refer to Mayr *et al.* [43] and Shippy *et al.* [45].

5.16. Anisotropy

The main concern for the application of the boundary element technique to anisotropic elastic analysis is with the appropriate fundamental solution. The starting boundary integral equation can be represented as before (Eq. (5.77)) on condition that the elastic constants for the fundamental problem be the same as those in the actual one.

Following a general procedure presented by John [47], Vogel and Rizzo [48] have arrived at an integral representation for the fundamental three-dimensional displacements of the form (see also Synge [49])

$$u_{ij}^*(\xi, x) = \frac{1}{8\pi^2 r} \oint_{|\varrho|=1} K_{ij}^{-1}(\boldsymbol{\varrho})\, ds \tag{5.185}$$

in which the line integral is to be taken over a unit circle lying in the plane perpendicular to $\mathbf{r}$ (difference between the position vectors for x and ξ) and centered at ξ. Also, the function $K_{ij}^{-1}(\boldsymbol{\varrho})$ is defined as the inverse of the characteristic matrix K_{ij} given by

$$K_{ij} = C_{ijkl}\, \varrho_k\, \varrho_l, \tag{5.186}$$

where C_{ijkl} is the anisotropic tensor of elastic constants which plays the role of expression (5.31) in the anisotropic case.

As indicated by Vogel and Rizzo, the contour integral presented in Eq. (5.185) can only be performed analytically for some particular cases. It is, however, dependent on the orientation of $\mathbf{r}$ only, and consequently, nonsingular. Therefore, numerical integration techniques can easily be employed. In a recent publication, Wilson and Cruse [50] have studied this procedure in the light of computer efficiency and proposed the following representation:

$$G_{ij}(v_1, v_2) = \oint_{|\varrho|=1} K_{ij}^{-1}(\boldsymbol{\varrho})\, ds, \tag{5.187}$$

where v_1 and v_2 define the orientation of the vector **r**.

Substitution of Eq. (5.187) into Eq. (5.185) yields

$$u_{ij}^* = \frac{1}{8\pi^2 r} G_{ij} \tag{5.188}$$

which also facilitates the computation of the corresponding derivatives with respect to the coordinates of x required for the definition of the fundamental tractions. These derivatives can be written as

$$u_{ij,k}^* = \frac{r_{,k}}{8\pi^2 r^2} G_{ij} + \frac{1}{8\pi^2 r} G_{ij,\alpha} v_{\alpha,k}, \qquad \alpha = 1, 2. \tag{5.189}$$

Guidelines for efficient computation of the integrals involved in expressions (5.185) and (5.189) can be found in Ref. [50], these can be followed for general three-dimensional anisotropic solids. For the case of transversely isotropic materials, the complete fundamental solution can be written in closed form as indicated by Pan and Chou for infinite [51] and semi-infinite [52] medium problems. The half-space transversely isotropic solution was also discussed by Kobayashi and Nishimura [53].

An application of the boundary element technique to plane anisotropic problems was presented by Rizzo and Shippy [54]. Here, as indicated by them, closed form fundamental solutions can be obtained [55], and the implementation of the orthotropic case was presented in detail, including some practical examples. This fundamental solution is presented below for completeness.

Under the assumption of plane stress distribution in a homogeneous orthotropic material, Hooke's law can be written as

$$\begin{aligned} u_{1,1} &= s_{11}\sigma_{11} + s_{12}\sigma_{22}, \\ u_{2,2} &= s_{12}\sigma_{11} + s_{22}\sigma_{22}, \\ u_{1,2} + u_{2,1} &= s_{66}\sigma_{12}. \end{aligned} \tag{5.190}$$

The fundamental displacements due to a unit concentrated load at the origin of the coordinate system is given by

$$\begin{aligned} u_{11}^* &= K_a [\sqrt{\alpha_1} A_2^2 \ln \bar{r}_1 - \sqrt{\alpha_2} A_1^2 \ln \bar{r}_2], \\ u_{12}^* &= u_{21}^* = -K_a A_1 A_2 (\theta_1 - \theta_2), \\ u_{22}^* &= -K_a \left[\frac{1}{\sqrt{\alpha_1}} A_1^2 \ln \bar{r}_1 - \frac{1}{\sqrt{\alpha_2}} A_2^2 \ln \bar{r}_2 \right], \end{aligned} \tag{5.191}$$

where

$$\begin{aligned}
K_a &= \frac{1}{2\pi(\alpha_1-\alpha_2)\,s_{22}}, \\
\alpha_1+\alpha_2 &= \frac{1}{s_{22}}(2s_{12}+s_{66}), \\
\alpha_1\alpha_2 &= s_{11}s_{22}, \\
A_i &= s_{12}-\alpha_i s_{22},
\end{aligned} \tag{5.192}$$

and

$$\begin{aligned}
\bar{r}_i &= \sqrt{x_1^2+\frac{x_2^2}{\alpha_i}}, \\
\theta_i &= \arctan\,(x_2/x_1\sqrt{\alpha_i}).
\end{aligned} \tag{5.193}$$

In the above expressions α_1 and α_2 are supposed to be real and positive numbers (this is the case for a large number of materials).

The fundamental tractions can be written as follows

$$\begin{aligned}
p_{11}^* &= K_a\left[\frac{A_1}{\sqrt{\alpha_2}\,\bar{r}_2^2}-\frac{A_2}{\sqrt{\alpha_1}\,\bar{r}_1^2}\right]x_k n_k, \\
p_{12}^* &= K_a\left[M_1\frac{A_2}{\bar{r}_1^2}-M_2\frac{A_1}{\bar{r}_2^2}\right], \\
p_{21}^* &= K_a\left[M_1\frac{A_1}{\alpha_1\bar{r}_1^2}-M_2\frac{A_2}{\alpha_2\bar{r}_2^2}\right], \\
p_{22}^* &= K_a\left[\frac{A_1}{\sqrt{\alpha_1}\,\bar{r}_1^2}-\frac{A_2}{\sqrt{\alpha_2}\,\bar{r}_2^2}\right]x_k n_k,
\end{aligned} \tag{5.194}$$

in which

$$M_i = \sqrt{\alpha_i}\,x_1 n_2 - \frac{1}{\sqrt{\alpha_i}}\,x_2 n_1. \tag{5.195}$$

The two-dimensional (plane strain and plane stress) fundamental solution for elastic orthotropic materials was also presented by Tomlin [56] who followed an earlier work by Lekhnitskii [57].

Example 5.8. By employing the fundamental solution given above (expressions (5.191)–(5.195)), Rizzo and Shippy [54] analyzed the problem of a circular orthotropic ring, fixed on its outer boundary and subjected to a uniform shear traction p_0 on its inner boundary (see Fig. 5.41). The discretization adopted consisted of 48 equally sized constant boundary elements for each circular boundary and the material constants were

$$s_{11} = 8.33\times10^{-6}\ \text{in}^2/\text{lb}\,,$$

$$s_{22} = 16.67 \times 10^{-6}\,\text{in}^2/\text{lb}\,,$$
$$s_{12} = -0.6 \times 10^{-6}\,\text{in}^2/\text{lb}\,,$$
$$s_{66} = 143 \times 10^{-6}\,\text{in}^2/\text{lb}$$

which correspond to a type of plywood [57].

The problem was solved for a series of ratios r_i/r_0 (Fig. 5.41) with the inner radius being kept constant. The inner hoop stress distribution for the cases $r_i/r_0 = 3/1000$, 3/7, 3/4 is depicted in Fig. 5.42. The first case is seen to approximately correspond to an infinite plate with a circular hole and has been solved analytically in Ref. [57]. This limiting case compares well with the analytical solution as indicated in Table 5.3.

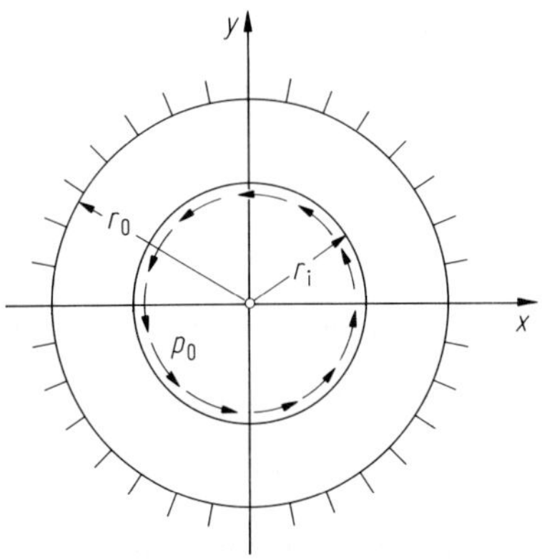

Fig. 5.41. Circular orthotropic ring under uniform shear traction p_0

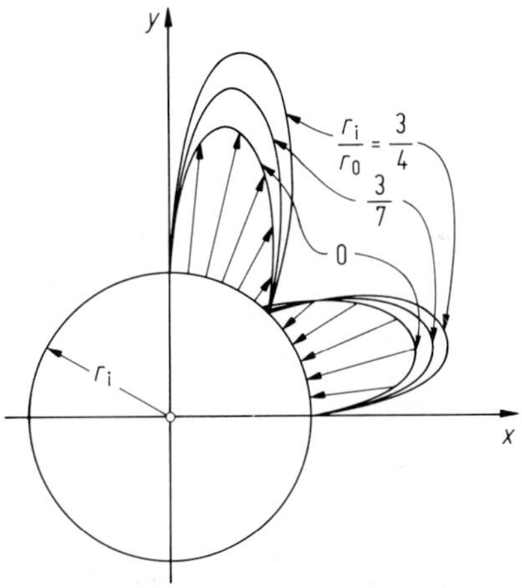

Fig. 5.42. Hoop stress distribution around the inner surfaces of circular rings

Table 5.3. Hoop stress distribution around boundary of the hole.

Angle from the x axis (deg)	Stress/p_0	
	Numerical	Analytical
0	0	0
7.5	−0.85	−0.84
15.0	−1.16	−1.15
22.5	−1.05	−1.05
30.0	−0.77	−0.78
37.5	−0.44	−0.44
45.0	−0.08	−0.08
52.5	0.29	0.29
60.0	0.69	0.69
67.5	1.11	1.11
75.0	1.45	1.45
87.5	1.28	1.31
90.0	0	0

References

1. Love, A. E. H., *A Treatise on the Mathematical Theory of Elasticity*, Dover, New York, 1944.
2. Ford, H., and Alexander, J. M., *Advanced Mechanics of Materials*, 2nd ed., Ellis Horwood, Chichester, 1977.
3. Malvern L. E., *Introduction to the Mechanics of a Continuous Medium*, Prentice–Hall, Englewood Cliffs, N. J., 1969.
4. Fung, Y. C., *Foundations of Solid Mechanics*, Prentice-Hall, Englewood Cliffs, N. J., 1965.
5. Timoshenko, S. P., and Goodier, J. N., *Theory of Elasticity*, 3rd ed., McGraw-Hill, Tokyo, 1970.
6. Nayak, G. C., and Zienkiewicz, O. C., Convenient form of stress invariants for plasticity, Proc. Am. Soc. Civil Engrs., J. Struct. Div. **98,** 949–954 (1972).
7. Kellogg, O. D., *Foundations of Potential Theory*, Springer Verlag, Berlin, 1929.
8. Somigliana, C., Sopra l'equilibrio di un corpo elastico isotropo, *Il Nuovo Ciemento* 17–19 (1886).
9. Brebbia, C. A., *The Boundary Element Method for Engineers*, Pentech Press, London; Halstead Press, New York, 1978 (second edition, 1980).
10. Brebbia, C. A., and Walker, S., *Boundary Element Techniques in Engineering*, Butterworths, London, 1980.
11. Melan, E., Der Spannungszustand der durch eine Einzelkraft im Innern beanspruchten Halbscheibe, Z. Angew. Math. Mech. **12,** 343–346 (1932).
12. Mindlin, R. D., Force at a point in the interior of a semi-infinite solid, Physics **7,** 195–202 (1936).
13. Nakaguma, R. K., Three dimensional elastostatics using the boundary element method, Ph.D. Thesis, University of Southampton, 1979.
14. Telles, J. C. F., and Brebbia, C. A., Boundary element solution for half-plane problems, Int. J. Solids Structures **17,** 1149–1158 (1981).
15. Zabreyko, P. P., *et al., Integral Equations – A Reference Text*, Noordhoff, Amsterdam, 1975.
16. Riccardella, P. C., An implementation of the boundary integral technique for planar problems in elasticity and elastoplasticity, Report No. SM-73-10, Dept. Mech. Engng., Carnegie Mellon Univ., Pittsburg, 1973.
17. Hartmann, F., Computing the C-matrix in non-smooth boundary points, in *New Developments in Boundary Element Methods* (C. A. Brebbia, Ed.), pp. 367–379, Butterworths, London, 1980. CML Southampton, 1983.
18. Rizzo, F. J., and Shippy, D. J., An advanced boundary integral equation method for three dimensional thermoelasticity, Int. J. Numerical Methods Engng. **11,** 1753–1768, (1977).
19. √ Cruse, T. A., Mathematical foundations of the boundary integral equation method in solid mechanics, Report No. AFOSR-TR-77-1002, Pratt and Whitney Aircraft Group, 1977.
20. √ Jaswon, M. A., and Symm, G. T., *Integral Equation Methods in Potential Theory and Elastostatics*, Academic Press, London, 1977.
21. Cruse, T. A., and Vanburen, W., Three-dimensional elastic stress analysis of a fracture specimen with an edge crack, Int. J. Fracture Mech. **7,** 1–15 (1971).
22. Lachat, J. C., A further development of the boundary integral technique for elastostatics, Ph.D. Thesis, University of Southampton, 1975.
23. Telles, J. C. F., and Brebbia, C. A., On the application of the boundary element method to plasticity, Appl. Math. Modelling **3,** 466–470 (1979).
24. Telles, J. C. F., Mansur, W. J., and Halbritter, A. L., The boundary element method applied to two dimensional linear elasticity, in *Proc. 2nd Symposium on Computational Systems in Civil Engineering*, pp. 303–314, CESP, São Paulo, 1978.
25. Chaudonneret, M., On the discontinuity of the stress vector in the boundary integral equation method for elastic analysis, in *Recent Advances in Boundary Element Methods* (C. A. Brebbia, Ed.), pp. 185–194, Pentech Press, London, 1978.
26. Cruse, T. A., An improved boundary integral equation method for three dimensional elastic stress analysis, Comp. Structures **4,** 741–754 (1974).

27. Cruse, T. A., and Wilson, R. B., Advanced applications of boundary integral equation methods, Nuclear Engng. Design **46,** 223–234 (1978).
28. Cruse, T. A., and Meyers, G. J., Two dimensional fracture mechanics analysis. J. Struct. Div. Proc. ASCE **103,** 309–320 (1977).
29. Brebbia, C. A., and Nakaguma, R., Applications of boundary elements in the analysis of offshore structures, in *Proc. Brazil Ofsshore/77 Rio de Janeiro,* Pentech Press, London, GULF Publications, Houston, 1978.
30. Boissenot, J. M., Lachat, J. C., and Watson, J., Etude par equations integrals d'une eprouvette C.T. 15, Rev. Phys. Appl. **9,** 611 (1974).
31. Cruse, T. A., Two dimensional BIE fracture mechanics analysis, in *Proc. 1st Int. Seminar on Recent Advances in Boundary Element Methods* (C. A. Brebbia, Ed.), Pentech Press, London, 1978.
32. Poulos, H. G., and Davis, E. H., *Elastic Solutions for Soil and Rock Mechanics,* Wiley, New York, 1974.
33. Jeffery, G. B., Plane stress and plane strain in bipolar coordinates, Trans. Roy. Soc. (London), Ser. A **221,** 265–293 (1920).
34. Mindlin, R. D., Stress distribution around a hole near the edge of a plate under tension, Proc. Soc. Exptl. Stress. Anal. **5,** 56–68 (1948).
35. Watson, J. O., The solution of boundary integral equations of three-dimensional elastostatics for infinite regions. Paper presented at the 1st Int. Seminar on Recent Advances in Boundary Element Methods, University of Southampton, 1978.
36. Stippes, M., and Rizzo, F. J., A Note on the body force integral of classical elastostatics, Z. Angew. Math. Phys. **28,** 339–341 (1977).
37. Danson, D. J., A boundary element formulation of problems in linear isotropic elasticity with body forces, in *Boundary Element Methods* (C. A. Brebbia, Ed.), pp. 105–122, Springer-Verlag, Berlin, 1981.
38. Cruse, T. A., Boundary integral equation method for three dimensional elastic fracture mechanics, Report No. AFOSR-TR-75-0813, Pratt and Whitney Aircraft Group, (1975).
39. Kermanidis, T., A numerical solution for axially symmetrical elasticity problems, Int. J. Solids Structures **11,** 493–500 (1975).
40. Cruse, T. A., Snow, D. W., and Wilson, R. B., Numerical solutions in axisymmetric elasticity. Comput. Structures **7,** 445–451 (1977).
41. Erdelyi, A., *et al., Higher Transcendental Functions,* Baethman Manuscript Project, Vol. 1, McGraw-Hill, New York, 1953.
42. Mayr, M., and Neuretier, W., Ein Numerisches Verfahren zur Lösung des Axialsymmetrischen Torsionsproblems, Ingenieur-Archiv. **46,** 137–142 (1977).
43. Mayr, M., Drexler, W., and Kuhn, G., A semianalytical boundary integral approach for axisymmetric elastic bodies with arbitrary boundary conditions, Int. J. Solids Structures **16,** 863–871 (1980).
44. Mayr, M., On the numerical solution of axisymmetric elasticity problems using an integral equation approach, Mech. Res. Com. **3,** 393–398 (1976).
45. Shippy, D. J., Rizzo, F. J., and Nigan, R. K., A boundary integral equation method for axisymmetric elastostatic bodies under arbitrary surface loads, in *Proc. 2nd Int. Symp. on Innovative Numerical Analysis in Appl. Engng. Sci.* (R.P. Shaw *et al.,* Eds), University of Virginia Press, Charlottesville, 1980.
46. Rizzo, F. J., Gupta, A. K., and Wu, Y., A boundary integral equation method for torsion of variable diameter circular shafts and related problems, in *Proc. 2nd Int. Symp. on Innovative Numerical Analysis in Appl. Engng. Sci.* (R.P. Shaw *et al.,* Eds), University of Virginia Press, Charlottesville, 1980.
47. John, F., *Plane Waves and Spherical Means Applied to Partial Differential Equations,* Interscience Publishers, New York, 1955.
48. Vogel, S. K., and Rizzo, F. J., An integral equation formulation of three dimensional anisotropic elastostatic boundary value problems, J. Elasticity **3,** 203–216 (1973).
49. Synge, J. L., *The Hypercircle in Mathematical Physics,* Cambridge University Press, Cambridge, 1957.
50. Wilson, R. B., and Cruse, T. A., Efficient implementation of anisotropic three dimensional boundary-integral equation stress analysis, Int. J. Numerical Methods Engng. **12,** 1383–1397 (1978).

51. Pan, Y. C., and Chou, T. W., Point force solution for an infinite transversely isotropic solid, Trans. ASME, J. Appl. Mech. **43,** 608–612 (1976).
52. Pan, Y. C., and Chou, T. W., Green's function solutions for semi-infinite transversely isotropic materials, Int. J. Engng. Sci. **17,** 545–551 (1979).
53. Kobayashi, S., and Nishimura, N., Green's tensors for elastic half-spaces: An application of boundary integral equation method. Mem. Faculty Engng, Kyoto Univ. **42,** 228–241 (1980).
54. Rizzo, F. J., and Shippy, D. J., A method for stress determination in plane anisotropic bodies, J. Composite Materials **4,** 36–61 (1970).
55. Green, A. E., A note on stress systems in aelotropic materials, Philos. Mag. **34,** 416–418 (1943).
56. Tomlin, G. R., Numerical analysis of continuum problems of zoned anisotropic media, Ph.D. Thesis, Southampton University, 1972.
57. Lekhnitskii, S. G., *Theory of Elasticity of an Anisotropic Elastic Body,* Holden-Day, San Francisco, 1963.
58. Alarcon, E., Brebbia, C., and Dominguez, J., The boundary element method in elasticity. Int. J. Mech. Sci. **20,** 625–639 (1978).

Chapter 6

Boundary Integral Formulation for Inelastic Problems

6.1. Introduction

Although applications of integral equations in elasticity were already known in the 1960s, it was only during the last decade that the first papers on nonlinear material problems appeared. The first publication on this subject was due to Swedlow and Cruse [1] in 1971. The article was concerned with the generalization of the strain hardening elastoplastic constitutive equations, previously presented by the first author, to compressible and anisotropic plastic flow, and presented an extended form of Somigliana's identity including plastic strain rates. In addition, the starting boundary integral equation for the direct boundary element formulation was first introduced, for three-dimensional problems, but examples were not shown nor was the integral expression for internal stresses given. The authors, however, pointed out the existence of a domain integral which accounts for the plastic strains contribution to the formulation.

This early work was taken up by Riccardella [3] in 1973, who implemented the von Mises yield criterion (isotropic hardening) for two-dimensional problems using piecewise constant interpolation for the plastic strains. The complete integral expression for stresses at internal points was not presented due to the author's recognition of a singularity in the plastic strain integral. Instead, this apparent difficulty was correctly avoided by first integrating analytically the plastic strain term and then obtaining the derivatives also in closed form. A direct consequence of the procedure was that interpolation functions other than constant could not be easily implemented. By using a rather cumbersome implicit solution technique, some examples were solved and the author concluded that, although not entirely successful, the results were encouraging. This work deserves considerable credit, not only for being the first of its kind, but also because it laid the numerical basis for much of the work that followed. Linear boundary elements, for instance, were first presented together with the analytical expressions for the free term. Also, the closed form integrals for the plastic strain term remained the only correct expressions available until recently.

During the same period, Mendelson [2] presented and discussed different integral formulations for elastoplastic problems, namely, indirect, direct, and a direct biharmonic formulation therein called semidirect approach. Partial solutions to some elastoplastic examples were presented, including a trivial closed form expression for the torsion problem of a circular shaft and some early numerical results for an edge-notched beam under pure bending. The latter was solved by using the so-called semidirect formulation. By contrast with the previous references, the direct formulation was presented including the integral expressions for the internal stresses (two- and three-dimensional problems). Such expressions, however, were later seen to be incorrect due to the way in which the plastic strain term was considered.

In 1975 an extension of the above work was presented by Mendelson and Albers [4]. In this paper the numerical results for the torsion problem of a bar with square cross section were presented within the context of the direct formulation (warping function) and the deformation theory of plasticity. Ideal plasticity and strain hardening were considered, and a comparison of results with finite difference solutions indicated the powerfulness of the technique. The paper also produced some further results for the beam problem presented before, but in addition to the complete solution obtained by the semidirect formulation, an attempt to apply the direct procedure was presented with inconclusive results.

Two years later Mukherjee [5] presented a theoretical paper concerned with the proper care in reducing the three-dimensional direct boundary element formulation to the plane strain case. In this work he partially corrected the equations presented in Refs. [2, 4] and produced modified versions for the kernels of the plastic strain integrals. Such modifications are entirely based on the incompressibility of the plastic strains and, consequently, cannot be valid for plasticity problems in which plastic dilation is allowed (plastic potentials type Druker–Prager or Mohr–Coulomb) to occur. An application of the formulation to obtain closed form solutions for some simple problems was also discussed in [16] by the same author and a coworker.

Still in 1977, Chaudonneret [11] used a direct boundary element formulation for the viscoplastic analysis of a notched plate. In her study, original constitutive equations were employed and a confirmation of the results was obtained experimentally. Also, the integral equations presented were based on an "initial stress" form of the viscoplastic strains influence and the numerical implementation was carried out using linear boundary elements and constant rectangular cells for integrating the nonlinear term. It is worth mentioning that Riccardella's procedure for obtaining the internal stresses appears to have been used since the corresponding integral expressions presented in the paper were still not correct.

The following year saw a major contribution towards the proper inelastic boundary element formulation; Bui [6] presented a paper in which he points out the appropriate concept (originally due to Mikhlin [17]) for the derivative of the singular integral of the inelastic term. Here, the three-dimensional integral expressions are discussed and the author indicates the existence of a free term in the integral equation for internal stresses which was not considered in any of the previous publications. This free term, however, does not ease the numerical implementation because the associated domain integral (inelastic term) still has to be evaluated in the principal value sense. Nevertheless, it was the very first time the correct integral expression was proposed.

Recognition of the above work led Mukherjee and Kumar [10] to adopt the procedure previously described by Riccardella. In this paper they succeeded in performing time-dependent inelastic analysis of some plane stress examples using power law creep and the recently developed constitutive relations due to Hart (metallic media). The solution procedure employed was a predictor–corrector time integration scheme coupled with piecewise constant spatial interpolation for both boundary unknowns and inelastic strains.

One year later (1979), Telles and Brebbia [7] produced the complete boundary element formulation for three- and two-dimensional plasticity problems. The correct expressions for internal stresses were given, including the proper derivatives of the singular domain integrals. In their work, emphasis was given to the numerical implementation of the integral equations and a simple procedure for numerically computing the principal value of the plastic strain integrals, together with the corresponding free terms, was proposed. Such a procedure is based on the application of a uniform plastic strain field to the discretized integral equations and allows for the implementation of higher-order internal cells. This work did not show any solutions for engineering examples, but the possibility of correctly employing higher-order representation for the inelastic strains was demonstrated for the first time.

As it is seen, the 1970s saw a great deal of controversy with respect to the correctness of the boundary element formulation. In the beginning of the present decade, however, the technique was already capable of solving many practical problems using more sophisticated numerical implementations. As early as 1980, for instance, Telles and Brebbia [8] employed the direct boundary element method to solve some elastoplastic problems in two dimensions (plane stress and plane strain). An "initial strain" form of the inelastic term was considered and the formulation was capable of handling incompressible plastic strains using the isotropic von Mises yield criterion with strain hardening, ideal plasticity, and strain softening behavior. The numerical implementation was accomplished by using linear interpolation functions for both, boundary elements and internal cells. In this work the potentialities of boundary elements for inelastic analysis were highlighted by comparing the results with finite element solutions for the same problems.

In another publication by the same authors [9], an "initial stress" formulation was introduced with four different yield criteria (Tresca, Mises, Mohr–Coulomb, and Drucker–Prager). The possibility of plastic dilation was therefore considered. In addition, alternative direct boundary element formulations were also discussed; namely, initial strain, initial stress, and fictitious tractions and body forces approach. Among the different applications presented is the geotechnical problem of a deep tunnel which clearly demonstrates the suitability of boundary elements for inelastic infinite medium problems.

An improvement in an early formulation was presented by Morjaria and Mukherjee [12], where the implementation previously described by the second author and colleague [10] was made more efficient by employing linear boundary elements and an Euler-type time integration scheme. The inelastic term, however, was still spatially interpolated in constant piecewise form. In this publication some further examples (plane stress) are solved, including a plate with an elliptic cutout; comparisons with the previous attempt reveal substantial improvement in computer efficiency.

In 1981 the first successful formulations employing fundamental solutions that satisfy particular boundary conditions are introduced. Telles and Brebbia [13] implement the half-plane singular solution in the context of the different formulations previously proposed by them. In this work finite and semi-infinite plasticity problems are solved with great resolution, and this is achieved without boundary discretization over the traction-free surface of the semiplane. Another interesting implementation was presented by Morjaria and Mukherjee [14] where an indirect and biharmonic boundary element formulation is presented in conjunction with the fundamental solution for planar bodies with cutouts (circular and elliptic). This biharmonic formulation is applied to solve two examples including the challenging problem of a cracked plate simulated by a narrow elliptic cutout.

A further development is also presented by Telles and Brebbia [15] where they introduce a viscoplastic boundary element implementation which is capable of handling plasticity, creep, and viscoplasticity in a unified approach. In this reference the Perzyna's constitutive model for elastic/viscoplastic material behavior is adopted with four different yield criteria. The solution routine employed is a simple Euler time integration scheme with time step limiting considerations. The examples shown illustrate the capabilities of boundary elements in these classes of time-dependent nonlinear problems.

This chapter presents the general boundary integral formulation for inelastic problems which is then applied in two subsequent chapters. Chapter 7 is concerned with the application of the inelastic boundary element equations to solve plasticity problems. The yield condition presented in Section 6.2 is extended to general continuum problems and the von Mises yield criterion [18–22, 26–29] is first introduced in conjunction with the initial strain equations. Also, a solution algorithm [21] based on these expressions is presented and discussed in detail, including a series of examples and comparisons with existing results.

In order to extend the range of applications, general stress-strain relations for post yield behavior are introduced in incremental form. This is accomplished by considering four different yield criteria, namely [34], Tresca, von Mises, Mohr–Coulomb, and Drucker–Prager. Such relations are seen to be particularly useful when the initial stress equations are employed; hence two different algorithms for stepwise plasticity solutions are presented and implemented for the initial stress formulations. At the end of Chapter 7 examples and comparisons with alternative solutions are presented and these also include applications of the half-plane fundamental solution.

Chapter 8 is mainly devoted to the application of the boundary element technique to time-dependent nonlinear material problems. The uniaxial models presented in Section 6.2 are employed in equivalent or effective form to generate the constitutive equations and a unified procedure, capable of handling viscoplasticity and creep, is presented. In addition, the procedure is also applicable to simulating pure elastoplastic solutions through the consideration of long-term load increments followed by stationary conditions. The last part of Chapter 8 deals with the analysis of no-tension materials such as concrete, rocks, and others. For these problems a redistribution of stresses occurs during loading and the solution technique is similar to the one applied for plasticity.

6.2. Inelastic Behavior of Materials

In the theory of elasticity reviewed in Section 5.1, there were two controlling factors: complete recovery of the unstrained configuration when the loads are removed and the dependence of the deformations only on the final stresses, not on the previous load history or strain path. In plasticity and inelasticity, in general, these two factors are not realized.

Plasticity is defined as a property which enables a material to be deformed continuously and permanently without rupture during the application of stresses exceeding the elastic limit of the material. Thus, residual strains are expected to occur on removal of the load and, furthermore, the final deformation depends not only on the final stresses, but also on the path stress history from the beginning of yield.

The problem of formulating physical relations describing the actual behavior of a material during plastic flow is a very complex one. This complexity is due to the nonlinearity and irreversibility of the deformation processes and to a number of phenomena which occur only after the material becomes plastic. The yield characteristics of many materials, for instance, are modified by the rate of straining, with the resistance to deformation increasing markedly with the speed of loading (viscous effect). On the other hand, creep of metals is one example where deformations will occur (usually at elevated temperatures) with extended periods of time under constant stress. In order to simplify the present discussion, some possible approximate diagrams are considered that may represent the behavior of a specimen stressed in simple tension or compression.

An elastic perfectly plastic material is shown in Fig. 6.1. Here, as the stress in the loaded specimen is increased, from O towards A, an elastic recoverable strain takes place until the stress reaches the value $\sigma = Y$, when a plastic strain is superimposed and further deformation will occur under constant yield stress. If, after a point B has been reached, the specimen is unloaded, the path O–A–B is not retraced due to the irreversibility of plastic deformation, but the stress point will follow the line B–C parallel to O–A. Stressing the specimen in compression will therefore lead to point C for which the compressive yield stress $\sigma = -Y$ is attained. Thereafter the specimen deforms under constant value of yield stress and point D may be reached allowing for the entire cycle to be repeated.

A more complex situation occurs when hardening/softening effects are taken into account. This can be seen in Fig. 6.2 where simplified linear hardening is characterized by a constant modulus E_T. After reaching point A for which $\sigma = Y$, a further increase of stress is now required to induce further deformation. When the

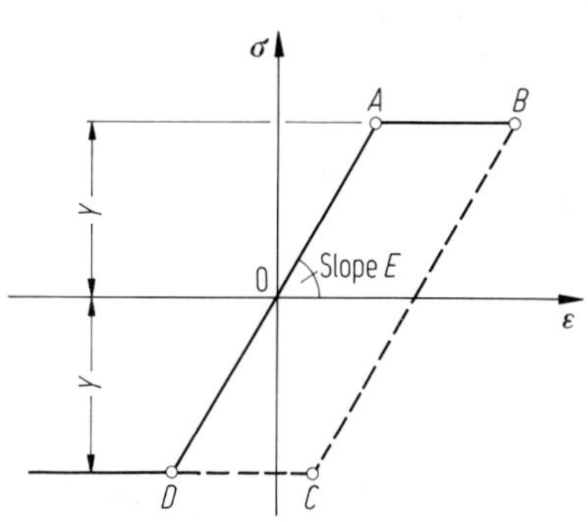

Fig. 6.1. Uniaxial stress-strain diagram for an elastic perfectly plastic material

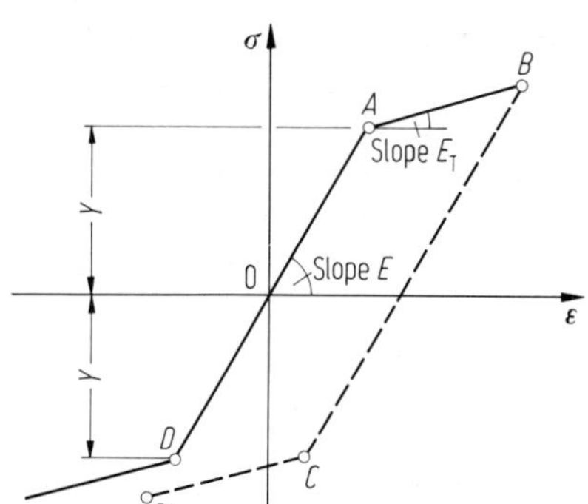

Fig. 6.2. Uniaxial stress-strain diagram for a hardening material

specimen is unloaded from point B, the stress point moves along the line B–C as before, but it is known from experiments that the compressive yield stress will vary depending on the previous deformation history; thus $|\sigma_B| \neq |\sigma_C|$, in general, and this is reffered to in the literature as the Bauschinger effect.

There are several simplified models used to describe the Bauschinger effect. At one extreme it is assumed that the elastic unloading range will be double the initial yield stress (kinematic hardening). Hence,

$$\sigma_C = \sigma_B - 2Y . \tag{6.1}$$

At the other extreme there is the isotropic hardening theory which assumes that the mechanism that produces hardening acts equally in tension and compression; thus,

$$\sigma_C = -\sigma_B . \tag{6.2}$$

Actually, neither theory accurately represents the hardening effects in reverse loading. The kinematic model, though more accurate in this situation, tends to overcorrect for the Bauschinger effect [21] and the isotropic model does not take into account such anisotropic behavior. The latter, however, is mathematically simpler and, consequently, has been most frequently used. Furthermore, the drawback involved in the isotropic hardening theory can be overcome by making use of the fraction model [35], also known as the overlay model [36]. In this model a material particle is considered to be composed of various portions which can be represented by subelements connected in parallel, showing isotropic hardening behavior in plastic deformation. By assigning different properties to each subelement and assuming that all subelements are subjected to the same total strain, the proper material behavior can be simulated as closely as possible, including the Bauschinger effect.

If only one subelement is chosen, the isotropic hardening theory is obtained. However, if necessary, the model can also describe kinematic behavior by making a suitable choice of the number of subelements, their size, and isotropic hardening rules. This means that the kinematic model is no longer needed and, consequently, attention will be given in the subsequent parts of this book to the isotropic hardening theory. Thus, from now on $\sigma \geqq 0$ is always implied for simplicity.

Assuming that the total strain ε is subdivided into an elastic strain ε^e and a plastic strain ε^p, one obtains

$$\varepsilon = \varepsilon^e + \varepsilon^p , \tag{6.3}$$

where

$$\varepsilon^e = \frac{\sigma}{E} . \tag{6.4}$$

With reference to Fig. 6.3 it is seen that pure elastic behavior is obtained for initial loading when

$$\sigma - Y < 0 . \tag{6.5}$$

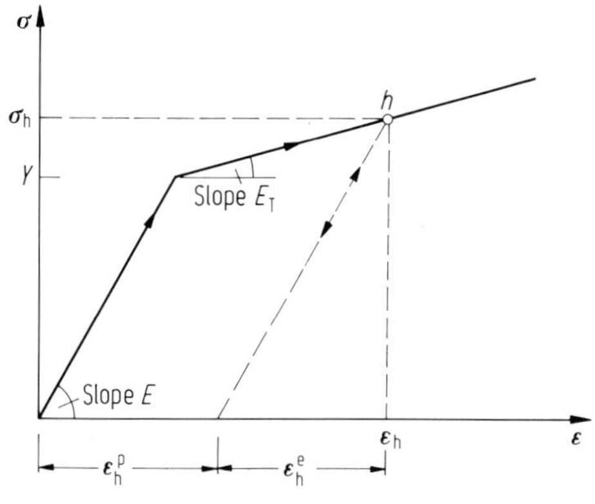

Fig. 6.3. Uniaxial stress-strain diagram showing elastic and plastic strains

Once σ exceeds Y, however, this condition changes such that σ is tested against the yield stress σ_0 as follows:

$$\sigma - \sigma_0 < 0\,, \tag{6.6}$$

where σ_0 has the initial value Y and varies according to a certain rule as plasic flow progresses. For the case depicted in Fig. 6.3 it is easily seen that

$$\sigma_0 = Y + \frac{E_T}{1 - E_T/E}\,\varepsilon^p\,. \tag{6.7}$$

In order to keep the present discussion sufficiently general, the above expression can be related to the work hardening hypothesis by assuming that σ_0 is a function of a hardening parameter k which represents the total plastic work, namely,

$$k = \int \sigma\, d\varepsilon^p\,. \tag{6.8}$$

Hence,

$$\sigma_0 = g\left(\int \sigma\, d\varepsilon^p\right) \tag{6.9}$$

and

$$\frac{dg}{dk} = \frac{1}{\sigma}\frac{dg}{d\varepsilon^p} = \frac{H'}{\sigma}\,, \tag{6.10}$$

where H' is the slope of the uniaxial curve replotted as stress versus plastic strain.

Equation (6.7), which corresponds to linear work hardening, can therefore be written

$$\sigma_0 = Y + H'\,\varepsilon^p \tag{6.11}$$

in which for this case H' is a constant given by

$$H' = \frac{E_T}{1 - E_T/E}\,. \tag{6.12}$$

Recalling the condition presented in (6.6), plastic behavior is possible if the following condition or criterion is satisfied:

$$F(\sigma, k) = \sigma - \sigma_0 = 0\,, \tag{6.13}$$

where $F(\sigma, k)$ is a yield function subjected to the restriction

$$F(\sigma, k) \leqq 0\,. \tag{6.14}$$

It was mentioned before that some materials present pronounced rate-dependent plastic behavior. Within the context of the classical or inviscid theory of plasticity, however, time independence is a basic assumption and this makes a simultaneous description of plastic and rheologic effects impossible. Such a unified description is the object of the viscoplastic theory.

Every material shows more or less pronounced viscous properties. In some problems these properties can be neglected without any real effect in the results, but in other problems this influence may be essential and the important feature of the inelastic behavior is the time dependence of the deformation process. Thus, in such cases, the inelastic strains will depend on the time stress history as well as on the path stress history. Consequently, different results will be obtained for different loading paths and different durations of the loading processes.

One of the more general inelastic models is the elastic/viscoplastic one due to Perzyna [30–32]. This model assumes that the material exibits viscous properties in the plastic region only, which means that $F < 0$ represents a pure elastic behavior. Moreover, the yield criterion on Eq. (6.13) is still valid as an initial condition, now designated as the static yield criterion. In spite of these common features, viscoplasticity allows for

$$F(\sigma, k) > 0 \tag{6.15}$$

which is impossible in the so-called inviscid theory of plasticity.

The uniaxial plastic strain for rate-dependent plastic materials is given in rate form as

$$\dot{\varepsilon}^p = \gamma \left\langle \Phi\left(\frac{F}{\sigma_0}\right)\right\rangle\,, \tag{6.16}$$

where the dot indicates time derivative, γ is a material parameter possibly function of time, temperature, etc., and

$$\begin{aligned} \left\langle \Phi\left(\frac{F}{\sigma_0}\right)\right\rangle &= 0 \qquad \text{for} \quad F \leqq 0 \\ &= \Phi\left(\frac{F}{\sigma_0}\right) \qquad \text{for} \quad F > 0\,. \end{aligned} \tag{6.17}$$

The function Φ is selected from experimental results and different types have been proposed [30], e.g.,

$$\Phi(X) = X^n , \qquad \Phi(X) = X , \qquad \Phi(X) = \exp X - 1 ,$$
$$\Phi(X) = \sum_{\alpha=1}^{N} A_\alpha (\exp X^\alpha - 1) , \qquad \Phi(X) = \sum_{\alpha=1}^{N} B_\alpha X^\alpha . \tag{6.18}$$

Equation (6.16) clearly indicates that the rate of increase of the inelastic strain is a function of the excess stress above the static yield criterion. This function of the excess stress generates the viscoplastic strain rate according to a predetermined viscosity law which is better illustrated by means of the rheological model of Fig. 6.4.

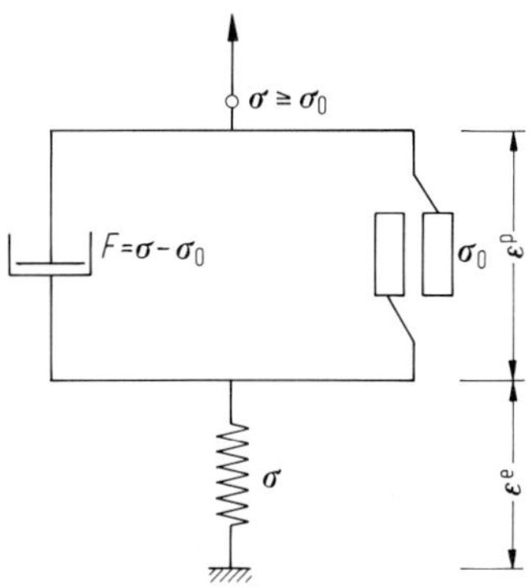

Fig. 6.4. Rheological model for elastic/viscoplastic behavior

In this mechanical model, the friction slider is assumed to sustain all the stress σ up to $\sigma = \sigma_0$, when it then becomes active and slides for $\sigma > \sigma_0$. When this happens, the excess stress $\sigma - \sigma_0$ is carried by the (possible nonlinear) dashpot which generates the viscoplastic strain. The elastic part of the total strain is instantaneously given by the elastic spring.

It should be noticed that, in general, the dashpot and the slider may have properties that depend on the viscoplastic strain ($H' \neq 0$). Thus, after some time under constant applied σ, the slider tends to become rigid again and an asymptotic static configuration ($\dot{\varepsilon}^p = 0$) is achieved providing satisfaction of the static yield criterion.

In order to demonstrate the equivalence of the rheological model and Eq. (6.16), consider the equilibrium condition ($\sigma \geqq \sigma_0$)

$$\sigma = F + \sigma_0 , \tag{6.19}$$

where F represents the stress acting on the dashpot and σ_0 is the part that corresponds to the friction slider.

The stress in the viscous dashpot is related to the viscoplastic strain rate as follows:

$$F = \mu \, \dot{\varepsilon}^p = \mu (\dot{\varepsilon} - \dot{\varepsilon}^e) , \tag{6.20}$$

where μ denotes the damping parameter of the dashpot.

Substituting Eq. (6.20) into Eq. (6.19) and rearranging gives

$$\dot{\varepsilon} = \dot{\varepsilon}^e + \frac{1}{\mu}(\sigma - \sigma_0) \tag{6.21}$$

which reads

$$\dot{\varepsilon} = \dot{\varepsilon}^e + \dot{\varepsilon}^p \, . \tag{6.22}$$

Hence,

$$\dot{\varepsilon}^p = \frac{1}{\mu}(\sigma - \sigma_0) \tag{6.23}$$

which corresponds to Eq. (6.16) if

$$\mu = \frac{\sigma_0}{\gamma} \tag{6.24}$$

and

$$\Phi\left(\frac{F}{\sigma_0}\right) = \frac{F}{\sigma_0} \, . \tag{6.25}$$

It is interesting to study some closed form solutions to Eq. (6.21). For simplicity, assume that $H' = 0$ ($\sigma_0 = Y$) and that the uniaxial model is subjected to a constant total strain rate. Thus, Eq. (6.21) becomes

$$\dot{\varepsilon} = \frac{\dot{\sigma}}{E} + \frac{\gamma}{Y}(\sigma - Y) \tag{6.26}$$

and leads to the following linear differential equation

$$\dot{\sigma} + \frac{\gamma E}{Y}\sigma = E(\dot{\varepsilon} + \gamma) \, , \tag{6.27}$$

where $\dot{\varepsilon} =$ constant.

The solution of Eq. (6.27) is given by

$$\sigma = Y\left(\frac{\dot{\varepsilon}}{\gamma} + 1\right) + C \exp\left(-\frac{\gamma E}{Y} t\right) \tag{6.28}$$

in which t denotes time and C depends on the initial conditions.

If at $t = 0$, $\varepsilon = Y/E$ and $\varepsilon^p = 0$, the following expression for the stress arises:

$$\sigma = \frac{Y\dot{\varepsilon}}{\gamma}\left[1 - \exp\left(-\frac{\gamma E}{Y} t\right)\right] + Y \tag{6.29}$$

which can also be written in terms of strain instead of time as (see Fig. 6.5).

$$\sigma = \frac{Y\dot{\varepsilon}}{\gamma}\left\{1 - \exp\left[\frac{\gamma}{\dot{\varepsilon}}\left(1 - \frac{E\varepsilon}{Y}\right)\right]\right\} + Y \, . \tag{6.30}$$

Alternatively, one can assume that initially an instantaneous $\varepsilon = (Y/E)\,(\dot{\varepsilon}/\gamma + 1)$ is applied and then the total strain increases at a constant rate. In this case the above expression greatly simplifies and the stress remains constant throughout, i.e.,

$$\sigma = Y\left(\frac{\dot{\varepsilon}}{\gamma} + 1\right). \tag{6.31}$$

This case is illustrated in Fig. 6.6.

It is instructive now to point out an important distinction between the inviscid theory of plasticity and the viscoplastic theory adopted here. For pure plasticity, the yield condition presented in Eq. (6.13) leads to a necessary condition for plastic behavior to occur. Once the stress point satisfies the equation $F = 0$, a loading criterion can be defined ($H' > 0$), depending on what happens next, i.e.,

$F = 0$, $\dot{\sigma} < 0$ represents unloading (leads to an elastic state);
$F = 0$, $\dot{\sigma} > 0$ represents loading (leads to another plastic state).

In viscoplasticity, however, the case $F > 0$ exists and consequently viscoplastic behavior will continue to occur completely independent of whether $\dot{\sigma} \gtrless 0$.

An interesting feature of the elastic/viscoplastic model is that for slow incremental loading processes, the results obtained by the classical theory of plasticity are approached (provided the stationary state $F = 0$ is possible). This has been mentioned before when describing the rheological model and is indeed observable in Fig. 6.6. When this is the case, clearly the function Φ and the parameter γ become immaterial, the latter acting just as a time scale factor which renders time a fictitious variable.

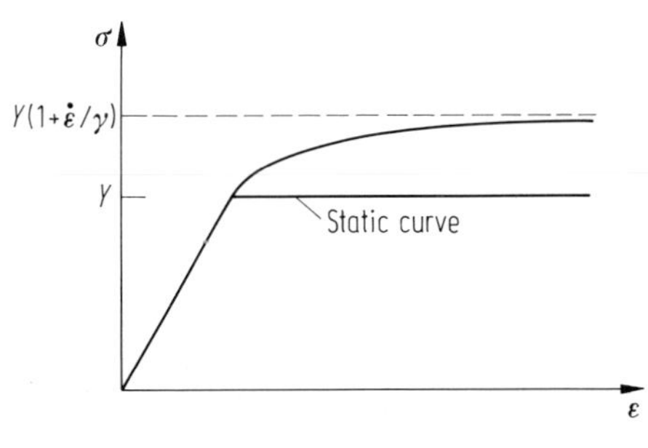

Fig. 6.5. Umaxial stress-strain curve for $\dot{\varepsilon}$ equal constant ($t = 0$, $\varepsilon = Y/E$)

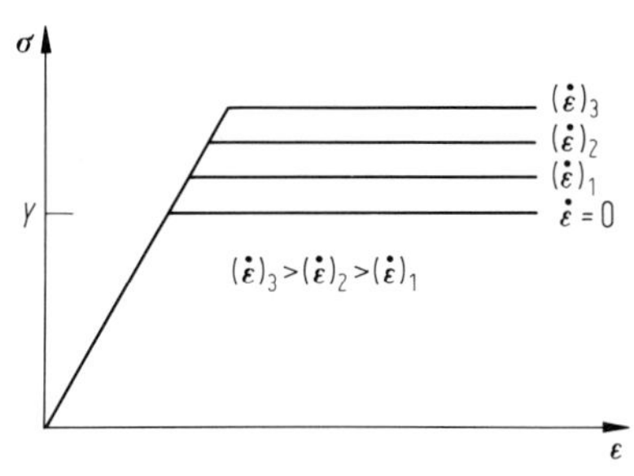

Fig. 6.6. Uniaxial stress-strain curve for $\dot{\varepsilon}$ equal constant ($t = 0$, $\varepsilon = \dfrac{Y}{E}\left(\dfrac{\dot{\varepsilon}}{\gamma} + 1\right)$ instantaneously applied)

Such features can be better explained by rewriting Eq. (6.16) in the form ($F \geqq 0$)

$$\dot{\varepsilon}^p = \gamma\, \Phi \left(\frac{F}{\sigma_0}\right) \tag{6.32}$$

which after rearranging gives

$$F = \sigma_0\, \Phi^{-1} \left(\frac{\dot{\varepsilon}^p}{\gamma}\right) \tag{6.33}$$

or

$$F = \sigma_0\, \Phi^{-1} \left(\frac{\dot{\varepsilon} - \dot{\sigma}/E}{\gamma}\right). \tag{6.34}$$

For slow incremental loading processes, the rates become vanishingly small along the loading path, thus $F = 0$ is approximately attained throughout. In practical terms, one can think of a discrete loading program in which sufficiently small load increments are applied instantaneously. After each load increment, the load is kept constant and a stationary state is allowed to occur (i.e., the friction slider "locks" again in the rheological model). In this fashion, the complete loading path can be followed with the statical yield condition being satisfied at a number of discrete points along the path.

In the simple uniaxial behavior discussed here, increments of any size can be applied because the result at the final point is always the same. This is not the same for continuum problems; in such cases, stress redistribution usually occurs, hence the same stress path may not be obtained. Consequently, small increments must always be kept in mind for the general case.

With reference to the mechanical model of Fig. 6.4, it is immediately apparent that on removal of the dashpot – assuming $\mu = 0$ (i.e., $\gamma \to \infty$) – a pure elastoplastic problem is simulated and only instantaneous response is obtained. Here, the restriction $\sigma \leqq \sigma_0$ is readily found necessary to maintain equilibrium. Another useful simulation can be obtained by assuming that instead of the dashpot, the friction slider is removed (i.e., $\sigma_0 = 0$). In this case, the mechanical model retains its rheological properties and corresponds to the well-known Maxwell model where a linear dashpot is associated in series with a spring. Therefore, by assuming nonlinear properties to the dashpot, the so-called secondary or steady creep of metals [21, 23–25] can be equally represented in this comprehensive model. This matter will be dealt with in what follows.

There is experimental evidence that some metals, usually at elevated temperature, deform continuously with time under constant load. This phenomenon is designated creep and the time-dependent strain originated in the process is called creep strain. A typical uniaxial curve of creep strain (ε^c) versus time under constant load is shown in Fig. 6.7. The first part, AB, where the creep rate decreases rapidly, is known as primary or transient creep. This portion is usually recoverable with time after unloading. The second stage, BC, is associated with a constant creep rate and consequently called steady or secondary creep. In this stage, creep leaves permanent strain. The final stage, CD, known as tertiary creep, is characterized by a rapid increase in the creep rate and leads quickly to rupture. Tertiary creep is greatly affected by the reduction in the cross-sectional area at large strains. This fact, allied to the usual short duration of the primary stage, generally leads to interest in the secondary creep only, though the primary cannot always be neglected.

In constant stress tests, it is customary to represent the creep strain by a general equation of the form

$$\varepsilon^c = g(\sigma, t, T), \tag{6.35}$$

where T is temperature.

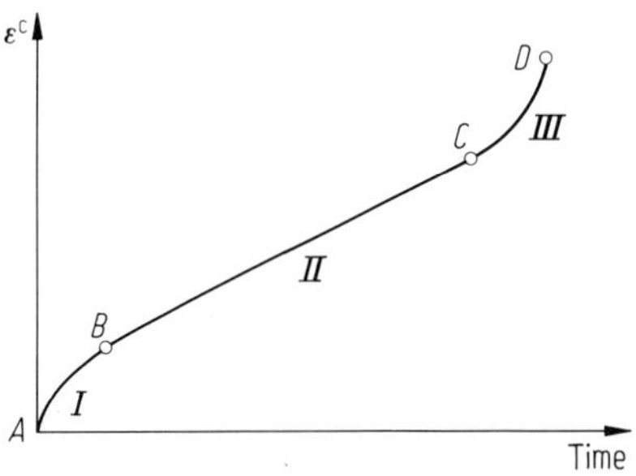

Fig. 6.7. Typical uniaxial creep curve under constant load

A good review of the different types of relations proposed for Eq. (6.35) is given in [24]. In tests of short duration, primary creep predominates. A commonly used expression to represent this primary creep is

$$\varepsilon^c = B\,\sigma^n\,t^{k+1}. \tag{6.36}$$

For the secondary part, the following representation has been preferred:

$$\varepsilon^c = K\,\sigma^m\,t, \tag{6.37}$$

where B, K, m, n, and k are temperature-dependent material parameters.

Generalization of the above equations to include time varying stress is a questionable assumption commonly made. Here, it is the strain rate at any time which is of interest. Thus, Eq. (6.36) gives the time hardening expression

$$\dot{\varepsilon}^c = (k+1)\,B\,\sigma^n\,t^k \tag{6.38}$$

and Eq. (6.37) reduces to the well-known Norton's law, i.e.,

$$\dot{\varepsilon}^c = K\,\sigma^m. \tag{6.39}$$

Equation (6.39) seems to be acceptable for materials which only exhibits secondary creep and has been widely applied in many practical problems. Note that this equation together with the elastic strain rate ($\dot{\sigma}/E$) can be simulated by the nonlinear Maxwell model mentioned before.

For short-term problems, Eq. (6.38) can be substituted by its strain hardening counterpart. This can be done by expressing t from Eq. (6.36) as a function of ε^c and σ, and then substituting the result in expression (6.38). The final expression is

$$\dot{\varepsilon}^c = (k+1)\, B^{1/(k+1)}\, \sigma^{n/(k+1)}\, (\varepsilon^c)^{k/(k+1)}\,. \tag{6.40}$$

For constant stress the above equation is the same as Eq. (6.38), but for time varying stress, different results will be obtained. Experimental data seem to agree better with the strain hardening approach. This is true especially for very short time tests [24]. A shortcoming of both relations is that they do not predict the reversal of the creep strain after unloading. Here, the use of overlay type models [35, 36] appears to be promising.

Throughout this brief exposition only $\sigma \geqq 0$ was considered. It should be kept in mind that for $\sigma < 0$ a negative creep or plastic strain ($|\sigma| \geqq \sigma_0$) will be generated instead. The expressions remain valid if only the absolute values are considered. The reason for keeping $\sigma \geqq 0$ will be more evident in Chapters 7 and 8 where the generalization for multiaxial stress states will be presented and the relations discussed here will be readily applied in equivalent or effective form.

In many practical applications, materials such as concrete and rocks can be idealized as being capable of sustaining only compressive stresses and straining without resistence in tension. Such an idealization is now commonly referred to in the literature as a no-tension behavior [37]. In this case the material can be considered in one of the following ways [38]: (i) as a material which cannot withstand any tension as shown in Fig. 6.8a, loading and unloading for zero tension occur along a horizontal straight line with another straight line representing the linearly elastic property for compression. The material behavior is elastic; (ii) by assuming that the material behaves as a plastic one for zero tension. This will produce a different unloading path as shown in Fig. 6.8b.

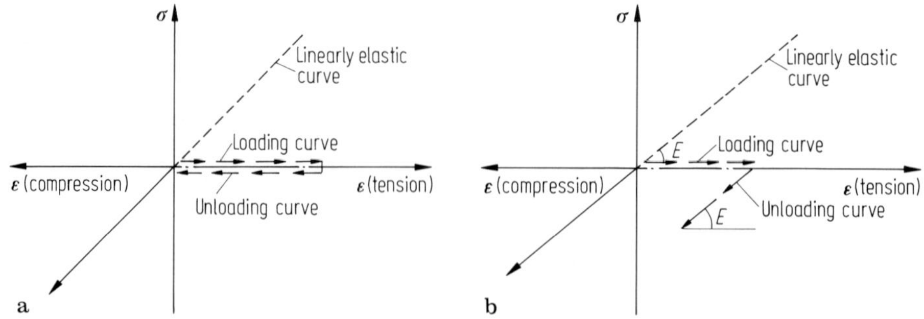

Fig. 6.8. **a** No-tension as a path-independent problem. **b** No-tension as a path-dependent problem

6.3. Governing Equations

In the present section the basic differential equations for continuum inelastic problems are introduced. In order to keep a unified notation, these equations are presented in rate form. This is a natural procedure for time-dependent problems such as viscoplasticity and creep. For classical plasticity, it should be emphasized that pure incremental quantities could be equally used since the relations are homogeneous in time due to the lack of time-dependent effects. Plasticity, however, can be associated to a time-like parameter which is in fact independent of the time scale.

Within the context of small strain theory, the total strain rate for inelastic problems is assumed to be represented by

$$\dot{\varepsilon}_{ij} = \frac{1}{2}\,(\dot{u}_{i,j} + \dot{u}_{j,i}) = \dot{\varepsilon}^{e}_{ij} + \dot{\varepsilon}^{a}_{ij}\,, \tag{6.41}$$

where $\dot{\varepsilon}^{e}_{ij}$ and $\dot{\varepsilon}^{a}_{ij}$ are respectively the elastic and inelastic parts of the total strain rate tensor.

Herein, by inelastic strains one means any kind of strain field which can be considered as "initial strains", i.e.,

$$\dot{\varepsilon}^{a}_{ij} = \dot{\varepsilon}^{p}_{ij} + \dot{\varepsilon}^{c}_{ij} + \dot{\varepsilon}^{T}_{ij} + \dot{\varepsilon}^{I}_{ij}\,, \tag{6.42}$$

where $\dot{\varepsilon}^{p}_{ij}$ is the plastic or viscoplastic strain rate, $\dot{\varepsilon}^{c}_{ij}$ is the creep strain rate, $\dot{\varepsilon}^{T}_{ij}$ is the thermal strain rate, and $\dot{\varepsilon}^{I}_{ij}$ is the initial strain rate due to other causes.

The equilibrium conditions presented in Eq. (5.3) can now be written in rate form as

$$\dot{\sigma}_{ij,i} + \dot{b}_j = 0\,. \tag{6.43}$$

Equation (6.43) is valid in the interior of the body. The same condition when applied to the boundary surface leads to the following rate version of Eq. (5.5),

$$\dot{p}_i - \dot{\sigma}_{ij}\, n_j = 0\,, \tag{6.44}$$

where n_j represents the direction cosines of the outward normal to the boundary of the body.

If inelastic strains are considered as initial strains, the application of Hooke's law to the elastic part of the total strain rate tensor results in the following expression for the stress rate components:

$$\dot{\sigma}_{ij} = 2G\,(\dot{\varepsilon}_{ij} - \dot{\varepsilon}^{a}_{ij}) + \frac{2G\nu}{1-2\nu}\,(\dot{\varepsilon}_{ll} - \dot{e})\,\delta_{ij} \tag{6.45}$$

in which $\dot{e} = \dot{\varepsilon}^{a}_{kk}$, i.e., inelastic dilatational strain rate.

The above expression can be rewritten in terms of initial stresses

$$\dot{\sigma}_{ij} = 2G\,\dot{\varepsilon}_{ij} + \frac{2G\,\nu}{1-2\nu}\,\dot{\varepsilon}_{kk}\,\delta_{ij} - \dot{\sigma}^{a}_{ij}\,, \tag{6.46}$$

where $\dot{\sigma}_{ij}^{a}$ represents the components of the "initial stresses" given by

$$\dot{\sigma}_{ij}^{a} = 2G\,\dot{\varepsilon}_{ij}^{a} + \frac{2G\nu}{1-2\nu}\,\dot{e}\,\delta_{ij}\,. \tag{6.47}$$

The substitution of Eq. (6.45) into Eqs. (6.43) and (6.44) together with Eq. (6.41) gives [23]

$$\dot{u}_{j,ll} + \frac{1}{1-2\nu}\,\dot{u}_{l,lj} = 2\left(\dot{\varepsilon}_{ij,i}^{a} + \frac{\nu}{1-2\nu}\,\dot{e}_{,j}\right) - \frac{\dot{b}_j}{G} \tag{6.48}$$

and

$$\dot{p}_i + 2G\left(\dot{\varepsilon}_{ij}^{a}\,n_j + \frac{\nu}{1-2\nu}\,\dot{e}\,n_i\right) = \frac{2G\nu}{1-2\nu}\,\dot{u}_{l,l}\,n_i + G(\dot{u}_{i,j} + \dot{u}_{j,i})\,n_j\,. \tag{6.49}$$

Equation (6.48) is an extended form of the Navier equation presented in Eq. (5.32) and Eq. (6.49) represents its traction boundary conditions (see Eq. (5.33)). The above expressions can alternatively be written in the following form:

$$\dot{u}_{j,ll} + \frac{\nu}{1-2\nu}\,\dot{u}_{l,lj} = -\frac{\dot{\hat{b}}_j}{G} \tag{6.50}$$

and

$$\dot{\hat{p}}_i = \frac{2G\nu}{1-2\nu}\,\dot{u}_{l,l}\,n_i + G(\dot{u}_{i,j} + \dot{u}_{j,i})\,n_j\,, \tag{6.51}$$

where $\dot{\hat{b}}_j$ and $\dot{\hat{p}}_i$ are pseudo-body-forces and pseudo-tractions given by

$$\dot{\hat{b}}_j = \dot{b}_j - 2G\left(\dot{\varepsilon}_{ij,i}^{a} + \frac{\nu}{1-2\nu}\,\dot{e}_{,j}\right) = \dot{b}_j - \dot{\sigma}_{ij,i}^{a} \tag{6.52}$$

and

$$\dot{\hat{p}}_i = \dot{p}_i + 2G\left(\dot{\varepsilon}_{ij}^{a}\,n_j + \frac{\nu}{1-2\nu}\,\dot{e}\,n_i\right) = \dot{p}_i + \dot{\sigma}_{ij}^{a}\,n_j\,. \tag{6.53}$$

One can notice that Eq. (6.50) represents a set of three quasi-linear partial differential equations for the displacement rates (inelastic terms appear on the right-hand side). Therefore, provided the inelastic strain rates are known, one can still apply the fundamental solutions presented in Chapter 5.

Expressions (6.41)–(6.53) have been presented for three-dimensional bodies. For plane problems, these equations can also be used $(i, j, k, l = 1, 2)$ with $\dot{e} = \dot{\varepsilon}_{11}^{a} + \dot{\varepsilon}_{22}^{a} + \dot{\varepsilon}_{33}^{a}$ in plane strain and ν replaced by $\bar{\nu} = \nu/(1+\nu)$ with $\dot{e} = \dot{\varepsilon}_{11}^{a} + \dot{\varepsilon}_{22}^{a}$ in plane stress.

Different procedures for the boundary element solution of the above equations will be presented in this book. The various formulations will be seen to stem from the equations introduced in this section and consequently the terms initial strain and initial stress will be broadly used to indicate their corresponding integral equations. This remark may appear unnecessary, but it is here included to avoid confusion with some early finite element formulations where the terms initial strain/stress were used to indicate the way in which plastic strain increments are calculated from the constitutive equations [39]. In these formulations, the so-called initial strain is unable to handle ideal plasticity. This restriction, of course, does not apply to the formulations presented here.

6.4. Boundary Integral Formulation

In Chapter 5 we have seen how the integral formulation could be obtained using weighted residual techniques. The advantage of using a weighted residual procedure is that one can start from the beginning with the idea of finding a numerical solution to the actual problem. Thus, the technique brings some physical insight into the numerical solution of the differential equations and more importantly, since it is general, a unified procedure capable of relating the boundary element method to other numerical methods (such as finite elements and finite differences) is obtained.

The basic steps for this procedure will be outlined in what follows, for further details the reader is referred to Chapters 1 and 2 where a complete discussion about the technique is presented.

We seek an approximate solution to the equilibrium equation presented in Section 6.3 (Eq. (6.43))

$$\dot{\sigma}_{ij,i} + \dot{b}_j = 0 \tag{6.54}$$

with boundary conditions

$$\begin{aligned} \dot{u}_i &= \dot{\bar{u}}_i \quad \text{on } \Gamma_1, \\ \dot{p}_i &= \dot{\bar{p}}_i \quad \text{on } \Gamma_2, \end{aligned} \tag{6.55}$$

where $\Gamma_1 + \Gamma_2 = \Gamma$.

For an assumed solution, $\dot{u}_j$, the error introduced can be minimized by writing the following weighted residual statement:

$$\int_\Omega (\dot{\sigma}_{jk,j} + \dot{b}_k)\, u_k^*\, d\Omega = \int_{\Gamma_2} (\dot{p}_k - \dot{\bar{p}}_k)\, u_k^*\, d\Gamma + \int_{\Gamma_1} (\dot{\bar{u}}_k - \dot{u}_k)\, p_k^*\, d\Gamma\,, \tag{6.56}$$

where u_k^* and p_k^* correspond to the displacements and surface tractions of the weighting field. Note that

$$p_k^* = \sigma_{jk}^*\, n_j \tag{6.57}$$

with n_j being the direction cosines of the outward normal to the boundary of the body.

If the same material constants (E, G, and ν) are valid for the approximating and the weighting fields, the first term in Eq. (6.56) can be integrated by parts to give

$$\begin{aligned} & -\int_\Omega \dot{\sigma}_{jk}\, \varepsilon^*_{jk}\, d\Omega + \int_\Omega \dot{b}_k\, u^*_k\, d\Omega \\ & = -\int_{\Gamma_2} \dot{\bar{p}}_k\, u^*_k\, d\Gamma - \int_{\Gamma_1} \dot{p}_k\, u^*_k\, d\Gamma + \int_{\Gamma_1} (\dot{\bar{u}}_k - \dot{u}_k)\, p^*_k\, d\Gamma\,. \end{aligned} \tag{6.58}$$

Recalling expression (6.46), one obtains

$$\dot{\sigma}_{ij} = \dot{\sigma}^e_{ij} - \dot{\sigma}^a_{ij} \tag{6.59}$$

where $\dot{\sigma}^e_{ij} = C_{ijkl}\, \dot{\varepsilon}_{kl}$ (see expression (5.31)).

Expression (6.59) can be substituted into Eq. (6.58) as follows:

$$\begin{aligned} & -\int_\Omega \dot{\sigma}^e_{jk}\, \varepsilon^*_{jk}\, d\Omega + \int_\Omega \dot{\sigma}^a_{jk}\, \varepsilon^*_{jk}\, d\Omega + \int_\Omega \dot{b}_k\, u^*_k\, d\Omega \\ & = -\int_{\Gamma_2} \dot{\bar{p}}_k\, u^*_k\, d\Gamma - \int_{\Gamma_1} \dot{p}_k\, u^*_k\, d\Gamma + \int_{\Gamma_1} (\dot{\bar{u}}_k - \dot{u}_k)\, p^*_k\, d\Gamma\,, \end{aligned} \tag{6.60}$$

and again the first term can be integrated by parts to give

$$\begin{aligned} & \int_\Omega \sigma^*_{jk,j}\, \dot{u}_k\, d\Omega + \int_\Omega \dot{\sigma}^a_{jk}\, \varepsilon^*_{jk}\, d\Omega + \int_\Omega \dot{b}_k\, u^*_k\, d\Omega \\ & = -\int_{\Gamma_2} \dot{\bar{p}}_k\, u^*_k\, d\Gamma - \int_{\Gamma_1} \dot{p}_k\, u^*_k\, d\Gamma + \int_{\Gamma_2} \dot{u}_k\, p^*_k\, d\Gamma + \int_{\Gamma_1} \dot{\bar{u}}_k\, p^*_k\, d\Gamma\,. \end{aligned} \tag{6.61}$$

The above equation can now be written in general form as

$$\int_\Omega b^*_k\, \dot{u}_k\, d\Omega = \int_\Gamma u^*_k\, \dot{p}_k\, d\Gamma - \int_\Gamma p^*_k\, \dot{u}_k\, d\Gamma + \int_\Omega u^*_k\, \dot{b}_k\, d\Omega + \int_\Omega \varepsilon^*_{jk}\, \dot{\sigma}^a_{jk}\, d\Omega \tag{6.62}$$

where the substitution $\sigma^*_{jk,j} = -\, b^*_k$ was made.

Proceeding as in Chapter 5, one can assume that the weighting field is the solution to the fundamental problem (Eq. (5.54)) which allows for b^*_k to be given by expression (5.49). Thus, for each unit point load e_i, the following equation is obtained:

$$\begin{aligned} \dot{u}_i(\xi) = & \int_\Gamma u^*_{ij}(\xi, x)\, \dot{p}_j(x)\, d\Gamma(x) - \int_\Gamma p^*_{ij}(\xi, x)\, \dot{u}_j(x)\, d\Gamma(x) \\ & + \int_\Omega u^*_{ij}(\xi, x)\, \dot{b}_j(x)\, d\Omega(x) + \int_\Omega \varepsilon^*_{jki}(\xi, x)\, \dot{\sigma}^a_{jk}(x)\, d\Omega(x)\,. \end{aligned} \tag{6.63}$$

Equation (6.63) is the inelastic counterpart of Eq. (5.53). Consequently, as discussed before, if the half-space fundamental solution is adopted, Γ can be substituted by Γ' in the second boundary integral. An interesting feature of the initial stress form is that, in contrast with the initial strain equation, the reduction to two-dimensional problems is accomplished by simply keeping the subscripts with a range of two. In both cases, however, the specialization for $\xi \to \Gamma$ can be performed as in pure elastic analysis. Therefore, the following expression is obtained:

$$c_{ij}(\xi)\,\dot{u}_j(\xi) + \int_\Gamma p^*_{ij}(\xi, x)\,\dot{u}_j(x)\,d\Gamma(x) = \int_\Gamma u^*_{ij}(\xi, x)\,\dot{p}_j(x)\,d\Gamma(x) + \int_\Omega u^*_{ij}(\xi, x)\,\dot{b}_j(x)\,d\Omega(x) + \int_\Omega \varepsilon^*_{jki}(\xi, x)\,\dot{\sigma}^a_{jk}(x)\,d\Omega(x) \qquad (6.64)$$

where the last integral can be replaced by

$$\int_\Omega \sigma^*_{jki}(\xi, x)\,\dot{\varepsilon}^a_{jk}(x)\,d\Omega(x) \qquad (6.65)$$

for the initial strain formulation.

It is worth mentioning that the initial stress and initial strain equations are entirely equivalent. This will be proved in Section 6.6 where a complete discussion about the alternative boundary element formulations is presented.

6.5. Internal Stresses

Of fundamental importance for the stepwise solution of nonlinear material problems is the calculation of stresses at internal points. In order to combine both accuracy and computational efficiency, it has been demonstrated by Telles and Brebbia [13] that the proper integral equation should be used in preference to computing displacements at internal points and differentiating them numerically as it is done in finite differences or finite elements.

The correct integral equations for stresses at internal points have been presented in previous papers by the present authors. Since its derivation requires the derivative of the singular integral of the inelastic term and this had often led to incorrect expressions in the past [2, 5, 11, 16], a proper procedure for obtaining these equations is presented in this section [15]. It is the authors' belief that this will enlighten the general concept originally due to Mikhlin [17] which has been applied in Refs. [6, 7].

In order to simplify the presentation, only the Kelvin fundamental solution will be considered in conjunction with the initial stress equation. Also, from now on the initial notation will be somewhat simplified allowing Eq. (6.63) to be written as

$$\dot{u}_i = \int_\Gamma u^*_{ij}\,\dot{p}_j\,d\Gamma - \int_\Gamma p^*_{ij}\,\dot{u}_j\,d\Gamma + \int_\Omega u^*_{ij}\,\dot{b}_j\,d\Omega + \int_\Omega \varepsilon^*_{jki}\,\dot{\sigma}^a_{jk}\,d\Omega\,. \qquad (6.66)$$

From the application of Hooke's law to the elastic part of the total strain rate tensor comes the following expression for the stress rates (see Eq. (6.46))

$$\dot{\sigma}_{ij} = G\left(\frac{\partial \dot{u}_i}{\partial x_j} + \frac{\partial \dot{u}_j}{\partial x_i}\right) + \frac{2G\nu}{1-2\nu}\frac{\partial \dot{u}_k}{\partial x_k}\,\delta_{ij} - \dot{\sigma}^a_{ij}. \qquad (6.67)$$

Stresses at points located within the body can be computed by substituting Eq. (6.66) into Eq. (6.67) on condition that the space derivatives present in Eq. (6.67) be taken with reference to the coordinates of the load point. As in the elastic case, such differentiation can be applied directly to the fundamental solution tensors for the first three integrals of Eq. (6.66). However, the last one needs further examination; in a more formal representation, this integral should be written in the following form

$$V_i = \lim_{\varepsilon \to 0} \int_{\Omega_\varepsilon} \varepsilon^*_{jki}\, \dot{\sigma}^a_{jk}\, d\Omega\,, \tag{6.68}$$

where Ω_ε arises from Ω by removing a ball of radius ε centered at the load point ξ.

The proper expression for the derivative of V_i can therefore be written

$$\frac{\partial V_i}{\partial x_m} = \lim_{\varepsilon \to 0} \left\{ \frac{\partial}{\partial x_m} \int_{\Omega_\varepsilon} \varepsilon^*_{jki}\, \dot{\sigma}^a_{jk}\, d\Omega \right\}. \tag{6.69}$$

For simplicity and without loss of generality, the two-dimensional case will be carried out. In this case ε represents the radius of a circle and one can define a cylindrical coordinate system $(\bar{r}, \bar{\theta})$ based at the point $0 \equiv \xi$ as shown in Fig. 6.9a. In this system of coordinates the tensor ε^*_{jki} can be represented by

$$\frac{1}{r(\bar{r}, \bar{\theta})}\, \psi_{jki}(\varphi) \tag{6.70}$$

where for the case depicted in Fig. 6.9a one has $r(\bar{r}, \bar{\theta}) = \bar{r}$ and $\varphi(\bar{r}, \bar{\theta}) = \bar{\theta}$, but if a small increment in the rectangular coordinate x_m of the singular point is given, not only r and φ become different from $\bar{r}$ and $\bar{\theta}$ but also $\bar{\Gamma}_\varepsilon$ is shifted (see Fig. 6.9b), indicating their dependence on the coordinates of the load point.

Expression (6.69) is now of the form

$$\frac{\partial V_i}{\partial x_m} = \int_0^{2\pi} \lim_{\varepsilon \to 0} \left\{ \frac{\partial}{\partial x_m} \int_{\bar{\varepsilon}}^{R(\bar{\theta})} \frac{\psi_{jki}}{r}\, \dot{\sigma}^a_{jk}\, \bar{r}\, d\bar{r} \right\} d\bar{\theta} \tag{6.71}$$

which allows for the application of the Leibnitz formula* to the term between brackets, i.e.,

$$\begin{aligned} &\frac{\partial}{\partial x_m} \int_{\bar{\varepsilon}}^{R} \frac{\psi_{jki}}{r}\, \dot{\sigma}^a_{jk}\, \bar{r}\, d\bar{r} \\ &= \int_{\bar{\varepsilon}}^{R} \frac{\partial}{\partial x_m} \left(\frac{\psi_{jki}}{r} \right) \dot{\sigma}^a_{jk}\, \bar{r}\, d\bar{r} - \frac{\psi_{jki}}{r(\bar{\varepsilon}, \bar{\theta})}\, \dot{\sigma}^a_{jk}\, \bar{\varepsilon}\, \frac{\partial \bar{\varepsilon}}{\partial x_m}\,. \end{aligned} \tag{6.72}$$

---

* The Leibnitz formula gives

$$\frac{d}{d\alpha} \int_{\varphi_1(\alpha)}^{\varphi_2(\alpha)} F(x, \alpha)\, dx = \int_{\varphi_1(\alpha)}^{\varphi_2(\alpha)} \frac{\partial F}{\partial \alpha}\, dx - F(\varphi_1, \alpha) \frac{d\varphi_1}{d\alpha} + F(\varphi_2, \alpha) \frac{d\varphi_2}{d\alpha}\,.$$

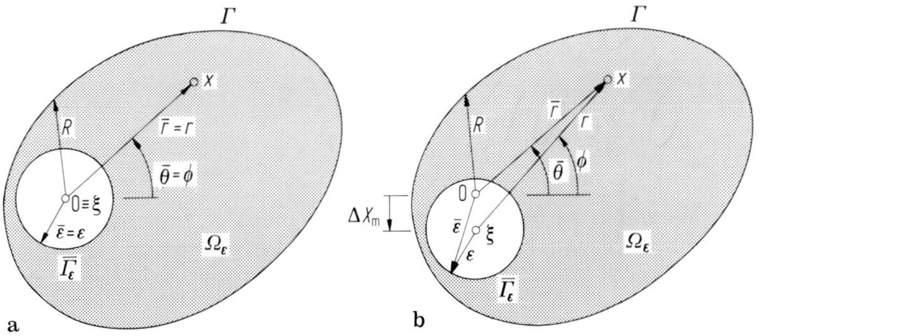

Fig. 6.9. Cylindrical coordinate system based at $0 \equiv \xi$ (**a**) and effect of increment in the rectangular coordinate x_m of ξ (**b**)

Taking into consideration that $0 \equiv \xi$ and $r(\bar{\varepsilon}, \bar{\theta}) = \varepsilon$, the substitution of expression (6.72) in (6.71) gives

$$\frac{\partial V_i}{\partial x_m} = \int_0^{2\pi} \lim_{\varepsilon \to 0} \int_{\varepsilon}^{R(\varphi)} \frac{\partial}{\partial x_m}\left(\frac{\psi_{jki}}{r}\right) \dot{\sigma}^a_{jk}\, r\, dr\, d\varphi - \dot{\sigma}^a_{jk}(\xi) \int_0^{2\pi} \psi_{jki} \cos(r, x_m)\, d\varphi, \tag{6.73}$$

where $\dot{\sigma}^a_{jk}(\xi)$ represents the value of the initial stress rate at the singular point.

Finally, one has to study the existence of the first integral in expression (6.73) which can be further written as

$$\begin{aligned} &\int_0^{2\pi} \lim_{\varepsilon \to 0} \int_{\varepsilon}^{R} \frac{\partial}{\partial x_m}\left(\frac{\psi_{jki}}{r}\right) \dot{\sigma}^a_{jk}\, r\, dr\, d\varphi \\ &\quad = \int_0^{2\pi} \lim_{\varepsilon \to 0} \left\{ \bar{\psi}_{jkim} \int_{\varepsilon}^{R} (\dot{\sigma}^a_{jk} - \dot{\sigma}^a_{jk}(\xi)) \frac{1}{r}\, dr \right\} d\varphi \\ &\quad + \dot{\sigma}^a_{jk}(\xi) \int_0^{2\pi} \bar{\psi}_{jkim} \ln(R)\, d\varphi \\ &\quad - \lim_{\varepsilon \to 0} \left\{ \dot{\sigma}^a_{jk}(\xi) \ln(\varepsilon) \int_0^{2\pi} \bar{\psi}_{jkim}\, d\varphi \right\}, \end{aligned} \tag{6.74}$$

where $\bar{\psi}_{jkim}(\varphi) = r^2\, \partial(\psi_{jki}/r)/\partial x_m$, and it is assumed that $\dot{\sigma}^a_{jk}$ satisfies a Hölder condition [40, 41] at ξ such that

$$\left| \dot{\sigma}^a_{jk}(x) - \dot{\sigma}^a_{jk}(\xi) \right| \leqq A\, r^{\alpha}, \tag{6.75}$$

where A and α are positive constants.

Clearly, the first two terms on the right-hand side of Eq. (6.74) are convergent and the last one requires that

$$\int_0^{2\pi} \bar{\psi}_{jkim}\, d\varphi = 0; \tag{6.76}$$

a condition which is fulfilled by an intrinsic property of $\bar{\psi}_{jkim}$. Therefore, the proof is complete and one can transform expression (6.73) back to the rectangular coordinate system as follows:

$$\frac{\partial V_i}{\partial x_m} = \int_\Omega \frac{\partial \varepsilon^*_{jki}}{\partial x_m} \dot{\sigma}^a_{jk}\, d\Omega - \dot{\sigma}^a_{jk}(\xi) \int_{\Gamma'_1} \varepsilon^*_{jki}\, r_{,m}\, d\Gamma \tag{6.77}$$

in which the first integral is to be interpreted in the sense of Cauchy principal value, Γ'_1 defines a circle of unit radius centered at the load point, and $r_{,m}$ is the derivative of r with reference to the coordinates of the field point. Note that $r_{,m} = -\partial r/\partial x_m$.

It is worth mentioning that the derivative of the body force rate integral can be investigated by the same procedure. In this case, due to the weaker singularity of u^*_{ij}, the free term (corresponding to the Γ'_1 integral) vanishes when $\varepsilon \to 0$.

Expression (6.77) is also valid for three-dimensional problems with Γ'_1 representing the surface of a unit sphere. In both cases the corresponding Γ'_1 integral can be computed in closed form and directly substituted in Eq. (6.67). In addition, since σ^*_{jki} and ε^*_{jki} present singularities of the same order, the same concept applies for the initial strain formulation. Therefore, the complete set of equations for two- and three-dimensional problems (Kelvin) is presented in the next section.

6.6. Alternative Boundary Element Formulations

In this section different formulations using the Kelvin fundamental solutions are discussed.

6.6.1. Initial Strain

The adoption of an initial strain formulation for three-dimensional inelastic problems leads to

$$c_{ij}\, \dot{u}_j = \int_\Gamma u^*_{ij}\, \dot{p}_j\, d\Gamma - \int_\Gamma p^*_{ij}\, \dot{u}_j\, d\Gamma + \int_\Omega u^*_{ij}\, \dot{b}_j\, d\Omega + \int_\Omega \sigma^*_{jki}\, \dot{\varepsilon}^a_{jk}\, d\Omega\,. \tag{6.78}$$

Equation (6.78) is assumed valid for any location of the load point (interior or boundary points) provided c_{ij} and the second boundary integral on the right-hand side are properly interpreted as known from the elastic application of the technique.

Under this assumption, the stress rates at interior points can be computed by use of expressions (6.41) and (6.45). The derivative of Eq. (6.78) yields ($c_{ij} = \delta_{ij}$)

$$\frac{\partial \dot{u}_i}{\partial x_m} = \int_\Gamma \frac{\partial u^*_{ij}}{\partial x_m} \dot{p}_j\, d\Gamma - \int_\Gamma \frac{\partial p^*_{ij}}{\partial x_m} \dot{u}_j\, d\Gamma + \int_\Omega \frac{\partial u^*_{ij}}{\partial x_m} \dot{b}_j\, d\Omega$$

$$+ \int_\Omega \frac{\partial \sigma^*_{jki}}{\partial x_m} \dot{\varepsilon}^a_{jk}\, d\Omega - \dot{\varepsilon}^a_{jk} \int_{\Gamma'_1} \sigma^*_{jki}\, r_{,m}\, d\Gamma \tag{6.79}$$

where the fourth integral is to be interpreted in the principal value sense and the last integral is to be performed over the surface of a unit sphere centered at the singular point. Note that the derivatives are taken with respect to the load point; as before these are indicated explicitly to differentiate from the comma notation which is taken with reference to the field point.

The last integral in Eq. (6.79) can now be computed

$$-\dot{\varepsilon}^a_{jk} \int_{\Gamma'_1} \sigma^*_{jki}\, r_{,m}\, d\Gamma = \frac{1}{15(1-\nu)} [(8-10\nu)\, \dot{\varepsilon}^a_{im} - (1-5\nu)\, \dot{\varepsilon}^a_{ll}\, \delta_{im}]. \tag{6.80}$$

In what follows the reader is referred to the end of this section for the components of the new tensors related to the fundamental solutions.

The above expressions together with Eqs. (6.41) and (6.45) allow for the determination of the internal stresses

$$\begin{aligned}\dot{\sigma}_{ij} = &\int_\Gamma u^*_{ijk}\, \dot{p}_k\, d\Gamma - \int_\Gamma p^*_{ijk}\, \dot{u}_k\, d\Gamma + \int_\Omega u^*_{ijk}\, \dot{b}_k\, d\Omega \\ &+ \int_\Omega \sigma^*_{ijkl}\, \dot{\varepsilon}^a_{kl}\, d\Omega - \frac{2G}{15(1-\nu)} [(7-5\nu)\, \dot{\varepsilon}^a_{ij} + (1+5\nu)\, \dot{\varepsilon}^a_{ll}\, \delta_{ij}]\end{aligned} \tag{6.81}$$

where the last two terms represent the influence of the inelastic strains.

For plane strains the procedure is analogous, the only difference being the fact that the inelastic strain rate integrals still have to take into consideration the work performed in the third direction ($\sigma^*_{33i}\, \dot{\varepsilon}^a_{33}$). This effect is easily incorporated through some particular assumptions such as incompressibility of the inelastic strains (valid for creep and plasticity of metals) or the isolated case of thermal strains. These cases lead to

$$c_{ij}\, \dot{u}_j = \int_\Gamma u^*_{ij}\, \dot{p}_j\, d\Gamma - \int_\Gamma p^*_{ij}\, \dot{u}_j\, d\Gamma + \int_\Omega u^*_{ij}\, \dot{b}_j\, d\Omega + \int_\Omega \hat{\sigma}^*_{jki}\, \dot{\varepsilon}^a_{jk}\, d\Omega \tag{6.82}$$

in which

$$\hat{\sigma}^*_{jki} = \sigma^*_{jki} + \frac{2\nu\, \delta_{jk}\, r_{,i}}{4\pi(1-\nu)\, r} \qquad \text{if } \dot{e} = 0 \qquad \text{(see Section 6.3)} \tag{6.83}$$

or

$$\hat{\sigma}^*_{jki} = \sigma^*_{jki} - \frac{\nu\, \delta_{jk}\, r_{,i}}{4\pi(1-\nu)\, r} \qquad \text{if } \dot{\varepsilon}^a_{jk} = \dot{\varepsilon}^T_{jk} = \alpha\, \dot{T}\, \delta_{jk}, \tag{6.84}$$

where α is the coefficient of linear thermal expansion and $\dot{T}$ is the temperature rate.

The corresponding internal stress rates are computed by

$$\dot{\sigma}_{ij} = \int_\Gamma u^*_{ijk}\, \dot{p}_k\, d\Gamma - \int_\Gamma p^*_{ijk}\, \dot{u}_k\, d\Gamma + \int_\Omega u^*_{ijk}\, \dot{b}_k\, d\Omega + \int_\Omega \hat{\sigma}^*_{ijkl}\, \dot{\varepsilon}^a_{kl}\, d\Omega + f_{ij}(\dot{\varepsilon}^a_{kl}) \tag{6.85}$$

where the last integral is to be taken in the Cauchy principal value sense and if $\dot{e} = 0$

$$\begin{aligned}\hat{\sigma}^*_{ijkl} &= \sigma^*_{ijkl} + \frac{G}{2\pi(1-\nu)\, r^2} [4\nu\, r_{,i}\, r_{,j}\, \delta_{kl} - 2\nu\, \delta_{ij}\, \delta_{kl}], \\ f_{ij} &= -\frac{G}{4(1-\nu)} [2\, \dot{\varepsilon}^a_{ij} + (1-4\nu)\, \dot{\varepsilon}^a_{ll}\, \delta_{ij}].\end{aligned} \tag{6.86}$$

For pure thermal strains one has

$$\hat{\sigma}^*_{ijkl} = \sigma^*_{ijkl} - \frac{G}{2\pi(1-\nu)\,r^2}\,[2\,\nu\,r_{,i}\,r_{,j}\,\delta_{kl} - \nu\,\delta_{ij}\,\delta_{kl}],$$

$$f_{ij} = -\frac{G(1+\nu)}{1-\nu}\,\alpha\,\dot{T}\,\delta_{ij}. \tag{6.87}$$

Plane stress problems can also be solved by using Eqs. (6.82) and (6.85) with $\hat{\sigma}^*_{jki} = \sigma^*_{jki}$, $\hat{\sigma}^*_{ijkl} = \sigma^*_{ijkl}$, ν replaced by $\bar{\nu}$ in all $(\,)^*$ tensors and the free term being given by

$$f_{ij} = -\frac{G}{4(1-\bar{\nu})}\,[2\,\dot{\varepsilon}^a_{ij} + \dot{\varepsilon}^a_{ll}\,\delta_{ij}]. \tag{6.88}$$

6.6.2. Initial Stress

In order to discuss the initial stress formulation, let us merely study the plastic strain rate integral presented in Eq. (6.78). Recalling expressions (5.59) and (5.60) we see that

$$\int_\Omega \sigma^*_{jki}\,\dot{\varepsilon}^a_{jk}\,d\Omega = \int_\Omega C_{jkrs}\,\varepsilon^*_{rsi}\,\dot{\varepsilon}^a_{jk}\,d\Omega; \tag{6.89}$$

by simple inspection of expression (5.31) we can make

$$C_{jkrs} = C_{rsjk}. \tag{6.90}$$

Moreover,

$$C_{rsjk}\,\dot{\varepsilon}^a_{jk} = \dot{\sigma}^a_{rs}, \tag{6.91}$$

where $\dot{\sigma}^a_{rs}$ was given in expression (6.47).

Thus,

$$\int_\Omega \sigma^*_{jki}\,\dot{\varepsilon}^a_{jk}\,d\Omega = \int_\Omega \varepsilon^*_{jki}\,\dot{\sigma}^a_{jk}\,d\Omega. \tag{6.92}$$

Hence the initial stress equation is seen to be equivalent to the initial strain. For plane strain problems the demonstration follows the same pattern and the corresponding expression is

$$\int_\Omega \hat{\sigma}^*_{jki}\,\dot{\varepsilon}^a_{jk}\,d\Omega = \int_\Omega \varepsilon^*_{jki}\,\dot{\sigma}^a_{jk}\,d\Omega. \tag{6.93}$$

In both cases the internal stresses can be computed by

$$\dot{\sigma}_{ij} = \int_\Gamma u^*_{ijk}\,\dot{p}_k\,d\Gamma - \int_\Gamma p^*_{ijk}\,\dot{u}_k\,d\Gamma + \int_\Omega u^*_{ijk}\,\dot{b}_k\,d\Omega + \int_\Omega \varepsilon^*_{ijkl}\,\dot{\sigma}^a_{kl}\,d\Omega + g_{ij}(\dot{\sigma}^a_{kl}) \tag{6.94}$$

where the integral of the initial stress term is in the principal value sense and the expressions for the free term are of form

$$g_{ij} = -\frac{1}{15(1-\nu)}[(7-5\nu)\,\dot{\sigma}^a_{ij} + (1-5\nu)\,\dot{\sigma}^a_{ll}\,\delta_{ij}] \qquad \text{for three dimensions} \tag{6.95}$$

and

$$g_{ij} = -\frac{1}{8(1-\nu)}[2\,\dot{\sigma}^a_{ij} + (1-4\nu)\,\dot{\sigma}^a_{ll}\,\delta_{ij}] \qquad \text{for two-dimensional plane strain.} \tag{6.96}$$

It is worth noting that for plane strains the initial stress integrals do not require the contribution of the work performed in the third direction, nor do the particular assumptions concerning $\dot{\varepsilon}^a_{ij}$ need to be made. This is because $\varepsilon^*_{33i} = 0$ and the effect of $\dot{\varepsilon}^a_{33}$ is already included into the components of $\dot{\sigma}^a_{ij}$. As a consequence, plane stress problems can be handled by the plane strain expressions with the replacement of ν by $\bar{\nu}$ being the only modification.

6.6.3. Fictitious Tractions and Body Forces

The last integral presented in Eq. (6.78) can be written in terms of the derivatives of u^*_{ij} as follows

$$\int_\Omega \sigma^*_{jki}\,\dot{\varepsilon}^a_{jk}\,d\Omega = \int_\Omega \left\{G\,(u^*_{ij,k} + u^*_{ik,j}) + \frac{2G\nu}{1-2\nu}\,u^*_{il,l}\,\delta_{jk}\right\}\dot{\varepsilon}^a_{jk}\,d\Omega \tag{6.97}$$

which after integrating by parts gives the identity

$$\int_\Omega \sigma^*_{jki}\,\dot{\varepsilon}^a_{jk}\,d\Omega = \int_\Gamma u^*_{ij}\,2G\left(\dot{\varepsilon}^a_{jk}\,n_k + \frac{\nu}{1-2\nu}\,\dot{\varepsilon}^a_{ll}\,n_j\right)d\Gamma$$

$$-\int_\Omega u^*_{ij}\,2G\left(\dot{\varepsilon}^a_{jk,k} + \frac{\nu}{1-2\nu}\,\dot{\varepsilon}^a_{ll,j}\right)d\Omega\,. \tag{6.98}$$

The substitution of expression (6.98) in Eq. (6.78) gives as a result

$$c_{ij}\,\dot{u}_j = \int_\Gamma u^*_{ij}\,\dot{\hat{p}}_j\,d\Gamma - \int_\Gamma p^*_{ij}\,\dot{u}_j\,d\Gamma + \int_\Omega u^*_{ij}\,\dot{\hat{b}}_j\,d\Omega\,, \tag{6.99}$$

where $\dot{\hat{b}}_j$ and $\dot{\hat{p}}_j$ were given in expressions (6.52) and (6.53), respectively.

Therefore, we have arrived at an inelastic formulation in which traction and body force rates are fictitious (depend on the inelastic strains), but the displacements are the actual ones. In order to apply Eq. (6.99) one has to be aware that although it looks like the elastic application of the boundary element technique, the internal stresses still have to be computed by use of Eqs. (6.41) and (6.45), i.e.,

$$\dot{\sigma}_{ij} = \int_\Gamma u^*_{ijk}\,\dot{\hat{p}}_k\,d\Gamma - \int_\Gamma p^*_{ijk}\,\dot{u}_k\,d\Gamma + \int_\Omega u^*_{ijk}\,\dot{\hat{b}}_k\,d\Omega - C_{ijkl}\,\dot{\varepsilon}^a_{kl}. \tag{6.100}$$

Another feature of this formulation is that in contrast with the two previous approaches, it needs computation of space derivatives of the inelastic strains (see expression (6.52)). This may be considered as a disadvantage for numerical implementation (since constant interpolation is ruled out), but, nevertheless, it is a valid procedure for formulating the BEM to inelastic problems.

Tensors related to fundamental solution

Finally, the new tensors related to the fundamental solutions that appeared in this section are of the following form:

$$\sigma^*_{ijkl} = \frac{G}{2\alpha\pi(1-\nu)\,r^\beta}\{\beta(1-2\nu)(\delta_{ij}\,r_{,k}\,r_{,l} + \delta_{kl}\,r_{,i}\,r_{,j})$$
$$+\beta\nu(\delta_{li}\,r_{,j}\,r_{,k} + \delta_{jk}\,r_{,l}\,r_{,i} + \delta_{ik}\,r_{,l}\,r_{,j} + \delta_{jl}\,r_{,i}\,r_{,k}) - \beta\gamma\,r_{,i}\,r_{,j}\,r_{,k}\,r_{,l}$$
$$+(1-2\nu)(\delta_{ik}\,\delta_{lj} + \delta_{jk}\,\delta_{li}) - (1-4\nu)\,\delta_{ij}\,\delta_{kl}\}, \tag{6.101}$$

$$\varepsilon^*_{ijkl} = \frac{1}{4\alpha\pi(1-\nu)\,r^\beta}\{(1-2\nu)(\delta_{ik}\,\delta_{lj} + \delta_{jk}\,\delta_{li} - \delta_{ij}\,\delta_{kl} + \beta\,\delta_{ij}\,r_{,k}\,r_{,l})$$
$$+\beta\nu(\delta_{li}\,r_{,j}\,r_{,k} + \delta_{jk}\,r_{,l}\,r_{,i} + \delta_{ik}\,r_{,l}\,r_{,j} + \delta_{jl}\,r_{,i}\,r_{,k})$$
$$+\beta\,\delta_{kl}\,r_{,i}\,r_{,j} - \beta\gamma\,r_{,i}\,r_{,j}\,r_{,k}\,r_{,l}\}, \tag{6.102}$$

where $\alpha = 2, 1$; $\beta = 3, 2$; $\gamma = 5, 4$ for three-dimensions and plane strain, respectively.

6.7. Half-Plane Formulations

The extension of the elastic half-plane boundary-element formulation to inelastic problems follows the same procedure as in the Kelvin implementation.

If we consider the inelastic strains to be incompressible, the starting equation for the initial strain formulation is given by

$$c_{ij}\,\dot{u}_j = \int_\Gamma u^*_{ij}\,\dot{p}_j\,d\Gamma - \int_{\Gamma'} p^*_{ij}\,\dot{u}_j\,d\Gamma + \int_\Omega u^*_{ij}\,\dot{b}_j\,d\Omega + \int_\Omega \hat{\sigma}^*_{jki}\,\dot{\varepsilon}^a_{jk}\,d\Omega \tag{6.103}$$

in which for plane strain problems the complementary part of the tensor that multiplies the inelastic strain rates is

$$\hat{\sigma}^c_{jki} = \sigma^c_{jki} - \nu\,\sigma^c_{lli}\,\delta_{jk}, \tag{6.104}$$

whereas for plane stress

$$\hat{\sigma}^c_{jki} = \sigma^c_{jki}. \tag{6.105}$$

Equation (6.103) is valid for any location of the load point on condition that c_{ij} and the integral over Γ' be properly interpreted as discussed in Chapter 5.

By suitably modifying the inelastic strain rate integral, an initial stress equation without the condition of incompressibility of the inelastic strains can be equally obtained for plane strains,

$$c_{ij}\,\dot{u}_j = \int_{\Gamma} u^*_{ij}\,\dot{p}_j\,d\Gamma - \int_{\Gamma'} p^*_{ij}\,\dot{u}_j\,d\Gamma + \int_{\Omega} u^*_{ij}\,\dot{b}_j\,d\Omega + \int_{\Omega} \varepsilon^*_{jki}\,\dot{\sigma}^a_{jk}\,d\Omega\,, \tag{6.106}$$

where, as explained in Section 6.6 plane stress problems can be dealt with by simply modifying the Poisson ratio.

In order to accurately compute the stress rates at interior points, the derivatives of Eq. (6.103) are combined to produce the expressions for the total strain rates and then substituted into Eq. (6.45). Here one notices that, due to the nonsingular nature of the complementary tensors, the derivatives of the inelastic strain rate integral create exactly the same singularities obtained for the single Kelvin implementation. Hence, for plane strains one has

$$\begin{aligned}\dot{\sigma}_{ij} = &\int_{\Gamma} u^*_{ijk}\,\dot{p}_k\,d\Gamma - \int_{\Gamma'} p^*_{ijk}\,\dot{u}_k\,d\Gamma + \int_{\Omega} u^*_{ijk}\,\dot{b}_k\,d\Omega\\ &+ \int_{\Omega} \hat{\sigma}^*_{ijkl}\,\dot{\varepsilon}^a_{kl}\,d\Omega + f_{ij}(\dot{\varepsilon}^a_{kl})\end{aligned} \tag{6.107}$$

in which the inelastic strain integral is to be computed in the principal value sense and f_{ij} is the same free term obtained for the Kelvin formulation, i.e.,

$$f_{ij} = -\frac{G}{4(1-\nu)}\left[2\,\dot{\varepsilon}^a_{ij} + (1-4\nu)\,\dot{\varepsilon}^a_{ll}\,\delta_{ij}\right]. \tag{6.108}$$

In addition,

$$\hat{\sigma}^c_{ijkl} = \sigma^c_{ijkl} - \nu\,\sigma^c_{ijmm}\,\delta_{kl} \tag{6.109}$$

and

$$\sigma^c_{ijkl} = G\left(\frac{\partial\sigma^c_{kli}}{\partial x_j} + \frac{\partial\sigma^c_{klj}}{\partial x_i}\right) + \frac{2G\nu}{1-2\nu}\,\frac{\partial\sigma^c_{klm}}{\partial x_m}\,\delta_{ij} \tag{6.110}$$

where the derivatives are taken with reference to the load point and the expressions for u^c_{ijk} and p^c_{ijk} were given in Ref. [42].

An interesting feature of the half-plane implementation is that if the problem to be analyzed satisfies the traction-free condition ($\dot{p}_k = 0$) over some part of the boundary $\Gamma - \Gamma'$, stresses at points located along this part of the boundary can be computed as if they were internal points. In order to validate Eq. (6.107) for such cases, only the expression of f_{ij} needs to be modified to take into consideration the limiting case $x_1(\xi) = 0$. This expression can be easily obtained as follows [13]: let us assume a semicircular free body, of radius ϱ, whose straight boundary is contained by the surface of the half-plane (see Fig. 6.10). If body forces are not considered, the application of a uniform plastic strain field ($\bar{\varepsilon}^p_{ij}$) to this body will only produce displacements; internal stresses and tractions remain zero throughout the process.

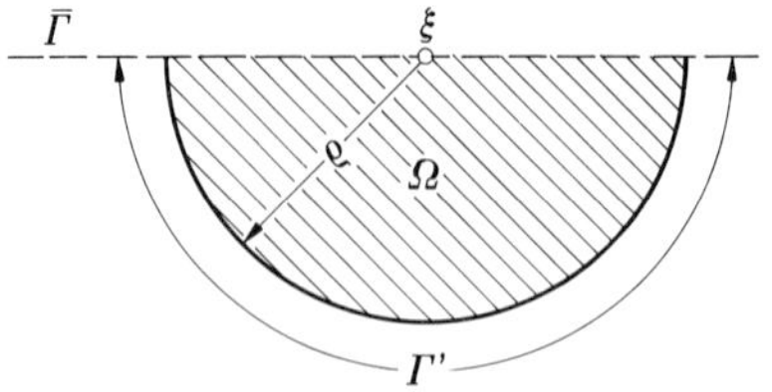

Fig. 6.10. Semicircular free body under a constant plastic strain field

The application of Eq. (6.107) to represent the stresses at the center ξ of the semicircle leads directly to

$$\int_{\Gamma'} p^*_{ijk}\,\bar{u}_k\,d\Gamma = \bar{\varepsilon}^p_{kl} \int_\Omega \hat{\sigma}^*_{ijkl}\,d\Omega + f_{ij}(\bar{\varepsilon}^p_{kl}); \tag{6.111}$$

moreover, from the condition of existence of the principal value (see expression (6.76)), one can prove that

$$\int_\Omega \hat{\sigma}^*_{ijkl}\,d\Omega = 0\,. \tag{6.112}$$

Hence,

$$f_{ij}(\bar{\varepsilon}^p_{kl}) = \int_{\Gamma'} p^*_{ijk}\,\bar{u}_k\,d\Gamma \tag{6.113}$$

where the relevant boundary displacements (neglecting rigid-body movements) can be computed by [7]

$$\bar{u}_i = \varrho\,(\bar{\varepsilon}^p_{ij} - \nu\,\bar{\varepsilon}^p_{kk}\,\delta_{ij})\,n_j \tag{6.114}$$

in which n_j represents the direction cosines of the outward normal to the curved boundary.

Equation (6.113) therefore provides the required expression for f_{ij} when $x_1(\xi) = 0$:

$$\begin{aligned} f_{11} &= f_{12} = 0\,,\\ f_{22} &= -\frac{G}{2(1-\nu)}\,(\dot{\varepsilon}^a_{22} - \dot{\varepsilon}^a_{11})\,. \end{aligned} \tag{6.115}$$

For the initial stress formulation the procedure is entirely similar and the equation equivalent to Eq. (6.107) is of the following form:

$$\begin{aligned} \dot{\sigma}_{ij} = &\int_\Gamma u^*_{ijk}\,\dot{p}_k\,d\Gamma - \int_{\Gamma'} p^*_{ijk}\,\dot{u}_k\,d\Gamma + \int_\Omega u^*_{ijk}\,\dot{b}_k\,d\Omega\\ &+ \int_\Omega \varepsilon^*_{ijkl}\,\dot{\sigma}^a_{kl}\,d\Omega + g_{ij}(\dot{\sigma}^a_{kl})\,, \end{aligned} \tag{6.116}$$

where ε^c_{ijkl} is obtained from Eq. (6.110) by substituting ε^c_{jki} for σ^c_{jki},

$$g_{ij} = -\frac{1}{8(1-\nu)}\,[2\,\dot{\sigma}^a_{ij} + (1-4\,\nu)\,\dot{\sigma}^a_{ll}\,\delta_{ij}] \qquad \text{for } x_1(\xi) > 0 \tag{6.117}$$

and

$$g_{11} = g_{12} = 0,$$
$$g_{22} = -\frac{1}{4(1-\nu)}[\dot{\sigma}^a_{22} - \dot{\sigma}^a_{11}] \qquad \text{for } x_1(\xi) = 0. \tag{6.118}$$

Plane stress problems can be handled by Eqs. (6.107) and (6.116) if Poisson's ratio is modified as before, $\hat{\sigma}^*_{ijkl} = \sigma^*_{ijkl}$, and expression (6.108) is replaced by

$$f_{ij} = -\frac{G}{4(1-\bar{\nu})}[2\,\dot{\varepsilon}^a_{ij} + \dot{\varepsilon}^a_{ll}\,\delta_{ij}]. \tag{6.119}$$

Notice that expressions (6.115), (6.117), and (6.118) still remain valid.

6.8. Spatial Discretization

The spatial discretization of the equations previously presented is illustrated in this section for two-dimensional problems. The boundary of the body is assumed to be represented by surface elements as discussed in Chapter 5 and the part of the domain in which nonzero inelastic strains are expected to develop is discretized using internal cells for integration. The body force term, though not causing any difficulty for implementation, is not considered for simplicity.

The discretization of the boundary integrals has been thoroughly discussed in Section 5.8 and the same procedure can be followed here. Consequently, emphasis will be given to the domain integrals of the inelastic terms.

It is instructive to start by assuming an initial strain formulation in which the inelastic strains are incompressible. Thus, let us now concentrate attention to the equation

$$c_{ij}\,\dot{u}_j + \int_\Gamma p^*_{ij}\,\dot{u}_j\,d\Gamma = \int_\Gamma u^*_{ij}\,\dot{p}_j\,d\Gamma + \int_\Omega \hat{\sigma}^*_{jki}\,\dot{\varepsilon}^a_{jk}\,d\Omega \tag{6.120}$$

where the appropriate alterations for half-plane problems are implied.

For the domain discretization of Eq. (6.120) the Cartesian coordinates $\boldsymbol{x}$ of points located within each cell Ω_j can be represented by the following equation

$$\boldsymbol{x} = \bar{\boldsymbol{\Psi}}^T \boldsymbol{x}^m, \tag{6.121}$$

where $\bar{\boldsymbol{\Psi}}$ represents the interpolation functions and $\boldsymbol{x}^m$ the coordinates of some special points which define the geometry of the cell.

The inelastic strain rates are assumed to be interpolated within the cell in the form

$$\dot{\boldsymbol{\varepsilon}}^{a(j)} = \bar{\boldsymbol{\Phi}}^T \dot{\boldsymbol{\varepsilon}}^{a,n} \tag{6.122}$$

in which $\bar{\boldsymbol{\Phi}}$ stands for the interpolation functions and $\dot{\boldsymbol{\varepsilon}}^{a,n}$ for the values of the inelastic strain rates at a certain number of stress points (equivalent to nodal points in two-dimensional finite elements).

Assuming N boundary elements and M internal cells, the discretized version of Eq. (6.120) for a boundary node ξ_i is given by

$$\begin{aligned} \boldsymbol{c}(\xi_i)\,\dot{\boldsymbol{u}}(\xi_i) + \sum_{j=1}^{N}\left(\int_{\Gamma_j} \boldsymbol{p}^*\,\boldsymbol{\Phi}^T d\Gamma\right)\dot{\boldsymbol{u}}^n \\ = \sum_{j=1}^{N}\left(\int_{\Gamma_j} \boldsymbol{u}^*\,\boldsymbol{\Phi}^T d\Gamma\right)\dot{\boldsymbol{p}}^n + \sum_{j=1}^{M}\left(\int_{\Omega_j} \hat{\boldsymbol{\sigma}}^*\,\bar{\boldsymbol{\Phi}}^T d\Omega\right)\dot{\boldsymbol{\varepsilon}}^{a,n} \end{aligned} \tag{6.123}$$

and the same form is valid for an internal stress point $\xi_i(\boldsymbol{c} = \boldsymbol{I})$.

For general purposes it is convenient to compute the cell integrals by using a suitable numerical quadrature scheme; e.g., for triangular cells, Hammer's integration formulas can be used:

$$\int_{\Omega_j} \hat{\boldsymbol{\sigma}}^*\,\bar{\boldsymbol{\Phi}}^T d\Omega = \sum_{k=1}^{K} |\boldsymbol{J}|_k\, w_k\,(\hat{\boldsymbol{\sigma}}^*\,\bar{\boldsymbol{\Phi}}^T)_k, \tag{6.124}$$

where K is the number of integration points, w_k is the associated weighting factor, and $|\boldsymbol{J}|$ is the Jacobian of the coordinate transformation which allows for the representation of the interpolation functions in terms of a homogeneous coordinate system (η_1, η_2). Note that these integrals present integrable singularities when the singular node ξ_i lies on the cell Ω_j. Thus, some special care must be taken in such cases.

The application of Eq. (6.123) to all boundary nodes generates the following matrix relationship:

$$\boldsymbol{H}\dot{\boldsymbol{U}} = \boldsymbol{G}\dot{\boldsymbol{P}} + \boldsymbol{D}\,\dot{\boldsymbol{\varepsilon}}^a, \tag{6.125}$$

where matrices $\boldsymbol{H}$ and $\boldsymbol{G}$ are the same as those obtained for elastic analysis and matrix $\boldsymbol{D}$ is due to the inelastic strain integral.

Computation of stress rates at internal points follows a similar procedure. Here, the equation equivalent to Eq. (6.120) is given by

$$\dot{\sigma}_{ij} = \int_{\Gamma} u^*_{ijk}\,\dot{p}_k\,d\Gamma - \int_{\Gamma} p^*_{ijk}\,\dot{u}_k\,d\Gamma + \int_{\Omega} \hat{\sigma}^*_{ijkl}\,\dot{\varepsilon}^a_{kl}\,d\Omega + f_{ij}(\dot{\varepsilon}^a_{kl}) \tag{6.126}$$

where as before the appropriate alterations for half-plane problems are implied.

Application of Eq. (6.126) in discretized form leads to

$$\begin{aligned} \dot{\boldsymbol{\sigma}}(\xi_i) = \sum_{j=1}^{N}\left(\int_{\Gamma_j} {}'\boldsymbol{u}^*\,\boldsymbol{\Phi}^T d\Gamma\right)\dot{\boldsymbol{p}}^n - \sum_{j=1}^{N}\left(\int_{\Gamma_j} {}'\boldsymbol{p}^*\,\boldsymbol{\Phi}^T d\Gamma\right)\dot{\boldsymbol{u}}^n \\ + \sum_{j=1}^{M}\left(\int_{\Omega_j} {}'\hat{\boldsymbol{\sigma}}^*\,\bar{\boldsymbol{\Phi}}^T d\Omega\right)\dot{\boldsymbol{\varepsilon}}^{a,n} + \boldsymbol{c}'(\xi_i)\,\dot{\boldsymbol{\varepsilon}}^a(\xi_i) \end{aligned} \tag{6.127}$$

for a stress point ξ_i.

As before, numerical integration schemes can be used for integrating over the cells. In this case, however, when the singular point ξ_i lies on the cell Ω_j some of the integrals are only possible in the principal value sense. Here we recall a general procedure devised by Telles and Brebbia [7] which provides indirect means of computing such principal values for any kind of interpolation functions or cells shape. This procedure is based on the application of a constant inelastic strain field to the discretized integral equations. For simplicity, and without loss of generality, the use of triangular cells will be used to illustrate the idea. Thus, let us consider the part of the domain Ω which is represented by all the adjacent cells connected to the singular stress point ξ_i (see Fig. 6.11).

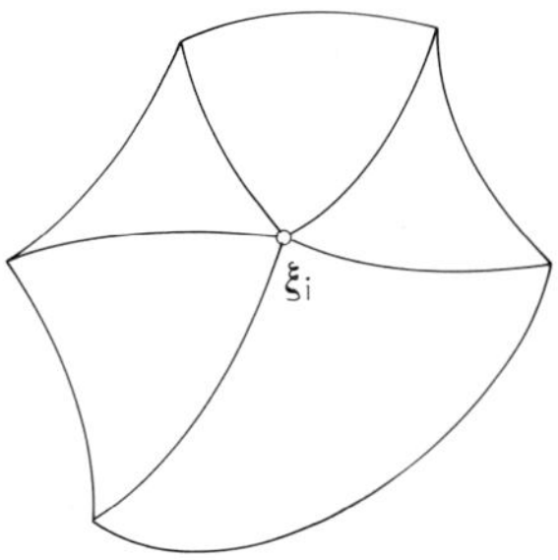

Fig. 6.11. Internal region corresponding to all the adjacent cells connected to the stress point ξ_i

The application of a constant inelastic strain field $\bar{\varepsilon}^a_{ij}$ to the reduced free body depicted in Fig. 6.11 will produce zero stresses and tractions. Consequently, Eq. (6.126) in discretized fashion can be employed to represent the zero stresses at the internal point ξ_i. This leads to the following equation

$$\sum_{j=1}^{\bar{M}} \left(\int_{\Omega_j} {}'\hat{\boldsymbol{\sigma}}^* \, \bar{\boldsymbol{\Phi}}^T d\Omega \right) \bar{\boldsymbol{\varepsilon}}^{a,n} + \boldsymbol{c}'(\xi_i) \, \bar{\boldsymbol{\varepsilon}}^a = \sum_{j=1}^{\bar{N}} \left(\int_{\hat{\Gamma}_j} {}'\boldsymbol{p}^* \, \boldsymbol{\Phi}^T d\Gamma \right) \bar{\boldsymbol{u}}^n, \tag{6.128}$$

where $\bar{M}$ corresponds to the reduced number of cells and similarly and $\bar{N}$ corresponds to the number of fictitious boundary elements $\hat{\Gamma}_j$ that represent the sides of the cells which are opposite to ξ_i (note that $\bar{M} \neq \bar{N}$ in general). Also, for plane strain one has [7]

$$\bar{u}_i = (\bar{\varepsilon}^a_{ij} - \nu \, \bar{\varepsilon}^a_{kk} \, \delta_{ij}) \, \Delta x_j, \tag{6.129}$$

and for plane stress

$$\bar{u}_i = \bar{\varepsilon}^a_{ij} \, \Delta x_j \tag{6.130}$$

in which Δx_j is the difference between the x_j coordinate of the field point and the reference point ξ_i.

One can easily verify that completely independent of the interpolation functions or cells shape adopted, each cell will always have three principal value integrals to be computed for each stress component (i.e., the terms that multiply $\dot{\varepsilon}^a_{ij}(\xi_i)$ in Eq. (6.127)). Therefore, since $\bar{\boldsymbol{u}}^n$ is computed analytically, such principal values together with the free term $\boldsymbol{c}'(\xi_i)$ are obtained by simply applying Eq. (6.128) to represent three independent sets of constant inelastic strains of the form $\bar{\varepsilon}^a_{ij} = \delta_{1i} \, \delta_{1j}$, $\bar{\varepsilon}^a_{ij} = 1 - \delta_{ij}$, and $\bar{\varepsilon}^a_{ij} = \delta_{2i} \, \delta_{2j}$.

It should be noted that for computational purposes the procedure can be applied to each cell at a time; i.e., after performing the allowable integrals over the cell, the same routine that performs the boundary integrals is called to integrate over the opposite side of the cell. The appropriate operations are then carried out and the total set of integrals becomes ready to be assembled in complete form, including not only the principal values but also the partial contribution of the corresponding terms of $\boldsymbol{c}'$.

Equation (6.127) when applied to all internal stress points yields

$$\dot{\boldsymbol{\sigma}} = \boldsymbol{G}'\,\dot{\boldsymbol{P}} - \boldsymbol{H}'\,\dot{\boldsymbol{U}} + (\boldsymbol{D}' + \boldsymbol{C}')\,\dot{\boldsymbol{\varepsilon}}^a\,, \tag{6.131}$$

where $\boldsymbol{C}'$ is a well-defined matrix that represents the free terms (last term in Eq. (6.127)) and $\boldsymbol{D}'$ is due to the inelastic strain integral. Matrices $\boldsymbol{H}'$ and $\boldsymbol{G}'$ correspond to the boundary integrals in a similar fashion to $\boldsymbol{H}$ and $\boldsymbol{G}$.

Note that Eq. (6.131) is only valid for internal stress points and possibly for stress points over the traction-free part of the boundary $\Gamma - \Gamma'$. Therefore, in order to compute the stress rates at boundary nodes, different expressions must be used. Such expressions are obtained by means of strain-displacement relationships and traction rate values along each boundary element. They do not require any integration and their elastic form was presented in Chapter 5. Thus, by employing an entirely similar procedure, the stress rates at boundary nodes can be obtained with reference to the local coordinate system $\bar{x}_i$ by (plane strain)

$$\begin{aligned}
\dot{\bar{\sigma}}_{11} &= \frac{1}{1-\nu}\,(2G\,\dot{\bar{\varepsilon}}_{11} + \nu\,\dot{\bar{\sigma}}_{22}) + 2G\left(\frac{\nu}{1-\nu}\,\dot{\bar{\varepsilon}}^a_{22} - \dot{\bar{\varepsilon}}^a_{11}\right),\\
\dot{\bar{\sigma}}_{12} &= \dot{\bar{p}}_1\,,\\
\dot{\bar{\sigma}}_{22} &= \dot{\bar{p}}_2\,,
\end{aligned} \tag{6.132}$$

where for plane stress the first equation is replaced by

$$\dot{\bar{\sigma}}_{11} = \frac{1}{1-\bar{\nu}}\,(2G\,\dot{\bar{\varepsilon}}_{11} + \bar{\nu}\,\dot{\bar{\sigma}}_{22}) + \frac{2G}{1-\bar{\nu}}\,\dot{\bar{\varepsilon}}^a_{11}\,. \tag{6.133}$$

From the above it is seen that such simple expressions can be assembled into Eq. (6.131) and computation of stress rates at all stress points becomes possible in a unified manner. Note that after performing the required coordinate transformations (local to global axes), the above expressions are ready to be assembled into the corresponding global matrices. Here, the contribution of adjacent elements to the common boundary nodes should be automatically averaged for nondouble nodes.

It is obvious that what has been discussed also applies for the initial stress equations. In this case, Eqs. (6.125) and (6.131) are replaced by

$$\boldsymbol{H}\dot{\boldsymbol{U}} = \boldsymbol{G}\dot{\boldsymbol{P}} + \boldsymbol{Q}\,\dot{\boldsymbol{\sigma}}^a \tag{6.134}$$

and

$$\dot{\boldsymbol{\sigma}} = \boldsymbol{G}'\,\dot{\boldsymbol{P}} - \boldsymbol{H}'\,\dot{\boldsymbol{U}} + (\boldsymbol{Q}' + \boldsymbol{E}')\,\dot{\boldsymbol{\sigma}}^a \tag{6.135}$$

in which the part of $\boldsymbol{E}'$ that corresponds to the integral equations stands for the free terms g_{ij} and the equivalent part of $\boldsymbol{Q}'$ together with $\boldsymbol{Q}$ are due to the initial stress integrals. Notice that the principal values of $\boldsymbol{Q}'$ can still be computed by applying a state of constant initial stresses in much the same way as it was shown for the initial strain equation. In addition, computation of stress rates at selected boundary nodes is included into Eq. (6.135) by following the procedure previously indicated with the first equation in (6.132) replaced by

$$\dot{\bar{\sigma}}_{11} = \frac{1}{1-\nu}\,(2G\,\dot{\bar{\varepsilon}}_{11} + \nu\,\dot{\bar{\sigma}}_{22}) + \frac{\nu}{1-\nu}\,\dot{\bar{\sigma}}_{22}^{a} - \dot{\bar{\sigma}}_{11}^{a} \tag{6.136}$$

where for plane stress ν is replaced by $\bar{\nu}$. Note that the condition of incompressibility of the inelastic strains is no longer included.

In order to minimize the computer effort for the solution of inelastic problems, let us reexamine Eqs. (6.125) and (6.131). For a well-posed problem, a sufficient number of tractions and boundary displacements needs to be prescribed. The unknowns are then reordered, leading to

$$\boldsymbol{A}\,\dot{\boldsymbol{Y}} = \dot{\boldsymbol{F}} + \boldsymbol{D}\,\dot{\boldsymbol{\varepsilon}}^{a} \tag{6.137}$$

and, similarly,

$$\dot{\boldsymbol{\sigma}} = -\,\boldsymbol{A}'\,\dot{\boldsymbol{Y}} + \dot{\boldsymbol{F}}' + \boldsymbol{D}^{*}\,\dot{\boldsymbol{\varepsilon}}^{a}\,, \tag{6.138}$$

where $\boldsymbol{D}^{*} = \boldsymbol{D}' + \boldsymbol{C}'$ and the contribution of the prescribed values is included in vectors $\dot{\boldsymbol{F}}$ and $\dot{\boldsymbol{F}}'$.

From the multiplication of Eq. (6.137) by $\boldsymbol{A}^{-1}$ comes

$$\dot{\boldsymbol{Y}} = \boldsymbol{K}\,\dot{\boldsymbol{\varepsilon}}^{a} + \dot{\boldsymbol{M}}\,, \tag{6.139}$$

where

$$\boldsymbol{K} = \boldsymbol{A}^{-1}\boldsymbol{D} \tag{6.140}$$

and

$$\dot{\boldsymbol{M}} = \boldsymbol{A}^{-1}\dot{\boldsymbol{F}}\,. \tag{6.141}$$

Substituting Eq. (6.139) into Eq. (6.138) yields

$$\dot{\boldsymbol{\sigma}} = \boldsymbol{B}\,\dot{\boldsymbol{\varepsilon}}^{a} + \dot{\boldsymbol{N}} \tag{6.142}$$

in which

$$\boldsymbol{B} = \boldsymbol{D}^{*} - \boldsymbol{A}'\boldsymbol{K} \tag{6.143}$$

and

$$\dot{\boldsymbol{N}} = \dot{\boldsymbol{F}}' - \boldsymbol{A}'\dot{\boldsymbol{M}}\,. \tag{6.144}$$

Note that the elastic solution to the rate problem is given by the vectors $\dot{\boldsymbol{M}}$ and $\dot{\boldsymbol{N}}$.

From the above it is seen that Eq. (6.142) represents a single recursive expression which relates stress rates at selected boundary nodes and internal points to the corresponding inelastic strains and the elastic solution. Also, this expresssion is now independent of the boundary equation (6.139) and provides a useful means of solving nonlinear material problems. This will be the subject of Chapter 7.

In terms of efficiently programming, it should be noted that initially matrix $\boldsymbol{A}$ is assembled in the array of matrix $\boldsymbol{B}$ and that once the system of equations is solved, Eq. (6.142) is generated in such a way that only matrix $\boldsymbol{B}$ is actually formed. Thus, only $\boldsymbol{K}$ and $\boldsymbol{B}$ require storage.

Finally, the same matrix manipulations can be performed in the initial stress equations (6.134) and (6.135). In this case, however, a slight modification in Eq. (6.135) has proven to be convenient for elastoplastic problems and this will be discussed in Chapter 7. The initial stress equations are also employed in Chapter 8 where problems concerning viscoplasticity and creep are discussed.

6.9. Internal Cells

The implementation of boundary elements and internal cells can be accomplished by following the procedures previously discussed and using the interpolation functions described in Chapter 3. The internal cell integrals can be carried out by using numerical integration formulas or, in some simple cases, analytically. For more complex implementations, analytical integration becomes very difficult and numerical or partially numerical procedures appear to be the only viable alternatives.

A semianalytical integration scheme has been proposed by Telles and Brebbia [8] which is particularly attractive when the singular node or stress point coincides with one of the cell points. Therefore, this integration scheme will be outlined below for triangular cells with linear interpolation functions.

Consider the triangular cell shown in Fig. 6.12. For this cell the interpolation functions are expressed in terms of a homogeneous coordinate system (η_1, η_2) and the Jacobian $|\boldsymbol{J}|$ indicated in expression (6.124) is simply twice the area of the triangle. In addition, the interpolation functions are given by (see Eq. (6.122))

$$\bar{\boldsymbol{\Phi}}^T = [\boldsymbol{I}\eta_1 \;\; \boldsymbol{I}\eta_2 \;\; \boldsymbol{I}\eta_3] , \tag{6.145}$$

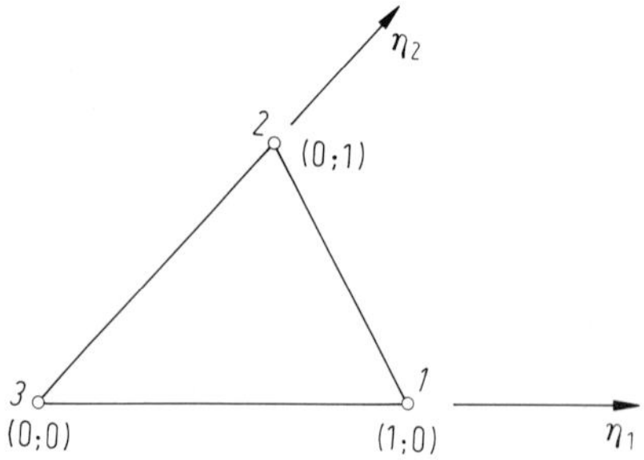

Fig. 6.12. Triangular cell and definition of intrinsic coordinate system (η_1, η_2)

where $\eta_3 = 1 - \eta_2 - \eta_1$ and $\boldsymbol{I}$ is the identity matrix of order three. Note that the relation between η_α and the Cartesian coordinate system (x, y) has been presented in Chapter 3.

For the total computation of matrix $\boldsymbol{D}$ each cell will contribute with 2×9 matrices of the following form:

$$\boldsymbol{d} = \int_{\Omega_\Delta} \{[\hat{\boldsymbol{\sigma}}^* \eta_1 \;\; \hat{\boldsymbol{\sigma}}^* \eta_2 \;\; \hat{\boldsymbol{\sigma}}^* \eta_3]\} \, d\Omega \tag{6.146}$$

where

$$\hat{\boldsymbol{\sigma}}^* = \begin{bmatrix} \hat{\sigma}^*_{111} & 2\hat{\sigma}^*_{121} & \hat{\sigma}^*_{221} \\ \hat{\sigma}^*_{112} & 2\hat{\sigma}^*_{122} & \hat{\sigma}^*_{222} \end{bmatrix}. \tag{6.147}$$

To illustrate the present semianalytical integration scheme, the case when the singular node coincides with one of the cell points will be studied. Let us consider the typical expression

$$\boldsymbol{d}^\alpha = \int_{\Omega_\Delta} \hat{\boldsymbol{\sigma}}^* \eta_\alpha \, d\Omega \,, \tag{6.148}$$

where $\boldsymbol{d}^\alpha$ represents the 2×3 submatrix α of $\boldsymbol{d}$.

In order to perform the integral, one can define a cylindrical coordinate system (r, φ) based at the singular point γ as shown in Fig. 6.13. In this system of coordinates the tensor $\hat{\sigma}^*_{jki}$ (Kelvin) can be represented by

$$\frac{1}{r} \psi_{jki} \tag{6.149}$$

in which ψ_{jki} is a function of φ only. Expression (6.148) can then be written

$$\boldsymbol{d}^\alpha = \lim_{\varepsilon \to 0} \int_{\varphi_1}^{\varphi_2} \int_{\varepsilon}^{R(\varphi)} \boldsymbol{\psi} \, \bar{\eta}_\alpha \, dr \, d\varphi \,, \tag{6.150}$$

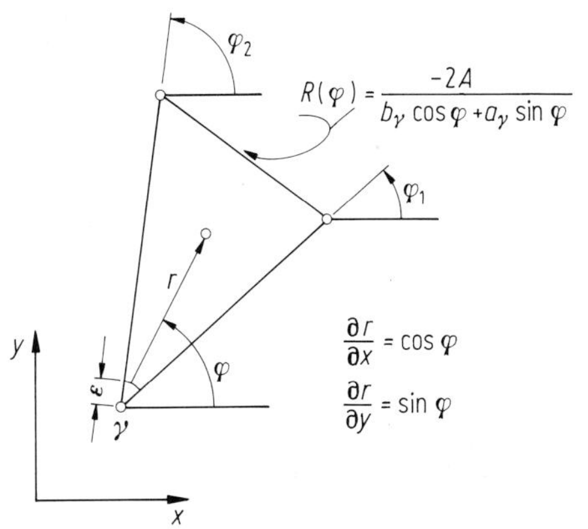

Fig. 6.13. Cylindrical coordinate system based at the singular point γ

where $\bar{\eta}_\alpha$ is the interpolation function with reference to the (r, φ) system:

$$\bar{\eta}_\alpha = {}^\gamma\eta_\alpha + \frac{r}{2A}(b_\alpha \cos\varphi + a_\alpha \sin\varphi)\,, \tag{6.151}$$

${}^\gamma\eta_\alpha$ being the value of the interpolation function at the singular point γ, i.e.,

$$\begin{aligned} {}^\gamma\eta_\alpha &= 0 \qquad \text{for } \alpha \neq \gamma \\ &= 1 \qquad \text{for } \alpha = \gamma\,. \end{aligned} \tag{6.152}$$

Notice that a_α, b_α and A have been defined in Chapter 3.

Since $\boldsymbol{\psi}$ does not depend on r, one can easily integrate with respect to this variable, and take the limit for $\varepsilon \to 0$. The following expressions are then obtained:

$$\boldsymbol{d}^\alpha = A \int_{\varphi_1}^{\varphi_2} \boldsymbol{\psi} \left\{ \frac{b_\alpha \cos\varphi + a_\alpha \sin\varphi}{(b_\gamma \cos\varphi + a_\gamma \sin\varphi)^2} \right\} d\varphi \qquad \text{for } \alpha \neq \gamma \tag{6.153}$$

and

$$\boldsymbol{d}^\alpha = -A \int_{\varphi_1}^{\varphi_2} \boldsymbol{\psi} \left\{ \frac{1}{b_\gamma \cos\varphi + a_\gamma \sin\varphi} \right\} d\varphi \qquad \text{for } \alpha = \gamma\,. \tag{6.154}$$

The advantage of integrating analytically with respect to r is now evident as very simple expressions are achieved and the singularity of the $\hat{\sigma}^*_{jki}$ tensor is removed. Integration with respect to φ does not present any problems and one can use standard one-dimensional Gaussian quadrature formulas. This can be done by simply expressing the variable φ as

$$\varphi = \frac{\eta}{2}(\varphi_2 - \varphi_1) + \frac{1}{2}(\varphi_2 + \varphi_1)\,, \tag{6.155}$$

where η is defined on the interval $[-1, 1]$.

Computation of matrix $\boldsymbol{D}'$ follows the same pattern with the difference that each cell contributes with 3×9 matrices of the form

$$'\boldsymbol{d} = \int_{\Omega_\Delta} \{[{}'\hat{\boldsymbol{\sigma}}^* \eta_1 \; {}'\hat{\boldsymbol{\sigma}}^* \eta_2 \; {}'\hat{\boldsymbol{\sigma}}^* \eta_3]\}\, d\Omega\,, \tag{6.156}$$

where

$$'\hat{\boldsymbol{\sigma}}^* = \begin{bmatrix} \hat{\sigma}^*_{1111} & 2\hat{\sigma}^*_{1112} & \hat{\sigma}^*_{1122} \\ \hat{\sigma}^*_{1211} & 2\hat{\sigma}^*_{1212} & \hat{\sigma}^*_{1222} \\ \hat{\sigma}^*_{2211} & 2\hat{\sigma}^*_{2212} & \hat{\sigma}^*_{2222} \end{bmatrix}\,. \tag{6.157}$$

Typically one can consider

$$'\boldsymbol{d}^\alpha = \int_{\Omega_\Delta} {}'\hat{\boldsymbol{\sigma}}^* \eta_\alpha\, d\Omega\,. \tag{6.158}$$

As before, in order to illustrate the semianalytical process of integration, the case when the singular stress point coincides with one of the cell points will be studied. In this case, expression (6.158) presents a further difficulty for $\alpha = \gamma$, which means that integration is only possible in the principal value sense taking into consideration the contribution of all the adjacent cells connected to γ.

Introducing the cylindrical coordinate system of Fig. 6.13, the Kelvin tensor $\hat{\sigma}^*_{ijkl}$ can be written

$$\frac{1}{r^2}\,\psi_{ijkl}\,, \tag{6.159}$$

where ψ_{ijkl} is a function of φ only.

Expression (6.158) turns to be represented by

$$'\boldsymbol{d}^\alpha = \lim_{\varepsilon\to 0} \int_{\varphi_1}^{\varphi_2} \int_{\varepsilon}^{R(\varphi)} {}'\psi\,\frac{\bar{\eta}_\alpha}{r}\,dr\,d\varphi\,. \tag{6.160}$$

For $\alpha \neq \gamma$, after integrating with respect to r and performing the limit for $\varepsilon \to 0$, expression (6.160) gives

$$'\boldsymbol{d}^\alpha = -\int_{\varphi_1}^{\varphi_2} {}'\psi \left\{\frac{b_\alpha\cos\varphi + a_\alpha\sin\varphi}{b_\gamma\cos\varphi + a_\gamma\sin\varphi}\right\} d\varphi\,. \tag{6.161}$$

When $\alpha = \gamma$, integration with respect to r gives

$$'\boldsymbol{d}^\alpha = \lim_{\varepsilon\to 0} \int_{\varphi_1}^{\varphi_2} {}'\psi \left\{\ln\left(\frac{-2A}{b_\gamma\cos\varphi + a_\gamma\sin\varphi}\right) - 1 - \frac{\varepsilon}{2A}(b_\gamma\cos\varphi + a_\gamma\sin\varphi) - \ln(\varepsilon)\right\} d\varphi\,. \tag{6.162}$$

To calculate the principal value let us first consider the following part of (6.162):

$$\omega = -\int_{\varphi_1}^{\varphi_2} {}'\psi\,[1 + \ln(\varepsilon)]\,d\varphi\,. \tag{6.163}$$

It is easily verified that in order to compute the contribution of all the adjacent cells connected to γ, one simply has to modify the integration limits in expression (6.163). This gives, due to an intrinsic property of ψ_{ijkl} (see expression (6.76)),

$$\omega = -[(1 + \ln(\varepsilon)]\int_0^{2\pi} {}'\psi\,d\varphi = \boldsymbol{0}\,. \tag{6.164}$$

Once the singular term has dropped, one can consider the rest of expression (6.162) and take the limit for $\varepsilon \to 0$. The following expression then arises:

$$'\boldsymbol{d}^\alpha = \int_{\varphi_1}^{\varphi_2} {}'\psi\,\ln\left(\frac{-2A}{b_\gamma\cos\varphi + a_\gamma\sin\varphi}\right) d\varphi\,. \tag{6.165}$$

As before, one dimensional Gaussian quadrature formulas can now be used to integrate expressions (6.161) and (6.165).

The general case when the singular node or stress point γ is not coincident with one of the cell points follows the same process. Here, the use of a semianalytical integration procedure is not so important since the integrals are always regular. Nevertheless, this procedure is still recommended in order to save computer time.

Notice that since the Kelvin tensors ε^*_{jki} and ε^*_{ijkl} can be cast into the form presented in Eqs. (6.149) and (6.159), the same integration scheme can be applied to the initial stress formulation. In addition, in order to optimize the procedure, the angle $(\varphi_2 - \varphi_1)$ can be used as an error-controlling factor for the numerical part of the integral. In practical terms, a reduced number of less than five integration points is found to be required.

The above scheme has been extensively used in computer programs and has proved to be very efficient.

With reference to the half-plane formulations, it has been shown (see Appendix B) that the complementary part of the expressions present no singularities when $c > 0$. Consequently, simple quadrature formulas can be used for the corresponding domain integrals. The limiting case $c = 0$ has been seen to produce singularities of the same order as those present in Kelvin's, therefore the two parts of the fundamental solution are added up and the integrals are properly computed using the same integration scheme normally employed for the Kelvin part.

6.10. Axisymmetric Case

Axisymmetric inelastic problems can be dealt with by following exactly the same procedures introduced in this chapter and using the appropriate fundamental solution presented in Section 5.15. In order to illustrate the procedure, the initial strain approach will be taken further. In this case, leaving aside the torsion terms, the inelastic strain rates are included into the analysis through the last domain integral of the equation presented below (see Eq. (5.182):

$$\begin{aligned} &c_{ij}(\xi)\,\dot{u}_j(\xi) + 2\pi \int_\Gamma p^*_{ij}(\xi, x)\,\dot{u}_j(x)\,r(x)\,d\Gamma(x) \\ &= 2\pi \int_\Gamma u^*_{ij}(\xi, x)\,\dot{p}_j(x)\,r(x)\,d\Gamma(x) + 2\pi \int_\Omega u^*_{ij}(\xi, x)\,\dot{b}_j(x)\,r(x)\,d\Omega(x) \\ &\quad + 2\pi \int_\Omega \sigma^*_{\alpha\beta i}(\xi, x)\,\dot{\varepsilon}^a_{\alpha\beta}(x)\,r(x)\,d\Omega(x)\,, \qquad i, j = r, z;\ \alpha, \beta = r, \varphi, z\,. \end{aligned} \tag{6.166}$$

In addition, the internal stresses can be computed as

$$\begin{aligned} \dot{\sigma}_{\alpha\beta}(\xi) &= 2\pi \int_\Gamma u^*_{\alpha\beta k}(\xi, x)\,\dot{p}_k(x)\,r(x)\,d\Gamma(x) \\ &\quad - 2\pi \int_\Gamma p^*_{\alpha\beta k}(\xi, x)\,\dot{u}_k(x)\,r(x)\,d\Gamma(x) + 2\pi \int_\Omega u^*_{\alpha\beta k}(\xi, x)\,\dot{b}_k(x)\,r(x)\,d\Omega(x) \\ &\quad + 2\pi \int_\Omega \sigma^*_{\alpha\beta\gamma\varrho}(\xi, x)\,\dot{\varepsilon}^a_{\gamma\varrho}(x)\,r(x)\,d\Omega(x) + h_{\alpha\beta}(\dot{\varepsilon}^a_{\gamma\varrho})\,, \\ &\qquad k = r, z;\ \alpha, \beta, \gamma, \varrho = r, \varphi, z\,, \end{aligned} \tag{6.167}$$

where the last domain integral is in the principal value sense and the terms of $h_{\alpha\beta}$ can be computed by employing limiting expressions for $\sigma^*_{\alpha\beta i}$ near the singularity. As discussed in Section 5.15 for the c_{ij} coefficient, the expressions for $h_{\alpha\beta}$ are also equivalent to the plane strain case, now without the inclusion of the contribution of the third direction into the other two, i.e.

$$\begin{aligned}
h_{rr} &= -\frac{G}{4(1-\nu)}[3\,\dot{\varepsilon}^a_{rr} + \dot{\varepsilon}^a_{zz} + 4\,\nu\dot{\varepsilon}^a_{\varphi\varphi}] \\
h_{rz} &= -\frac{G}{2(1-\nu)}\,\dot{\varepsilon}^a_{rz} \\
h_{zz} &= -\frac{G}{4(1-\nu)}[3\,\dot{\varepsilon}^a_{zz} + \dot{\varepsilon}^a_{rr} + 4\,\nu\dot{\varepsilon}^a_{\varphi\varphi}] \\
h_{\varphi\varphi} &= -\frac{G}{4(1-\nu)}[8\,\dot{\varepsilon}^a_{\varphi\varphi} + 4\nu(\dot{\varepsilon}^a_{rr} + \dot{\varepsilon}^a_{zz})].
\end{aligned} \tag{6.168}$$

References

1. Swedlow, J. L., and Cruse, T. A., Formulation of boundary integral equations for three-dimensional elastoplastic flow, Int. J. Solids Structures **7**, 1673–1683 (1971).
2. Mendelson, A., Boundary integral methods in elasticity and plasticity, Report No. NASA TN D-7418, NASA, 1973.
3. Riccardella, P. C., An implementation of the boundary integral technique for planar problems in elasticity and elastoplasticity, Report No. SM-73-10, Dept. Mech. Engng., Carnegie Mellon University, Pittsburg, 1973.
4. Mendelson, A., and Albers, L. U., Application of boundary integral equations to elasto-plastic problems, in *Boundary Integral Equation Method: Computational Applications in Applied Mechanics* (T. A. Cruse and F. J. Rizzo, Eds.), pp. 47–84, ASME, New York, 1975.
5. Mukherjee, S., Corrected boundary integral equations in planar thermoelastoplasticity, Int. J. Solids Structures, **13**, 331–335 (1977).
6. Bui, H. D., Some remarks about the formulation of three-dimensional thermoelastoplastic problems by integral equations, Int. J. Solids Structures **14**, 935–939 (1978).
7. Telles, J. C. F., and Brebbia, C. A., On the application of the boundary element method to plasticity, Appl. Math. Modelling **3**, 466–470 (1979).
8. Telles, J. C. F., and Brebbia, C. A., The boundary element method in plasticity, in *New Developments in Boundary Element Methods* (C. A. Brebbia, Ed.), pp. 295–317, CML Publications. Southampton, 1980; Appl. Math. Modelling **5**, 275–281 (1981).
9. Telles, J. C. F., and Brebbia, C. A., Elastoplastic boundary element analysis, in *Proc. Europe – U.S. Workshop on Nonlinear Finite Element Analysis in Structural Mechanics* (W. Wunderlich *et al.*, Eds.), pp. 403–434, Ruhr-University Bochum, Ruhr, Springer-Verlag Berlin, 1980.
10. Mukherjee, S., and Kumar, V., Numerical analysis of time dependent inelastic deformation in metallic media using the boundary integral equation method, Trans. ASME J. Appl. Mech. **45**, 785–790 (1978).
11. Chaudonneret, M., Méthode des équations intégrales appliquées a la résolution de problèmes de viscoplasticité, J. Mécanique Appl **1**, 113–132 (1977).
12. Morjaria, M., and Mukherjee, S., Improved boundary integral equation method for time dependent inelastic deformation in metals, Int. J. Numerical Methods Engng. **15**, 97–111 (1980).

13. Telles, J. C. F., and Brebbia, C. A., Boundary elements: New developments in elastoplastic analysis, Appl. Math. Modelling **5,** 376–382 (1981).
14. Morjaria, M., and Mukherjee, S., Numerical analysis of planar time dependent inelastic deformation of plates with cracks by the boundary element method, Int. J. Solids Structures **17,** 127–143 (1981).
15. Telles, J. C. F., and Brebbia, C. A., Elastic/viscoplastic problems using boundary elements, Int. J. Mech. Sci., **24,** 605–618 (1982).
16. Kumar, V., and Mukherjee, S., A boundary integral equation formulation for time-dependent inelastic deformation in metals, Int. J. Mech. Sci. **19,** 713–724 (1977).
17. Mikhlin, S. G., Singular integral equations, Amer. Math. Soc. Trans. Series 1, **10,** 84–197 (1962).
18. Hill, R., *The Mathematical Theory of Plasticity,* Clarendon Press, Oxford, 1950.
19. Ford, H., and Alexander, J. M., *Advanced Mechanics of Materials,* 2nd ed., Ellis Horwood, Chichester, 1977.
20. Prager, W., and Hodge, P. G., *Theory of Perfectly Plastic Solids,* Dover, New York, 1968.
21. Mendelson, A., *Plasticity: Theory and Application,* Macmillan, New York, 1968.
22. Chen, W. F., *Limit Analysis and Soil Plasticity,* Elsevier, Amsterdam, 1975.
23. Lin, T. H., *Theory of Inelastic Structures,* Wiley, New York, 1968.
24. Penny, R. K., and Marriott, D. L., *Design for Creep,* McGraw-Hill, London, 1971.
25. Odqvst, F. K. G., *Mathematical Theory of Creep and Creep Rupture,* 2nd ed., Clarendon Press, Oxford, 1974.
26. Malvern, L. E., *Introduction to the Mechanics of a Continuous Medium,* Prentice-Hall, Englewood Cliffs, N.J., 1969.
27. Fung, Y. C., *Foundations of Solid Mechanics,* Prentice-Hall, Englewood Cliffs, N.J., 1965.
28. Johnson, W., and Mellor, P. B., *Plasticity for Mechanical Engineers,* Van Nostrand, London, 1962.
29. Olszak, W., Mroz, Z., and Perzyna, P., *Recent Trends in the Development of the Theory of Plasticity,* Pergamon Press, Oxford; PWN, Warsaw, 1963.
30. Perzyna, P., Fundamental problems in viscoplasticity, Advan. Appl. Mech. **9,** 243–377 (1966).
31. Perzyna, P., The constitutive equations for rate sensitive plastic materials, Quart. Appl. Math. **20,** 321–332 (1963).
32. Olszak, W., and Perzyna, P., Stationary and non stationary visco-plasticity, in *Inelastic Behaviour of Solids,* (Kanninen, *et al.,* Eds.), pp.53–75, McGraw-Hill, New York, 1970.
33. Timoshenko, S. P., and Goodier, J. N., *Theory of Elasticity,* 3rd ed., McGraw-Hill, Tokyo, 1970.
34. Nayak, G. C., and Zienkiewicz, O. C., Convenient form of stress invariants for plasticity, Proc. Amer. Soc. Civil Engrs. J. Struct. Div. **98,** 949–954 (1972).
35. Besseling, J. F., A theory of elastic, plastic and creep deformations of an initially isotropic material showing anisotropic strain hardening, creep recovery, and secondary creep, Trans. ASCE, J. Appl. Mech. **25,** 529–536 (1958).
36. Owen, D. R. J., Prakash, A., and Zienkiewicz, O. C., Finite element analysis of nonlinear composite materials by use of overlay systems, Computer Structures **4,** 1251–1267 (1974).
37. Zienkiewicz, O. C., Valliappan, S., and King, I. P., Stress analysis of rock as a no-tension Material, Geotechnique **18,** 56–66 (1968).
38. Venturini, W. S., and Brebbia, C. A., The boundary element method for the solution of no-tension materials, in *Boundary Element Methods,* (C. A. Brebbia, Ed.), Springer-Verlag, Berlin, 1981.
39. Zienkiewicz, O. C., *The Finite Element Method,* 3rd ed., McGraw-Hill, London, 1977.
40. Kellogg, O. D., *Foundations of Potential Theory,* Springer-Verlag, Berlin, 1929.
41. Zabreyko, P. P., *et al., Integral Equations – A Reference Text,* Noordhoff, Holland, 1975.
42. Telles, J. C. F., and Brebbia, C. A., Boundary element solution for half-plane problems, Int. J. Solids Structures **17,** 1149–1158 (1981).
43. Novati, G., and Brebbia, C. A., Boundary element formulation for geometrically nonlinear elastostatics, Appl. Math. Modelling **6,** 136–138 (1982).

Chapter 7

Elastoplasticity

7.1. Introduction

In this chapter the boundary element equations presented in Chapter 6 are employed to solve problems concerned with the inviscid or classical theory of plasticity. An application of the initial strain equations for incompressible plastic strains is first introduced in conjunction with the von Mises yield criterion and Mendelson's successive elastic solutions method [4]. This simple solution technique, also called "elastic predictor–radial corrector method" by Schreyer *et al.* [10], has proved to be very efficient and stable with reference to the load increment size. The initial stress equations, on the other hand, are more general and are here implemented to handle four different yield criteria (Tresca, Mises, Mohr–Coulomb, and Drucker–Prager) with two different iterative routines. The first is a pure incremental technique comparable to what was used by Zienkiewicz *et al.* [11] for finite elements. The second deals with accumulated values of the initial stresses in a similar fashion to the initial strain implementation.

Several examples are presented to illustrate the applicability of boundary elements to elastoplastic problems and these also include some geotechnical problems solved by using the half-plane fundamental solution.

7.2. Some Simple Elastoplastic Relations

In Section 6.2 it was demonstrated that uniaxial plastic behavior is only possible if the yield criterion given by expression (6.13) is satisfied. The expression is repeated here for completeness:

$$F(\sigma, k) = \sigma - \sigma_0 = 0\,. \tag{7.1}$$

This yield condition was seen to be valid to describe uniaxial yield behavior. For general stress states this sort of representation is generalized to handle any possible combination of stresses. In the present section, only the von Mises yield criterion will be considered and this can be written as [1–9]

$$F(\sigma_{ij}, k) = \sqrt{3J_2} - \sigma_0 = 0\,, \tag{7.2}$$

where J_2 is the second invariant of the stress deviator tensor (see Chapter 5, expression (5.15)) and as before k is a hardening parameter representing the total plastic work, i.e.,

$$k = W^p = \int \sigma_{ij}\, d\varepsilon_{ij}^p . \tag{7.3}$$

As discussed before, plasticity is a path-dependent phenomenon; therefore, it becomes necessary to compute the differentials or increments of plastic strain throughout the loading history and then obtain the accumulated strains by integration or summation. A suitable relation for the determination of the plastic strain increments is given by the well-known Prandtl–Reuss equations [1, 2, 4]

$$d\varepsilon_{ij}^p = S_{ij}\, d\lambda , \tag{7.4}$$

where $d\lambda$ is a proportionality factor which may vary throughout the loading history, but is always positive.

In addition, it is convenient to define an equivalent or effective stress and an equivalent or effective plastic strain increment as

$$\sigma_e = \sqrt{3 J_2} \tag{7.5}$$

and

$$d\varepsilon_e^p = \sqrt{\tfrac{2}{3}\, d\varepsilon_{ij}^p\, d\varepsilon_{ij}^p} . \tag{7.6}$$

Note that for the uniaxial case presented in Section 6.2, $\sigma_e = \sigma$ and $d\varepsilon_e^p = d\varepsilon^p$. Moreover, the von Mises yield criterion can now be written

$$\sigma_e - \sigma_0 = 0 \tag{7.7}$$

which is entirely equivalent to expression (7.1).

With reference to Eq. (7.4) it is seen that the proportionality factor $d\lambda$ can be expressed in terms of the equivalent forms σ_e and $d\varepsilon_e^p$ if we square both sides of the equation as follows:

$$d\varepsilon_{ij}^p\, d\varepsilon_{ij}^p = S_{ij}\, S_{ij}\, d\lambda^2 , \tag{7.8}$$

which leads to

$$(\tfrac{3}{2}\, d\varepsilon_e^p)^2 = (\sigma_e\, d\lambda)^2 \tag{7.9}$$

or

$$d\lambda = \frac{3}{2}\frac{d\varepsilon_e^p}{\sigma_e} . \tag{7.10}$$

For the initial strain implementaiton of the boundary element technique, the plastic strain increments are computed by using the above expressions as follows [4]: let us assume that a loading path is found to reach a given state of stresses and accumulated plastic strains ε_{ij}^p. When the load is increased by a small amount, the additional plastic strains produced are $\Delta\varepsilon_{ij}^p$ and the total strains are given by

$$\varepsilon_{ij} = \varepsilon_{ij}^e + \varepsilon_{ij}^p + \Delta\varepsilon_{ij}^p, \tag{7.11}$$

where ε_{ij}^e already includes the current load increment.

It is now convenient to define a modified total strain tensor of the form

$$\varepsilon'_{ij} = \varepsilon_{ij} - \varepsilon_{ij}^p \tag{7.12}$$

or

$$\varepsilon'_{ij} = \frac{1}{2G}\left(\sigma_{ij} - \frac{\nu}{1+\nu}\sigma_{kk}\,\delta_{ij}\right) + \Delta\varepsilon_{ij}^p, \tag{7.13}$$

where expression (7.13) is simply (see expression (5.28))

$$\varepsilon'_{ij} = \varepsilon_{ij}^e + \Delta\varepsilon_{ij}^p. \tag{7.14}$$

Expression (7.13) can also be written in deviatoric form as (note that $\Delta\varepsilon_{kk}^p = 0$)

$$e'_{ij} = \frac{S_{ij}}{2G} + \Delta\varepsilon_{ij}^p \tag{7.15}$$

in which

$$e'_{ij} = \varepsilon'_{ij} - \frac{\delta_{ij}}{3}\varepsilon'_{kk}. \tag{7.16}$$

Recalling the Prandtl–Reuss equations given by Eq. (7.4), expression (7.15) yields

$$e'_{ij} = \left(1 + \frac{1}{2G\Delta\lambda}\right)\Delta\varepsilon_{ij}^p. \tag{7.17}$$

By squaring both sides of Eq. (7.17) in a manner similar to what was done for expressions (7.8)–(7.10), the following relation arises:

$$1 + \frac{1}{2G\Delta\lambda} = \frac{\varepsilon_{et}}{\Delta\varepsilon_e^p}, \tag{7.18}$$

where

$$\varepsilon_{et} = \sqrt{\tfrac{2}{3}\,e'_{ij}\,e'_{ij}}. \tag{7.19}$$

Substituting Eq. (7.18) into Eq. (7.17) gives

$$\Delta\varepsilon_{ij}^p = \frac{\Delta\varepsilon_e^p}{\varepsilon_{et}}\,e'_{ij}. \tag{7.20}$$

From the above equation it is seen that in order to determine the actual magnitudes of the plastic strain increments, the equivalent plastic strain increment must be determined. Therefore, substituting the proportionality factor $\Delta\lambda$ given by Eq. (7.10) into expression (7.18) we obtain

$$1+\frac{\sigma_e}{3G\,\Delta\varepsilon_e^p}=\frac{\varepsilon_{et}}{\Delta\varepsilon_e^p} \tag{7.21}$$

which gives

$$\Delta\varepsilon_e^p=\varepsilon_{et}-\frac{\sigma_e}{3G}. \tag{7.22}$$

Since the condition expressed in Eq. (7.7) must be satisfied throughout the plastic process, σ_0 can be substituted for σ_e in Eq. (7.22),

$$\Delta\varepsilon_e^p=\varepsilon_{et}-\frac{\sigma_0}{3G}. \tag{7.23}$$

Note that σ_0 corresponds to the uniaxial yield stress after the application of the current load increment; consequently, it is still unknown. This term, however, can be approximated by a truncated Taylor series about the preceding value of σ_0 (i.e., before the load increment is applied) as follows

$${}^{k}\sigma_0={}^{k-1}\sigma_0+{}^{k-1}H'\,\Delta\varepsilon_e^p+\cdots, \tag{7.24}$$

where H' has been defined in Section 6.2.

Substituting Eq. (7.24) into Eq. (7.23) and solving for $\Delta\varepsilon_e^p$ one finally obtains

$$\Delta\varepsilon_e^p=\frac{3G\,\varepsilon_{et}-\sigma_0}{3G+H'}, \tag{7.25}$$

where the values of σ_0 and H' are computed before the load increment.

The equations discussed here have been presented for the general three-dimensional case. For two-dimensional problems these equations are properly modified to account for plane strain or plane stress as indicated in Appendix C. This allows one to work with accumulated values of tractions, displacements, and stresses in Eqs. (6.139) and (6.142). Equation (6.139) can now be written

$$\boldsymbol{Y}=\boldsymbol{K}(\boldsymbol{\varepsilon}^p+\Delta\boldsymbol{\varepsilon}^p)+\boldsymbol{M} \tag{7.26}$$

and Eq. (6.142)

$$\boldsymbol{\sigma}=\boldsymbol{B}(\boldsymbol{\varepsilon}^p+\Delta\boldsymbol{\varepsilon}^p)+\boldsymbol{N}, \tag{7.27}$$

where $\boldsymbol{\varepsilon}^p$ represents the accumulated plastic strains up to (but not including) the corresponding to the current load increment $\Delta\boldsymbol{\varepsilon}^p$ which are to be determined iteratively.

7.3. Initial Strain: Numerical Solution Technique

With reference to Eqs. (7.26) and (7.27) one notices that vector $\boldsymbol{M}$ represents the elastic solution to the boundary problem (tractions and displacements unknown) and that vector $\boldsymbol{N}$ stands for the corresponding stresses. Therefore, load at first yield can be calculated by taking the most highly stressed boundary node or internal point and comparing its equivalent stress $\sigma_e^{\max}$ with the uniaxial yield stress of the material. The incremental process starts by reducing this stress value with a load factor defined as

$$\lambda_0 = \frac{\sigma_0}{\sigma_e^{\max}}. \tag{7.28}$$

The load increment is then calculated and further values of the load factor are given by

$$\lambda_i = \lambda_{i-1} + \beta, \tag{7.29}$$

where $\beta = \lambda_0 \omega$, ω being the given value of the load increment with reference to load at first yield.

Equations (7.26) and (7.27) are now written as

$$\boldsymbol{Y} = \boldsymbol{K}(\boldsymbol{\varepsilon}^p + \Delta\boldsymbol{\varepsilon}^p) + \lambda_i \boldsymbol{M} \tag{7.30}$$

and

$$\boldsymbol{\sigma} = \boldsymbol{B}(\boldsymbol{\varepsilon}^p + \Delta\boldsymbol{\varepsilon}^p) + \lambda_i \boldsymbol{N}. \tag{7.31}$$

For a given value of λ_i, the plastic strain increment is determined iteratively at each selected boundary node and internal stress point as follows:

(a) Compute stress (Eq. (7.31)).
(b) Calculate
 ε'_{ij} (Eq. (7.13)),
 ε_{et} (Eq. (7.19)),
 $\Delta\varepsilon_e^p \geqq 0$ (Eq. (7.25)).
(c) Verify convergence, i.e., compare $\Delta\varepsilon_e^p$ calculated with its previous value.
(d) Compute new estimate of $\Delta\varepsilon_{ij}^p$ (Eq. (7.20)).
(e) Continue with next node or point and start with (b) until all nodes and points have been considered.
(f) Go to (a) for a new iteration.

Once convergence is obtained (within prescribed tolerance) for all selected nodes and points, $\Delta\boldsymbol{\varepsilon}^p$ is added to $\boldsymbol{\varepsilon}^p$ and its value is also used as an initial guess for the next load increment.

Note that for the whole incremental–iterative process to take place, only Eq. (7.31) is required. Equation (7.30) is used only after convergence is achieved and if boundary unknowns are requested by the user. Furthermore, matrices $\boldsymbol{K}$ and $\boldsymbol{B}$ as well as vectors $\boldsymbol{M}$ and $\boldsymbol{N}$ are generated only once at the beginning of the entire process, which represents a great saving in computer time.

7.3.1. Examples – Initial Strain Formulation

To outline the applicability of the formulation described in the previous sections, some examples were run, with linear elements and cells, and results have been compared with finite element and experimental analyses. Also, whenever possible, analytical solutions are included for checking the results.

Example 7.1. – Polystyrene Crazing Problem [12]. In order to study the effect of voids in polystyrene strength, this example was first run by Haward and Owen [13] using the finite element method. The geometry of the problem is given in Fig. 7.1 where boundary element and internal cell discretization is also shown. For the present comparison, plane strain approximation was used in both FEM and BEM analyses, the former using quadratic isoparametric elements as depicted in Fig. 7.2.

Ideal plasticity was assumed with

$$E = 42 \times 10^3 \text{ MN/m}^2,$$

$$\sigma_0 = Y = 105 \text{ MN/m}^2,$$

$$\nu = 0.33 .$$

Two loading conditions were considered: biaxial tension and uniaxial tension, both applied by prescribing displacements at the edges.

Figures 7.3a, b show the results obtained by both programs for the two loading cases. As can be seen, agreement between the different formulations has been obtained.

Example 7.2. – Plane Strain Punch [12]. This example consists of a rigid flat punch indented into a solid plane strain specimen (see Fig. 7.4). Finite element solu-

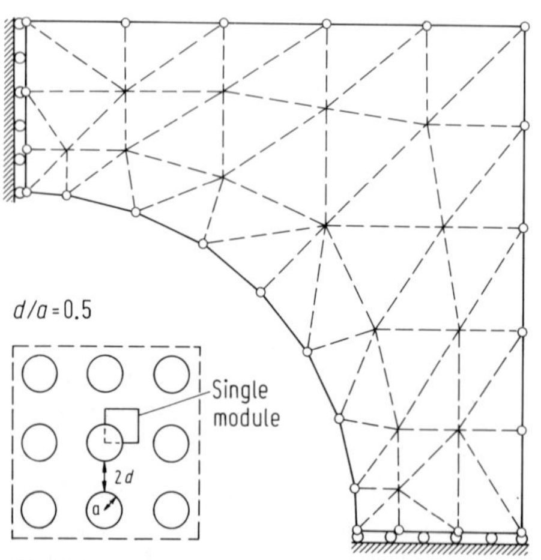

Fig. 7.1. Two-dimensional cylindrical void model and discretization used for BEM

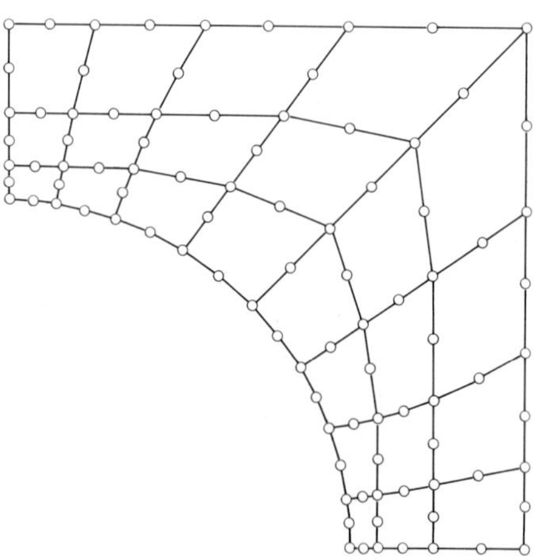

Fig. 7.2. Quadratic isoparametric finite element mesh used for polystyrene crazing problem

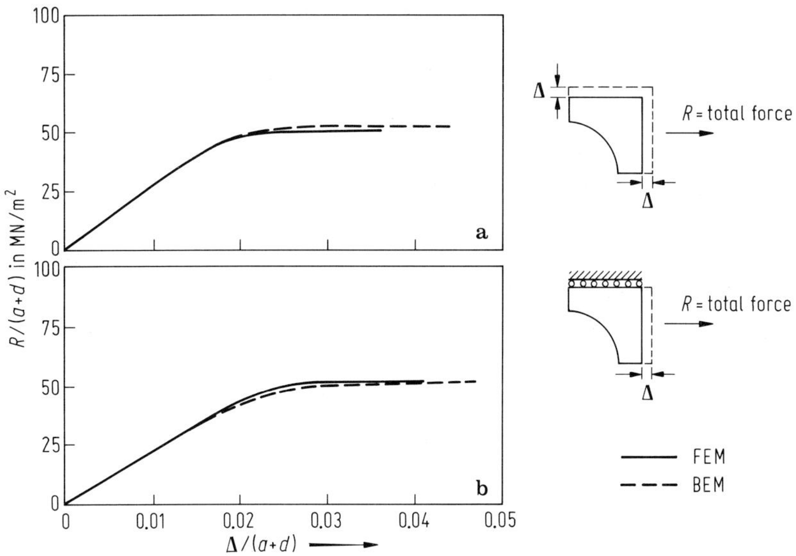

Fig. 7.3. Mean stress–strain curves for polystyrene crazing problem. **a** Biaxial tension, **b** uniaxial tension (fixed edge)

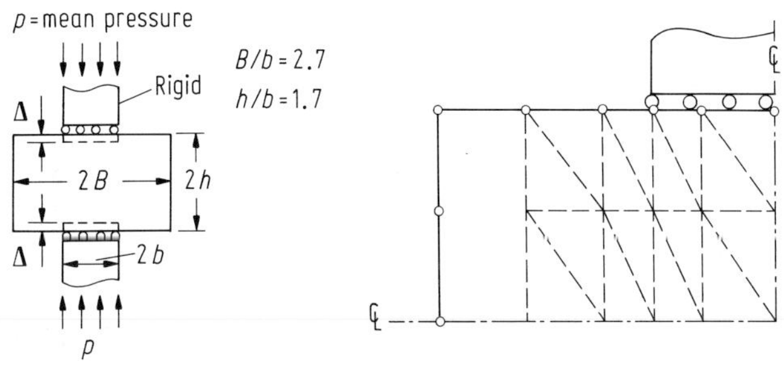

Fig. 7.4. Plane strain punch problem. Boundary element and internal cell discretization

tions with different material parameters were presented by Nayak and Zienkiewicz [14]. Boundary element results were calculated with the discretization shown in Fig. 7.4 (no boundary elements along symmetry axes) and by incrementing the rigid punch displacements.

For this comparison two different material properties were used; ideal plasticity ($H' = 0$) and strain softening ($H' = -0.1\,E$). Mean pressure–displacement curves shown in Fig. 7.5 exhibit close agreement between finite element and boundary element results, despite the rather coarse discretization employed for the boundary element solution.

Plastic zones obtained for the strain softening case using FEM and the mesh shown in Fig. 7.6a agree reasonably well with BEM plastified points as shown in Fig. 7.6b.

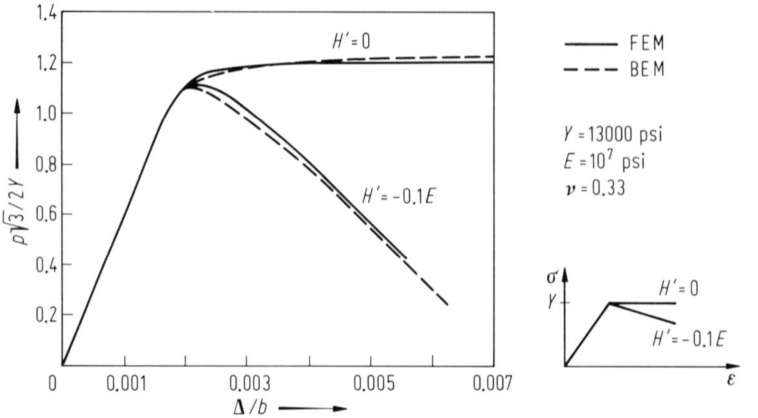

Fig. 7.5. Mean pressure – displacement curves for plane strain punch problem

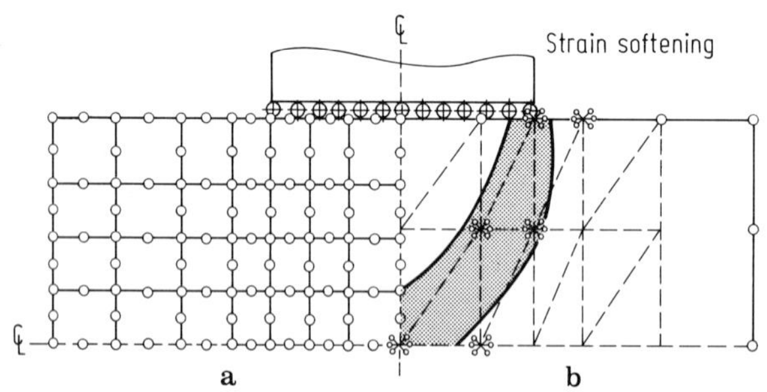

Fig. 7.6. Plane strain punch problem. **a** Quadratic isoparametric finite element mesh, **b** comparison between plastic zones obtained by FEM and plastified points obtained by BEM ($\Delta/b = 0.0052$)

Example 7.3. – Thick Cylinder [16]. In this example the plane strain expansion of a thick cylinder subjected to internal pressure is studied. Ideal plasticity is assumed with the following material parameters:

$$E = 12000 \text{ dN/mm}^2,$$
$$\sigma_0 = 24 \text{ dN/mm}^2,$$
$$\nu = 0.3.$$

Boundary element results computed without boundary discretization of the symmetry axes (see Fig. 7.7) are compared with the analytical solution produced by Hodge and White [3].

Radial displacements over the outer boundary and circumferential stress distribution (plastic front at $r' = 1.6a$) exhibit good agreement with the analytical solution as shown in Figs. 7.8 and 7.9, respectively.

The applications shown in the present section clearly indicate the potentiality of boundary elements for solving plasticity problems. In all the examples the load increment was kept between 5% and 25% of the load at first yield and it was verified that the successive elastic solutions procedure employed is very stable with reference to the load increment size. Consequently, this procedure is recommended for Mises material problems. In the next section the stress–strain relations will be presented in a more general form and different yield criteria will be included for the initial stress implementation.

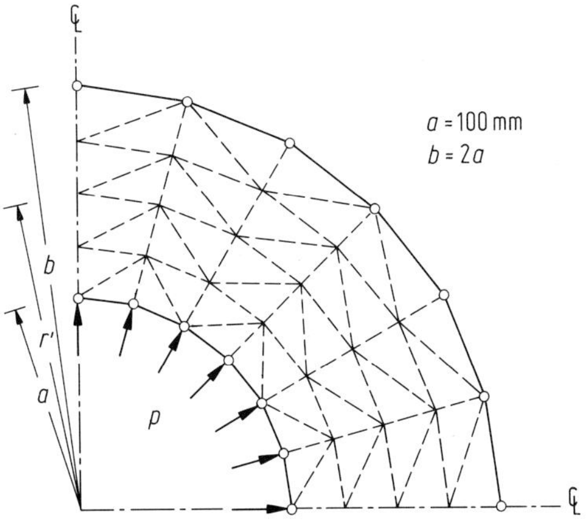

Fig. 7.7. Thick cylinder problem. Boundary element and internal cell discretization

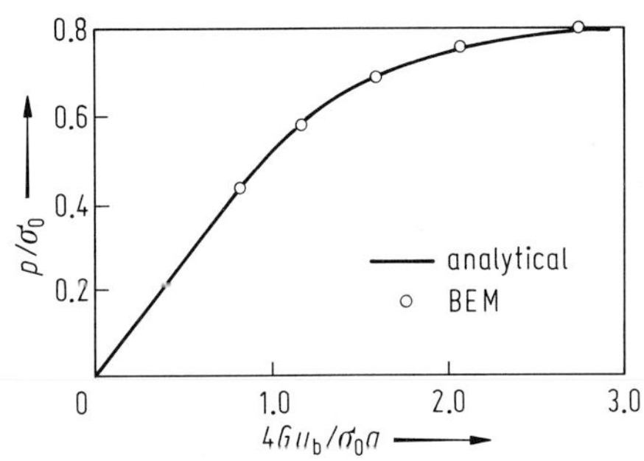

Fig. 7.8. Outer surface displacements for thick cylinder problem

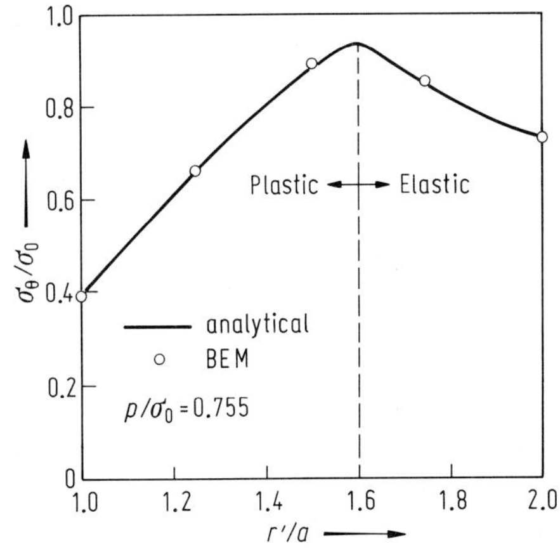

Fig. 7.9. Circumferential stress distribution in thick cylinder. Plastic front at $r' = 1.6\,a$

7.4. General Elastoplastic Stress–Strain Relations

For the formulation of a theory which models elastoplastic material deformation, three requirements have to be met:

(a) Explicit elastic relationship between stress and strain before the onset of plastic deformation.
(b) A yield criterion indicating the stress level at which plastic flow commences.
(c) Relationship between stress and strain for postyield behavior.

Requirement (a) has been thoroughly discussed in Chapter 5. Therefore only (b) and (c) will be considered here.

The yield criterion for isotropic hardening can be written in general form as

$$F(\sigma_{ij}, k) = 0, \tag{7.32}$$

where k is the work hardening parameter (see expression (7.3)) that gives the instantaneous position of the yield surface in the n-dimensional stress space.

On physical grounds, one can notice that the yield criterion is independent of the orientation of the coordinate system employed and should be a function of the three stress invariants. It is common to represent two of these invariants as functions of the deviatoric stresses (see Chapter 5):

$$I_1 = \sigma_{kk},$$
$$J_2 = \tfrac{1}{2} S_{ij} S_{ij},$$
$$J_3 = \tfrac{1}{3} S_{ij} S_{jk} S_{ki}.$$

In the present work, instead of J_3, the alternative stress invariant α, known as the Lode angle [15], was used. This invariant has been introduced in Section 5.1 and is given by

$$-\frac{\pi}{6} \leqq \alpha = \frac{1}{3} \sin^{-1}\left(-\frac{3\sqrt{3}}{2} \frac{J_3}{J_2^{3/2}}\right) \leqq \frac{\pi}{6}. \tag{7.33}$$

By using these stress invariants, different yield criteria can be applied, such as [15]

Tresca:

$$2\sqrt{J_2} \cos \alpha - \sigma_0 = 0; \tag{7.34}$$

Von Mises:

$$\sqrt{3J_2} - \sigma_0 = 0; \tag{7.35}$$

Mohr–Coulomb:

$$\frac{I_1}{3} \sin \varphi' + \sqrt{J_2}\left(\cos \alpha - \frac{1}{\sqrt{3}} \sin \alpha \sin \varphi'\right) - c' \cos \varphi = 0 \tag{7.36}$$

in which φ' is the angle of internal friction and c' is the cohesion of the material.

Drucker–Prager:

$$\alpha' I_1 + \sqrt{J_2} - K' = 0, \tag{7.37}$$

where

$$\alpha' = \frac{2\sin\varphi'}{\sqrt{3}(3-\sin\varphi')}, \qquad K' = \frac{6c'\cos\varphi'}{\sqrt{3}(3-\sin\varphi')}. \tag{7.38}$$

Mohr–Coulomb hypothesis may be simulated by the Drucker–Prager criterion in plane strains if α' and K' are written as [17]

$$\alpha' = \frac{\tan\varphi'}{(9+12\tan^2\varphi')^{1/2}}, \qquad K' = \frac{3c'}{(9+12\tan^2\varphi')^{1/2}}. \tag{7.39}$$

For our practical purposes, Eq. (7.32) can then be written

$$F(\sigma_{ij}, k) = f(\sigma_{ij}) - \psi(k) = 0 \tag{7.40}$$

where one can notice that $f(\sigma_{ij})$ is a scalar function of σ_{ij} which plays the role of an equivalent stress, designated by σ_e. As a consequence, we can define an equivalent plastic strain ε_e^p whose increment produces an increment in the plastic strain energy as follows:

$$\sigma_e\, d\varepsilon_e^p = \sigma_{ij}\, d\varepsilon_{ij}^p = dk. \tag{7.41}$$

Note that for the von Mises criterion $d\varepsilon_e^p$ defined above is given by expression (7.6).

In order to obtain the stress–strain relations for postyield behavior, let us first rewrite Eq. (6.45) in incremental form,

$$d\sigma_{ij} = C_{ijkl}(d\varepsilon_{kl} - d\varepsilon_{kl}^p). \tag{7.42}$$

Within the context of associated plasticity, the flow rule also known as normality principle [1, 6, 7], can be described by

$$d\varepsilon_{ij}^p = d\lambda \frac{\partial F}{\partial \sigma_{ij}}, \tag{7.43}$$

where $d\lambda$ is a proportionality factor, termed the plastic multiplier. It should be pointed out that here the Prandtl–Reuss equations can also be simulated for the von Mises criterion; $d\lambda$ however would no longer be represented by expression (7.10).

The substitution of Eq. (7.43) into Eq. (7.42) gives

$$d\sigma_{ij} = C_{ijkl}(d\varepsilon_{kl} - a_{kl}\, d\lambda) \tag{7.44}$$

in which

$$a_{kl} = \frac{\partial F}{\partial \sigma_{kl}} = \frac{\partial f}{\partial \sigma_{kl}}. \tag{7.45}$$

When plastic yielding occurs the stresses satisfy Eq. (7.40) which by differentiation gives

$$dF = a_{ij}\, d\sigma_{ij} - \frac{d\psi}{dk}\, dk = 0 \tag{7.46}$$

or, according to Eq. (7.3),

$$a_{ij}\, d\sigma_{ij} - \frac{d\psi}{dk}\, \sigma_{ij}\, d\varepsilon_{ij}^{p} = 0. \tag{7.47}$$

From the application of the normality principle to Eq. (7.47) results

$$a_{ij}\, d\sigma_{ij} - \frac{d\psi}{dk}\, \sigma_{ij}\, a_{ij}\, d\lambda = 0. \tag{7.48}$$

If we substitute Eq. (7.44) into the above equation and solve for $d\lambda$ we obtain

$$d\lambda = \frac{1}{\gamma'}\, a_{ij}\, C_{ijkl}\, d\varepsilon_{kl}, \tag{7.49}$$

where

$$\gamma' = a_{ij}\, C_{ijkl}\, a_{kl} + \frac{d\psi}{dk}\, \sigma_{ij}\, a_{ij}. \tag{7.50}$$

Before we go further the last term in expression (7.50) can be examined. It is easy to show that $f(\sigma_{ij})$ is a homogeneous function of degree one and this allows the application of Euler's theorem [18] as follows:

$$\sigma_{ij} \frac{\partial f}{\partial \sigma_{ij}} = f(\sigma_{ij}) = \sigma_e. \tag{7.51}$$

The substitution of Eqs. (7.51) and (7.41) into Eq. (7.50) yields

$$\gamma' = a_{ij}\, C_{ijkl}\, a_{kl} + \frac{d\psi}{d\varepsilon_e^{p}}, \tag{7.52}$$

where $d\psi/d\varepsilon_e^p = H'$ if ψ is defined as the uniaxial yield stress.

Equation (7.49) can now be used to substitute $d\lambda$ in (7.44) providing the required incremental stress–strain relations

$$d\sigma_{ij} = C_{ijkl}^{ep}\, d\varepsilon_{kl} \tag{7.53}$$

in which

$$C_{ijkl}^{ep} = C_{ijkl} - \frac{1}{\gamma'}\, C_{ijmn}\, a_{mn}\, a_{op}\, C_{opkl}. \tag{7.54}$$

For the application of the above relations to the initial stress formulation, a further modification has proved to be convenient. Let us adopt the following notation (see expression (6.59)):

$$d\sigma_{ij}^{e} = C_{ijkl}\, d\varepsilon_{kl}, \tag{7.55}$$

where $d\sigma_{ij}^{e}$ stands for the components of the elastic stress increments (i.e., these represent the stress increment values as if a pure elastic problem were being solved).

Equation (7.53) can then be written in the form:

$$d\sigma_{ij} = d\sigma_{ij}^{e} - \frac{1}{\gamma'}\, C_{ijmn}\, a_{mn}\, a_{kl}\, d\sigma_{kl}^{e} \tag{7.56}$$

which means that the true stresses can be computed from the corresponding elastic stresses in incremental form. In addition to this, the increments of initial stress presented in Eq. (6.47) can also be calculated by the relation

$$d\sigma_{ij}^{p} = d\sigma_{ij}^{e} - d\sigma_{ij} = \frac{1}{\gamma'}\, C_{ijmn}\, a_{mn}\, a_{kl}\, d\sigma_{kl}^{e}, \tag{7.57}$$

where $d\sigma_{ij}^{p}$ corresponds to $d\sigma_{ij}^{a}$ with $d\varepsilon_{ij}^{a} = d\varepsilon_{ij}^{p}$.

All the expressions presented in this section are valid for the three-dimensional case. For two-dimensional problems the reader is referred to Appendix C for further details.

By simply examining Eq. (6.46) we notice that Eq. (6.135) can be applied for the computation of $d\sigma_{ij}^{e}$ if the matrix $\boldsymbol{E}'$ is replaced by

$$\bar{\boldsymbol{E}} = \boldsymbol{E}' + \boldsymbol{I}, \tag{7.58}$$

where $\boldsymbol{I}$ is the identity matrix. This gives

$$d\boldsymbol{\sigma}^{e} = \boldsymbol{G}'\, d\boldsymbol{P} - \boldsymbol{H}'\, d\boldsymbol{U} + \boldsymbol{Q}^{*}\, d\boldsymbol{\sigma}^{p} \tag{7.59}$$

in which

$$\boldsymbol{Q}^{*} = \boldsymbol{Q}' + \bar{\boldsymbol{E}}. \tag{7.60}$$

Finally, it is worth mentioning that the problem of indeterminacy of the normality principle (see Eq. (7.43)) at the so-called "corners" of the yield surface (typical of Tresca and Mohr–Coulomb surfaces) can be overcome by adopting the simple procedure indicated in [14]. This consists of "rounding off" the corners whenever $|\alpha| > (\pi/6 - \pi/180)$ and consequently avoids the singularity which occurs when $|\alpha| = \pi/6$.

7.5. Initial Stress: Outline of Solution Techniques

In order to minimize the computer effort for the initial stress formulation, Eqs. (6.134) and (7.59) can be further manipulated as discussed in Section 6.8. This leads to

$$dY = R\, d\sigma^p + dM \tag{7.61}$$

and

$$d\sigma^e = S\, d\sigma^p + dN, \tag{7.62}$$

where

$$R = A^{-1} Q \tag{7.63}$$

and

$$S = Q^* - A' R. \tag{7.64}$$

Note that, as before, vectors dM and dN represent the elastic solution to the incremental problem (actual solution in absence of plasticity). Furthermore, Eqs. (7.61) and (7.62) remain valid if, instead of incremental loading, the total load is applied. The only reason to proceed incrementally is the constitutive equations presented in Eq. (7.53). This enables us to compute load at first yield by simply scaling down the total elastic solution by a load factor λ_0. The incremental process starts at this load level and further values of the load factor are given by expression (7.29).

For elastoplastic solutions, Eqs. (7.61) and (7.62) can therefore be applied as

$$Y = R(\sigma^p + \Delta\sigma^p) + \lambda_i M \tag{7.65}$$

and

$$\sigma^e = S(\sigma^p + \Delta\sigma^p) + \lambda_i N \tag{7.66}$$

or alternatively for pure incremental relations,

$$\Delta Y = R\,\Delta\sigma^p + \beta M \tag{7.67}$$

and

$$\Delta\sigma^e = S\,\Delta\sigma^p + \beta N, \tag{7.68}$$

where in both cases vectors M and N correspond to the application of the total load and $\Delta\sigma^p$ stands for the current initial stress increment.

For a typical load increment (i.e., a given value of λ_i), the initial stress increment can be determined iteratively at each selected boundary node or internal point exhibiting plastic behavior by two processes. The former is, in fact, a pure incremental procedure. Once the load increment βN has been applied, the initial stress increment corresponding to the solution of the elastic problem is computed and has to be applied back into the body, providing an elastic stress redistribution. This operation again generates a new initial stress field to be redistributed elastically, and so on. Iteration is halted when the contribution of the last initial stress increment can be neglected.

The above process is in essence comparable to what was presented in [11] for the finite element method and is summarized as follows:

(a) Compute elastic stress increment by Eq. (7.68) if first iteration is being performed or $\Delta\boldsymbol{\sigma}^e = \boldsymbol{S}\Delta\boldsymbol{\sigma}^p$ otherwise.
(b) Find true stress increment $\Delta\sigma_{ij}$ (Eq. (7.56)).
(c) Verify convergence, i.e., compare $\Delta\varepsilon_e^p$ calculated with its accumulated value obtained during the current load increment to see if it can be neglected.
(d) Calculate initial stress increment by $\Delta\sigma_{ij}^p = \Delta\sigma_{ij}^e - \Delta\sigma_{ij}$.
(e) Accumulate values of initial stress and true stress;
$\sigma_{ij}^p \Rightarrow \sigma_{ij}^p + \Delta\sigma_{ij}^p$,
$\sigma_{ij} \Rightarrow \sigma_{ij} + \Delta\sigma_{ij}$.
(f) Continue with next node or point and start with (b) until all nodes and points have been considered.
(g) Go to (a) for a new iteration.

Iterations are performed until convergence is obtained (within prescribed tolerance) at every selected node or point.

It is interesting to note that in order to avoid cumulative errors, $\Delta\boldsymbol{\sigma}^p$ obtained at the end of iterations is applied together with $\beta\boldsymbol{N}$ in Eq. (7.68) for the first iteration of the next load increment.

The second process, which proved to be less dependent on the load increment size but not always more economical [19], deals with accumulated values of the elastic stress in a similar fashion to the procedure adopted for the initial strain implementation.

The initial stress increment is kept separate from its accumulated value until convergence is obtained as follows:

(a) Compute elastic stress (Eq. (7.66)).
(b) Calculate elastic stress increment by $\Delta\sigma_{ij}^e = \sigma_{ij}^e - \sigma_{ij} - \sigma_{ij}^p$.
(c) Find true stress increment $\Delta\sigma_{ij}$ (Eq. (7.56)).
(d) Verify convergence, i.e., compare $\Delta\varepsilon_e^p$ with its previous value.
(e) Calculate new estimate of initial stress increment by $\Delta\sigma_{ij}^p = \Delta\sigma_{ij}^e - \Delta\sigma_{ij}$.
(f) Continue with next node or point and start with (b) until all nodes and points have been considered.

Once that convergence is obtained for all selected nodes and points, the true stress and initial stress increments are accumulated and the latter is also used as an initial guess for the next load increment.

Note that neither procedure requires computation of the boundary unknowns. Consequently, Eq. (7.65) need only be used to print the boundary unknowns once convergence is achieved. In addition, if the body to be analyzed is under an initial (*in situ*) stress field, these stresses are simply added to the total stress vector at the beginning of the entire process. In this case, load at first yield can no longer be computed by expression (7.28); nevertheless any approximated value of λ_0 (provided it corresponds to a pure elastic state) can be used to start the incremental process.

Before the application of the above algorithms to solve plasticity problems it should be pointed out that although solution procedures are incremental, finite-sized load increments are always prescribed and this may create some drifts of the stress level beyond the yield surface. If load increments are kept sufficiently small, this problem is practically eliminated, but if relatively large load increment sizes are to be permitted, special techniques of the type presented in Refs. [10, 14, 20] have been found necessary to maintain the stresses on the yield surface. Basically, such techniques make use of a subincremental procedure which subdivides the increment of elastic stress into a number of subincrements. Consequently, relation (7.56) is always applied for small subincrement sizes. Also, once all subincrements have been considered, the satisfaction of Eq. (7.40) is verified and the final excess stress (if any) which still violates the yield criterion is added to the initial stress increment.

7.5.1. Examples – Kelvin Implementation

Following the solution algorithms presented in this section, the results for a series of examples solved by the boundary element technique are now compared with analytical solutions where such solutions are available and with finite element results.

Example 7.4. – Notched Tensile Specimen [19]. This example is one of the very early plasticity problems solved by using the finite element technique. Plane stress and plane strain results have been presented in several papers, creating a good opportunity to compare the boundary element computations.

Material parameters are as follows:

$$E = 7000 \text{ dN/mm}^2,$$

$$\sigma_0 = 24.3 \text{ dN/mm}^2,$$

$$\nu = 0.2,$$

$$H' = 0 \text{ (von Mises yield criterion).}$$

Plane stress analysis was carried out using the discretization shown in Fig. 7.10. Note that symmetry was taken into consideration without boundary discretization of the symmetry axes. This is due to a direct condensation process which automatically integrates over reflected elements and cells in such a way that the size of the final matrices corresponds to the reduced number of boundary elements and internal points presented.

Figure 7.11 gives the load–displacement curve for this case. It is seen that the curve remains nearly straight until very close to the limit load, when a sharp bend then occurs. Such behavior was also observed by Yamada *et al.* [21] in a similar problem. The limit load achieved by the boundary elements ($2\sigma_a/\sigma_0 = 1.21$) coincides with the results presented by Nayak and Zienkiewicz [14] using four different finite elements to analyze the same problem. Their limit load was found to vary between $2\sigma_a/\sigma_0 = 1.19$ and 1.23, and simple triangular, isoparametric linear, quadratic and cubic elements were used, all four meshes with approximately 97 nodes.

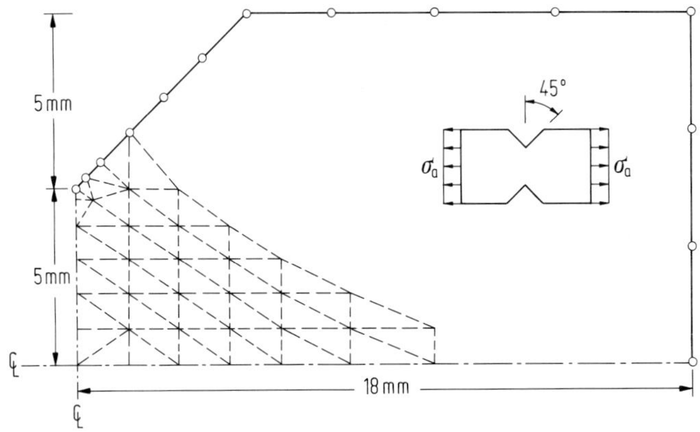

Fig. 7.10. Notched tensile specimen. Boundary element and internal cell discretization (plane stress case)

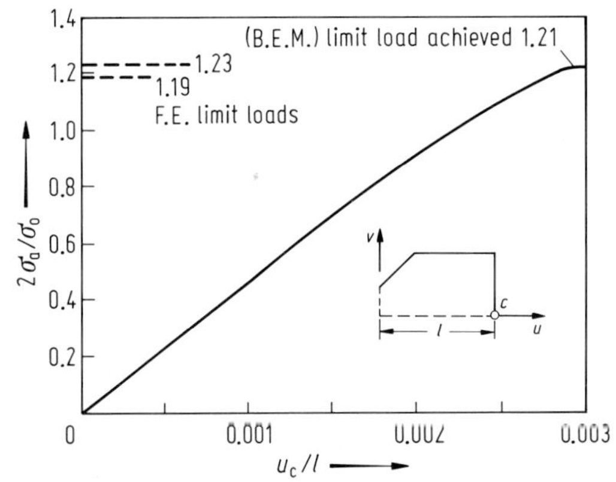

Fig. 7.11. Load–displacement curve for notched specimen in plane stress

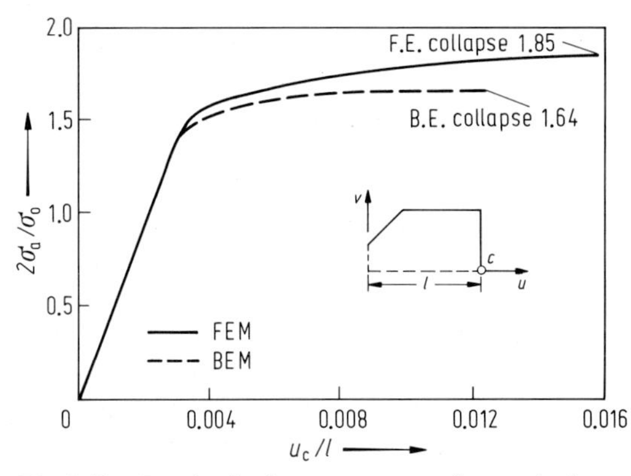

Fig. 7.12. Load–displacement curves for notched specimen in plane strain

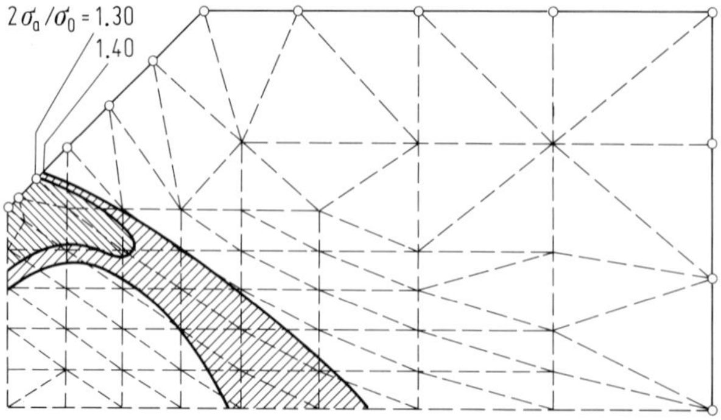

Fig. 7.13. Plastic zones obtained by BEM for different load levels (plane strain)

For the plane strain case, because of a large spread of plastic zone before limit load is achieved, the number of internal points and cells was increased from 33 and 51 to 59 and 97, respectively. The load – displacement curve is shown in Fig. 7.12 where the equivalent finite element results presented by Chen [5] are also given. The limit load obtained by BEM ($2\sigma_a/\sigma_0 = 1.64$) is below the value given by the finite element method ($2\sigma_a/\sigma_0 = 1.85$). But, as stated by Chen, bound theorems demonstrate that the maximum load should lie between 1.52 and 1.73, which supports the boundary element results.

Spread of plastic zones at lower load levels presented in Fig. 7.13 exhibits good agreement with finite element computations [5, 11] for the same problem.

Example 7.5. – Deep Circular Tunnel [19]. This example was selected to emphasize the advantages of boundary elements over "domain"-type techniques to solve infinite medium problems. A circular excavation studied by Reyes [22] and later by Baker *et al.* [23] with linear displacement triangular and simple quadrilateral finite elements, respectively, is compared with boundary element results.

The plane strain problem was analyzed under the Drucker – Prager simulation of the Mohr – Coulomb yield criterion (α' and K' given by expressions (7.39)) and by assuming the infinite domain to be initially subjected to a uniform stress field of 1 ksi vertical and 0.4 ksi in both horizontal directions ($K_0 = 0.4$). For the present study, external loads corresponding to the relaxation of this *in situ* stress field were applied over the surface of the opening.

The material (rock) was assumed to be perfectly plastic with

$$E = 500\ \text{ksi}\,,$$

$$c' = 0.28\ \text{ksi}\,,$$

$$\nu = 0.2\,,$$

$$\varphi' = 30°\,.$$

Boundary element and internal cell discretization is presented in Fig. 7.14 where the plastic zone on complete removal of the *in situ* stresses from the boundary of the cavity is also given.

Stresses along the horizontal section computed at the end of the relaxation process are compared with the corresponding results presented by Reyes and Baker in Fig. 7.15. Here, internal stresses outside the discretized region were calculated at simple internal points not connected to any internal cells.

It is important to note that the refinement of the two finite element meshes (about 253 nodes) should not lead to the differences in the σ_y values shown in Fig. 7.15. Although no reference was made by the authors, this discrepancy is probably due to the outer boundary conditions considered in the two analyses. The boundary element technique does not require any outer boundary discretization, but in order to study its influence in the results a quarter of a circle with radius

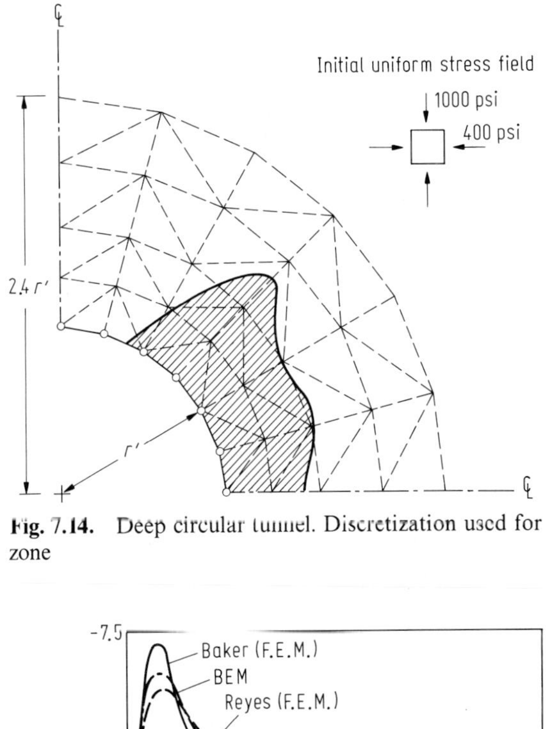

Fig. 7.14. Deep circular tunnel. Discretization used for BE results and total spread of plastic zone

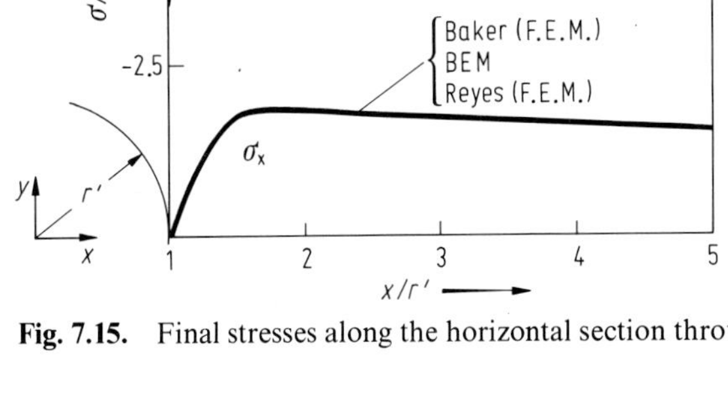

Fig. 7.15. Final stresses along the horizontal section through the medium

equal to nine times the radius of the cavity was discretized using six boundary elements; this is approximately the extent of the finite element meshes. The outer circle was then considered to be free to displace, giving as a result a better agreement with Baker computations. A second alternative was carried out by prescribing zero displacements over the outer boundary, leading now to improved agreement with Reyes results.

Example 7.6. – Rough Punch [19]. In this example the elastoplastic behavior of a square block compressed by two opposite perfectly rough rigid punches is studied. The problem is analyzed under plane strain condition and the material is considered to be perfectly plastic, obeying the von Mises yield criterion.

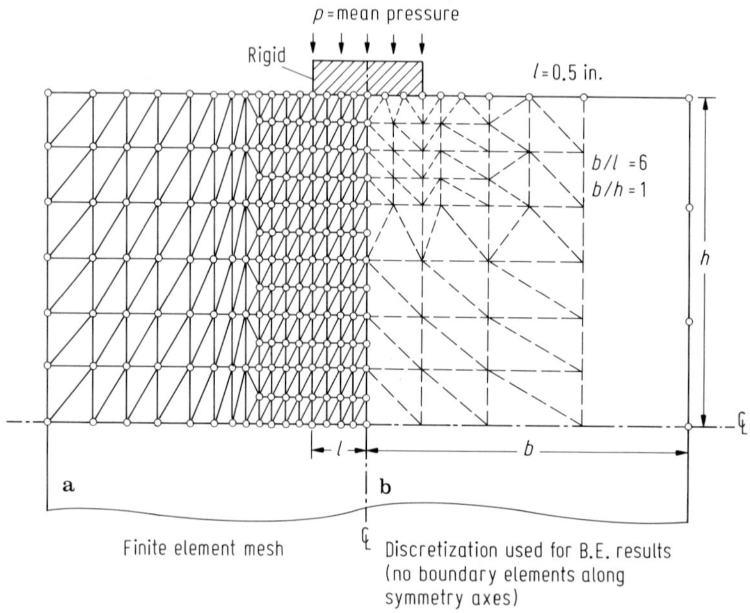

Fig. 7.16. Geometry of rough punch problem

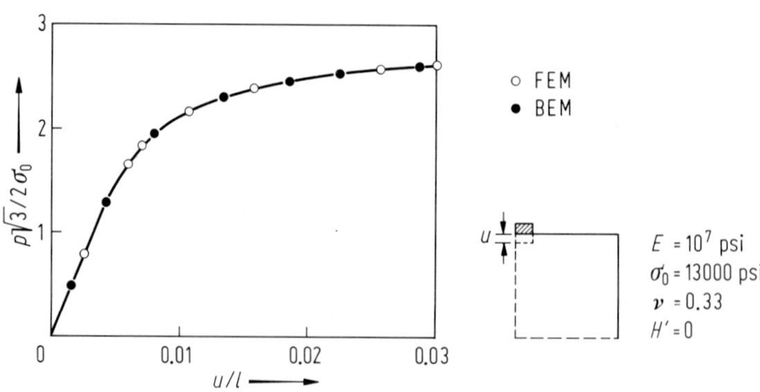

Fig. 7.17. Mean pressure–applied displacement curve for rough punch problem

By using a very refined mesh of 274 linear displacement triangular finite elements and 173 nodal points (see Fig. 7.16a), solutions to this problem were presented by Chen [5, 24]. The boundary element analysis was performed with the discretization shown in Fig. 7.16b, requiring less than one-third of the FE data to run the problem.

The indentation process was developed by prescribing the flat punch displacements leading to the average pressure–applied displacement curve presented in Fig. 7.17. As can be seen, agreement between the two analyses has been thoroughly obtained, both methods slightly exceeding (4%) the theoretical limit load $\sqrt{3}\,p/2\,\sigma_0 = 2.5$.

7.5.2. Examples – Half-Plane Implementation

In the present subsection the results of some applications of the half-plane implementation are compared with numerical and analytical solutions presented in the literature.

Example 7.7. – Strip Footing [16]. In this example the plane strain analysis of a flexible strip footing under uniform loading is presented. The finite soil stratum was discretized taking full advantage of both symmetry and free-surface condition, using the reduced number of 14 boundary elements and 42 internal points as shown in Fig. 7.18.

The soil was considered to be a perfectly plastic material, obeying the associated Mohr–Coulomb (M–C) criterion with

$$E = 30\,000 \text{ psi},$$
$$c' = 10 \text{ psi},$$
$$\nu = 0.3,$$
$$\varphi' = 20°.$$

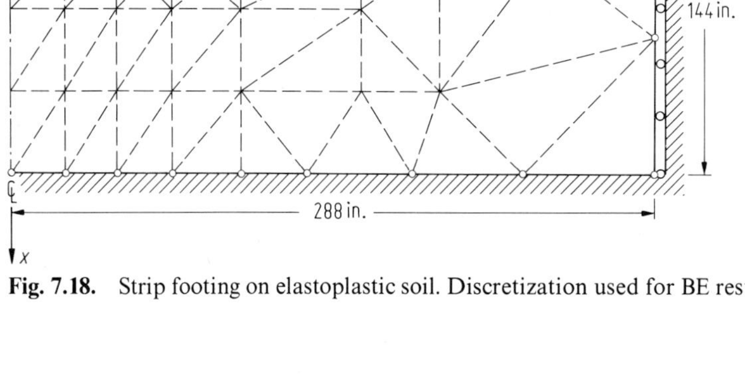

Fig. 7.18. Strip footing on elastoplastic soil. Discretization used for BE results

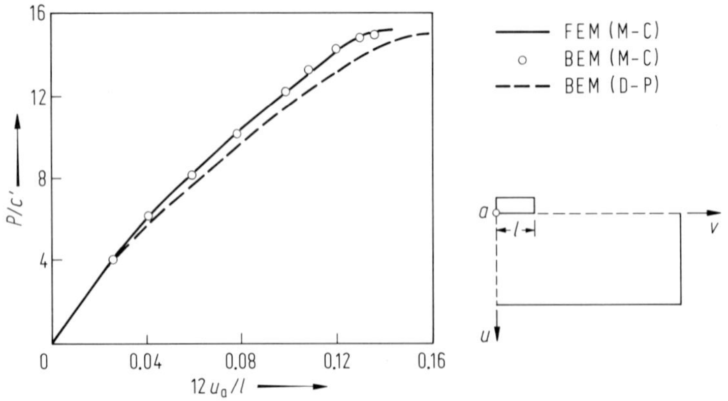

Fig. 7.19. Load–displacement curves for strip footing problem

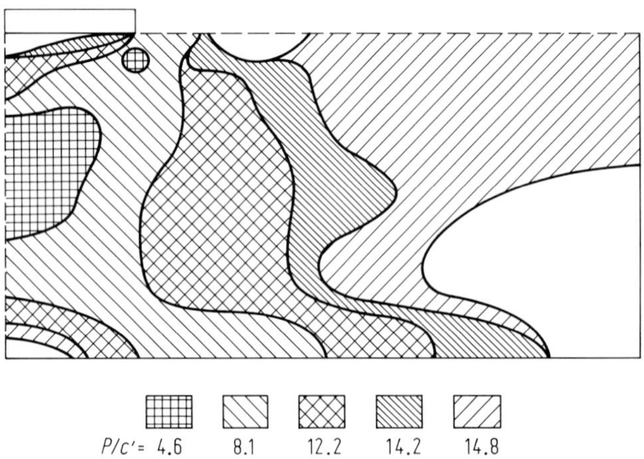

Fig. 7.20. Spread of plastic zones at different load levels. Mohr–Coulomb yield criterion

An alternative solution was also obtained by using the associated Drucker–Prager (D–P) yield criterion given by expressions (7.37) and (7.39).

Ground surface displacements are presented in Fig. 7.19. Also included is the equivalent M–C finite element solution obtained by Zienkiewicz *et al.* [25] using quadratic isoparametric elements with 121 nodal points. The collapse loads achieved by the boundary element and finite element techniques (M–C) are $p/c' = 14.9$ and $p/c' = 15.1$, respectively, which agree well with the Prandtl solution (Chen [5]) $p/c' = 14.8$. As for the D–P results, it is seen that although the displacements were larger, the maximum load obtained was still not far from the previous ones.

Zones of yielding defined by the M–C solution are shown in Fig. 7.20. These zones compare well with the reported finite element computations.

Example 7.8. – Shallow Tunnel [16]. In the last section the elastoplastic boundary element technique was applied to solve the problem of a deep circular excavation of radius r' in an infinite medium. The great advantages of the technique were then pointed out when comparing results with different finite element solutions. Here, an analogous problem is studied by considering the tunnel to be shallow, located within a semi-infinite domain and with its center at a depth of $5r'$.

As before, the rock-like material was assumed to follow the Drucker–Prager yield criterion (α' and K' as given in Eq. (7.39)), with the following characteristics:

$$E = 500 \text{ ksi},$$
$$c' = 0.28 \text{ ksi},$$
$$\nu = 0.2,$$
$$\varphi' = 30°.$$

In order to produce a more realistic analysis, the semi-infinite medium was assumed to be initially under the *in situ* linearly varying stress field given by

$$\sigma_v = \bar{\sigma}_v + \gamma h \qquad \text{(vertical stress)},$$
$$\sigma_h = 0.4\,\sigma_v \qquad \text{(horizontal stresses)},$$

where $\bar{\sigma}_v$ is a uniform pressure that may be due to an overburden of water or very weak material, γ is the unit weight of the rock, and h is the distance from the ground surface.

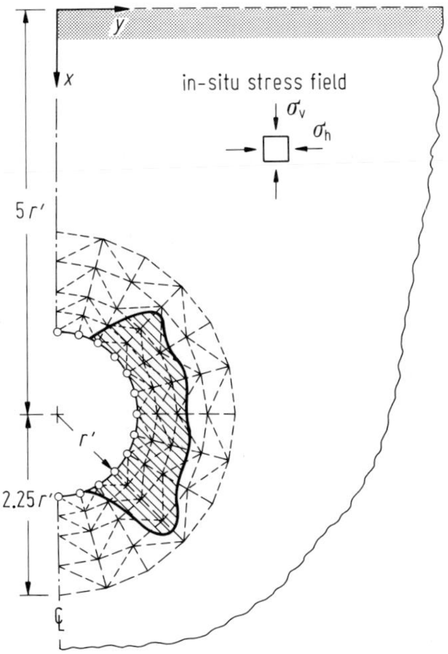

Fig. 7.21. Shallow circular tunnel problem. Discretization used for BE results and total spread of plastic zone

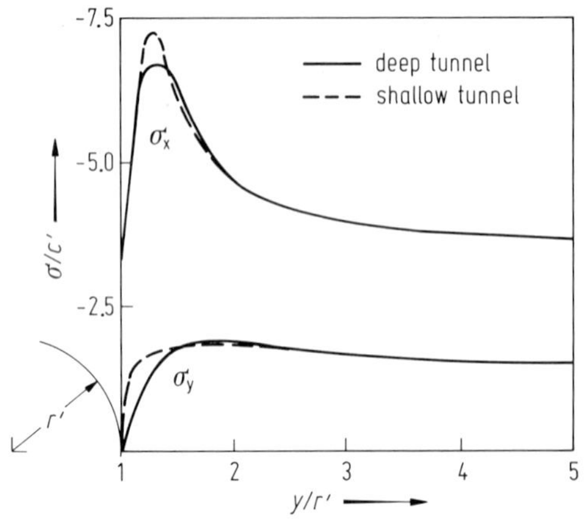

Fig. 7.22. Final stresses along the horizontal section through the medium

To simulate the stress state adopted for the deep tunnel problem ($\sigma_v = 1$ ksi) at the depth of the excavation axis, the following values were chosen:

$$\bar{\sigma}_v = 0.3 \text{ ksi},$$

$$\gamma = 8.9 \times 10^{-2} \text{ lb/in}^3,$$

$$r' = 131 \text{ ft}.$$

The plane strain analysis was carried out by applying increments of external loads, corresponding to the relaxation of the *in situ* stresses, over the boundary of the cavity. The discretization employed is depicted in Fig. 7.21 where the total extent of the plastic zone is also shown.

Final stresses along the horizontal section are presented in Fig. 7.22 with the equivalent results from the deep tunnel case included for comparison. Note that stress values outside the internal cell region were computed at simple internal points.

The above example clearly indicates the powerfulness of the half-plane implementation. Problems of this sort can only be satisfactorily solved by using this technique, which requires neither ground surface nor outer boundary discretization.

7.6. Comparison with Finite Elements

In this section a comparison between boundary element elastoplastic solutions and finite elements is made. The comparison is based on the main factors that affect the efficiency of the programs, i.e., computer time, accuracy of the solution, and simplicity of data structure. The boundary element program used here was developed by Telles in 1979 [26] and the finite element program was written by Owen and Hinton [27]. All the cases were run using the same ICL 2970 computer and the comparison was carried out through two typical examples [28]:

(i) perforated aluminum strip,
(ii) circular cavity under internal pressure.

Example 7.9. – Perforated Aluminum Strip. In this example a rectangular aluminum plate in plane stress and with a central hole is uniformly stressed at the ends. The finite element mesh using nine noded Lagrangian elements is shown in Fig. 7.23a and the boundary element discretization (linear interpolation for both boundary unknowns and plastic strains) is presented in Fig. 7.23b. The material properties are as follows:

$$
\begin{aligned}
E &= 70{,}000\ \mathrm{N/mm^2},\\
\nu &= 0.2,\\
Y &= 243\ \mathrm{N/mm^2},\\
H' &= 0.032\,E\ \text{(Mises criterion)}.
\end{aligned}
$$

Figure 7.24 presents the load–displacement curve for a point located on the loaded edge using boundary or finite elements. Both techniques give results of similar accuracy, but the differences in computer time and data preparation (Table 7.1) indicate the advantage of using boundary elements.

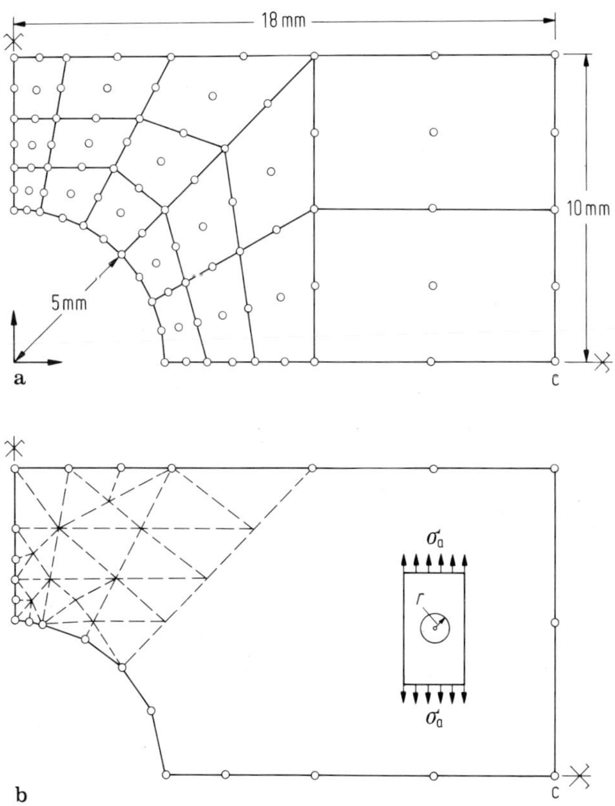

Fig. 7.23. **a** Finite element mesh for perforated strip. **b** Boundary element discretization for perforated strip

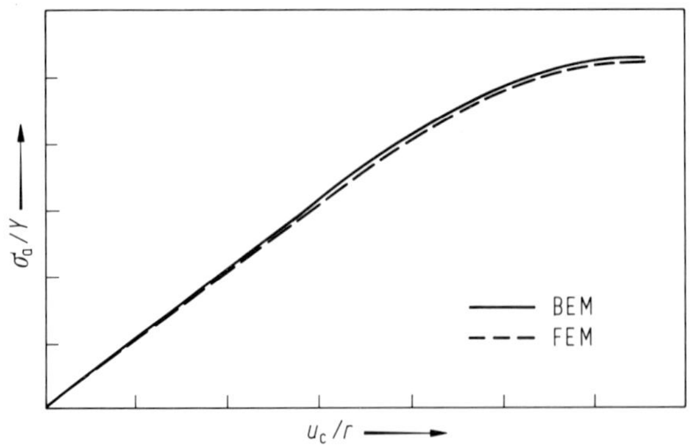

Fig. 7.24. Load–displacement curves for perforated strip problem

Notice that in order to obtain the same accuracy, the number of boundary element nodes required on the surface of the body is slightly less than the number of finite element nodes required. This is due to the good convergence properties of the "mixed-type" boundary element formulation.

Example 7.10. – Circular Cavity under Internal Pressure. Boundary elements are at their best for problems with infinite domain, such as the circular cavity under internal pressure. This is a plane strain problem and the difficulty arises, when using finite elements, of having to discretize the domain up to a certain distance. This distance was here taken to be 10 times the radius of the cavity. The nine noded finite element mesh is shown in Fig. 7.25a and the material was assumed to follow the von Mises yield criterion with the following properties:

$$E = 70{,}000\ \mathrm{N/mm^2},$$

$$\nu = 0.2,$$

$$\sigma_0 = 1000\ \mathrm{N/mm^2}\ \text{(ideal plasticity)}.$$

The problem was solved by finite elements, first assuming that the outer boundary was fixed and then free to displace. Here, in theory, the radial displacements at the cavity should lie between the displacement values found by the two assumptions. The problem was then analyzed using boundary elements, for which the infinite body can be properly taken into consideration. The discretization employed is shown in Fig. 7.25b. Figure 7.26 shows that the boundary element results for displacements lie between the two finite element solutions for the higher loads. However, for the first load increments, the BE displacements are slightly larger than those obtained by finite elements. This may be due to the well-known rigidity of the finite element results, especially in the largely elastic region. When the yielding zone lies near the cavity, the restraint conditions of the outer boundary have little effect on the radial displacements of the cavity, and the rigidity of the FE dominates the outcome. Further advantages of using BE can be seen in Table 7.1. Notice that for this problem the input data is greatly simplified, especially when

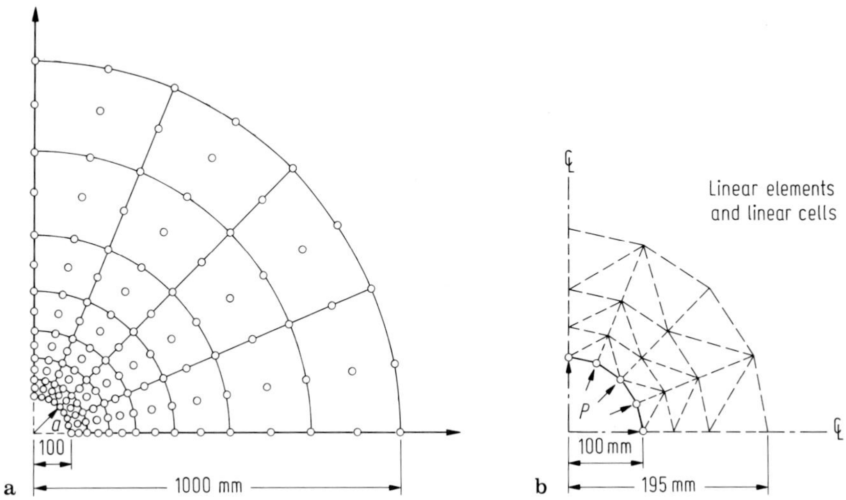

Fig. 7.25. **a** Finite element mesh; circular cavity. **b** Boundary element discretization; circular cavity

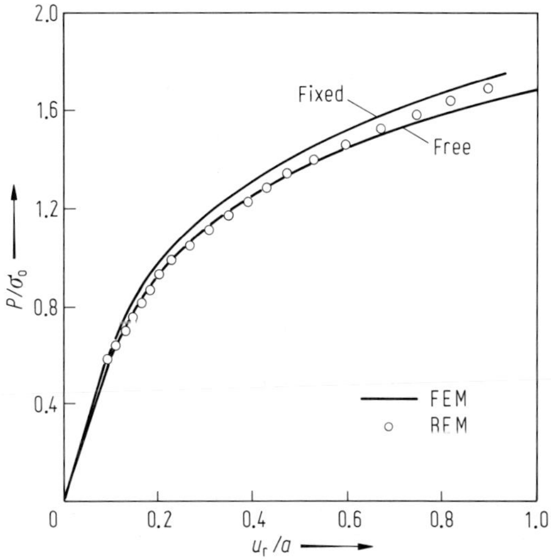

Fig. 7.26. Radial deflections versus loads; circular cavity

Table 7.1. Comparison of BE and FE solutions.

Problem	FE or BE	Total No. of nodes/points needed	Boundary or finite elements needed	CPU time using ICL 2970 (in units)
Perforated aluminum strip	FE BE	87 40	17 23	430 142
Circular cavity under internal pressure	FE (fixed) FE (free) BE	135 135 20	28 28 4	1212 1341 103

taking symmetry into consideration. The boundary element technique is also more efficient than finite elements in terms of computer time, and substantial savings can be achieved in this case.

References

1. Hill, R., *The Mathematical Theory of Plasticity*, Clarendon Press, Oxford, 1950.
2. Ford, H., and Alexander, J. M., *Advanced Mechanics of Materials*, 2nd ed., Ellis Horwood, Chichester, 1977.
3. Prager, W., and Hodge, P. G., *Theory of Perfectly Plastic Solids*, Dover, New York, 1968.
4. Mendelson, A., *Plasticity: Theory and Application*, Macmillan, New York, 1968.
5. Chen, W. F., *Limit Analysis and Soil Plasticity*, Elsevier, Amsterdam, 1975.
6. Malvern, L. E., *Introduction to the Mechanics of a Continuous Medium*, Prentice–Hall, Englewood Cliffs, N. J., 1969.
7. Fung, Y. C., *Foundations of Solid Mechanics*, Prentice–Hall, Englewood Cliffs, N. J., 1965.
8. Johnson, W., and Mellor, P. B., *Plasticity for Mechanical Engineers*, Van Nostrand, London, 1962.
9. Olszak, W., Mroz, A., and Perzyna, P., *Recent Trends in the Development of the Theory of Plasticity*, Pergamon Press, Oxford; PWN, Warsaw, 1963.
10. Schreyer, H. L., Kulak, R. F., and Kramer, J. M., Accurate numerical solutions for elastic-plastic models, Trans. ASME J. Pressure Vessel Technol. **101,** 226–234 (1979).
11. Zienkiewicz, O. C., Valliappan, S., and King, I. P., Elasto-plastic solutions of engineering problems, initial stress finite element approach, Int. J. Numerical Methods Engng. **1,** 75–100 (1969).
12. Telles, J. C. F., Brebbia, C. A., The boundary element method in plasticity, Appl. Math. Modelling **5,** 275–281 (1981).
13. Haward, R. N., and Owen, D. R. J., The yielding of a two dimensional void assembly in an organic glass, J. Materials Sci **8,** 1136–1144 (1973).
14. Nayak, G. C., and Zienkiewicz, O. C., Elasto-plastic stress analysis: a generalization for various constitutive relations including strain softening, Int. J. Numerical Methods Engng. **5,** 113–135 (1972).
15. Nayak, G. C., and Zienkiewicz, O. C., Convenient form of stress invariants for plasticity, Proc. Amer. Soc. Civil Engrs. J. Struct. Div. **98,** 949–954 (1972).
16. Telles, J. C. F., and Brebbia, C. A., Boundary elements: New developments in elastoplastic analysis, Appl. Math. Modelling **5,** 376–382 (1981).
17. Drucker, D. C., and Prager, W., Soil mechanics and plastic analysis or limit design, Quart. Appl. Math. **10,** 157–165 (1952).
18. Wylie, C. R., *Advanced Engineering Mathematics*, 4th ed., McGraw–Hill, Tokyo, 1975.
19. Telles, J. C. F., and Brebbia, C. A., Elastoplastic boundary element analysis, in *Proc. Europe–U.S. Workshop on Nonlinear Finite Element Analysis in Structural Mechanics* (W. Wunderlich *et al.*, Eds.), pp. 403–434, Springer-Verlag, Berlin, 1981.
20. Bushnell, D., A strategy for the solution of problems involving large deflections, plasticity and Creep, Int. J. Numerical Methods Engng. **11,** 683–708 (1977).
21. Yamada, Y., Yoshimura, N., and Sakurai, T., Plastic stress-strain matrix and its application for the solution of elastic-plastic problems by the finite element method, Int. J. Mech. Sci. **10,** 343–354 (1968).
22. Reyes, S. F., and Deere, D. U., Elastic-plastic analysis of underground openings by the finite element method, in *Proc. 1st Int. Congr. Rock Mechanics*, pp. 477–483, Lisbon, 1966.
23. Baker, L. E., Shandu, R. S., and Shieh, W. Y., Application of elastoplastic analysis in rock mechanics by finite element method, in *Proc. 11th Symp. Rock Mechanics*, (J. Somerton, Ed.), pp. 237–251, University of California, Berkeley, 1969.
24. Chen, A. C. T., and Chen, W. F., Constitutive equations and punch-indentation of concrete, Proc. Amer. Soc. Civil Engrs. J. Engng. Mech. Div. **101,** 889–906 (1975).

25. Zienkiewicz, O. C., Humpheson, C., and Lewis, R. W., Associated and non-associated visco-plasticity and plasticity in soil mechanics, Geotechnique **25,** 671–689 (1975).
26. Telles, J. C. F., On the application of the boundary element method to inelastic problems, Ph.D. Thesis, University of Southampton, 1981.
27. Owen, D. R. J., and Hinton, E., *Finite Elements in Plasticity: Theory and Practice,* Pineridge Press, Swansea, 1980.
28. Lee, K. N., A comparison of finite elements and boundary element solutions in plasticity M.Sc. Thesis, University of Southampton, 1982.
29. Maier, G., and Novati, G., Elastic-plastic boundary element analysis as a linear complementarity problem, Applied Mathematical Modelling, **7** (1983)

Chapter 8

Other Nonlinear Material Problems

8.1. Introduction

In the present chapter an application of the boundary element equations to viscoplasticity is presented. The procedure can be used for creep problems as well. The Perzyna [1–3] approach has been adopted since it is appropriate for computer applications and – as indicated in Chapter 6 – can be used to simulate pure elastoplastic solutions. The time-dependent solution is obtained by a simple Euler one-step procedure and some guidelines for the selection of the time step length are discussed. In addition, problems involving no-tension materials are also presented and illustrated by examples.

8.2. Rate-Dependent Constitutive Equations

In this chapter we shall restrict ourselves to the solution of either creep or elastic/viscoplastic problems in the sense described by Perzyna [1]. It is worth mentioning that transient or steady-state thermal strains could be equally considered by solving a coupled boundary element problem following the procedure presented in Chapter 4 for the thermal part of the problem.

With reference to Section 6.2, the static yield criterion for isotropic hardening can now be written in general form as

$$F(\sigma_{ij}, k) = 0\,, \tag{8.1}$$

where, as before, k represents a hardening parameter which dictates the position of the static yield surface in the nine-dimensional stress space. This condition can be better visualized in the form

$$f(\sigma_{ij}) = \psi(k) \tag{8.2}$$

in which $F = f - \psi$, and if the work hardening hypothesis is being adopted, k is given by expression (7.3).

One can notice that the conditon expressed in Eq. (8.1) or (8.2) does not differ from the corresponding yield condition for the so-called inviscid theory of plasticity. Therefore, the different expressions for F introduced in Section 7.4 can still be used. This encourages a further interpretation; let us designate the scalar function of $f(\sigma_{ij})$ by σ_e, as before. Such designation allows for the definition of the equivalent plastic strain rate as (see expression (7.41))

$$\sigma_e \, \dot{\varepsilon}^p_e = \sigma_{ij} \, \dot{\varepsilon}^p_{ij} = \dot{k} \, . \tag{8.3}$$

Following the generalized normality principle due to Perzyna [1–3], the visco-plastic strain rates are given by

$$\dot{\varepsilon}^p_{ij} = \gamma \left\langle \Phi\left(\frac{F}{\psi}\right)\right\rangle \frac{\partial F}{\sigma_{ij}} \, , \tag{8.4}$$

where γ, Φ, and the symbol $\langle\ \rangle$ have been defined and commented upon in Section 6.2.

Equation (8.4) can be further written as

$$\dot{\varepsilon}^p_{ij} = \gamma \left\langle \Phi\left(\frac{F}{\psi}\right)\right\rangle \frac{\partial f}{\partial \sigma_{ij}} \tag{8.5}$$

which after multiplying both sides by σ_{ij} gives

$$\sigma_{ij} \, \dot{\varepsilon}^p_{ij} = \gamma \quad \Phi\left(\frac{F}{\psi}\right)\Big\rangle \, \sigma_{ij} \frac{\partial f}{\partial \sigma_{ij}} \, . \tag{8.6}$$

Assuming that $f(\sigma_{ij})$ is homogeneous of degree one (a requirement satisfied by the yield criteria adopted here) and applying Euler's theorem [7] we obtain

$$\sigma_{ij} \, \dot{\varepsilon}^p_{ij} = \gamma \left\langle \Phi\left(\frac{F}{\psi}\right)\right\rangle f(\sigma_{ij}) \, . \tag{8.7}$$

Recalling definition (8.3), expression (8.7) can be finally represented by

$$\dot{\varepsilon}^p_e = \gamma \left\langle \Phi\left(\frac{F}{\psi}\right)\right\rangle , \tag{8.8}$$

a relation which for $F > 0$ leads to

$$f(\sigma_{ij}) = \psi(k) \left[1 + \Phi^{-1}\left(\frac{\dot{\varepsilon}^p_e}{\gamma}\right)\right] . \tag{8.9}$$

Equation (8.9) when compared to Eq. (8.2) clearly demonstrates the explicit dependence of the flow surface on the equivalent plastic strain rate.

As a further illustration, consider the definition

$$\dot{\varepsilon}_e = \frac{\dot{\sigma}_e}{E} + \dot{\varepsilon}^p_e \tag{8.10}$$

in which E is the Young's modulus and $\dot{\varepsilon}_e$ represents an equivalent measure of the total strain rate. Note that in uniaxial problems $\dot{\varepsilon}_e$, $\dot{\sigma}_e$, and $\dot{\varepsilon}^p_e$ become the actual total strain, stress, and plastic strain rates if ψ is defined as the uniaxial yield stress.

The flow surface can now be written

$$f(\sigma_{ij}) = \psi(k)\left[1 + \Phi^{-1}\left(\frac{\dot{\varepsilon}_e - \dot{\sigma}_e/E}{\gamma}\right)\right], \tag{8.11}$$

indicating the explicit dependence of $f(\sigma_{ij})$ on the rate of induced strains/stresses.

For creep problems the equivalent version of expression (8.8) is assumed to be (see Section 6.2)

$$\dot{\varepsilon}_e^c = K\,\sigma_e^m\,t^n\,, \tag{8.12}$$

where K is a material parameter and $\sigma_e = f(\sigma_{ij})$ represents the von Mises equivalent stress.

It is interesting to note that the time hardening function t^n can be removed from Eq. (8.12) by the following transformation [9]:

$$\bar{t} = \int_0^t \tau^n\,d\tau\,, \tag{8.13}$$

where $\bar{t}$ denotes a transformed time leading to

$$\frac{d\varepsilon_e^c}{d\bar{t}} = K\,\sigma_e^m\,. \tag{8.14}$$

This means that the problem can be solved in terms of a fictitious time which relates to the true time t by means of the inverse relation

$$t = [\bar{t}(n+1)]^{1/(n+1)}\,. \tag{8.15}$$

Different time hardening functions can be equally transformed by the above procedure assuming that Eq. (8.12) is taken from experimental analysis under constant stress.

The creep strain rates can therefore be written

$$\dot{\varepsilon}_{ij}^c = K\,\sigma_e^m\,\frac{\partial f}{\partial \sigma_{ij}}\,, \tag{8.16}$$

where the dot indicates derivative with respect to $\bar{t}$ if $n \neq 0$.

Equation (8.16) corresponds to the Prandtl–Reuss equations and can be cast into the form of Eq. (8.5). In both cases the corresponding initial stress rates can be computed by the simple relation

$$\dot{\sigma}_{ij}^a = \gamma\,\langle\Phi\rangle\,\bar{d}_{ij} \tag{8.17}$$

where

$$\bar{d}_{ij} = C_{ijkl}\,\frac{\partial f}{\partial \sigma_{kl}}\,. \tag{8.18}$$

Herein, for the boundary element implementation, the initital stress equations have been adopted since they present the advantage of handling compressible or incompressible inelastic strains in plane strain or plane stress problems with minor alterations. The relevant two-dimensional forms of the above relations are presented in Appendix C.

In order to apply Eqs. (6.134) and (6.135) to the solution of time-dependent inelastic problems, the manipulations introduced in Section 6.8 can be performed, resulting in the following matrix equations:

$$\dot{Y} = \boldsymbol{R}\,\dot{\boldsymbol{\sigma}}^a + \dot{\boldsymbol{M}} \tag{8.19}$$

and

$$\dot{\boldsymbol{\sigma}} = \boldsymbol{V}\,\dot{\boldsymbol{\sigma}}^a + \dot{\boldsymbol{N}}, \tag{8.20}$$

where vectors $\dot{\boldsymbol{M}}$ and $\dot{\boldsymbol{N}}$ are given in Eqs. (6.141) and (6.144), matrix $\boldsymbol{R}$ is defined in Eq. (7.63), and the new matrix $\boldsymbol{V}$ is given by

$$\boldsymbol{V} = \bar{\boldsymbol{Q}} - \boldsymbol{A}'\boldsymbol{R} \tag{8.21}$$

in which

$$\bar{\boldsymbol{Q}} = \boldsymbol{Q}' + \boldsymbol{E}'\,. \tag{8.22}$$

From the above it is seen that (see expression (8.17)) Eq. (8.20) represents a system of ordinary differential equations for stresses at selected boundary nodes and internal points which can be solved by standard methods (provided it satisfies the Lipschitz condition [4–6]), producing a unique solution to the time-dependent problem. A simple and efficient solution procedure for this matrix equation is the subject of the next section.

8.3. Solution Technique: Viscoplasticity

For the solution of the examples presented in this chapter, a simple Euler one-step procedure [8] has been adopted in the following fashion: let us assume a load factor $\lambda(t)$ which is considered to be a known function of time. Equations (8.19) and (8.20) can be integrated on time to give

$$Y = \boldsymbol{R}\,\boldsymbol{\sigma}^a + \lambda(t)\,\boldsymbol{M} \tag{8.23}$$

and

$$\boldsymbol{\sigma} = \boldsymbol{V}\,\boldsymbol{\sigma}^a + \lambda(t)\,\boldsymbol{N}, \tag{8.24}$$

where vectors $\boldsymbol{M}$ and $\boldsymbol{N}$ correspond to the elastic solution at some reference load level.

For the time-marching procedure, Eq. (8.24) is applied after each discrete time step ($\Delta t = {}^{k+1}t - {}^{k}t$) with the value of the initial stresses being computed at selected boundary nodes and internal points by the Euler formula

$$ {}^{k+1}\sigma_{ij}^{a} = {}^{k}\sigma_{ij}^{a} + \Delta t\, \gamma \langle {}^{k}\Phi \rangle\, {}^{k}\bar{d}_{ij} \,. \tag{8.25} $$

During this process one may have that $\lambda(t)$ is left constant for some time, creating a situation in which after a sufficient number of time steps has been applied, the values of $(\Delta t\, \dot{\varepsilon}_{e}^{p})$ or $({}^{k+1}\sigma_{e} - {}^{k}\sigma_{e})$ become vanishingly small everywhere. In such cases a stationary condition is deemed to have occurred and the time-marching scheme can be stopped.

It is interesting to note that the time integration procedure does not require computation of the boundary unknowns. Consequently, Eq. (8.23) need only be used to print the boundary unknowns at some requested time/load values.

The success of this simple time integration scheme is dependent on the proper selection of the time step lengths. It has been known for quite some time [9] that ideally small time steps should be applied in the early stages of the computation (i.e., after the application of the load or load increment) and that these can be increased in size as stationary or steady state is approached.

Following the experience of many authors [9–12] with different spatial discretization techniques (mainly finite elements), the time step size should be controlled by a relation between accumulated and rate value of some variables to produce the above-described automatic lengthening as the asymptotic state is achieved. This can be considered at each node or point as follows:

$$ \Delta t \leqq \eta_1 \frac{\varepsilon_e}{\dot{\varepsilon}_e^{a}} = \frac{\eta_1}{\dot{\varepsilon}_e^{a}} \left(\frac{\sigma_e}{E} + \varepsilon_e^{a} \right) \tag{8.26} $$

under the condition that

$$ {}^{k+1}\Delta t \leqq \eta_0\, {}^{k}\Delta t \,, \tag{8.27} $$

where η_1 and η_0 are problem-dependent parameters that should be chosen to compromise between computer time and accuracy. Normally, $0.01 \leqq \eta_1 \leqq 0.15$ and $1.2 \leqq \eta_0 \leqq 2$.

A drawback of relations (8.26) and (8.27) is that they do not guarantee complete stability of the explicit time integration scheme, particularly near the steady state which produces large time step values. Useful bounds for the maximum time step length have been presented by Cormeau [13] for perfectly viscoplastic materials, these can be seen to correspond to the pure relaxation problem

$$ \dot{\varepsilon}_{ij} = \frac{1}{2G} \left[\dot{\sigma}_{ij} - \frac{\nu}{1+\nu} \dot{\sigma}_{kk}\, \delta_{ij} \right] + \dot{\varepsilon}_{ij}^{a} = 0 \,. \tag{8.28} $$

Herein, in addition to the above-referred bounds, an approximate, yet general, limiting value for the time step has been adopted.

Recalling expression (8.10) in rearranged form one obtains

$$\dot{\sigma}_e = g(t, \sigma_e)\,, \tag{8.29}$$

where for $\dot{\varepsilon}_e = 0$

$$g(t, \sigma_e) = -E\,\gamma\langle\Phi\rangle\,. \tag{8.30}$$

Equation (8.29) in equivalent stress form can be used to study a bound [4] for the stability condition of the simple Euler procedure adopted here. Thus, the coupling of Eqs. (8.29) and (8.25) leads to the equivalent relation

$$^{k+1}\sigma_e = {}^k\sigma_e + \Delta t\, g(t, {}^k\sigma_e)\,. \tag{8.31}$$

Let us now accept that throughout the time-marching process truncation and roundoff errors have been committed. The global error at time ^kt is then given by

$$^k\varrho = {}^k\sigma_e - {}^k\hat{\sigma}_e\,, \tag{8.32}$$

where $\hat{\sigma}_e$ represents the value of σ_e obtained through an exact integration on time of Eq. (8.29) which corresponds to the exact solution of Eq. (8.24).

Assuming that the error ϱ is sufficiently small to allow a truncated Taylor's expansion of $g(t, \sigma_e)$ about $\hat{\sigma}_e$, we obtain

$$g(t, {}^k\sigma_e) = g(t, {}^k\hat{\sigma}_e) + {}^k\varrho\,\frac{\partial g(t, {}^k\hat{\sigma}_e)}{\partial\sigma_e} + \cdots\,. \tag{8.33}$$

The substitution of Eq. (8.32) for ^kt and ^{k+1}t in Eq. (8.31), together with Eq. (8.33) gives

$$^{k+1}\hat{\sigma}_e + {}^{k+1}\varrho = {}^k\hat{\sigma}_e + {}^k\varrho + \Delta t\left[g(t, {}^k\hat{\sigma}_e) - {}^k\varrho\,\gamma\,E\,{}^k\!\left(\frac{\partial\Phi}{\partial\sigma_e}\right)\right]. \tag{8.34}$$

If a stationary state is likely to have occurred,

$$^{k+1}\hat{\sigma}_e = {}^k\hat{\sigma}_e\,, \qquad g(t, {}^k\hat{\sigma}_e) = 0\,, \tag{8.35}$$

hence,

$$^{k+1}\varrho = {}^k\varrho\left[1 - \Delta t\,\gamma\,E\,{}^k\!\left(\frac{\partial\Phi}{\partial\sigma_e}\right)\right]. \tag{8.36}$$

In order to ensure that errors remain bounded (stability), one has that $|^{k+1}\varrho| \leqq |^k\varrho|$, which gives

$$\Delta t \leqq \Delta t_{\mathrm{CRIT}} = \frac{2}{\gamma(\partial\Phi/\partial\sigma_e)\,E}\,. \tag{8.37}$$

Taking into consideration that for work hardening viscoplastic materials

$$\frac{\partial \Phi}{\partial \sigma_e} = \Phi' \left[\frac{1}{\psi} - \frac{\sigma_e}{\psi^2} \frac{d\psi}{dk} \frac{dk}{d\sigma_e} \right], \tag{8.38}$$

where $\Phi' = d\Phi/d(F/\psi)$, one finally obtains

$$\Delta t \leqq \Delta t_{\mathrm{CRIT}} = \frac{2\,\psi^2}{\gamma\,\Phi'(E\,\psi + \sigma_e\,d\psi/d\varepsilon_e^p)} \tag{8.39}$$

in which for $\psi = \sigma_0$ one has $d\psi/d\varepsilon_e^p = H'$.

For creep problems the equivalent expression is

$$\Delta t_{\mathrm{CRIT}} = \frac{2}{K E\, m\, \sigma_e^{m-1}} \tag{8.40}$$

and if Eq. (8.12) is used instead, the term t^n should appear in the denominator, producing the same critical time step obtained by Cormeau [13] and Irons [14] when $\nu \to 0.5$.

In order to study the effect of hardening in the critical time step let us consider the case $\Phi' = 1$, $d\psi/d\varepsilon_e^p = H' =$ constant. In this case expression (8.39) simplifies as follows:

$$\Delta t_{\mathrm{CRIT}} = \frac{2\,\sigma_0^2}{\gamma\, E(Y + H'\,\varepsilon_e)}, \tag{8.41}$$

where the relation $\sigma_e = E(\varepsilon_e - \varepsilon_e^p)$ with $\varepsilon_e =$ constant was used. The above relation indicates that when $H' > 0$, the effect of hardening produces an initial reduction in the critical time step (when compared to the case $H' = 0$), and that as viscoplastic flow progresses this limit is increased with the square of σ_0. One can notice that this is not the case when $H' < 0$ (softening); here the time step limit is initially increased but diminishes as viscoplasticity develops, producing a reduction in the region of stability which must not be overlooked in such cases.

In the next section the results of some examples solved in the light of the theory presented in this chapter are compared with existing results taken from the literature.

8.4. Examples: Time-Dependent Problems

An interesting feature of the elastic/viscoplastic theory (see Section 6.2) is that if the load is applied in small increments, allowing for stationary conditions to be achieved after each load step, a pure elastoplastic solution is obtained. The question of how small these increments should be taken still remains an open question and is, in fact, problem dependent. In the first example presented here this feature is fully explored for solving a current elastoplastic problem. But in the second and third applications, the total load is applied in one step and two problems of the type power law creep and quasilinear viscoplastic are analyzed.

Example 8.1. – Deep Beam. In the first example the elastoplastic behavior of a simply supported deep beam under uniform load is studied by the viscoplastic boundary element technique. The discretization employed is shown in Fig. 8.1 and the material is assumed to obey the Tresca yield criterion with the following parameters:

$$E = 30 \times 10^6 \text{ psi},$$
$$\sigma_0 = 36 \times 10^3 \text{ psi},$$
$$\nu = 0.3,$$
$$H' = 0; \quad \Phi\left(\frac{F}{\psi}\right) = \frac{F}{\psi}; \quad \gamma = 1 \text{ sec}^{-1}.$$

This problem has been analyzed by Anand *et al.* [15] by using a mesh of 272 linear displacement triangular finite elements, which corresponds to 33% more elements on the boundary than the discretization used here [8].

A comparison of results is depicted in Fig. 8.2. where the load-midspan displacement curves, for both numerical techniques, are plotted together with the

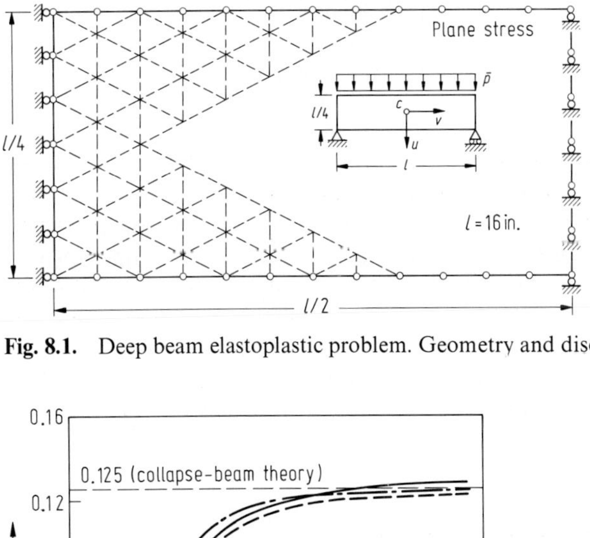

Fig. 8.1. Deep beam elastoplastic problem. Geometry and discretization used for BE results

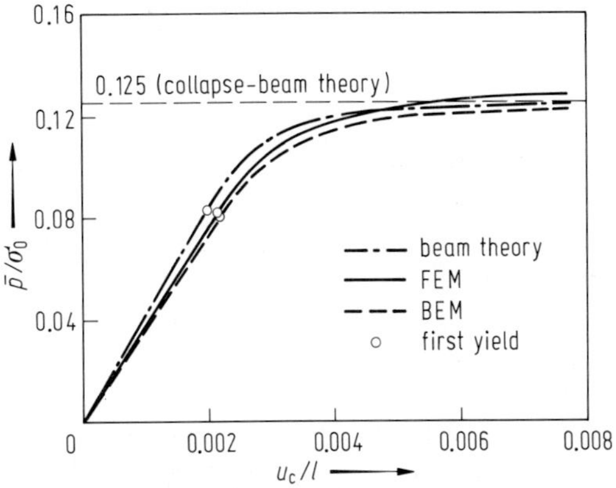

Fig. 8.2. Load–midspan displacement curves for deep beam problem

beam theory solution [16]. As can be seen, the boundary element solution asymptotically approaches the limit load obtained by the beam theory, whereas the finite element results slightly exceed this load level. A vanishing small difference is already noticed in the elastic results, with the BE technique predicting a lower load value for initial yield and larger displacements for the same load level. The plastic zones produced by both techniques were in good agreement with the beam theory and, therefore, are not shown here.

Example 8.2. – Thin Disc [8]. Accurate bounds for the creep problem of a thin disc with a central rigid insert under constant external edge load were produced by Sim [17]. These were obtained by direct time integration of the analytical solution and presented in dimensionless form using the so-called "reference stress" technique. In order to test the boundary element performance in the same problem, the following material parameters were chosen:

$$E = 17 \times 10^6 \text{ psi},$$
$$\nu = 0.33,$$
$$\dot{\varepsilon}_e^c = 5.8 \times 10^{-18}\, \sigma_e^{4.4} \qquad \text{(units: 1b, in, and s).}$$

The geometry and load value are given in Fig. 8.3 where the boundary element and internal cell discretization are also shown. Notice that improved axial symmetry was obtained by avoiding boundary discretization of the symmetry axes.

Radial displacements computed over the outer boundary are plotted against time for comparison with the solution bounds in Fig. 8.4. As expected, the boundary element technique produces a flat curve which lies within the narrow space between the two limiting lines taken from the reference. It is interesting to note that the slope of these parallel lines was calculated for an approximate stationary condition in which the variation of the displacement rates was 1%. Consequently,

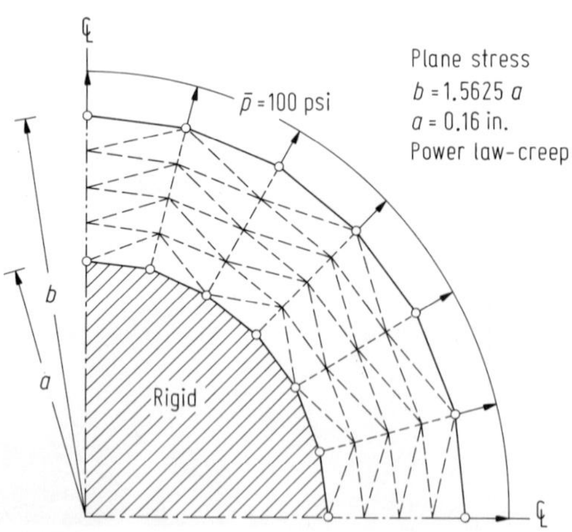

Fig. 8.3. Geometry of thin disc problem including boundary element and internal cell discretization

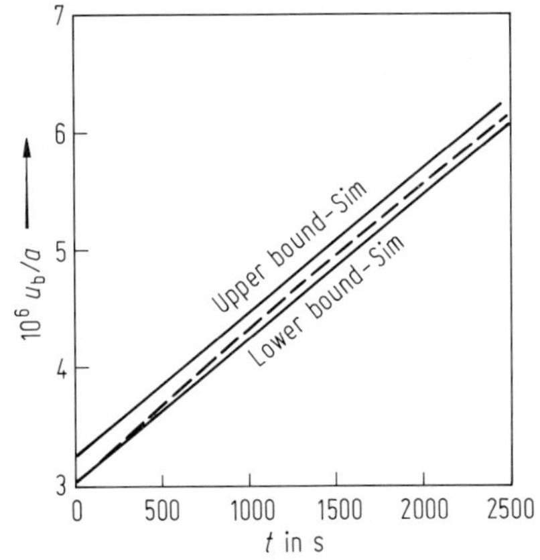

Fig. 8.4. Variations of outer boundary radial displacement with time for thin disc creep problem

the same stationarity criterion was adopted here, generating the final straight part of the curve.

Example 8.3. – Plate under Thermal Shrinkage [8]. In this example the analysis of a rectangular plate, bonded on one edge to a rigid support and subjected to a sudden uniform temperature drop is presented. The thermal shrinkage was assumed to be such that $\varepsilon_{ij}^T = -0.01\ \delta_{ij}$. The problem can be properly solved by prescribing tangential displacements corresponding to $\varepsilon_{ij} = -\varepsilon_{ij}^T$ over the fixed edge and computing the final displacements by simple superposition.

The material was assumed to be quasilinear ($\Phi(F/\psi) = F/\psi$) ideal viscoplastic, obeying the von Mises criterion.

Due to symmetry, only half the plate was discretized using 26 linear boundary elements and 17 internal points located in the region near to the restrained edge

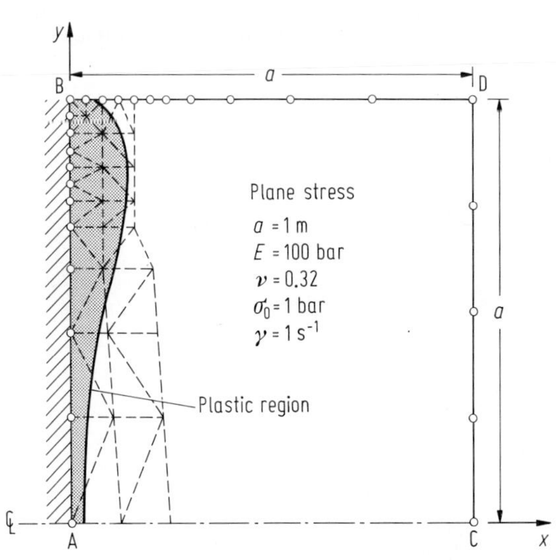

Fig. 8.5. Discretization for rectangular plate under thermal shrinkage and total extent of plastic region

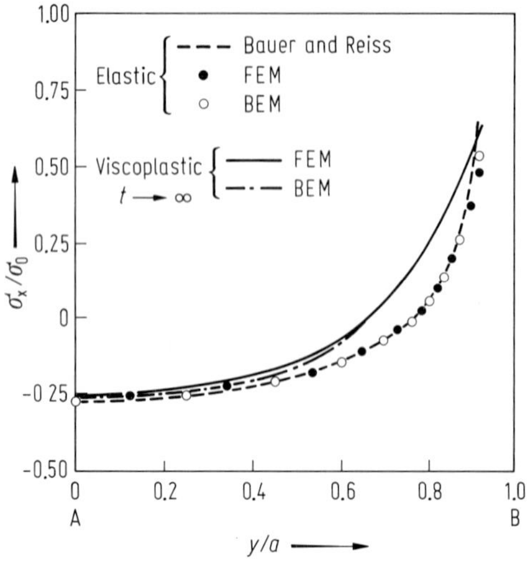

Fig. 8.6. Variation of σ_x over fixed edge for $t = 0$ and asymptotic state

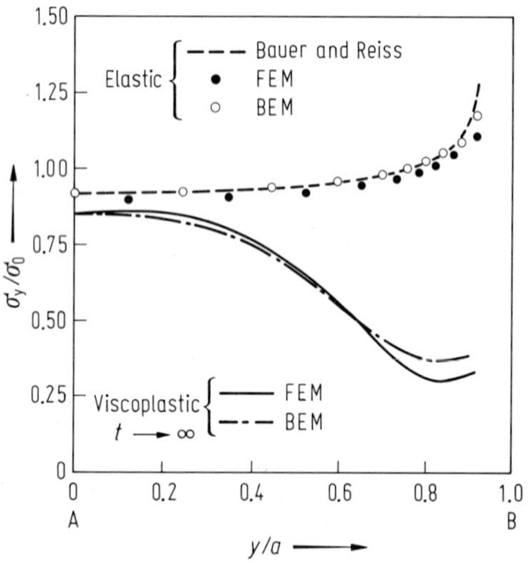

Fig. 8.7. Variation of σ_y over fixed edge for $t = 0$ and asymptotic state

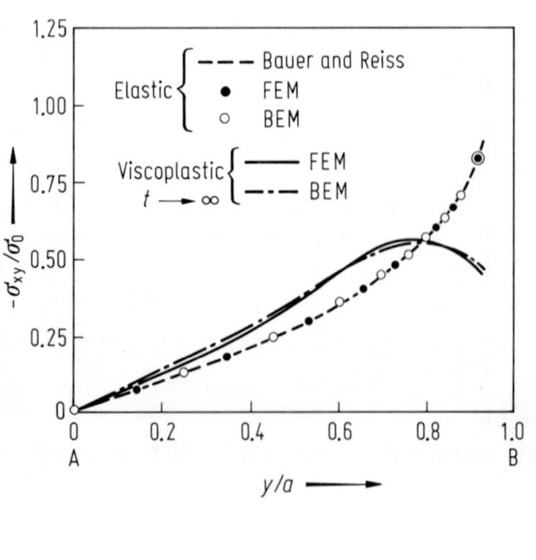

Fig. 8.8. Variation of σ_{xy} over fixed edge for $t = 0$ and asymptotic state

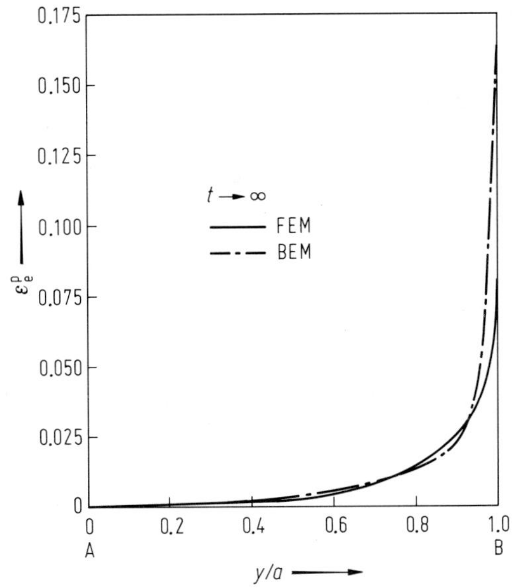

Fig. 8.9. Equivalent plastic strain distribution over bonded edge at stationary state

(linear cells) as shown in Fig. 8.5. Also included is the plastic zone produced by the instantaneous cooling process.

Finite element results for this problem have been presented by Zienkiewicz and Cormeau [11]. They used a mesh of 96 quadrilateral elements which was equivalent in size to the boundary element discretization over the bonded edge (A–B), but presented more refinement over the opposite edge (C–D).

An interesting comparison of results is depicted in Figs. 8.6–8.8 where the stresses computed at the fixed edge are shown for times $t = 0$ (elastic) and $t \to \infty$ when stationary condition is achieved. These include not only the FE and BE results, but also the sufficiently refined elastic finite difference solution produced by Bauer and Reiss [18], which provides a useful reference result for $t = 0$.

It is worth mentioning that neither method can predict the infinite value of the elastic stresses at the corners of the fixed edge. Consequently, a localized perturbation in the solutions is expected in the vicinity of corner B. Nevertheless, even though we neglect the results near to the singular node, one can notice that the boundary elements tend to produce a better representation of the singular behavior than the finite elements. This difference may be partly explained by the fact that the FE stresses were calculated at the Gauss points (2×2 integration) and is particularly apparent in Fig. 8.7 where the σ_y stresses are noticeably unequal over a large range.

A final comparison is presented in Fig. 8.9 in which the equivalent plastic strains computed by the boundary element technique are indicating a more severe concentration of plasticity near the corner than the finite element results.

8.5. No-Tension Materials

In the designing of tunnels and many underground structures, materials such as rock are often idealized as presenting a no-tension behavior. In this case, the rock surrounding the tunnel cannot withstand any tensile stresses produced by the rock excavation process. These sort of problems are particularly well suited to boundary elements and a good number of them has been recently solved [19, 20]. Basically, the procedure follows the first initial stress solution routine presented in Section 7.5, with the difference that now the initial stresses are computed by obtaining the principal stresses at each of the stress points and assigning zero value to the tensile values. This generates the initial stress field which has to be applied back into the body producing a stress redistribution. The iterative procedure is therefore the same indicated in Chapter 7 for the pure incremental initial stress solution.

In order to illustrate the present section some applications are now presented.

Example 8.4. – Hydroelectric Power Station. The deep tunnel depicted in Fig. 8.10 was first analyzed by Valliappan [21] using 500 linear displacement triangular finite elements. In a recent publication, Venturini and Brebbia [19] studied the same no-tension problem using the boundary element technique. The BE solution was obtained for the two cases presented in Figs. 8.10 and 8.11 (with and without prestressing forces) where the discretization employed – linear ele-

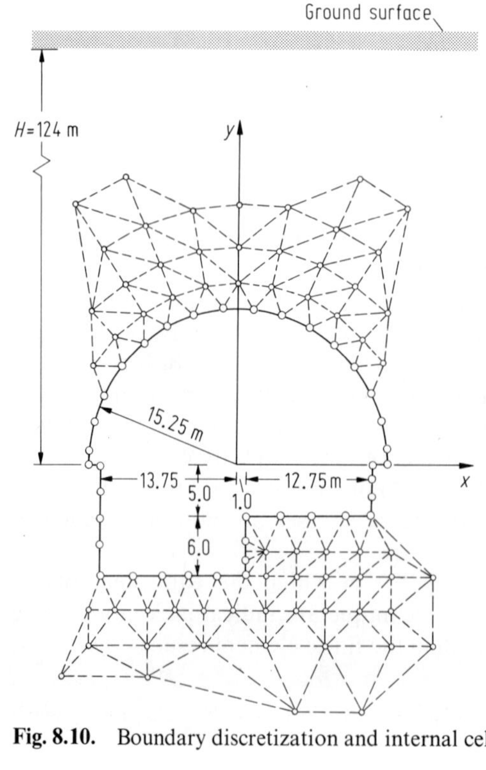

Fig. 8.10. Boundary discretization and internal cells

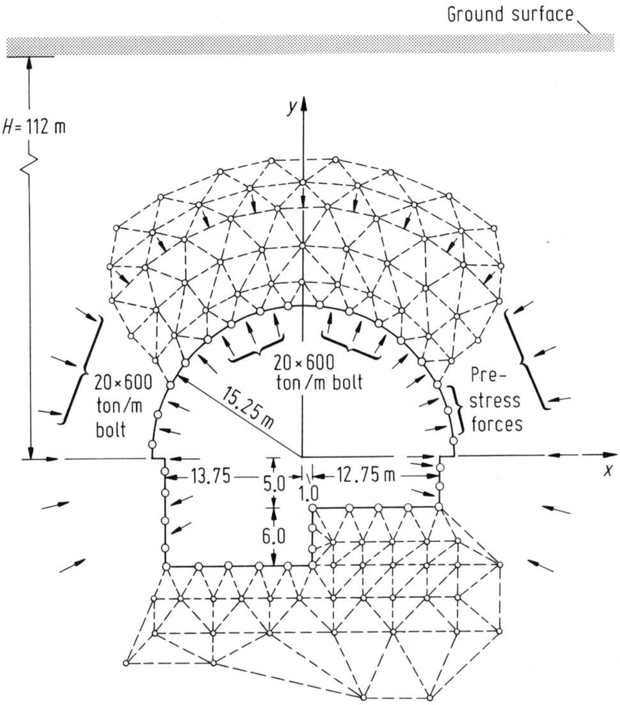

Fig. 8.11. Boundary discretization, internal cells and prestressing forces at the deep tunnel

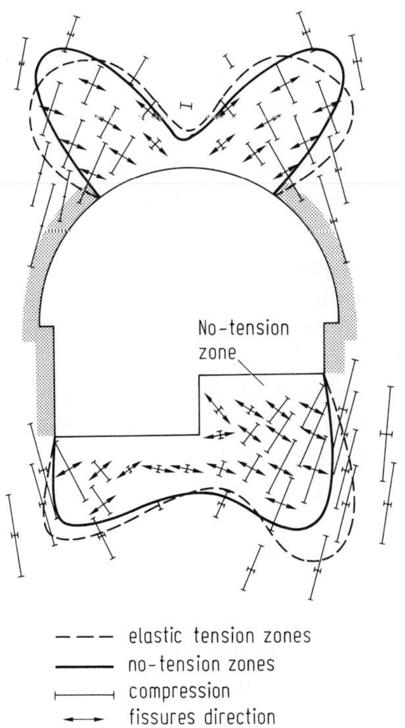

Fig. 8.12. Elastic and no-tension results. Tunnel without prestressing forces

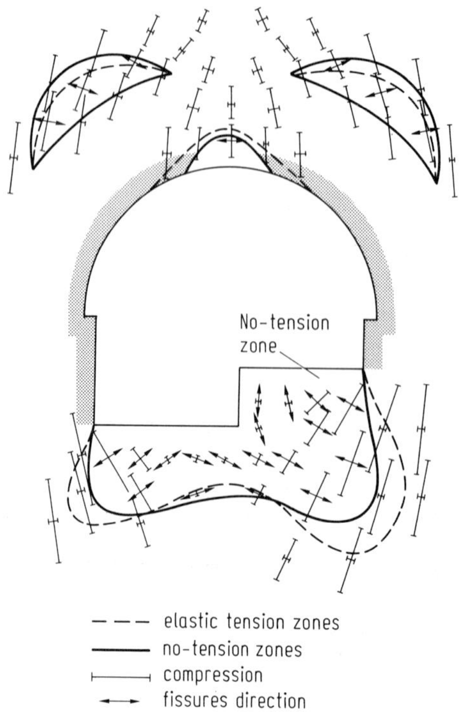

Fig. 8.13. Prestressed tunnel. Elastic and no-tension results

ments and linear cells – is also shown. For the first case the only applied forces are those corresponding to the removal of the material and Fig. 8.12 shows the initial elastic and final no-tension solutions.

Notice the large no-tension zone present over the roof which requires some additional structural component of prestressing to avoid the rock falling down.

The boundary discretization for the second case (with prestressing) is the same as before, but the internal cell configuration was changed to take into consideration the expected new tension-free zones. The prestressing forces shown in Fig. 8.11 were applied at 20 boundary points and 20 internal points which are distributed along circular paths.

The boundary element results are presented in Fig. 8.13, they show that the tension free zones over the roof have been considerably reduced. These results agree reasonably well with the finite element solution presented in [21]. They were, however, more efficiently obtained by the BE technique.

Example 8.5. – Lined Tunnel [20]. This example consists of a lined tunnel, under internal pressure, within the infinite medium. The analysis was carried out by assuming three subregions (see Fig. 8.14 and 8.15); the first is the infinite medium and is considered capable of sustaining tensile stresses, the second is the region where the internal cells are confined, i.e., no-tension region, and the third is assumed to represent the thick concrete lining (linear elastic). The first two regions represent

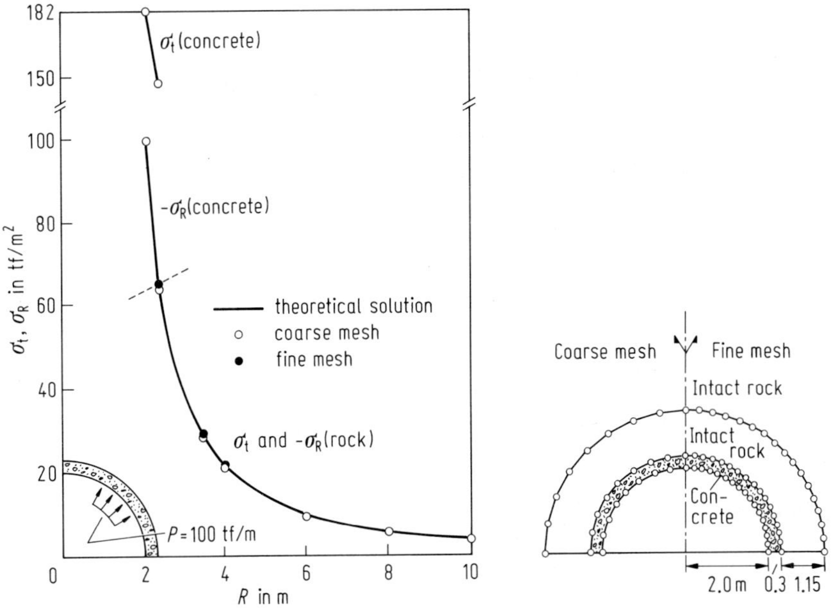

Fig. 8.14. Lined tunnel; boundary discretization and elastic results

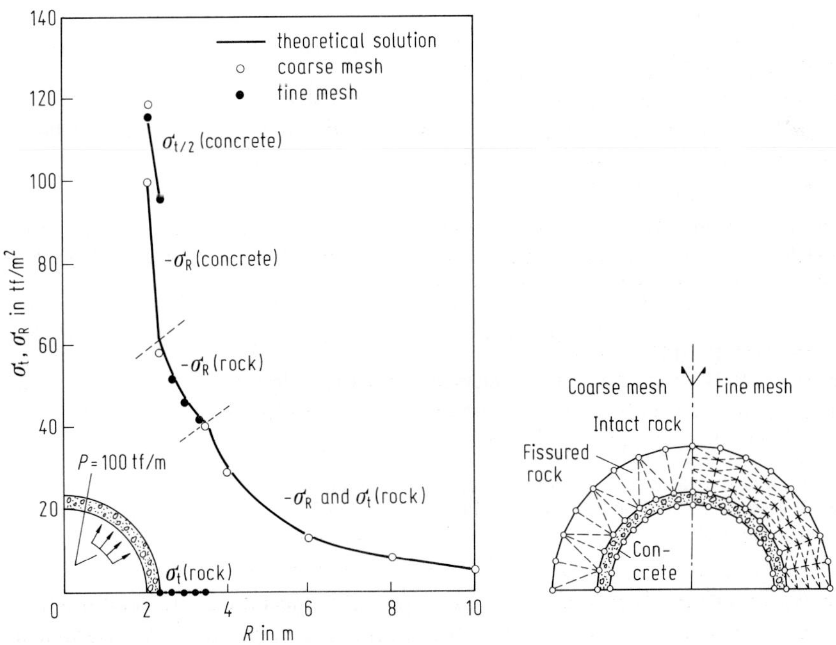

Fig. 8.15. Lined tunnel; cell meshes and no-tension results

the rock medium and have $\nu = 0.2$. The concrete of the lining is assumed to have $\nu = 0.15$ and the ratio between the elastic moduli of the concrete and rock is 2. The *in situ* stresses present in the rock mass are not included to allow a comparison between numerical and theoretical results.

The linear elastic results were first obtained by using the two different boundary discretizations shown in Fig. 8.14. These results were found to agree well with the theoretical solution [22], producing a maximum error of 1.7% and 0.3% for the two cases.

For the no-tension analysis, two different internal cell meshes were used with the coarsest boundary discretization. The boundary element solutions are presented in Fig. 8.15 where the theoretical solution taken from Ref. [23] is also shown. Notice that the agreement is still good with the maximum errors being less than 7.5% and 2.0% in both cases.

References

1. Perzyna, P., Fundamental problems in viscoplasticity, Advan. Appl. Mech. **9,** 243–377 (1966).
2. Perzyna, P., The constitutive equations for rate sensitive plastic materials, Quart. Appl. Math. **20,** 321–332 (1963).
3. Olszak, W., and Perzyna, P., Stationary and non stationary visco-plasticity, in *Inelastic Behaviour of Solids* (Kanninen *et al.*, Eds.), pp. 53–75, McGraw–Hill, New York, 1970.
4. Dahlquist, G., and Björck, A., *Numerical Methods,* Prentice–Hall, Englewood Cliffs, N.J., 1974.
5. Gear, C. W., *Numerical Initial Value Problems in Ordinary Differential Equations,* Prentice–Hall, Englewood Cliffs, N.J., 1971.
6. Lambert, J. D., *Computational Methods in Ordinary Differential Equations,* Wiley, London, 1973.
7. Wylie, C. R., *Advanced Engineering Mathematics,* 4th ed., McGraw–Hill, Tokyo, 1975.
8. Telles, J. C. F., and Brebbia, C. A., Elastic/viscoplastic problems using boundary elements, Int. J. Mech. Sci. **24,** 605–618 (1982).
9. Penny, R. K., The creep of spherical shells containing discontinuities, Int. J. Mech. Sci. **9,** 373–388 (1967).
10. Penny, R. K., and Hayhurst, D. R., The deformations and stresses in a stretched thin plate containing a hole during stress redistribution caused by creep, Int. J. Mech. Sci. **11,** 23–39 (1969).
11. Zienkiewicz, O. C., and Cormeau, I. C., Viscoplasticity and plasticity, an alternative for finite element solution of material nonlinearities, in *Proc. Colloque Meth. Calcul. Sci. Tech.,* pp. 171–199, IRIA, Paris, 1973.
12. Sutherland, W. H., AXICRIP – Finite element computer code for creep analysis of plane stress, plane strain and axisymmetric bodies, Nucl. Engng. Design **11,** 269–285 (1970).
13. Cormeau, I., Numerical stability in quasi-static elasto/viscoplasticity, Int. J. Numerical Methods Engng. **9,** 109–127 (1975).
14. Irons, B., and Treharne, G., A bound theorem in Eigenvalues and its practical applications, in *Proc. 3rd Conf. Matrix Meth. Struct. Mech.,* 245–254, Wright-Patterson A. F. B., USA Air Force, Ohio, 1971.
15. Anand, S. C., Lee, S. L., and Rossow, E. C., Finite element analysis based upon Tresca yield criterion, Ingenieur-Archiv **39,** 73–86 (1970).
16. Prager, W., and Hodge, P. G., *Theory of Perfectly Plastic Solids,* Dover, New York, 1968.
17. Sim, R. G., Reference results for plane stress creep behaviour, J. Mech. Engng. Sci. **14,** 404–410 (1972).

18. Bauer, F., and Reiss, E. L., On the numerical determination of shrinkage stresses, Trans. ASME J. Appl. Mech. **37,** 123–127 (1970).
19. Venturini, W. S., and Brebbia, C. A., The boundary element method for the solution of no-tension materials, in *Boundary Element Methods* (C. A. Brebbia, Ed.), Springer-Verlag, Berlin, 1981.
20. Venturini, W. S., and Brebbia, C. A., Boundary element formulation to solve no-tension problems in geomechanics, in *NATO Advanced Summer Institute on Numerical Methods in Geomechanics* (J. Martins, Ed.), D. Reidel Pub. Co., Holland, 1982.
21. Valliappan, S., Non-linear stress analysis of two-dimensional problems with special reference to rock and soil mechanics, Ph.D. Thesis, University College of Swansea, 1968.
22. Jaeger, C., *Rock Mechanics and Engineering,* Cambridge University Press, Cambridge, 1972.
23. Camargo, W. M., Projeto de túneis em maciço rochoso sob pressão hidrostática interna, Ph.D. Thesis, University of São Paulo, 1968.
24. Chaudonneret, M., Methode des equations integrales appliquees a la resolution de problemes de viscoplasticite. J. Mechanique Appliquee **1,** 113, 1977.
25. Mukherjee, S., and Kumar, V., Numerical analysis of time dependent inelastic deformations in metallic media using the boundary integral equation method. Trans. ASME, J. Appl. Mech. **45,** 785, 1978.
26. Morjaria, M., and Mukherjee, S., Improved boundary integral equation method for the time dependent inelastic deformation in metals. Int. J. Num. Meth. Engng. **15,** 97, 1980.
27. Brunet, M., Numerical analysis of viscoplasticity using the boundary element method, in *Boundary Element Methods in Engineering* (C. A. Brebbia, Ed.), Springer-Verlag, Berlin, 1982.

Chapter 9

Plate Bending

9.1. Introduction

This chapter deals with the boundary integral theory of plate bending and some of its applications using boundary elements. The usual assumptions of thin plate bending theory are reviewed and applied in a weighted residual manner, following the concepts presented by Washizu [1] and other authors [2]. Plates with transverse shear deformation present a much simpler formulation than the one for thin plates. This occurs because when shear deformation is included in the formulation, the displacements and rotations are independent of each other, while for thin plates they are not.

The first application of BEM to plate bending problems appears to be due to Jaswon and Maiti [3] who for simple loadings convert the problem into a biharmonic boundary-value problem and solve it as a function of two potentials. Maiti and Chakrabarty solved simply supported polygonal plates in 1974 [4]. Plates of arbitrary shapes were also solved by Niwa, Kobayashi, and Fukui at the same time [5]. At the end of the 1970s Altiero and Sikarskie [6] produced another important paper on plate bending, stressing the possibility of using solutions other than Green's function in an unbounded domain. Bezine [7–9] has written several contributions to the topic, culminating with Ref. [9] in which the concept of similarity is applied to the solution. Segedin and Bricknell [10] presented an integral method of solution for plates with reentry corners.

Important contributions to problems of plate bending analysis using integral equations have been made by Stern [11] who recently published a review of the general theory and applications [12]. Other papers concerned with applications can be seen in the latest proceedings of the International Conference on Boundary Element Methods. They include papers relating mixed principles to boundary elements [13], sandwich plate bending applications [14], large displacement approach [15], and stress concentration problems [16]. The governing equations of plate bending for integral equation formulations are reviewed in detail in Ref. [17].

9.2. Governing Equations

In plate bending the plane $x_1 - x_2$ is taken to coincide with the mean surface of the plate (Fig. 9.1). The thickness of the plate is called h. The applied forces are per unit area inside the plate and per unit of length along Γ. Forces can be given as direct forces or moments. The positive direction for moments and transverse shear forces is given in Fig. 9.2. The moments and shear forces are defined as

$$
\begin{aligned}
m_{11} &= \int_{-h/2}^{h/2} \sigma_{11} x_3 \, dx_3 , \qquad & q_1 &= \int_{-h/2}^{h/2} \sigma_{13} \, dx_3 \\
m_{22} &= \int_{-h/2}^{h/2} \sigma_{22} x_3 \, dx_3 , \qquad & q_2 &= \int_{-h/2}^{h/2} \sigma_{23} \, dx_3 \\
m_{12} = m_{21} &= \int_{-h/2}^{h/2} \sigma_{12} x_3 \, dx_3 . & &
\end{aligned}
\tag{9.1}
$$

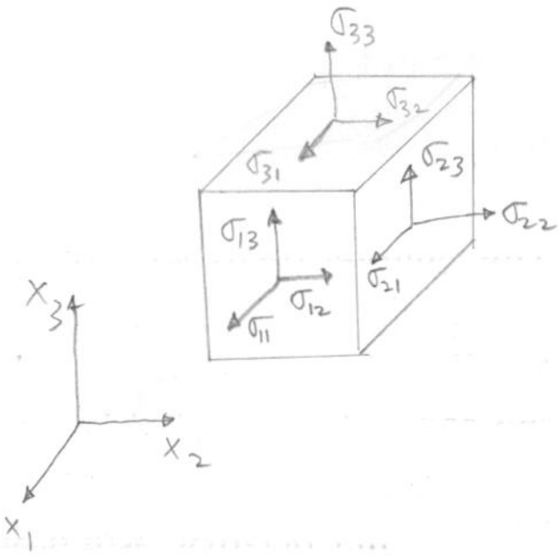

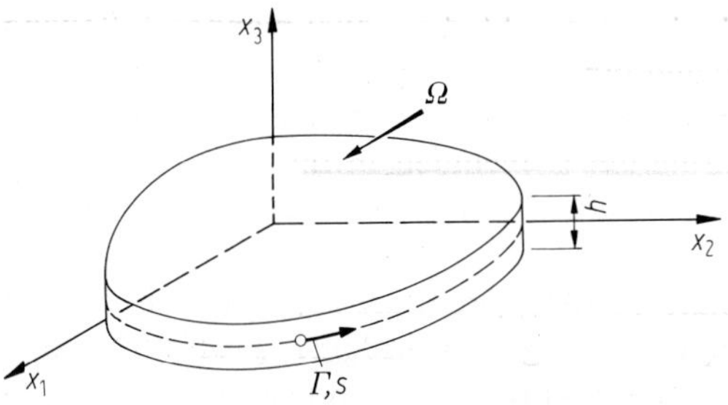

Fig. 9.1. Coordinate system

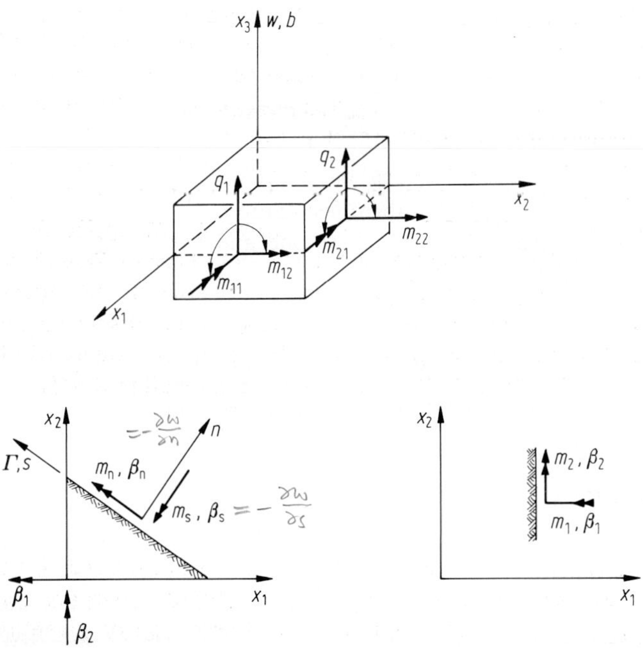

Fig. 9.2. Notation

If $n_i\,(i=1,2)$ are the cosines of the normal n with respect to the x_1 and x_2 directions, we can write the boundary moments as

$$\begin{aligned} m_1 &= n_1\, m_{11} + n_2\, m_{12}\,, \\ m_2 &= n_1\, m_{12} + n_2\, m_{22}\,. \end{aligned} \tag{9.2}$$

The equilibrium equations are given in terms of moments and vertical forces as shown in Fig. 9.2, i.e.,

$$q_{1,1} + q_{2,2} + b = 0\,, \qquad -q_1 + \frac{\partial m_{11}}{\partial x_1} + \frac{\partial m_{12}}{\partial x_2} = 0\,, \qquad -q_2 + \frac{\partial m_{22}}{\partial x_2} + \frac{\partial m_{12}}{\partial x_1} = 0\,. \tag{9.3}$$

The natural way of writing the moments on the boundary is in terms of the resultant couples on the boundary (Fig. 9.2). This gives

$$\begin{aligned} m_n &= n_1\, m_1 + n_2\, m_2 = m_{11}(n_1)^2 + m_{12}(2 n_1 n_2) + m_{22}(n_2)^2\,, \\ m_s &= -\,n_2\, m_1 + n_1\, m_2 = -(m_{11} - m_{22})\, n_1 n_2 + m_{12}(n_1^2 - n_2^2)\,. \end{aligned} \tag{9.4}$$

Notice that q_1 and q_2 of Eqs. (9.3) are the shearing forces per unit length as defined in Eq. 9.1. Equations (9.2) and (9.4) define the prescribed moments per unit length along the boundary. (m_n is the bending and m_s is the twisting moment.)

In this plate theory the vertical displacements defined as w can be related to the rotations. This is done by assuming that the generalized displacements vary linearly over the thickness. Then we can substitute them into the strain equations of elasticity and find that the assumption of negligible transverse shear deformation leads to

$$\beta_1 = -\frac{\partial w}{\partial x_1}\,, \qquad \beta_2 = -\frac{\partial w}{\partial x_2}\,. \tag{9.5}$$

Equations (9.5) represent constraints on β_1 and β_2 which are no longer independent variables. Consequently, the number of equilibrium equations (in principle three as shown in Eqs. (9.3)) is reduced to only the first one and the other two are identically satisfied by requiring that

$$q_1 = \frac{\partial m_{11}}{\partial x_1} + \frac{\partial m_{12}}{\partial x_2}\,, \qquad q_2 = \frac{\partial m_{22}}{\partial x_2} + \frac{\partial m_{21}}{\partial x_1}\,. \tag{9.6}$$

On a boundary with normal n one has the following force component:

$$q = n_1\, q_1 + n_2\, q_2\,. \tag{9.7}$$

9.3. Integral Equations

We can next write the weighted residual equations corresponding to the first equilibrium Eq. (9.3) and the following boundary conditions

Essential

$$w = \bar{w}, \qquad \beta_n = \bar{\beta}_n, \quad \beta_s = \bar{\beta}_s \quad \text{on } \Gamma_1, \tag{9.8}$$

Natural

$$q = \bar{q}, \qquad m_n = \bar{m}_n, \quad m_s = \bar{m}_s \quad \text{on } \Gamma_2.$$

β_s and β_n are the rotation components tangential and normal to the boundary. ($\beta_s = -\partial w/\partial s$; $\beta_n = -\partial w/\partial n$.)

The weighted residual statement corresponding to the first equation (9.3) and (9.8) is

$$\int_\Omega \{q_{1,1} + q_{2,2} + b\} w^* \, d\Omega = \int_{\Gamma_2} \{(m_n - \bar{m}_n) \beta_n^* + (m_s - \bar{m}_s) \beta_s^* + (q - \bar{q}) w^*\} \, d\Gamma$$
$$- \int_{\Gamma_1} \{(\beta_n - \bar{\beta}_n) m_n^* + (\beta_s - \bar{\beta}_s) m_s^* + (w - \bar{w}) q^*\} \, d\Gamma. \tag{9.9}$$

By integrating this equation by parts twice we obtain

$$\int_\Omega \{m_{11} \chi_{11}^* + 2m_{12} \chi_{12}^* + m_{22} \chi_{22}^*\} \, d\Omega - \int_\Omega b\, w^* \, d\Omega$$
$$= \int_{\Gamma_1} \{m_n \beta_n^* + m_s \beta_s^* + q\, w^*\} \, d\Gamma + \int_{\Gamma_2} \{\bar{m}_n \beta_n^* + \bar{m}_s \beta_s^* + \bar{q}\, w^*\} \, d\Gamma$$
$$+ \int_{\Gamma_1} \{(\beta_n - \bar{\beta}_n) m_n^* + (\beta_s - \bar{\beta}_s) m_s^* + (w - \bar{w}) q^*\} \, d\Gamma, \tag{9.10}$$

where the flexural strains χ_{ij} are defined as

$$\chi_{11} = -w_{,11}, \qquad \chi_{22} = -w_{,22}, \qquad \chi_{12} = -w_{,12}. \tag{9.11}$$

Notice that expression (9.10) is a generalized version of the principle of virtual work. If the virtual displacements are taken to be such that the displacements (and rotations) boundary conditions are identically satisfied as well as the homogeneous version of the virtual displacements, i.e.,

$$\beta_n^* = \beta_s^* = w^* = 0 \qquad \text{on } \Gamma_1, \tag{9.12}$$

one obtains the virtual displacement statement for a plate in bending, i.e.,

$$\int_\Omega (m_{11} \chi_{11}^* + 2m_{12} \chi_{12}^* + m_{22} \chi_{22}^*) \, d\Omega - \int_\Omega b\, w^* \, d\Omega$$
$$= \int_{\Gamma_2} (\bar{m}_n \beta_n^* + \bar{m}_s \beta_s^* + \bar{q}\, w^*) \, d\Gamma. \tag{9.13}$$

This formulation is well-known in finite element models.

We can now return to expression (9.10), and integrate by parts twice again using the stress-strain relationships for a homogeneous isotropic plate, i.e.,

$$
\begin{aligned}
m_{11} &= D(\chi_{11} + \nu\,\chi_{22})\,,\\
m_{22} &= D(\chi_{22} + \nu\,\chi_{11})\,,\\
m_{12} &= D(1-\nu)\,\chi_{12}\,,
\end{aligned}
\tag{9.14}
$$

where $D = Eh^3/12(1-\nu^2)$ is called the bending rigidity of the plate. The new domain integral gives the inverse of Eq. (9.9), i.e.,

$$
\begin{aligned}
&\int_\Omega \{q^*_{1,1} + q^*_{2,2}\}\, w\, d\Omega + \int_\Omega b\, w^*\, d\Omega = -\int_{\Gamma_1} \{m_n\, \beta^*_n + m_s\, \beta^*_s + q\, w^*\}\, d\Gamma\\
&- \int_{\Gamma_2} \{\bar{m}_n\, \beta^*_n + \bar{m}_s\, \beta^*_s + \bar{q}\, w^*\}\, d\Gamma + \int_{\Gamma_1} \{m^*_n\, \bar{\beta}_n + m^*_s\, \bar{\beta}_s + q^*\, \bar{w}\}\, d\Gamma\\
&+ \int_{\Gamma_2} \{m^*_n\, \beta_n + m^*_s\, \beta_s + q^*\, w\}\, d\Gamma
\end{aligned}
\tag{9.15}
$$

This is the starting statement for the boundary element solution of plate bending problems. If we write $\Gamma = \Gamma_1 + \Gamma_2$ for simplicity, Eq. (9.15) becomes

$$
\begin{aligned}
&\int_\Omega \left\{\frac{\partial^2 m^*_{11}}{\partial x_1^2} + 2\frac{\partial^2 m^*_{12}}{\partial x_1\, \partial x_2} + \frac{\partial^2 m^*_{22}}{\partial x_2^2}\right\} w\, d\Omega + \int_\Omega b\, w^*\, d\Omega\\
&= -\int_\Gamma \{m_n\, \beta^*_n + m_s\, \beta^*_s + q\, w^*\}\, d\Gamma + \int_\Gamma \{m^*_n\, \beta_n + m^*_s\, \beta_s + q^*\, w\}\, d\Gamma\,.
\end{aligned}
\tag{9.16}
$$

The problem reduces now to finding the *fundamental solution* for the plate. For plates of uniform rigidity, we have

$$
\frac{\partial^2 m^*_{11}}{\partial x_1^2} + 2\frac{\partial^2 m^*_{12}}{\partial x_1\, \partial x_2} + \frac{\partial^2 m^*_{22}}{\partial x_2^2} + b^* = -D\nabla^4 w^* + \Delta(\xi, x) = 0\,,
\tag{9.17}
$$

where ∇^4 is the biharmonic operator

$$
\nabla^4(\,) = \frac{\partial^4(\,)}{\partial x_1^4} + 2\frac{\partial^4(\,)}{\partial x_1^2\, \partial x_2^2} + \frac{\partial^4(\,)}{\partial x_2^4}
\tag{9.18}
$$

or in polar coordinates

$$
\nabla^4 w = \frac{1}{r}\frac{d}{dr}\left[r\frac{d}{dr}\left(\frac{1}{r}\frac{d}{dr}\left(r\frac{dw}{dr}\right)\right)\right].
\tag{9.19}
$$

The fundamental solution corresponding to Eq. (9.17) is the solution for a unit point load, and gives for the displacement

$$
w^* = \frac{r^2}{8\pi D}(\ln r)
\tag{9.20}
$$

Differentiating this expression one can obtain the rotations, moments, and shear forces corresponding to the fundamental solution.

Returning to Eq. (9.9) we find that some line integrals can be transformed through another integration by parts in order to work in terms of the vertical displacement w and the normal rotation β_n. Consider the terms in w and β_s, i.e.,

$$\int_{\Gamma_2} [(m_s - \bar{m}_s)\, \beta_s^* + (q - \bar{q})\, w^*]\, d\Gamma$$

$$= -[(m_s - \bar{m}_s)\, w^*]_{\Gamma_2} + \int_{\Gamma_2} \left[\left(q + \frac{\partial m_s}{\partial s} \right) - \left(\bar{q} + \frac{\partial \bar{m}_s}{\partial s} \right) \right] w^*\, d\Gamma\,. \qquad (9.21)$$

The term $q + \partial m_s/\partial s$ is called effective vertical shear force. Notice that the first term in Γ_2 is due to a possible difference in value at the ends of Γ_2. The effective shear force can be called

$$t = q + \frac{\partial m_s}{\partial s} \qquad (9.22)$$

Under this assumption, the variables under consideration reduce to four, i.e., t, m_n, β_n, and w. Taking into account the fundamental solution and the new boundary terms, Eq. (9.16) can now be written for any load point ξ_i as

$$\int_\Omega b\, w^* + \int_\Gamma (m_n \beta_n^* + t\, w^*)\, d\Gamma = c\, w_i + \int_\Gamma (m_n^*\, \beta_n + t^*\, w)\, d\Gamma\,, \qquad (9.23)$$

where for a smooth boundary $c = \frac{1}{2}$ and for an internal point $c = 1$, as previously seen.

This is the starting point for plate bending boundary element models. It should be noted that if the m_s value is discontinuous or the boundary has a corner as shown in Fig. 9.3, the importance of the first term on the right-hand side of Eq. (9.21) becomes obvious. At the point 0 in the figure we need to incorporate a corner force in our formulation given by

$$t_c = [(m_s)_{0^-} - (m_s)_{0^+}]\,. \qquad (9.24)$$

These forces and their inverses are easily incorporated by adding terms $t_c w^*$ and $t_c^*\, w$ to the left- and right-hand sides of Eq. (9.23), respectively, i.e.,

$$c\, w_i + \int_\Gamma (m^* \beta + t^* w)\, d\Gamma + \sum t_c^*\, w = \int_\Gamma (m\, \beta^* + t\, w^*)\, d\Gamma + \int_\Omega b\, w^*\, d\Omega + \sum t_c w^*\,, \qquad (9.25)$$

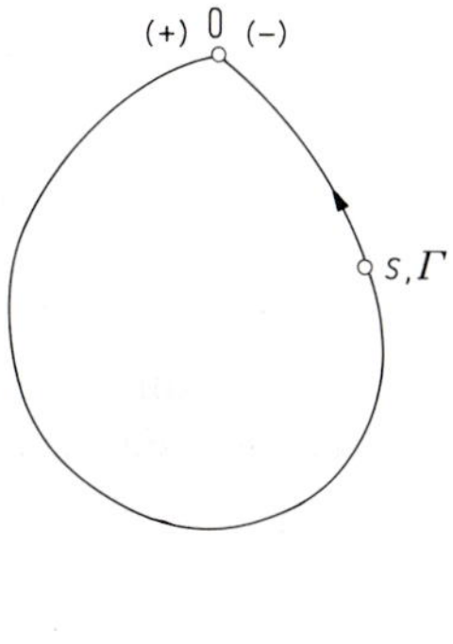

Fig. 9.3. Corner point

where the summation is carried out for as many corners as necessary. Notice that β_n is now β and m_n is m for simplicity. Notice also that we have now two unknowns per node, i.e., displacement w or effective shear force t, and rotation β or moment m. Hence we need another equation to solve the problem. This equation is given by finding the derivative of Eq. (9.25) with respect to the normal, which gives

$$c\,\theta_i + \int_\Gamma \left(\frac{\partial m^*}{\partial n}\,\beta + \frac{\partial t^*}{\partial n}\,w \right) d\Gamma + \sum \frac{\partial t_c^*}{\partial n}\,w$$

$$= \int_\Gamma \left(m\,\frac{\partial \beta^*}{\partial n} + t\,\frac{\partial w^*}{\partial n} \right) d\Gamma + \int_\Omega b\,\frac{\partial w^*}{\partial n}\,d\Omega + \sum t_c\,\frac{\partial w^*}{\partial n}\,. \tag{9.26}$$

In addition, we may need a further equation in certain corner problems where the values of the boundary moment may be discontinuous. This is equivalent to the extra equation mentioned in Chapter 5 for elastostatics. Another possibility is to use a double node or non-conforming elements at the corners. The interested reader is referred to Ref. [12] for a full discussion.

The above formulation can be easily extended to plates on elastic foundations supported by springs of the Winkler type. The effect of these springs can be introduced by considering the equilibrium equation,

$$q_{1,1} + q_{2,2} + b = -\,D\nabla^4 w + b = k\,w\,, \tag{9.27}$$

where k is the spring support per unit area. Hence, the elastic foundation introduces one more integral term, i.e.,

$$\int_\Omega k\,w\,w^*\,d\Omega \tag{9.28}$$

on the left-hand side of Eq. (9.25). The most elegant way of solving the problem is by finding the fundamental solution which will reduce the problem to a boundary-only solution, i.e., the solution of

$$D\nabla^4 w^* + k\,w^* = \Delta(\xi, x)\,. \tag{9.29}$$

Otherwise we can integrate numerically the term shown in expression (9.28) using internal cells.

Unfortunately, the fundamental solution for a plate supported on a half-space as reported by Woinowsky-Krieger in [18] is rather complex and consequently will not be given here.

9.3.1. Other Fundamental Solutions

In many cases it is interesting to use fundamental solutions different from the concentrated load given by Eq. (9.20). Solutions to plates under different boundary conditions and loads are given in many classical books, and can be used instead of the Green's function type of fundamental solution. In these cases we have a solution w^* which may not be singular and may also be due to some non concentrated force, e.g., the solution for

$$\frac{\partial^2 m_{11}^*}{\partial x_1^2} + 2\frac{\partial^2 m_{12}^*}{\partial x_1\,\partial x_2} + \frac{\partial^2 m_{22}^*}{\partial x_2^2} + b^* = 0 \tag{9.30}$$

which gives the following generalized integral statement:

$$\begin{aligned}&\int_\Omega b^* w\, d\Omega + \int_\Gamma (t^* w + m^* \beta)\, d\Gamma + \sum t_c^* w \\ &= \int_\Omega b w^*\, d\Omega + \int_\Gamma (t w^* + m\, \beta^*)\, d\Gamma + \sum t_c w^* .\end{aligned} \tag{9.31}$$

Notice that as the load is no longer a concentrated one, a new integral appears on the left-hand side which represents the work of the fundamental applied forces on the actual vertical displacements.

Altiero and Sikarskie [6] have used the solution of a clamped circular plate as their fundamental solution. Their application illustrates how the boundary element method can be generalized and that it is not always necessary to use fundamental solutions corresponding to an unbounded domain. Antes [19] has generalized these ideas to study the behavior of circular cylindrical shells using known stress function solutions.

9.4. Applications

Example 9.1. In Ref. [12] Stern analyzed a uniformly loaded square plate under two sets of boundary conditions: (i) simply supported (hinged) along the entire boundary and (ii) clamped on three edges and simply supported on the fourth. For each problem each edge has been subdivided into 4, 8, or 16 elements, corresponding to a total of 80, 144, or 272 degrees of freedom, respectively.

For the simply supported boundary conditions, the displacements and bending moments are zero along the edges, hence the unknowns are the normal rotation and total shear plus the concentrated force at each corner. Because of symmetry, results are presented on half the edge as shown in Fig. 9.4. Comparing these results with analytical ones [18], the solution appears to converge well. The corner reactions found by Stern were $t/b\, l^2 = 0.0641$ compared to 0.065 of Ref. [18].

Similar conclusions were achieved for the case of the plate with three edges clamped and one simply supported (Figs. 9.5a, b) where boundary element results are compared with a finite difference solution obtained by Moody [20].

Example 9.2. The following example is taken from Bezine [7] who solved square plates under different boundary conditions. We will consider three of those conditions, i.e., clamped plate, simply supported plate, and plate with three free edges and the fourth one clamped; all with a concentrated load at the center and $\nu = 0.3$. He divided the plate boundary into 6 or 12 equal-length constant elements per edge, and did not take symmetry into consideration, which gives a total of 24 or 48 boundary elements (Fig. 9.6).

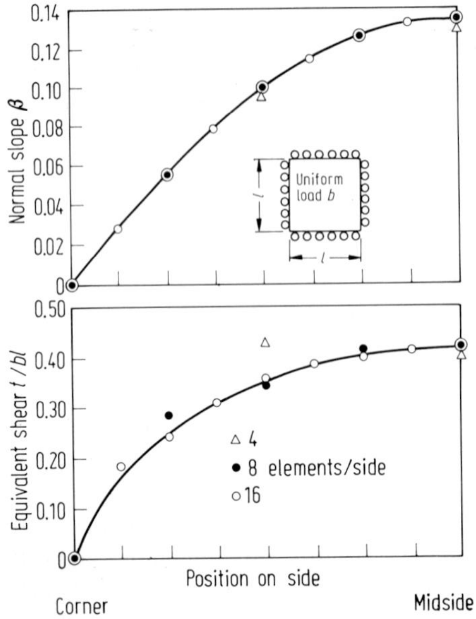

Fig. 9.4. Results for uniformly loaded simply supported plate

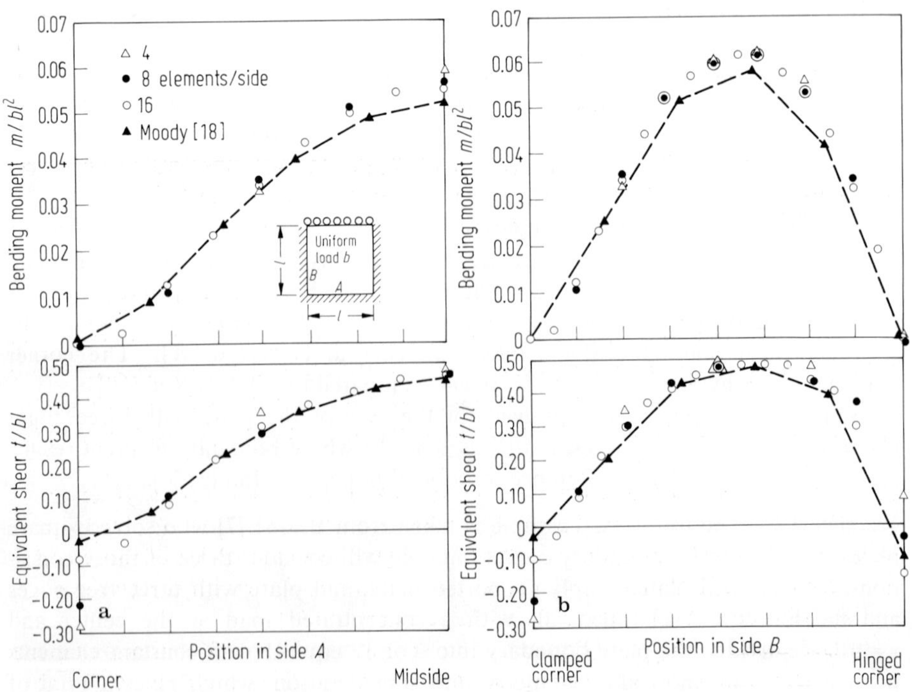

Fig. 9.5. Results for uniformly loaded, clamped simply supported plate

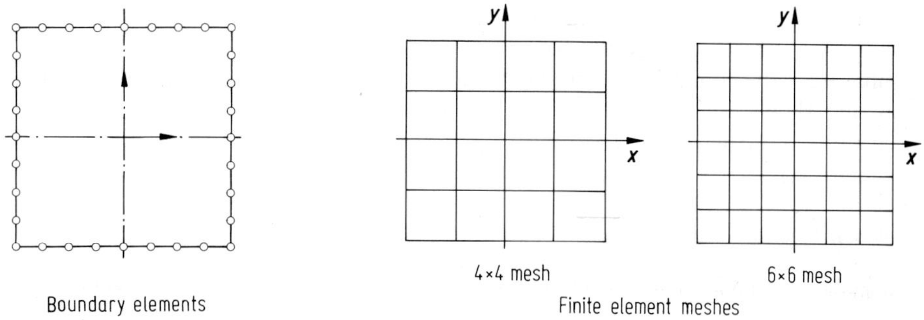

Fig. 9.6. Boundary elements and finite element meshes

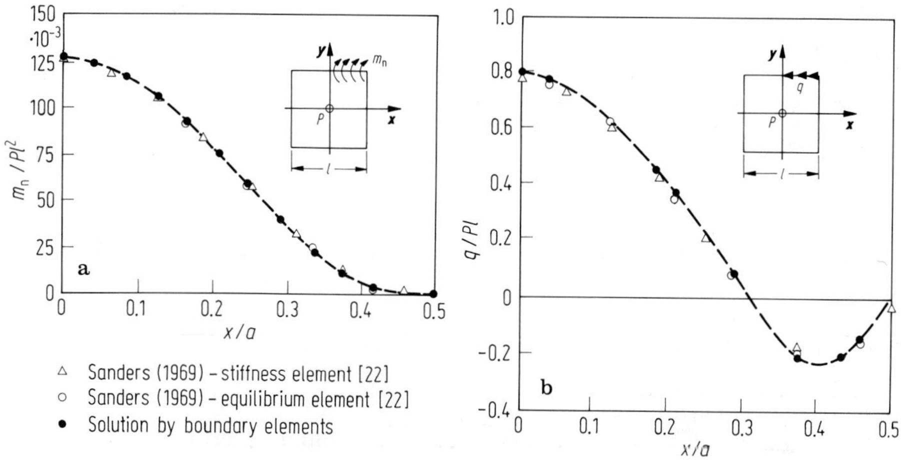

Fig. 9.7. **a** Normal bending moment m_n along a half-edge for a clamped plate with a load P at its center. **b** Transverse shear force q along a half-edge for a clamped plate with a load P at its center

Results are presented for deflections, normal slopes, bending moment, and transverse shear force at some points on the boundary and the deflection is also computed at several points inside the plate. The numerical results are compared with finite elements, using the 12 degrees of freedom element due to Melosh [21]. The finite element mesh is shown in Fig. 9.6.

Clamped Plate: For this case the exact deflection at the center is given in Ref. [18] and compared with the BEM and FEM solutions in Table 9.1. Notice the accuracy of the BEM solution. Results for bending moment and equivalent shear along the edge (Fig. 9.7) are also presented and compared with those obtained by Sanders using equilibrium and stiffness finite elements [22].

Simply Supported Plate: The center deflections for a simply supported plate are also shown in Table 9.1. The agreement between analytical solutions of Fabre [23] and boundary elements is excellent.

Plate with Three Free Edges and the Fourth Edge Clamped (cantilever plate): There is no theoretical solution for this problem and hence the boundary element results are only compared with the finite element solutions. The center deflections are shown in Table 9.1; Figs. 9.8a, b compare the deflections and normal derivatives on the boundary, and Fig. 9.8c shows the deflections along the symmetry axis.,

Table 9.1. Deflection of plate center for different boundary conditions.

Deflection divided by $10^{-1}\,Fl^2/D$	Boundary element solution		Analytical solution	Finite element solution	
	24 elements	48 elements		6 × 6 grids	4 × 4 grids
Clamped plate	0.05564	0.05599	0.05605	0.05911	0.06134
Simply supported plate	0.1159	0.1159	0.1160	0.1197	0.1233
Cantilever plate	0.4756	0.4628		0.4708	0.4709

Example 9.3. Altiero and Sikarskie [6] solved the case of a clamped triangular plate under a uniformly distributed load (Fig. 9.9). The plate was embedded inside a fictitious circular plate for which the Green function is known, i.e., the circular clamped plate. Hence the fundamental solution is not the unbounded one of Eq. (9.20) but the one given by Timoshenko and Woinowsky-Krieger [18] for this type of circular plates.

The transverse load integral was calculated defining a grid over the plate domain rather than attempting to convert it into boundary integrals. The number of constant boundary elements was 60 and results were presented along the y axis (Fig. 9.10). The deflection and bending moment results are reassembled and the extrapolated results at the base of the triangle were reported to compare favorably with those in Ref. [18] which were obtained using finite differences. At the center of the plate, the moments were as indicated in Table 9.2.

Table 9.2. Bending moments at center of the plate.

	Integral Equation	Ref. [18]
m_x	0.0179	0.0196
m_y	0.0180	0.0191

The most interesting contribution of Altiero and Sikarskie is that they pointed out the possibility of using a Green's function other than the one for an unbounded domain.

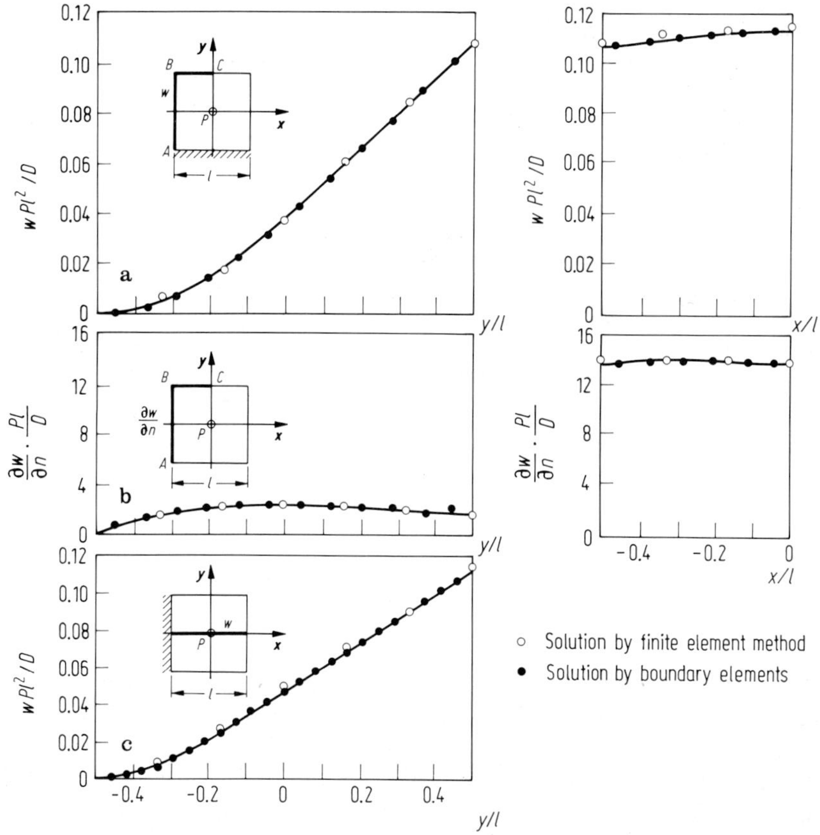

Fig. 9.8. **a** Deflection w along the free edges AB and BC for a cantilever plate with a load P at its center. **b** Slope $\partial w/\partial n$ along the free edges AB and BC for a cantilever plate with a load P at its center. **c** Deflection w along the axis of symmetry for a cantilever plate with a load P at its center

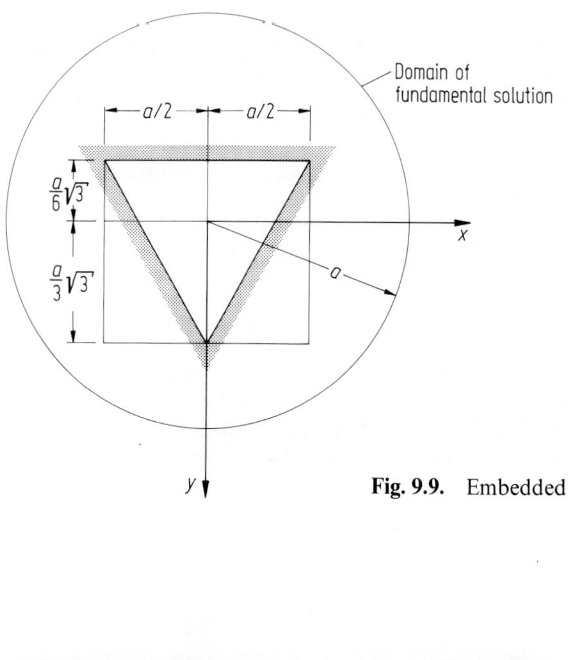

Fig. 9.9. Embedded triangular plate

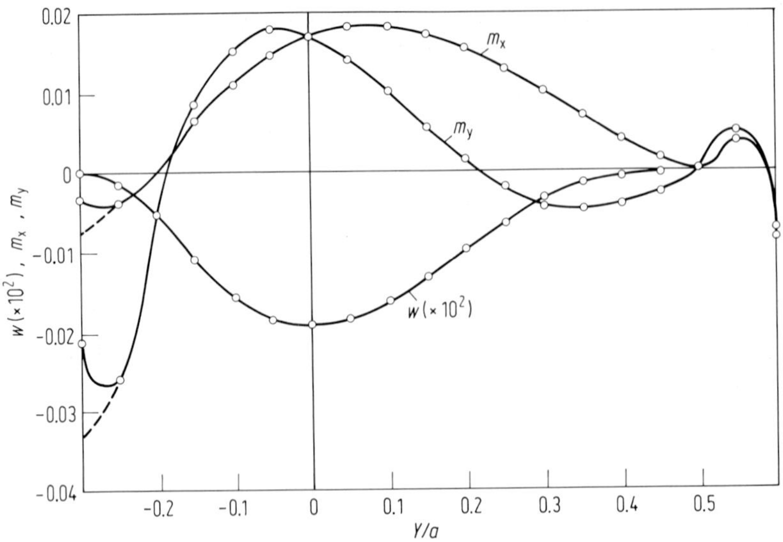

Fig. 9.10. Deflections and bending moments for triangular plate

References

1. Washizu, K., *Variational Methods in Elasticity and Plasticity*, 2nd ed., Pergamon Press, New York, 1975.
2. Tottenham, H., and Brebbia, C., (Eds.), *Finite Element Techniques in Structural Mechanics.* Stress Analysis Publishers, Southampton, England, 1970.
3. Jaswon, M. A., and Maiti, M., An integral formulation of plate bending problems, J. Engng. Math. **2,** 83–93 (1968).
4. Maiti, M., and Chakrabarty, S. K., Integral equation solutions for simply supported polygonal plates, Int. J. Engng. Sci. **12,** 793–806 (1974).
5. Niwa, Y., Kobayashi, S., and Fukui, T., An application of the integral equation method to plate bending, Faculty Engng., Kyoto Univ. **36** (Pt. 2), 140–158 (1974).
6. Altiero, N. J., and Sikarskie, D. L., A boundary integral method applied to plates of arbitrary plan form, Computer and Structures **9,** 163–168 (1978).
7. Bezine, G. P., and Gamby, D. A., A new integral equation formulation for plate bending problems, in *Recent Advances in Boundary Element Methods* (C. A. Brebbia, Ed.), Pentech Press, London, 1978.
8. Bezine G. P., Boundary integral formulation for plate flexure with arbitrary boundary conditions, Mech. Res. Commun. **5,** (1978).
9. Bezine, G. P., Application of similarity of results of new boundary integral equations for plate flexure problems, Appl. Math. Modelling **5** (1981).
10. Segedin, C. M., and Brickell, D. G. A., Integral equation method for corner plates, J. Struct. Div. Proc. ASCE **94,** 41–52 (1968).
11. Stern, M., A general boundary integral formulation for the numerical solution of plate bending problems, Int. J. Solids. Structures **15,** 769–782 (1979).
12. Stern, M., Boundary integral equations for bending of thin plates, in *Progress in Boundary Element Methods, Vol. 2,* Pentech Press, London; Springer-Verlag, New York, 1983.
13. Weeën, F. V., Application of the direct boundary element method to Reissner's plate model, in *Boundary Elements in Engineering* (C. Brebbia, Ed.), Springer-Verlag, Berlin, 1982.

14. Kamiya, N., Sawaki, Y., and Nakamura, Y., Boundary element nonlinear bending analysis of clamped sandwich plates and shells, in *Boundary Elements in Engineering* (C. Brebbia, Ed.), Springer-Verlag, Berlin, 1982.
15. Tanaka, N., Integral equation approach to small and large displacements of thin elastic plates, in *Boundary Elements in Engineering* (C. Brebbia, Ed.), Springer-Verlag, Berlin, 1982.
16. Kim, J. W., On the computation of the stress intensity factors in elastic plate flexure via boundary integral equations, in *Boundary Elements in Engineering* (C. Brebbia, Ed.), Springer-Verlag, Berlin, 1982.
17. Danson, D. J., Plate bending analysis using the boundary element method, Technical Report, Computational Mechanics Centre, Southampton, England, 1981.
18. Timoshenko, S., and Woinowsky-Krieger, S., *Theory of Plates and Shells,* McGraw–Hill, New York, 1959.
19. Antes, H., On boundary integral equations for circular cylindrical shells, in *Boundary Element Methods* (C. Brebbia, Ed.), Springer-Verlag, Berlin, 1981.
20. Moody, N. T., Moments and reactions for rectangular plates, US Department of Interior, Bureau of Reclamation, Engineering Monograph 27, 1960.
21. Melosh, R. J., Basis for derivation of matrices by the direct stiffness method, AIAA J. **1,** 1631 (1963).
22. Sanders, G., Applications de la methode des elements finis a la flexion des plaques, Ph.D. Thesis, University of Liege, 1969.
23. Fabre, H., Numerical solution of integro differential and singular integral equations for plate bending problems, J. Elasticity **6,** 39–56 (1976).

Chapter 10

Wave Propagation Problems

10.1. Introduction

The phenomenon of wave propagation is frequently encountered in a variety of engineering disciplines. For instance, in the design of antennas, it is important to know the interaction with electromagnetic waves. In earthquake analysis, knowledge of the elastodynamic wave propagation is essential. Problems of radiation and scattering of water waves are common features in fluid mechanics where they appear in the design of harbors, breakwaters, off-shore structures, etc.

This chapter deals with boundary integral solutions to the transient scalar wave equation

$$\nabla^2 u(x, t) - \frac{1}{c^2}\frac{\partial^2 u(x, t)}{\partial t^2} = 0 \tag{10.1}$$

with prescribed boundary and initial conditions. The constant factor c is the wave propagation speed (or celerity). It was realized a long time ago that this problem can be recast into an integral equation for the unknown potential function u [1–4]. For three-dimensional problems, this derivation dates back to Kirchhoff [5]. In the context of the BEM, numerical formulations for the solution of this integral equation have been presented by Friedman and Shaw [6], Shaw [7], Groenenboom [8, 9], and Mansur and Brebbia [10, 11].

For harmonic waves (i.e., assuming a time dependence for u in the form $e^{-i\omega t}$), the problem is more conveniently described in the frequency domain where it reduces to the solution of the Helmholtz equation

$$\nabla^2 u(x) + \varkappa^2 u(x) = 0 \tag{10.2}$$

in which ω is the frequency and $\varkappa = \omega/c$ is the wave number.

Equation (10.2) is the governing equation for problems involving the so-called expansion waves [1] and is more applicable to the case of scattering and radiation of acoustic waves by rigid obstacles [12, 13]. We will restrict our discussion on harmonic problems to the case of water waves, assuming an irrotational motion of an inviscid incompressible fluid governed by Laplace's equation. However, a particular configuration of this problem to be discussed in Section 10.5 mathematically reduces it to the solution of a two-dimensional Helmholtz equation, thus providing a complete analogy with the problem of acoustic waves.

10.2. Three-Dimensional Water Wave Propagation Problems

In this section, we discuss in detail the problem of wave-structure interaction for three-dimensional bodies of arbitrary geometry. We consider the linear diffraction problem which arises when the wave height is assumed sufficiently small for linear wave theory to apply [14, 15]. Then, the problem reduces to the determination of the velocity potential u which satisfies Laplace's equation

$$\nabla^2 u = 0 \tag{10.3}$$

within the fluid region. This is subject to the following boundary conditions (Fig. 10.1):

$$\frac{\partial^2 u}{\partial t^2} + g\frac{\partial u}{\partial x_3} = 0 \qquad \text{at } x_3 = 0\,, \tag{10.4}$$

$$\eta = -\frac{1}{g}\left(\frac{\partial u}{\partial t}\right)_{x_3=0}\,, \tag{10.5}$$

$$\frac{\partial u}{\partial x_3} = 0 \qquad \text{at } x_3 = -d\,, \tag{10.6}$$

$$\frac{\partial u}{\partial n} = 0 \qquad \text{at the body surface } \Gamma\,, \tag{10.7}$$

where g is the gravitational constant, η is the free-surface elevation and n denotes the outward (to the fluid) normal to the body surface. Equations (10.4) and (10.5) derive from the linearized kinematic and dynamic free-surface boundary condi-

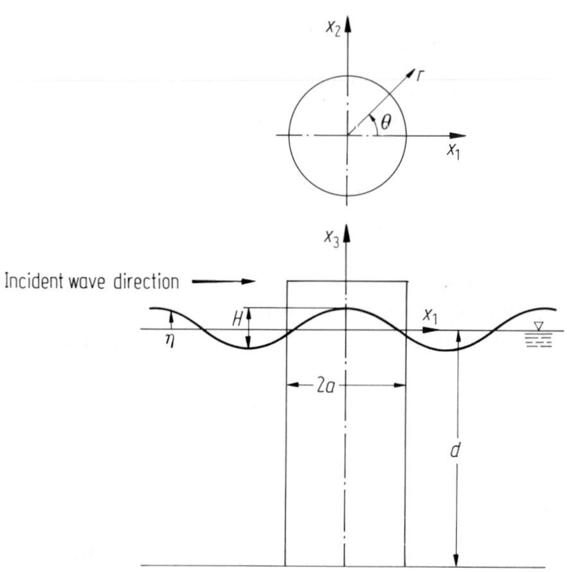

Fig. 10.1. Wave diffraction around an obstacle: definition sketch

tions, while Eqs. (10.6) and (10.7) correspond to the kinematic boundary conditions at the seabed and at the body surface, respectively.

Generally, the velocity potential u is represented as the sum of the incident wave and the scattered wave potentials,

$$u = u_I + u_s . \tag{10.8}$$

Furthermore, in order to ensure that u_s has the correct behavior in the far field, we impose the Sommerfeld radiation condition [16]

$$\lim_{r \to \infty} r^{1/2} \left[\frac{\partial u_s}{\partial r} - i \varkappa u_s \right] = 0 , \tag{10.9}$$

where r is the radial ordinate.

The expression of the velocity potential in the form of Eq. (10.8), involving a separation into undisturbed incident wave and scattered wave components, constitutes the basis of diffraction theory. The incident wave potential itself satisfies Eqs. (10.3)–(10.6) and is specified in complex form as

$$u_I = -\frac{igH}{2\omega} \frac{\cosh[\varkappa(x_3 + d)]}{\cosh(\varkappa d)} e^{i(\varkappa x_1 - \omega t)} \tag{10.10}$$

where H is the wave height and ω the wave angular frequency. Note that $\omega = 2\pi/T$ and $\varkappa = 2\pi/L$, L being the wave length and T its period.

Because all the equations of the problem are linear, the potential u_s also satisfies Eqs. (10.3)–(10.6), as well as the radiation condition (10.9). The body surface boundary condition may be written

$$\frac{\partial u_s}{\partial n} = -\frac{\partial u_I}{\partial n} \quad \text{on } \Gamma , \tag{10.11}$$

thus providing for the dependence of u_s on u_I. Equations (10.3)–(10.6) applied to u_s, together with conditions (10.9) and (10.11), define the problem in terms of u_s. Since the flow is considered to be harmonic in time, the dependence on time may be separated and the velocity potential u expressed as

$$u(x, t) = \mathrm{Re}\,[u'(x)\, e^{-i\omega t}] \tag{10.12}$$

in which u' is generally complex and $u' = u'_I + u'_s$.

Lamb [1] has shown through the application of Green's theorem that the velocity potential associated with the flow about a body can be described by a continuous distribution of point wave sources over the immersed surface of the body. Thus, the potential at any point in the fluid region may be expressed as

$$u'_s(x) = \frac{1}{4\pi} \int_\Gamma \sigma(\xi)\, u^*(\xi, x)\, d\Gamma(\xi) . \tag{10.13}$$

This expression is the starting point for the indirect BEM formulation, as discussed in Section 2.3 (Eq. (2.51)).

In order to determine the unknown source density distribution, we take the derivative of Eq. (10.13) in the direction of the outward normal to Γ as the point x is taken to Γ and apply the kinematic boundary condition on the body surface, yielding the following boundary integral equation (see [17] and Chapter 2, Eq. (2.52)):

$$-\frac{1}{2}\sigma(x)+\frac{1}{4\pi}\int_\Gamma \sigma(\xi)\frac{\partial u^*(\xi, x)}{\partial n(x)}\,d\Gamma(\xi)=-\frac{\partial u_I'(x)}{\partial n(x)}. \tag{10.14}$$

Once the above equation is solved for the source density distribution, values of u_s' at any boundary point or any point in the fluid region can be calculated by direct quadrature via Eq. (10.13).

Alternatively, a direct BEM formulation can also be applied to solve for the scattered velocity potential u_s', in which case the boundary integral equation is of the form

$$c(\xi)\,u_s'(\xi)=-\int_\Gamma\left[u_s'(x)\frac{\partial u^*(\xi, x)}{\partial n(x)}+u^*(\xi, x)\frac{\partial u_I'(x)}{\partial n(x)}\right]d\Gamma(x)\,. \tag{10.15}$$

This expression is similar to Eq. (2.69), its solution providing the unknown values of u_s' along Γ.

Although a fundamental solution of the type $u^*=1/R$ can be employed for the solution of the problem, this requires discretization not only of the surface Γ but also of the seabed, the free surface, and the truncation of the lateral boundaries at a finite distance where the radiation condition may be applied.

The most common numerical methods of solution to the problem just described [15, 18–22] employ a fundamental solution (or Green's function) u^* which satisfy all the boundary conditions of the problem, Eqs. (10.4), (10.5), (10.6), and (10.9), with the exception of the kinematic condition on the body surface, Eq. (10.7). In short, employ a function u^* which is of the form

$$u^*=\frac{1}{R}+\bar{u}^*\,, \tag{10.16}$$

where R is the distance between the points ξ and x and $\bar{u}^*$ represents a regular function satisfying $\nabla^2\bar{u}^*=0$ throughout the fluid region (see Section 2.11).

The particular expression for u^* appropriate to the boundary-value problem posed was developed by John [23] and may be expressed in integral form as

$$u^*=\frac{1}{R}+\frac{1}{R'}+2\int_0^\infty\frac{(\mu+v)\exp(-\mu d)\cosh[\mu(x_3(x)+d)]\cosh[\mu(x_3(\xi)+d)]}{\mu\sinh(\mu d)-v\cosh(\mu d)}J_0(\mu r)\,d\mu$$
$$-\,i\,C_0\cosh[\varkappa(x_3(x)+d)]\cosh[\varkappa(x_3(\xi)+d)]\;J_0(\varkappa r)\,, \tag{10.17}$$

where J_0 is the Bessel function of the first kind of zero order

$$R=\{[x_1(\xi)-x_1(x)]^2+[x_2(\xi)-x_2(x)]^2+[x_3(\xi)-x_3(x)]^2\}^{1/2},$$
$$R'=\{[x_1(\xi)-x_1(x)]^2+[x_2(\xi)-x_2(x)]^2+[x_3(\xi)+2d+x_3(x)]^2\}^{1/2},$$

$$r = \{[x_1(\xi) - x_1(x)]^2 + [x_2(\xi) - x_2(x)]^2\}^{1/2},$$

$$\nu = \varkappa \tanh(\varkappa d) = \omega^2/g,$$

$$C_0 = \frac{2\pi(\nu^2 - \varkappa^2)}{(\varkappa^2 - \nu^2)\, d + \nu},$$

and the Cauchy principal value of the integral is taken.

An alternative series form of u^* is given by [23]

$$u^* = -\, i\, C_0 \cosh[\varkappa(x_3(x) + d)] \cosh[\varkappa(x_3(\xi) + d]\, H_0^{(1)}(\varkappa r) + 4 \sum_{m=1}^{\infty} C_m \cos[\mu_m(x_3(x) + d)] \cos[\mu_m(x_3(\xi) + d)]\, K_0(\mu_m r), \tag{10.18}$$

where K_0 denotes the modified Bessel function of the second kind of zero order, $H_0^{(1)}$ is the Hankel function of the first kind of zero order, and

$$C_m = \frac{\mu_m^2 + \nu^2}{(\mu_m^2 + \nu^2)\, d - \nu}$$

in which μ_m are the real positive roots of the equation

$$\mu_m \tan(\mu_m d) + \nu = 0 \tag{10.19}$$

taken in ascending order.

The derivative $\partial u^*/\partial n$ may be evaluated from

$$\frac{\partial u^*}{\partial n} = \frac{\partial u^*}{\partial x_1} n_1 + \frac{\partial u^*}{\partial x_2} n_2 + \frac{\partial u^*}{\partial x_3} n_3, \tag{10.20}$$

where $\partial u^*/\partial x_i$ are determined through a straightforward differentiation of u^* and n_i are the direction cosines of the normal n to the surface Γ.

The numerical procedures for the evaluation of the integrals over each boundary element are discussed in several references (for instance, [15, 18–22]). Usually, constant quadrilateral boundary elements were employed. As a guideline on the choice of the form of the fundamental solution to be used, Eq. (10.17) or (10.18), Garrison and Chow [18] recommend that when $\varkappa r > 0.1$ the series form converges sufficiently rapidly for it to be employed, whereas for $\varkappa r < 0.1$ the integral form is more convenient.

Before closing this section, it should be pointed out that one difficulty that the integral equation approach sometimes encounters is that it breaks down at certain "irregular" wave frequencies [23] which generally correspond to wavelengths of the order of or less than the characteristic size of the body. This behavior has been described by Murphy [24] and is an inherent property of Eq. (10.14): at these frequencies there is no unique solution to the integral equation and so the problem cannot be solved by employing this integral formula. However, this difficulty is not too serious because the relatively short wavelengths corresponding to the irregular frequencies are not usually critical in design [15].

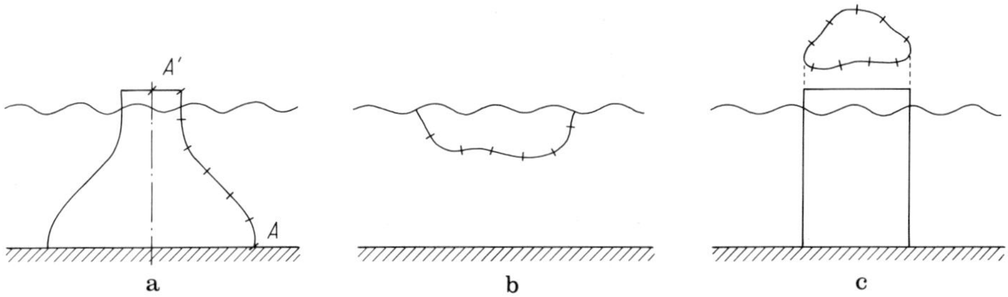

Fig. 10.2. Particular configurations: **a** vertical axisymmetric bodies; **b** horizontal cylinders of arbitrary section; **c** vertical cylinders of arbitrary section

The approach described in this section is generally costly to run in the computer because of the complexity of the fundamental solution. In the following sections we shall discuss alternative methods which, although more economical, are restricted to particular configurations (Fig. 10.2).

Example 10.1. Garrison and Chow's [18] comparison between numerical and experimental results for the case of a submerged tank are reproduced in Figs. 10.3, 10.4, and 10.5. Figures 10.3 and 10.4 present results for the horizontal and vertical force coefficients, respectively, for various depth ratios d/a (note that the height of the tank is approximately equal to a). It may be noted the decrease of wave force with depth and with decreasing values of the wavelength.

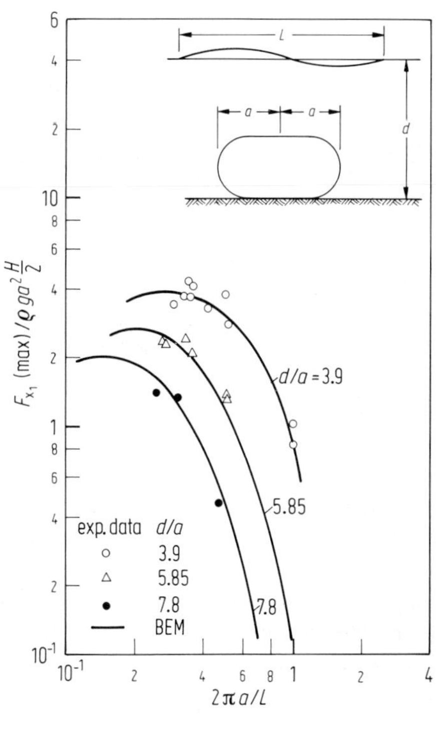

Fig. 10.3. Horizontal force coefficient

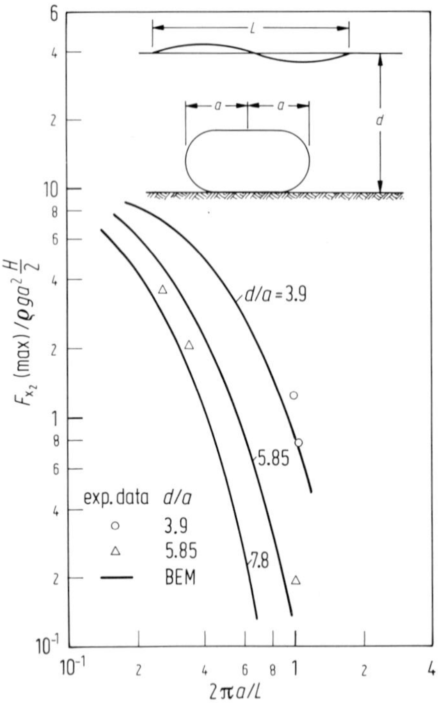

Fig. 10.4. Vertical force coefficient

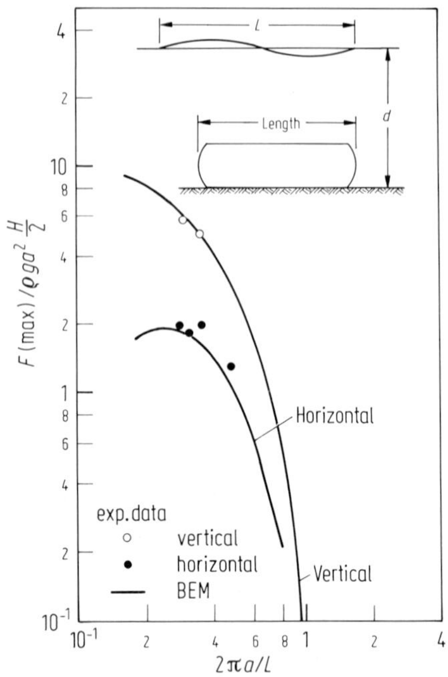

Fig. 10.5. Horizontal and vertical force coefficients for $d = 3.9$

Figure 10.5 shows horizontal and vertical force coefficients for the case when the tank is set with its longest dimension (approximately equal to 6a) perpendicular to the wave crests. By comparison of the results with that of Fig. 10.3, it is apparent that the maximum value of the horizontal force coefficient is reduced by approximately 50%. At larger values of $2\pi a/L$, both the horizontal and vertical force coefficients decrease quite rapidly.

10.3. Vertical Axisymmetric Bodies

The case of vertical axisymmetric bodies (Fig. 10.2a) has been treated by Black [25] and Fenton [26]. Black developed an axisymmetric fundamental solution on the basis of a method outlined by Morse and Feshbach [4]. Subsequently, Fenton demonstrated that this axisymmetric fundamental solution may be obtained using basically the same idea employed in Section 2.13, i.e., by writing the three-dimensional fundamental solution given by John [23] in cylindrical polar coordinates and integrating it analytically with respect to the circumferential direction. The main difference now is that, although the body is axisymmetric, it is subject to arbitrary (nonaxisymmetric) boundary conditions. Thus, the series form (10.18) of John's fundamental solution (more convenient to work with because the independent variables are separated) is initially expressed as a Fourier series in terms of θ (refer to Fig. 10.1) as [26]

$$u^* = \sum_{j=0}^{\infty} \left(\sum_{m=0}^{\infty} G_{jm} \right) (2 - \delta_{j0}) \cos [j (\theta (x) - \theta (\xi))] \tag{10.21}$$

with

$$\begin{aligned} G_{j0} &= - i\, C_0 \cosh [\varkappa (x_3 (x) + d)] \cosh [\varkappa (x_3 (\xi) + d)]\, J_j \binom{\varkappa R}{\varkappa r} H_j^{(1)} \binom{\varkappa r}{\varkappa R} , \\ G_{jm} &= 4\, C_m \cos [\mu_m (x_3 (x) + d)] \cos [\mu_m (x_3 (\xi) + d)]\, K_j \binom{\mu_m r}{\mu_m R} I_j \binom{\mu_m R}{\mu_m r} \\ & \qquad \text{for } m \geqq 1 , \\ R &= [x_1 (\xi)^2 + x_2 (\xi)^2]^{1/2}, \\ r &= [x_1 (x)^2 + x_2 (x)^2]^{1/2}. \end{aligned}$$

In the above equation, δ_{j0} is the Kronecker delta, I_j denotes the modified Bessel function of the first kind of jth order and the notation used in Eq. (10.18) is applicable apart from the variables R and r, which have been redefined. The upper value of the alternative argument is used if $r \geqq R$ and the lower, otherwise.

We now use the boundary integral equation (10.14) together with the Fourier expansion of the fundamental solution (10.21) to obtain a Fourier series, the coefficients of which are the integral equations valid on the arc AA′ of Fig. 10.2a.

The source density σ is a function of position on the body and may be written as $\sigma(\bar{\Gamma}, \theta)$ according to the notation of Section 2.13, where $\bar{\Gamma}$ specifies a point on the curve AA′. Since the flow is symmetric about the x_1 axis, we may expand σ as an even Fourier series in θ:

$$\sigma(\bar{\Gamma}, \theta) = \sum_{l=0}^{\infty} \sigma_l(\bar{\Gamma}) \cos l\, \theta . \tag{10.22}$$

The term involving the incident potential u_I' in Eq. (10.14) may likewise be expressed as an even Fourier series:

$$\frac{\partial u_I'}{\partial n} = \sum_{l=0}^{\infty} \bar{u}_l \cos l\, \theta , \tag{10.23}$$

where the values of $\bar{u}_l$ are given by Fenton [26].

Substituting Eqs. (10.21) – (10.23) into Eq. (10.14) and noting that $d\Gamma = R\, d\theta\, d\bar{\Gamma}$ (see Eq. (2.155)), the surface integral in Eq. (10.14) can now be integrated with respect to $\theta(\xi)$ to give an even Fourier series in $\theta(x)$ with each coefficient involving only an integral over the boundary contour $\bar{\Gamma}$. All of the Fourier coefficients for each value of l can now be equated to obtain an infinite number of one-dimensional integral equations, each independent of $\theta(x)$, giving

$$- \sigma_l(\bar{\Gamma}) + \int_{\bar{\Gamma}} \sigma_l(\bar{\Gamma}) \left\{ \sum_{m=0}^{\infty} \frac{\partial G_{lm}}{\partial n} \right\} R\, d\bar{\Gamma} = - 2\, \bar{u}_l \qquad \text{for } l = 0, 1, 2, \ldots . \tag{10.24}$$

Each such equation can be solved in the usual boundary elements manner. Expressions for the coefficients of the matrices involved in the solution (see Section 2.6) are given by Fenton [26]. These expressions are very long and consist of the sum of several series, but all series converge very quickly and all terms are finite.

Finally, Fenton [26] developed expressions for the forces and moments on the body in terms of the velocity potential. Of these components, the vertical force involves only the case $l=0$, the horizontal force and overturning moment (in the x_1-x_3 plane) involve only the case $l=1$, and the three remaining orthogonal components of force and moment are zero. Thus only the two one-dimensional integral equations corresponding to $l=0$ and $l=1$ need be solved to determine the wave loads on the body (but not the detailed pressure distribution around the body).

Example 10.2. This example is taken from Au [27] who solved some wave diffraction problems using Fenton's fundamental solution in the context of the direct boundary element method (Section 2.4). The example analyzes a submerged hemispherical tank of radius 10 m fixed on the bottom of the sea. The water depth was taken as 30 m and the wave amplitude is 2 m. The surface of the hemisphere

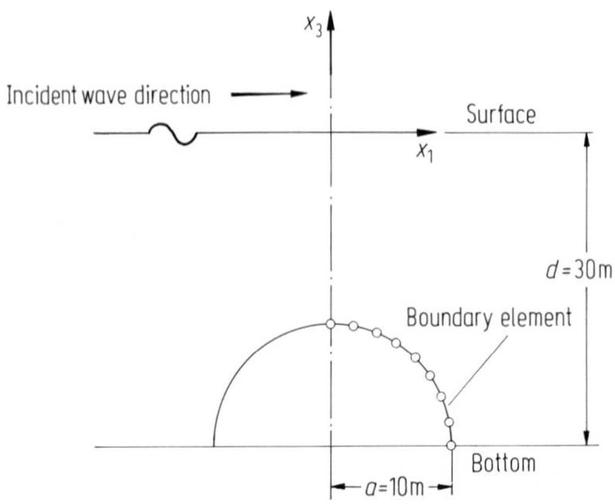

Fig. 10.6. Submerged hemispherical tank

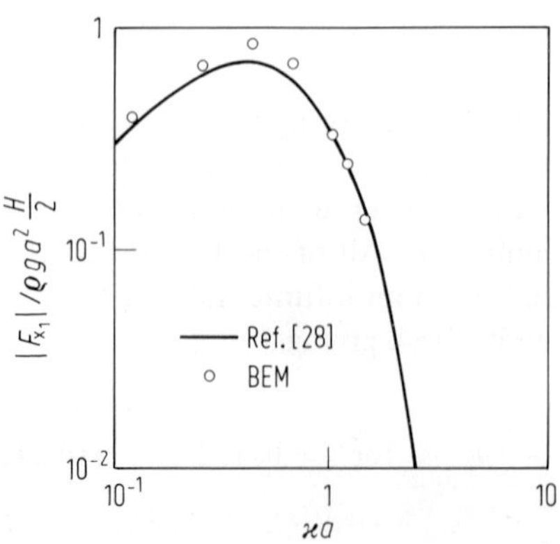

Fig. 10.7. Horizontal force coefficient for $d/a=3$

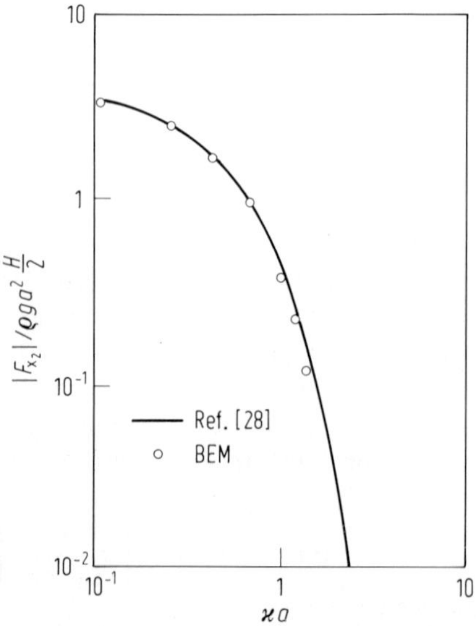

Fig. 10.8. Vertical force coefficient for $d/a=3$

was discretized into nine constant boundary elements each of them subtending an angle of 10° to the center of the sphere (Fig. 10.6). Results for the horizontal and vertical force coefficients are presented in Figs. 10.7 and 10.8, respectively, compared with analytical results given by Garrison *et al.* [28].

10.4. Horizontal Cylinders of Arbitrary Section

Another particular case of practical interest is the two-dimensional wave motion in the $x_1 - x_3$ plane past an infinite horizontal cylinder whose axis is parallel to the x_2 axis (Fig. 10.2b). Research effort has been directed towards two main aspects of the problem: the evaluation of wave reflection and transmission coefficients associated with the body's behavior as a breakwater [29, 30] and the treatment of moored and freely floating bodies undergoing motion [31, 32].

The governing equations of the problem are essentially the same (Eqs. (10.3)–(10.7)) apart from the radiation condition (10.9) which adopts a slightly different form [16]:

$$\frac{\partial u}{\partial r} = \frac{i\,\omega}{c}\,u \qquad \text{for } r \to \infty\,. \tag{10.25}$$

The two-dimensional fundamental solution which satisfies all boundary conditions apart from that on Γ is given by John [23] as

$$u^* = \ln\left(\frac{R\,R'}{d^2}\right) - 2\int_0^\infty \left\{\frac{(\mu+\nu)\exp(-\mu\, d)\cosh[\mu\,(x_3(x)+d)]\cosh[\mu\,(x_3(\xi)+d)]\cos(\mu\, r)}{\mu[\mu\sinh(\mu\, d) - \nu\cosh(\mu\, d)]} + \frac{\exp(-\mu\, d)}{\mu}\right\} d\mu\,, \tag{10.26}$$

where the notation following Eq. (10.17) applies with $x_2(\xi) - x_2(x) = 0$ here, and the integral is understood to be a Cauchy principal value.

The series form of u^* is

$$u^* = \frac{2\pi\,\nu}{\varkappa[\nu\, d + \sinh^2(\varkappa\, d)]}\cosh[\varkappa\,(x_3(x)+d)]\cosh[\varkappa\,(x_3(\xi)+d)]\sin(\varkappa\, r) - 2\pi\sum_{m=1}^{\infty}\frac{C_m}{\mu_m}\cos[\mu_m(x_3(x)+d)]\cos[\mu_m(x_3(\xi)+d)]\exp(-\mu_m r) \tag{10.27}$$

in which the notation following Eq. (10.18) applies. According to Naftzger and Chakrabarti [29], the series form of u^* can be evaluated numerically much more efficiently than expression (10.26) when the value of r is large.

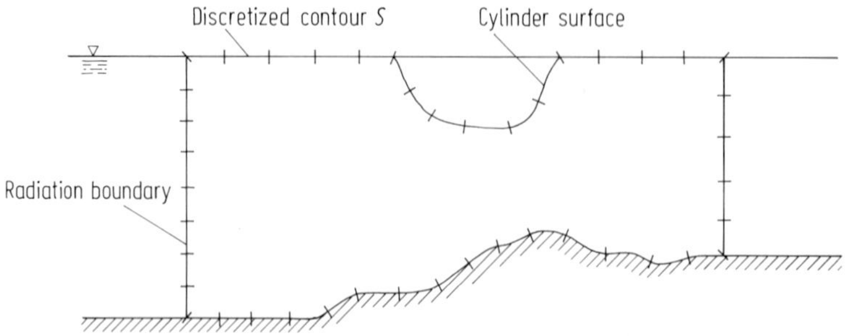

Fig. 10.9. Boundary discretization of truncated domain

The method adopted in [30–32] employs the fundamental solution $u^* = \ln(1/R)$ with the boundary integral equation (10.15), with Γ replaced by the boundary S of the truncated domain depicted in Fig. 10.9. The advantages of this method are that it may incorporate a variable depth in the vicinity of the body, and that the very simple logarithmic fundamental solution is used. The main disadvantage is that a relatively long boundary needs to be discretized. Still, very good results have been reported in the above-mentioned references.

Example 10.3. Naftzger and Chakrabarti [29] analyzed a fully submerged cylinder and compared their numerical results with that of Ogilvie [33] who obtained a semi-closed-form solution for the case of infinitely deep water. The results for the forces are presented in Figs. 10.10–10.12 in terms of $\varkappa a$ and $\varkappa h$, where a is the cylinder radius and h the depth of its center. The linear horizontal and vertical forces on the cylinder are equal to each other for all values of $\varkappa a$ and $\varkappa h$. A value of $\varkappa a = 2$ was adopted and it may be noted that the limiting value of $\varkappa h$ at which the cylinder cuts the still water level is $\varkappa h = \varkappa a = 2$. As Ogilvie [33] pointed out, the clearance between the top of the cylinder and the still water surface must be greater than the wave amplitude, if the linear theory is to have meaning.

Figure 10.10 presents a comparison where the value of $d/a = 4$ was employed. Since $\varkappa d > \pi$, the effect of the bottom on the incident wave is negligible. The two solutions are seen to be virtually the same for $\varkappa h < 6$. As the cylinder approaches the bottom, the bottom will have some effect on the results. In this particular case, however, the amplitude of the forces becomes so small that the difference between the two solutions is not evident from the plots.

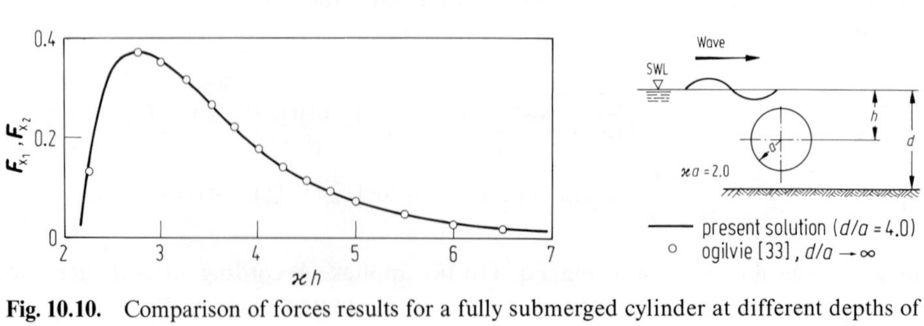

Fig. 10.10. Comparison of forces results for a fully submerged cylinder at different depths of submergence

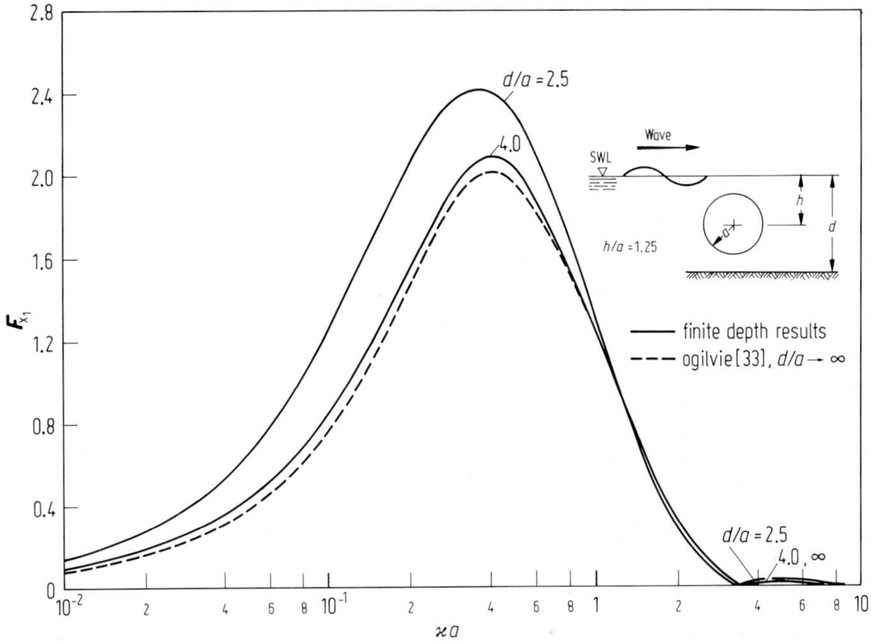

Fig. 10.11. Normalized maximum horizontal force

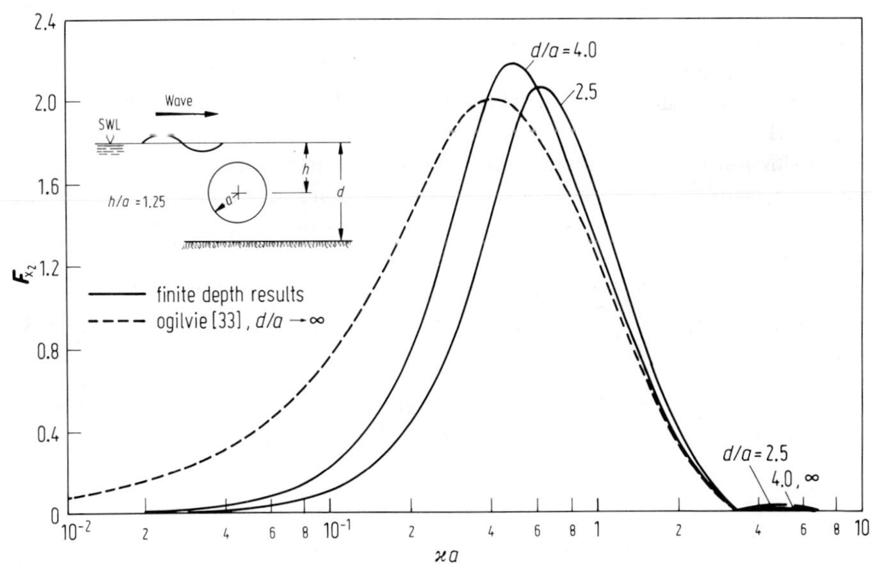

Fig. 10.12. Normalized maximum vertical force

To illustrate the effect of the bottom, results are presented in Figs. 10.11 and 10.12 for $h/a = 1.25$ and $d/a = 2.5$ and 4.0. Note that while the shape of the curves is similar to that for $d/a \to \infty$, the horizontal and vertical forces are no longer equal.

10.5. Vertical Cylinders of Arbitrary Section

This case (Fig. 10.2c) is equivalent to the problem of resonance of harbors of arbitrary shape. Harbor oscillation occurs due to waves arriving from the open sea into the harbor. These waves are partly reflected by the boundaries of the harbor and part of the waves is trapped inside the harbor. These waves produce resonance if the frequencies of the various incident and reflected waves happen to coincide with one or more of the free oscillatory modes of the harbor.

The present situation is much simpler to treat than the general three-dimensional problem since the surface integral equation of the three-dimensional problem is reduced to a line integral equation taken over the horizontal section of the body or bodies. The governing Laplace's equation of the three-dimensional problem is reduced to a two-dimensional Helmholtz equation. The corresponding fundamental solution which satisfies both the Helmholtz equation and the radiation condition (10.25) is particularly simple and may be evaluated rapidly.

As pointed out in Section 10.1, this problem is mathematically equivalent to the problem of scattering and radiation of acoustic waves. In this respect, the first application of a boundary integral formulation for the solution of the Helmholtz equation was derived by Banaugh and Goldsmith [12]. Other applications include that of Shaw [13] also for acoustic waves, Hwang and Tuck [34] and Lee [35] for harbor resonance, Isaacson [36], Harms [37], and Au and Brebbia [38] for diffraction of water waves.

The reduction of the problem to a Helmholtz equation for the case of water waves is accomplished by a separation of variables approach in which the scattered potential u_s is taken to have a hyperbolic cosine variation with depth (see Eq. (10.12))

$$u_s(x, t) = -\frac{i\,g\,H}{2\,\omega}\,\frac{\cosh\,[\varkappa\,(x_3 + d)]}{\cosh\,(\varkappa\,d)}\,u'_s(x_1, x_2)\,e^{-i\omega t}. \tag{10.28}$$

This directly satisfies the seabed and linearized free-surface boundary conditions. Substituting expression (10.28) into the Laplace equation for u_s gives

$$\nabla^2 u'_s + \varkappa^2 u'_s = 0. \tag{10.29}$$

The transformation of the above equation into a boundary integral equation was already carried out in Chapter 3, Example 3.4, using a weighted residual approach. This gives

$$c\,(\xi)\,u'_s(\xi) = -\int_\Gamma \left\{u'_s(x)\,\frac{\partial u^*\,(\xi, x)}{\partial n\,(x)} + u^*\,(\xi, x)\,\frac{\partial u'_I(x)}{\partial n\,(x)}\right\}\,d\Gamma\,(x), \tag{10.30}$$

where the fundamental solution u^* is of the form

$$u^* = \frac{i}{4} H_0^{(1)}(\varkappa r) \tag{10.31}$$

and the incident wave potential is (see Eq. (10.10))

$$u'_I = e^{i\varkappa x_1}. \tag{10.32}$$

Note that if the incident waves approach the body obliquely, say at an angle α to the x_1 axis, the incident wave potential may be written [15, 38]

$$u'_I = \exp\left[i\,\varkappa(x_1 \cos\alpha + x_2 \sin\alpha)\right]. \tag{10.33}$$

Mention is made in passing that the integral representation (10.30) of the Helmholtz equation is a classical one, known as Weber's equation [3].

The numerical solution of Eq. (10.30) is discussed in detail by several authors, notably [12, 34, 38] where analytical and numerical expressions for the evaluation of the integrals over each boundary element are given.

Extension of the formulation described in this section to the case of harbors consisting of several connected basins, each of which of a uniform but different depth, using the method of subregions discussed in Section 2.8 have been presented by Lee and Raichlen [39], Mattioli [40], and Rahman [41].

Problems involving harbors of variable depth have been studied through a combination of the boundary element method with other numerical techniques in [42–44]. The combination of the BEM with the finite element method is fully discussed in Chapter 13.

Example 10.4. This example, taken from Au and Brebbia [38], analyzes a vertical cylinder of elliptical cross-section (Fig. 10.13). An analytical solution of the problem was presented by Goda and Yoshimura [45] who solved the Helmholtz equation by using a separation of variables approach.

The boundary element results [38] were obtained using 32 constant or 16 quadratic elements with smaller elements near the major axis of the ellipse (Fig. 10.13) to take into consideration the more rapid change of the slope at that posi-

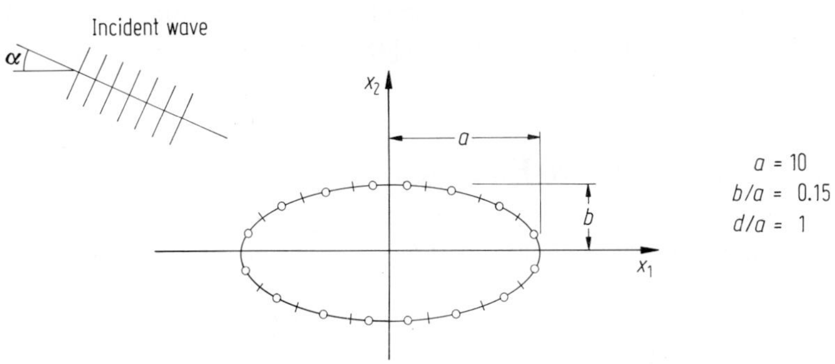

Fig. 10.13. Vertical cylinder of elliptical cross-section

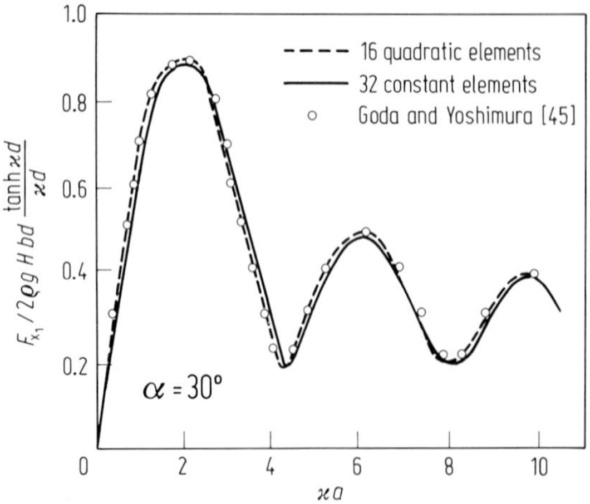

Fig. 10.14. F_{x_1} force on elliptical cylinder

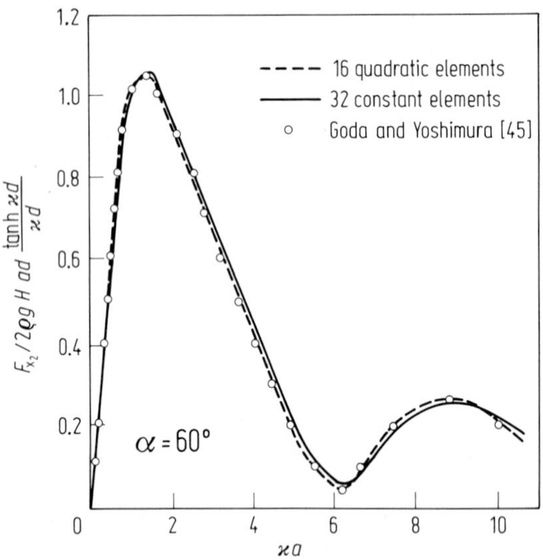

Fig. 10.15. F_{x_2} force on elliptical cylinder

tion. The forces were obtained for two different approach angles for the incident waves, i.e., $\alpha = 30°$ and $\alpha = 60°$. Results are plotted in Figs. 10.14 and 10.15; good agreement was found with the results presented in [45].

10.6. Transient Scalar Wave Equation

We shall now proceed to study boundary integral solutions to the transient scalar wave equation

$$\nabla^2 u(x, t) - \frac{1}{c^2} \frac{\partial^2 u(x, t)}{\partial t^2} = 0, \qquad x \in \Omega \tag{10.34}$$

with boundary conditions of the types

$$
\begin{aligned}
u(x, t) &= \bar{u}(x, t), && x \in \Gamma_1, \\
q(x, t) &= \frac{\partial u(x, t)}{\partial n(x)} = \bar{q}(x, t), && x \in \Gamma_2
\end{aligned}
\tag{10.35}
$$

and initial conditions at $t = t_0$,

$$
\begin{aligned}
u(x, t) &= u_0(x, t_0), && x \in \Omega, \\
\frac{\partial u(x, t)}{\partial t} &= \left[\frac{\partial u(x, t)}{\partial t}\right]_0, && x \in \Omega.
\end{aligned}
\tag{10.36}
$$

As in the case of the diffusion equation (Chapter 4), the problem represented by Eq. (10.34) with boundary conditions (10.35) and initial conditions (10.36) can also be recast into an integral equation for the unknown function u. Although Laplace transform formulations (see Section 4.2) can be employed in order to perform this transformation ([46] and Chapter 11), we shall concentrate on formulations which involve time-dependent fundamental solutions.

We start by writing the following weighted residual statement for the problem:

$$
\begin{aligned}
&\int_{t_0}^{t_F}\int_{\Omega}\left\{\nabla^2 u(x, t) - \frac{1}{c^2}\frac{\partial^2 u(x, t)}{\partial t^2}\right\} u^*(\xi, x, t_F, t)\, d\Omega(x)\, dt \\
&= \int_{t_0}^{t_F}\int_{\Gamma_2} [q(x, t) - \bar{q}(x, t)]\, u^*(\xi, x, t_F, t)\, d\Gamma(x)\, dt \\
&\quad - \int_{t_0}^{t_F}\int_{\Gamma_1} [u(x, t) \quad \bar{u}(x, t)]\, q^*(\xi, x, t_F, t)\, d\Gamma(x)\, dt,
\end{aligned}
\tag{10.37}
$$

where $q^*(\xi, x, t_F, t) = \partial u^*(\xi, x, t_F, t)/\partial n(x)$.

Integrating by parts twice the Laplacian with respect to space and twice the time derivative with respect to time, we obtain

$$
\begin{aligned}
&\int_{t_0}^{t_F}\int_{\Omega}\left\{\nabla^2 u^*(\xi, x, t_F, t) - \frac{1}{c^2}\frac{\partial^2 u^*(\xi, x, t_F, t)}{\partial t^2}\right\} u(x, t)\, d\Omega(x)\, dt \qquad (10.38) \\
&+ \frac{1}{c^2}\left\{\int_{\Omega}\left[u(x, t)\frac{\partial u^*(\xi, x, t_F, t)}{\partial t} - \frac{\partial u(x, t)}{\partial t} u^*(\xi, x, t_F, t)\right] d\Omega\right\}_{t=t_0}^{t=t_F} \\
&= -\int_{t_0}^{t_F}\int_{\Gamma} q(x, t)\, u^*(\xi, x, t_F, t)\, d\Gamma(x)\, dt + \int_{t_0}^{t_F}\int_{\Gamma} u(x, t)\, q^*(\xi, x, t_F, t)\, d\Gamma(x)\, dt.
\end{aligned}
$$

The fundamental solution u^* possesses the property

$$
\nabla^2 u^*(\xi, x, t_F, t) - \frac{1}{c^2}\frac{\partial^2 u^*(\xi, x, t_F, t)}{\partial t^2} = -\Delta(\xi, x)\,\Delta(t_F, t)
\tag{10.39}
$$

and it also satisfies a causality condition [4]

$$u^*(\xi, x, t_F, t) = 0 \qquad \text{whenever } c\,(t_F - t) < |x - \xi|. \tag{10.40}$$

In order to avoid ending the integrations exactly at the peak of a Dirac delta function, we may add to the upper limit of the integrals an arbitrarily small quantity ε. Thus, we have that the second integral on the left-hand side of Eq. (10.38) is identically zero for the upper limit value due to the causality condition (10.40). Taking the limit as $\varepsilon \to 0$ and accounting for condition (10.39) yields the equation [4, 10]

$$\begin{aligned} u(\xi, t_F) &+ \int_{t_0}^{t_F} \int_\Gamma u(x, t)\, q^*(\xi, x, t_F, t)\, d\Gamma(x)\, dt \\ &= \int_{t_0}^{t_F} \int_\Gamma q(x, t)\, u^*(\xi, x, t_F, t)\, d\Gamma(x)\, dt \\ &- \frac{1}{c^2} \int_\Omega \left\{ u_0(x, t_0) \left[\frac{\partial u^*(\xi, x, t_F, t)}{\partial t} \right]_0 - \left[\frac{\partial u(x, t)}{\partial t} \right]_0 u^*(\xi, x, t_F, t_0) \right\} d\Omega(x) \end{aligned} \tag{10.41}$$

which, on taking the point ξ to the boundary, gives the boundary integral equation

$$\begin{aligned} c(\xi)\, u(\xi, t_F) &+ \int_{t_0}^{t_F} \int_\Gamma u(x, t)\, q^*(\xi, x, t_F, t)\, d\Gamma(x)\, dt \\ &= \int_{t_0}^{t_F} \int_\Gamma q(x, t)\, u^*(\xi, x, t_F, t)\, d\Gamma(x)\, dt \\ &- \frac{1}{c^2} \int_\Omega \left\{ u_0(x, t_0) \left[\frac{\partial u^*(\xi, x, t_F, t)}{\partial t} \right]_0 - \left[\frac{\partial u(x, t)}{\partial t} \right]_0 u^*(\xi, x, t_F, t_0) \right\} d\Omega(x), \end{aligned} \tag{10.42}$$

the integrals being taken in the Cauchy principal value sense. One should note the analogy between this equation and the boundary integral equation (4.34) for diffusion problems.

10.7. Three-Dimensional Problems: The Retarded Potential

The fundamental solution of the wave equation in three dimensions is given in terms of a retarded time. Due to the finiteness of the wave propagation velocity (sound or light speed) the influence between the fields at two points separated in space will not be instantaneous. The time lag which depends on the distance between the source and the field points is called the retardation.

The fundamental solution is of the form [4, 47]

$$u^* = \frac{\Delta(t, t_R)}{4 \pi r}, \tag{10.43}$$

where r is the distance between the points ξ and x and $t_R = t_F - r/c$ is the retardation. The normal derivative of u^* along Γ is given by [8, 9]

$$q^* = \frac{\partial u^*}{\partial n} = -\frac{1}{4\pi r}\left[\frac{\Delta(t, t_R)}{r} - \frac{1}{c}\frac{\partial \Delta(t, t_R)}{\partial t}\right] n_r \tag{10.44}$$

in which $n_r = \partial r/\partial n$.

The time integrals in Eq. (10.42) can then be evaluated analytically as follows:

$$\int_{t_0}^{t_F} q(x, t)\, u^*(\xi, x, t_F, t)\, dt = \frac{1}{4\pi r}\int_{t_0}^{t_F} q(x, t)\,\Delta(t, t_R)\, dt = \frac{1}{4\pi r} q(x, t_R), \tag{10.45}$$

$$\begin{aligned}\int_{t_0}^{t_F} u(x, t)\, q^*(\xi, x, t_F, t)\, dt &= -\frac{n_r}{4\pi r}\left[\frac{1}{r}\int_{t_0}^{t_F} u(x, t)\,\Delta(t, t_R)\, dt\right.\\ &\quad\left. - \frac{1}{c}\int_{t_0}^{t_F} u(x, t)\frac{\partial \Delta(t, t_R)}{\partial t}\, dt\right]\\ &= -\frac{n_r}{4\pi r}\left\{\frac{1}{r} u(x, t_R) + \frac{1}{c}\left[\frac{\partial u(x, t)}{\partial t}\right]_{t=t_R}\right\},\end{aligned} \tag{10.46}$$

the result of the last integral being obtained through integration by parts.

Substituting the results of Eqs. (10.45) and (10.46) into Eq. (10.42) the boundary integral equation for the three-dimensional case becomes

$$\begin{aligned}c(\xi)\, u(\xi, t_F) &= \frac{1}{4\pi}\int_\Gamma \frac{1}{r}\left[q(x, t_R) + n_r\left\{\frac{1}{r} u(x, t_R) + \frac{1}{c}\left[\frac{\partial u(x, t)}{\partial t}\right]_{t=t_R}\right\}\right] d\Gamma(x)\\ &\quad + \frac{1}{4\pi}\left\{t N_0 + \frac{\partial}{\partial t}[t M_0]\right\},\end{aligned} \tag{10.47}$$

where M_0 and N_0 are respectively the mean value of u_0 and $[\partial u/\partial t]_0$ over a spherical surface with center at ξ and with a variable radius $c\,t$ [10]. It can be shown that the value of the coefficient $c(\xi)$ is the same as for Laplace's equation [8, 9].

Equation (10.47) can be viewed as an expression of the well-known Kirchhoff integral equation [1–5]. The necessary procedures for numerical solution of this equation are discussed in detail in [8, 9]. A special feature of the three-dimensional analysis is that no time integration is required. The same does not apply for the two-dimensional case as it is explained in the next section.

Example 10.5. Groenenboom [8] has presented some preliminary results of a computer code developed on the basis of the theory derived in this section. The specific example studied was that of wave propagation for a point source in a three-dimensional rectangular box. This problem was chosen because an analytical solution for it can be found by using the method of images.

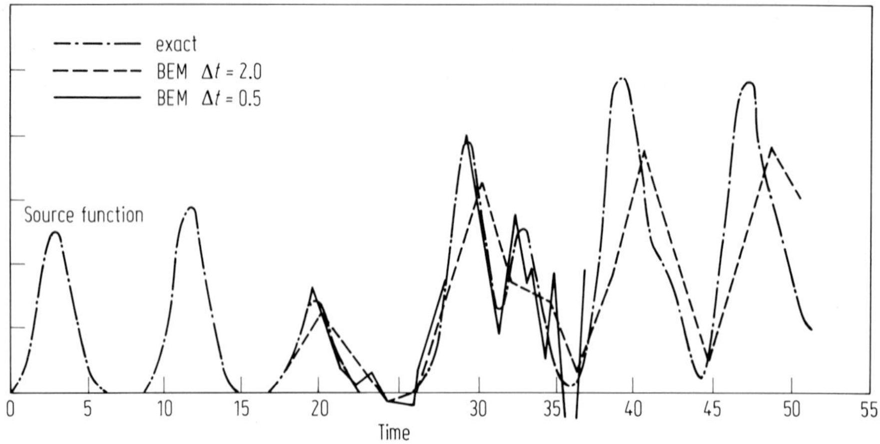

Fig. 10.16. Potential of a point source in a three-dimensional box

The dimensions of the box are 3×4×5 and its surfaces were divided into 376 constant quadrilateral boundary elements of dimensions 0.5×0.5. A short pulsed Gaussian distribution was taken for the source function.

Since the potential u in Eq. (10.47) is expressed solely in terms of quantities at previous values of time, a time-stepping technique was employed in the numerical solution. Results at a boundary point are presented in Fig. 10.16 for two time step values ($\Delta t = 0.5$ and 2 units of time) and compared with the exact solution. For the largest time step value the reproduction of the exact potential is, of course, not very accurate for such a fast transient, and a small delay in the response may be observed. For $\Delta t = 0.5$, the reproduction of the secondary and the third reflections is quite good, but at later times the solution seems to become unstable. Groenenboom [8] concluded that for such a crude mesh, the results obtained for this strong transient were encouraging.

10.8. Two-Dimensional Problems

The fundamental solution for the two-dimensional transient scalar wave equation is of the form [4, 47]

$$u^* = \frac{c}{2\pi\,[c^2(t_F - t)^2 - r^2]^{1/2}}\,H[c\,(t_F - t) - r], \tag{10.48}$$

where H is the Heaviside function. The normal derivative of u^* along Γ is given by [9, 10]

$$q^* = \frac{\partial u^*}{\partial n} = \left\{\frac{c\,r}{2\pi\,[c^2(t_F - t)^2 - r^2]^{3/2}}\,H[c\,(t_F - t) - r] - \frac{c}{2\pi\,[c^2(t_F - t)^2 - r^2]^{1/2}}\,\Delta\,[c\,(t_F - t) - r]\right\} n_r. \tag{10.49}$$

Since the two-dimensional fundamental solution is given in terms of a Heaviside (step) function rather than a delta function, as in the three-dimensional case, the influence of a source function at a point x on the potential at (ξ, t_F) is no longer restricted to the value at the retarded time but has to be integrated from the initial time t_0 up to the actual time t_F. Thus, time-marching schemes of the type discussed in Chapter 4 should be employed on the numerical solution of the corresponding boundary integral equation.

Two distinct types of singularities occur in the integrands of Eq. (10.42). The first one occurs in the domain integrals when $r = 0$ or in the boundary integrals when both r and $c(t_F - t)$ are simultaneously null. The second type of singularity appears for points located at the front of the wave represented by the fundamental solution, i.e., in the boundary integrals when $r = c(t_F - t)$ or in the domain integral when $r = c\,t_F$. In spite of these singularities, the numerical integration of Eq. (10.42) does not present any special difficulties [10] and some results of analyses using this formulation have already been presented by Mansur and Brebbia [11].

References

1. Lamb, H., *Hydrodynamics,* 6th ed., Cambridge University Press, Cambridge, 1932.
2. Stratton, J. A., *Electromagnetic Theory,* McGraw–Hill, New York, 1941.
3. Baker, B. B., and Copson, E. T., *The Mathematical Theory of Huygen's Principle,* 2nd ed., Oxford University Press, Oxford, 1953.
4. Morse, P. M., and Feshbach, H., *Methods of Theoretical Physics,* McGraw–Hill, New York, 1953.
5. Kirchhoff, G., Zur Theorie der Lichtstrahlen, Berliner Ber. 641, 1882.
6. Friedman, M. B., and Shaw, R. P., Diffraction of a plane shock wave by an arbitrary rigid cylindrical obstacle, Trans. ASME J. Appl. Mech. **29,** 40–46 (1962).
7. Shaw, R. P., An outer boundary integral equation applied to transient wave scattering in an inhomogeneous medium, Trans. ASME J. Appl. Mech. **42,** 147–152 (1975).
8. Groenenboom, P. H. L., The application of boundary elements to steady and unsteady potential fluid flow problems in two and three dimensions, in *Boundary Element Methods* (C. A. Brebbia, Ed.), Springer-Verlag, Berlin, 1981.
9. Groenenboom, P. H. L., Wave propagation phenomena, in *Progress in Boundary Elements, Vol. 2,* Pentech Press, London, Springer-Verlag, New York, 1982.
10. Mansur, W. J., and Brebbia, C. A., Formulation of the boundary element method for transient problems governed by the scalar wave equation, Appl. Math. Modelling **6,** 307–312 (1982).
11. Mansur, W. J., and Brebbia, C. A., Application of the boundary element method to solve the transient scalar wave equation, in *Boundary Elements in Engineering* (C. A. Brebbia, Ed.), Springer-Verlag, Berlin, 1982.
12. Banaugh, R. P., and Goldsmith, W., Diffraction of steady acoustic waves by surfaces of arbitrary shape, J. Acoust. Soc. Amer. **35,** 1590–1601 (1963).
13. Shaw, R. P., An integral equation approach to acoustic radiation and scattering, in *Topics in Ocean Engineering, Vol. II* (C. Bretschneider, Ed.), Gulf Publishing Co., Houston, 1970.
14. Brebbia, C. A., and Walker, S., *Dynamic Analysis of Off-Shore Structures,* Butterworths, London, 1979.
15. Sarpkaya, T., and Isaacson, M., *Mechanics of Wave Forces on Off-Shore Structures,* Van Nostrand Reinhold, New York, 1981.
16. Sommerfeld, A., *Partial Differential Equations in Physics,* Academic Press, New York, 1949.

17. Wehausen, J. V., and Laitone, E. V., Surface waves, in *Encyclopedia of Physics* (S. Flügge, Ed.), Vol. 9, pp. 446–778, Springer-Verlag, Berlin, 1960.
18. Garrison, C. J., and Chow, P. Y., Wave forces on submerged bodies, J. Waterways Harbours Coastal Eng. Div. ASCE **98,** 375–392 (1972).
19. Hogben, N., Osborne, J., and Standing, R. G., Wave loading on off-shore structures – Theory and experiment, in *Proc. Symp. Ocean Eng.*, pp. 19–36, National Physical Laboratory, London, 1974.
20. Hogben, N., and Standing, R. G., Wave loads on large bodies, in *Proc. Int. Symp. on the Dynamics of Marine Vehicles and Structures in Waves,* pp. 258–277, Univ. College, London, 1974.
21. Faltinsen, O. M., and Michelsen, F. C., Motions of large structures in waves at zero Froude number, in *Proc. Int. Symp. on the Dynamics of Marine Vehicles and Structures in Waves,* pp. 91–106, Univ. College, London, 1974.
22. Garrison, C. J., Hydrodynamic loading on large off-shore structures. Three-dimensional source distribution methods, in *Numerical Methods in Off-shore Engineering* (O. C. Zienkiewicz, R. W. Lewis, and R. G. Stagg, Ed.), Wiley, Chichester, 1978.
23. John, F., On the motion of floating bodies II, Commun. Pure Appl. Math. **3,** 45–101 (1950).
24. Murphy, J. E., Integral equation failure in wave calculations, J. Waterways Port Coastal Ocean Div. ASCE **104,** 330–334 (1978).
25. Black, J. L., Wave forces on vertical axisymmetric bodies, J. Fluid Mech. **67,** 369–376 (1975).
26. Fenton, J. D., Wave forces on vertical bodies of revolution, J. Fluid Mech. **85,** 241–255 (1978).
27. Au, M. C., Application of boundary element methods in wave propagation studies, M.Sc. Thesis, Southampton University, Southampton, 1979.
28. Garrison, C. J., Rao, V. S., and Snider, R. H., Wave interaction with large submerged objects, in *Proc. Offshore Tech. Conf.*, Paper No. OTC 1278, 1970.
29. Naftzger, R. A., and Chakrabarti, S. K., Scattering of waves by two-dimensional circular obstacles in finite water depths, J. Ship Res. **23,** 32–42 (1979).
30. Finnigan, T. D., and Yamamoto, T., Analysis of semi-submerged porous breakwaters, in *Proc. Civil Engineering in the Oceans IV,* pp. 380–397, ASCE, San Francisco, 1979.
31. Yamamoto, T., and Yoshida, A., Elastic mooring of floating breakwaters, in *Proc. 7th Int. Harbour Congress,* Antwerp, 1978.
32. Ijima, T., Chou, C. R., and Yoshida, A., Method of analysis for two-dimensional water wave problems, in *Proc. 15th Coastal Eng. Conf.*, pp. 2717–2736, ASCE, Honolulu, 1976.
33. Ogilvie, T. F., First- and second-order forces on a cylinder submerged under a free surface, J. Fluid mech. **16,** 451–472 (1963).
34. Hwang, L. S., and Tuck, E. O., On the oscillations of harbours of arbitrary shape, J. Fluid Mech. **42,** 447–464 (1970).
35. Lee, J. J., Wave-induced oscillations in harbours of arbitrary geometry, J. Fluid Mech. **45,** 375–394 (1971).
36. Isaacson, M. de St., Vertical cylinders of arbitrary section in waves, J. Waterways Port Coastal Ocean Div. ASCE **104,** 309–324 (1978).
37. Harms, V. W., Diffraction of water waves by isolated structures, J. Waterways Port Coastal Ocean Div. ASCE **105,** 131–147 (1979).
38. Au, M. C., and Brebbia, C. A., Diffraction of water waves by vertical cylinders using boundary elements, Appl. Math. Modelling **7,** 106–114 (1983).
39. Lee, J. J., and Raichlen, F., Oscillations in harbour with connected basins, J. Waterways Harbours Coastal Engng. Div. ASCE **98,** 311–332 (1972).
40. Mattioli, F., Wave-induced oscillations in harbours of variable depth, Comput. Fluids **6,** 161–172 (1978).
41. Rahman, M., Numerical response of an arbitrarily shaped harbour, Appl. Math. Modelling **5,** 109–121 (1981).
42. Shaw, R. P., and Falby, W., FEBIE – A Combined finite element–boundary integral equation method, Comput. Fluids **6,** 153–160 (1978).

43. Zienkiewicz, O. C., Bettess, P., and Kelly, D. W., The finite element method for determining fluid loadings on rigid structures: Two- and three-dimensional formulations, in *Numerical Methods in Off-shore Engineering* (O. C. Zienkiewicz, R. W. Lewis, and R. G. Stagg, Eds.), Wiley, Chichester, 1978.
44. Mattioli, F., Element integral approach for water waves, Comput. Fluids **9,** 181–203 (1981).
45. Goda, Y. and Yoshimura, T., Wave force on a vessel tied at off-shore dolphins, in *Proc. 13th Coastal Eng. Conf.*, pp. 1723–1742, ASCE, Vancouver, 1972.
46. Cruse, T. A., and Rizzo, F. J., A direct formulation and numerical solution of the general elasto dynamic problem I and II, J. Math. Anal. Appl. **22,** 244–259, 341–355 (1968).
47. Stakgold, I., *Green's Functions and Boundary Value Problems,* Wiley, New York, 1979.

Chapter 11

Vibrations

11.1. Introduction

While the BEM has rapidly developed in many engineering fields, there are relatively few papers on elastodynamics. The first formulation and solution of the transient elastodynamic problem by combining the boundary element technique and Laplace transform is due to Cruse and Rizzo [4] and Cruse [5], who applied this method to solve a half-plane wave propagation problem. In 1978, Cole *et al.* [6] formulated the problem in terms of boundary integral equations in space and time, and solved them by time-stepping numerical schemes. The application was made to a simple antiplane wave propagation problem.

Manolis and Beskos [7] extended Cruse's work to steady-state dynamic stress concentration, together with a modification of the Laplace transform inversion scheme. One of the main problems of the approach is the numerical inversion of the Laplace transform, although some guide lines have been given in Ref. [8]. Dynamic stiffness of foundations on a half-space in steady-state motion were recently studied by Dominguez *et al.* [9, 10].

More recently some important papers have been published on elastodynamics, and the use of BEM to find eigenvalues and eigenvectors have been reported by Hutchinson [11], Niwa *et al.* [12], and Nardini [26].

11.2. Governing Equations

The governing equations presented here correspond to the small displacement theory for homogeneous, isotropic, linearly elastic materials. In addition, the same assumptions made in Chapter 5 regarding the regularity conditions at infinity will apply.

The body is defined by a domain Ω bounded by Γ as indicated in Fig. 11.1. We can denote the displacement components associated with any point x of co-ordinates x_i by the notation

$$u_i(x, t)\,. \tag{11.1}$$

Under these assumptions, the displacement equations of motion can be written [1–3]

$$(C_1^2 - C_2^2)\, u_{i,ij} + C_2^2\, u_{j,ii} + \frac{b_j}{\varrho} = \ddot{u}_j\,, \tag{11.2}$$

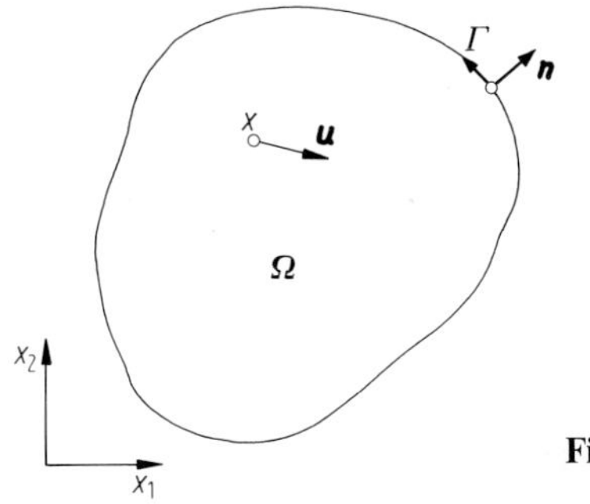

Fig. 11.1. Notation

where as before the commas indicate space derivatives and the dots indicate time derivatives. The propagation velocities of the dilatational and distortional waves are given as

$$C_1 = \sqrt{\frac{\lambda + 2\mu}{\varrho}}, \qquad C_2 = \sqrt{\frac{\mu}{\varrho}} \tag{11.3}$$

in which λ and μ are the Lamè constants given by

$$\lambda = \frac{E\nu}{(1+\nu)(1-2\nu)}, \qquad \mu = G = \frac{E}{2(1+\nu)} \tag{11.4}$$

and ϱ is the mass density of the material. The strains are obtained by

$$\varepsilon_{ij} = \tfrac{1}{2}(u_{i,j} + u_{j,i}) \tag{11.5}$$

and the material obeys Hooke's law, i.e.,

$$\sigma_{ij} = 2\mu\, \varepsilon_{ij} + \lambda\, \varepsilon_{kk} \delta_{ij}. \tag{11.6}$$

The fundamental boundary and initial conditions that have to be satisfied in elastodynamics are defined as follows:

Initial conditions

$$\begin{aligned} u_i(x,t) &= u_i^0(x), \\ \dot{u}_i(x,t) &= v_i^0(x) \qquad \text{for } t = t_0 \text{ in } \Omega + \Gamma, \end{aligned} \tag{11.7}$$

where $u_i^0(x)$ and $v_i^0(x)$ are prescribed functions;

displacement boundary conditions

$$u_i(x,t) = \bar{u}_i(x,t) \qquad \text{for } t > t_0 \text{ on } \Gamma_1; \tag{11.8}$$

traction boundary conditions

$$p_i(x,t) = \sigma_{ij} n_j = \bar{p}_i(x,t) \qquad \text{for } t > t_0 \text{ on } \Gamma_2 \tag{11.9}$$

in which the bar as usual indicates that the values are known and $\Gamma = \Gamma_1 + \Gamma_2$.

The stress and traction components can be expressed as functions of the displacements in the form

$$\sigma_{ij} = \mu (u_{i,j} + u_{j,i}) + \lambda\, u_{k,k}\, \delta_{ij}\,, \tag{11.10}$$

$$p_i = \mu \left(\frac{\partial u_i}{\partial n} + u_{j,i}\, n_j \right) + \lambda\, u_{j,j}\, n_i\,, \tag{11.11}$$

where n_j are the direction cosines of the outward normal to the boundary (Fig. 11.1).

By applying the divergence and curl operators to the vector form of Eq. (11.2), it can be shown [3] that in absence of body forces one obtains

$$\nabla^2 e = \frac{1}{C_1^2}\, \ddot{e} \tag{11.12}$$

and

$$\nabla^2 \boldsymbol{\omega} = \frac{1}{C_2^2}\, \ddot{\boldsymbol{\omega}} \tag{11.13}$$

where e is the dilatation ($e = \varepsilon_{kk}$) and $\boldsymbol{\omega}$ is the rotation vector whose components are related to the rotation tensor defined in Chapter 5 (Eq. (5.25)) by

$$\omega_1 = \omega_{23}\,, \qquad \omega_2 = \omega_{31}\,, \qquad \omega_3 = \omega_{12}\,. \tag{11.14}$$

Equations (11.12) and (11.13) indicate that the velocity C_1 is associated with the dilatational waves or P (primary) waves. The velocity C_2 is related to the equivoluminal or distortional waves also called S (secondary) waves. The propagation of these waves is a rather complex time-dependent phenomenon. In addition, when propagating waves meet a discontinuity, then reflection, refraction and diffraction are produced and the resultant wave motion is a superposition of all the components. For bodies with a surface of material discontinuity, there is a different type of plane waves called surface waves which propagate parallel to the surface and the disturbance decays exponentially with the distance from the surface. For a free surface they are called Rayleigh waves [13], the displacement of which is elliptic and counterclockwise moving in the plane of propagation. Surface waves with displacement perpendicular to the plane of propagation exist only in layered half-spaces. They are called Love's waves [14].

11.3. Time-Dependent Integral Formulation

The fundamental solution of equation (11.2) can be written as the response of an infinite medium to a unit impulse at time τ, i.e., a unit load concentrated at point ξ in the i direction and acting in an infinitely short time,

$$(C_1^2 - C_2^2)\, u^*_{ik,kj} + C_2^2\, u^*_{ij,kk} - \ddot{u}^*_{ij} = -\frac{\delta_{ij}}{\varrho}\, \Delta(\xi, x)\, \Delta(\tau, t)\,, \tag{11.15}$$

where $\Delta(\tau, t)$ represents the Dirac delta function of time.

The solution to the above equation can be written in terms of displacement components as follows [1]:

three-dimensional case

$$u_{ij}^*(\xi, x, t') = \frac{t'}{4\pi\varrho r^2}\left\{\frac{1}{r}(3r_{,i}r_{,j} - \delta_{ij})\left[H\left(t', \frac{r}{C_1}\right) - H\left(t', \frac{r}{C_2}\right)\right]\right.$$
$$\left. + r_{,i}r_{,j}\left[\frac{1}{C_1}\Delta\left(t', \frac{r}{C_1}\right) - \frac{1}{C_2}\Delta\left(t', \frac{r}{C_2}\right)\right] + \frac{\delta_{ij}}{C_2}\Delta\left(t', \frac{r}{C_2}\right)\right\}, \tag{11.16}$$

where H is the Heaviside or unit step function and $t' = t - \tau$;

two-dimensional case

$$u_{ij}^*(\xi, x, t') = \frac{1}{2\pi\varrho}\left\{\left(\frac{[2(t')^2 - (r^2/C_1^2)\, H(t', r/C_1)]}{[(t')^2 - (r^2/C_1^2)]^{1/2}}\right.\right.$$
$$\left. - \frac{[2(t')^2 - (r^2/C_2^2)]\, H(t', r/C_2)}{[(t')^2 - (r^2/C_2^2)]^{1/2}}\right)\frac{r_{,i}r_{,j}}{r^2}$$
$$- \left[H(t', r/C_1)\left((t')^2 - \frac{r^2}{C_1^2}\right)^{1/2}\right. \tag{11.17}$$
$$\left.\left. - H\left(t', \frac{r}{C_2}\right)\left((t')^2 - \frac{r^2}{C_2^2}\right)^{1/2}\right]\frac{\delta_{ij}}{r^2} + \frac{H(t', r/C_2)}{C_2^2[(t')^2 - (r^2/C_2^2)]^{1/2}}\,\delta_{ij}\right\}.$$

The corresponding stresses and tractions can be obtained by using expressions (11.10) and (11.11).

By using weighted residuals in time and space, one can obtain the boundary integral equation for the time-dependent formulation (the procedure is analogous to what was shown in Chapter 10):

$$c_{ij}u_j + \int_{t_0}^{\tau}\int_{\Gamma} p_{ij}^* u_j \, d\Gamma\, dt = \int_{t_0}^{\tau}\int_{\Gamma} u_{ij}^* p_j \, d\Gamma\, dt + \int_{t_0}^{\tau}\int_{\Omega} u_{ij}^* b_j \, d\Omega\, dt + \varrho\int_{\Omega}[u_{ij}^*\dot{u}_j]_{t_0}\, d\Omega$$
$$+ \varrho\int_{\Omega}[\dot{u}_{ij}^* u_j]_{t_0}\, d\Omega\,, \tag{11.18}$$

where on smooth surfaces ($\xi \in \Gamma$) $c_{ij} = \delta_{ij}/2$ and for interior points ($\xi \in \Omega$) $c_{ij} = \delta_{ij}$. This equation is the starting point for the boundary element formulation and can be discretized in time and space in similar fashion to the solution scheme indicated in Chapter 10 for the scalar wave problem. Its full formulation has only been attempted by Mansur in Ref. [15]. Other authors [4, 5, 7] have preferred to solve it using Laplace transform methods as described in the next section.

11.4. Laplace Transform Formulation

The Laplace transform of a function $f(t)$ is defined as

$$F(s) = L[f(t)] = \int_0^{\infty} f(t)e^{-st}\, dt\,, \tag{11.19}$$

where $f(t)$ is piecewise differentiable with a limited number of finite discontinuities and is conventionally regarded as zero for $t < 0$. The inversion formula is given by

$$f(t) = L^{-1}[F(s)] = \frac{1}{2\pi i} \int_{\gamma - i\infty}^{\gamma + i\infty} F(s)\, e^{st}\, ds \tag{11.20}$$

in which $\gamma > 0$ is greater than the real part of all the singularities of $F(s)$ and s is a complex number satisfying Re $(s) \geqq \gamma$. In the applications described in this section we shall mainly have available a number of sample points of F at the values of s for which we have performed our boundary element calculations. Consequently, numerical inversion techniques will be needed.

In order to remove the time dependence from the governing equations and facilitate their integral formulation, one can recast the set of equations using the Laplace transform. This gives, for Eq. (11.2),

$$(C_1^2 - C_2^2)\, U_{i,ij} + C_2^2\, U_{j,ii} + \frac{1}{\varrho} B_j - s^2 U_j + s\, u_j^0 + v_j^0 = 0\,, \tag{11.21}$$

where uppercase letters indicate transformed variables,

$$\begin{aligned} U_i(x, s) &= L[u_i(x, t)]\,, \\ B_i(x, s) &= L[b_i(x, t)]\,, \end{aligned} \tag{11.22}$$

and the initial velocity (v_j^0) together with the initial displacement (u_j^0) can be grouped with B_j in a modified body force term of the form

$$Q_j = B_j + \varrho(s\, u_j^0 + v_j^0)\,. \tag{11.23}$$

Hence, Eq. (11.21) becomes

$$(C_1^2 - C_2^2)\, U_{i,ij} + C_2^2\, U_{j,ii} - s^2 U_j + \frac{1}{\varrho} Q_j = 0 \tag{11.24}$$

which needs to be solved in conjunction with the transformed boundary conditions

$$\begin{aligned} U_j(x, s) &= \bar{U}_j(x, s) \qquad \text{on } \Gamma_1\,, \\ P_j(x, s) &= \bar{P}_j(x, s) \qquad \text{on } \Gamma_2\,. \end{aligned} \tag{11.25}$$

Following Doyle [16] and Cruse [4, 5], the fundamental solution for the Laplace transform of the dynamic equations can be written as

$$U_{ij}^*(\xi, x, s) = \frac{1}{\alpha \pi \varrho C_2^2} (\psi\, \delta_{ij} - \chi\, r_{,i}\, r_{,j})\,, \tag{11.26}$$

where for two-dimensional problems $\alpha = 2$ and

$$\psi = K_0\left(\frac{s\,r}{C_2}\right) + \frac{C_2}{s\,r}\left[K_1\left(\frac{s\,r}{C_2}\right) - \frac{C_2}{C_1}K_1\left(\frac{s\,r}{C_1}\right)\right],$$

$$\chi = K_2\left(\frac{s\,r}{C_2}\right) - \frac{C_2^2}{C_1^2}K_2\left(\frac{s\,r}{C_1}\right) \tag{11.27}$$

in which K_i is the modified Bessel function of the second kind and order i [17, 18].

For the three-dimensional case one has $\alpha = 4$ and

$$\psi = \frac{e^{-sr/C_2}}{r} + \left(\frac{C_2^2}{s^2 r^2} + \frac{C_2}{s\,r}\right)\frac{e^{-sr/C_2}}{r} - \frac{C_2^2}{C_1^2}\left(\frac{C_1^2}{s^2 r^2} + \frac{C_1}{s\,r}\right)\frac{e^{-sr/C_1}}{r},$$

$$\chi = \left(\frac{3\,C_2^2}{s^2 r^2} + \frac{3\,C_2}{s\,r} + 1\right)\frac{e^{-sr/C_2}}{r} - \frac{C_2^2}{C_1^2}\left(\frac{3\,C_1^2}{s^2 r^2} + \frac{3\,C_1}{s\,r} + 1\right)\frac{e^{-sr/C_1}}{r}. \tag{11.28}$$

The fundamental traction tensors P_{ij}^* can be obtained by substituting U_{ij}^* given in Eq. (11.26) into Eq. (11.11).

Using the weighted residuals as previously shown, we can obtain the transformed dynamic boundary integral statement for the transformed internal displacements, i.e.,

$$U_i = \int_\Gamma U_{ij}^*\, P_j\, d\Gamma - \int_\Gamma P_{ij}^*\, U_j\, d\Gamma + \int_\Omega U_{ij}^*\, Q_j\, d\Omega\,. \tag{11.29}$$

The internal stresses can then be obtained in terms of their transforms by substituting the above equation into the expression

$$L\,[\sigma_{ij}] = \varrho\left[(C_1^2 - 2\,C_2^2)\,\frac{\partial U_k}{\partial x_k}\,\delta_{ij} + C_2^2\left(\frac{\partial U_i}{\partial x_j} + \frac{\partial U_j}{\partial x_i}\right)\right], \tag{11.30}$$

where the derivatives are taken with reference to the coordinates of the load point. If we take the point ξ to the boundary Γ, we obtain the starting statement for the boundary element discretization, i.e.,

$$c_{ij}\, U_j + \int_\Gamma P_{ij}^*\, U_j\, d\Gamma = \int_\Gamma U_{ij}^*\, P_j\, d\Gamma + \int_\Omega U_{ij}^*\, Q_j\, d\Omega \tag{11.31}$$

in which for smooth boundaries $c_{ij} = \delta_{ij}/2$.

Equation (11.31) gives rise to the classical boundary element equations which can be solved for a series of values of the transform parameter s. Once this is done for a sufficient number of values of our transform parameter, a numerical inversion must be performed on the relevant variables U_i, P_i, and $L\,[\sigma_{ij}]$ to obtain the time-dependent solution. Different numerical inverse transform methods can be used; Cruse [5] adopted the procedure due to Papoulis [19], and Manolis and Beskos [7] gave preference to the method due to Durbin [20], which although more time-consuming, seems to produce improved results for late times. More information about numerical inverse transform techniques can be found in Ref. [8].

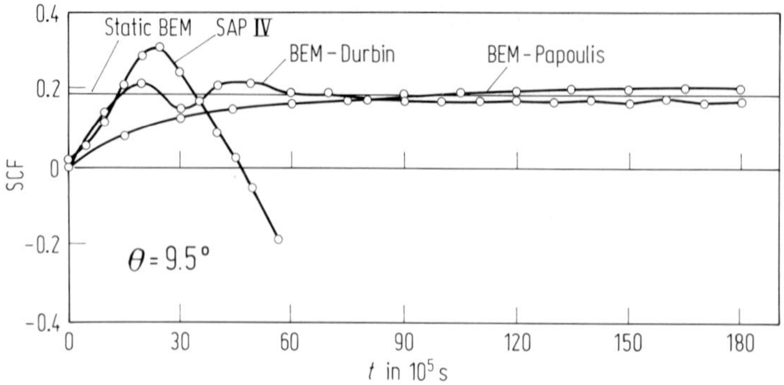

Fig. 11.2. Transient SCF history for a square hole under a plane shock (P) wave

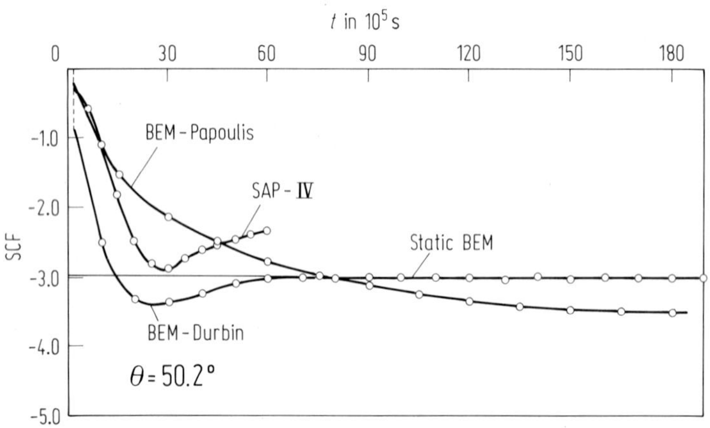

Fig. 11.3. Transient SCF history for a square hole under a plane shock (P) wave

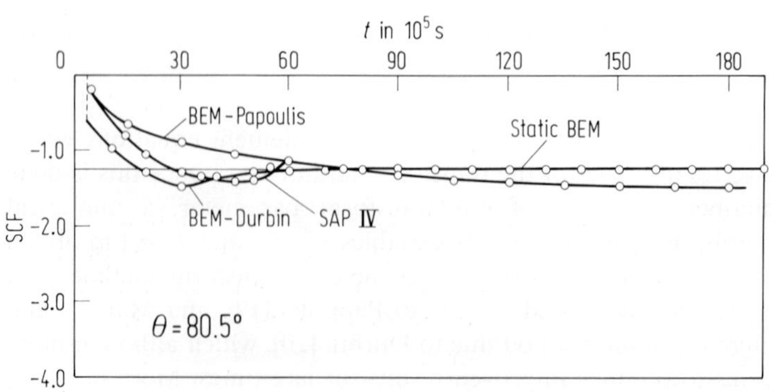

Fig. 11.4. Transient SCF history for a square hole under a plane shock (P) wave

Example 11.1. – Square Hole under a Plane Shock Wave. Some interesting comparisons between different boundary element solutions using the above-referred numerical inverse transform techniques and finite element results can be found in the paper by Manolis and Beskos [7]. The results for one of their examples are presented in Figs. 11.2–11.4. Those results correspond to a square cavity problem (plane strain) under the influence of a dilatational shock wave whose front is parallel to one of the cavity sides.

The material properties correspond to steel and in all the figures the stress concentration factor (SCF) in the tangent direction to the boundary is computed at different points along the boundary. The position of these points is defined by the angle θ which corresponds to a cylindrical coordinate system based at the center of the cavity. Also included are the results produced by the finite element program SAP IV [21] which seems to fail to predict a dynamic stress concentration factor higher than the static one for $\theta = 50.2°$.

11.5. Steady-State Elastodynamics

In many practical applications it is important to predict the dynamic behavior of a body or structure under harmonic excitation. The response is then a function of the exciting frequency and the initial conditions can be neglected assuming that a sufficiently long time has elapsed so that a steady-state is reached. This situation can be mathematically represented by taking the Fourier transform of the equations of motion.

The transformed variables can be represented by

$$\begin{aligned} U_i(x, \omega) &= \mathcal{F}[u_i(x, t)], \\ B_i(x, \omega) &= \mathcal{F}[b_i(x, t)] \end{aligned} \tag{11.32}$$

and the equations of motion now become

$$(C_1^2 - C_2^2)\, U_{i,ij} + C_2^2\, U_{j,ii} + \frac{1}{\varrho} B_j + \omega^2 U_j = 0 \tag{11.33}$$

with the transformed boundary conditions

$$\begin{aligned} U_i(x, \omega) &= \bar{U}_i(x, \omega) \qquad \text{on } \Gamma_1, \\ P_i(x, \omega) &= \bar{P}_i(x, \omega) \qquad \text{on } \Gamma_2. \end{aligned} \tag{11.34}$$

Notice that the initial conditions do not enter into the formulation.

As can be seen, Eq. (11.33) corresponds to Eq. (11.21) with s replaced by $i\,\omega$ and without the initial conditions. Consequently, the fundamental solutions presented in expressions (11.26)–(11.28) are valid for this formulation if we make $s = i\,\omega$. The corresponding boundary integral equation is therefore written as

$$c_{ij}\, U_j + \int_\Gamma P_{ij}^*\, U_j\, d\Gamma = \int_\Gamma U_{ij}^*\, P_j\, d\Gamma + \int_\Omega U_{ij}^*\, B_j\, d\Omega\,. \tag{11.35}$$

As before one can solve Eq. (11.35) for a sufficient number of values of ω and numerically invert the variable $U_j(x, \omega)$ to obtain the time-dependent displacements $u_j(x, t)$. A direct consideration of the magnitudes of $U_j(x, \omega)$ for different values of ω will give the natural modes of vibration of the elastic body. In fact, this is what makes such a procedure most attractive. For instance, if every variable of the problem can be written in the complex form

$$f(x, t) = F(x)\, e^{i\omega t}, \tag{11.36}$$

one obtains expression (11.33) as the governing equation and the solution for each prescribed circular frequency ω will give directly the response of the body in terms of the amplitudes U_i and P_i [7, 9, 10]. Some typical applications of this technique are presented in what follows; the first example is the analysis of an eigenvalue problem for an earth dam and the second indicates how the dynamic stiffness of foundations can be computed by this approach.

Example 11.2. – Earth Dam [22]. The behavior of an earth dam subject to ground motion is here studied under the assumption that its base is fixed to a perfectly rigid ground. This boundary condition, though unrealistic, was implemented to compare the results with a classical finite element solution [23]. The dimensions of the dam are shown in Fig. 11.5, where the finite element mesh used by Clough and Chopra [23] is also included.

Due to the boundary conditions, the rocking motion of the dam is expected to be negligible. Hence, the conventional shear wedge theory, in which the dam is idealized as a vertical shear beam of varying width, gives a reasonable approximation.

The boundary of the dam is divided into 30 constant boundary elements and 5 internal points are specified as shown in Fig. 11.6. Instead of horizontal ground accelerations applied to the dam base, linearly varying tractions in the opposite direction representing the inertia forces of the dam, are imposed on the upstream and downstream faces.

Figure 11.7 presents the variation of the horizontal motion at point B (node 20 in Fig. 11.6) over a frequency range of $\omega = 0$ to 9 rad/s. The fundamental

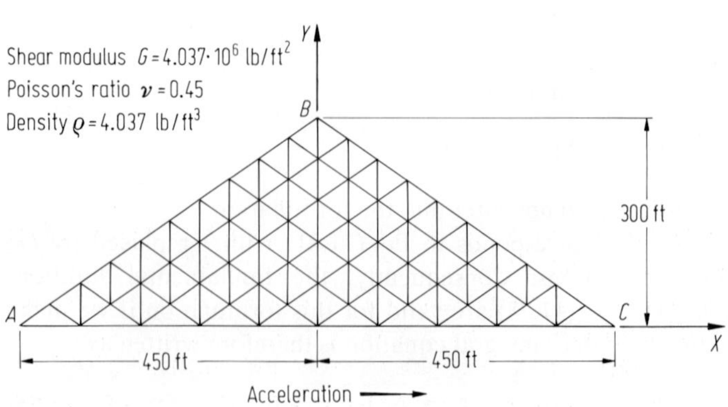

Fig. 11.5. Finite element mesh by Clough and Chopra

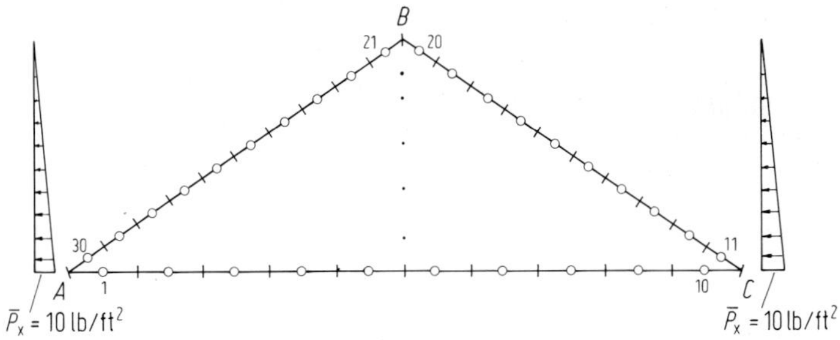

Fig. 11.6. Boundary element discretization

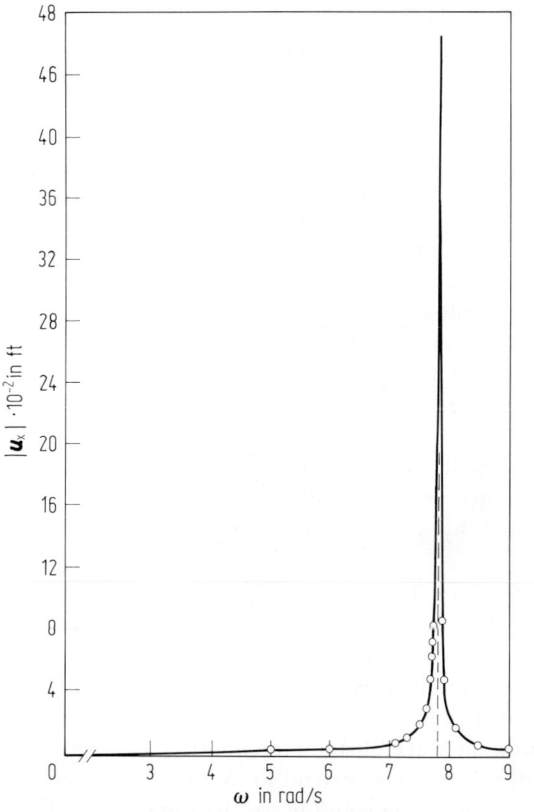

Fig. 11.7. Horizontal motion of point B versus frequency ω

frequency obtained from this plot is $\omega_1 = 7.78$ rad/s. The difference between this result and the finite element and shear wedge solutions is 0.9% and 2.89%, respectively (see Table 11.1). The greater discrepancy from the latter is expected, for the dam becomes stiffer under the constraint of no rocking motion.

The computed mode shape in normalized form is presented in Fig. 11.8, together with the result from the shear wedge theory for comparison. It can be seen that a good agreement has been obtained.

Table 11.1. Comparison of results for the fundamental frequency. ω in rad/s

BEM	FEM	Shear wedge theory
7.78	7.71	8.01

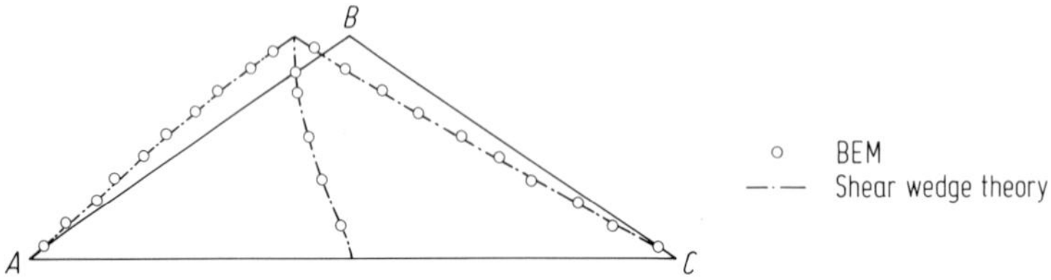

Fig. 11.8. Fundamental mode shape of earth dam

Example 11.3. – Dynamic Stiffness of Foundations [22]. In many engineering applications one needs to know the dynamic stiffness of the foundation. In most cases, the soil properties will not be uniform with depth – the shear modulus will normally increase – or a deposit of finite depth will be underlain by a much stiffer rock-like material. Nevertheless, the homogeneous isotropic linearly elastic half-space is a useful mathematical idealization and an application of the boundary element technique using this assumption was recently presented by Dominguez and Alarcón [9]. The results presented here have been found to be in close agreement with theirs.

The importance of boundary elements for modelling foundations is that it automatically takes into consideration the radiation damping and that the hysteretic damping can be simulated by simply introducing a complex shear modulus [24].

Part of the process in analyzing soil–structure interaction problems is the evaluation of the dynamic stiffnesses of foundations, i.e., the forces (or moments) required to produce unit dynamic motions (or rotations) of a massless rigid foundation on the half-space, with the other degrees of freedom being kept fixed.

Since the fundamental solution satisfies the infinite domain, only the boundary, or more precisely the soil–foundation interface plus the traction-free surface, need be discretized. In theory, the discretization should extend to infinity. However, for foundations resting on the surface, a reasonable solution can be obtained without any free-field element. It can be shown that all terms that represent the influence of the free-field elements are zero, except those responsible for the influence of the vertical motion on the horizontal, and vice versa. Nevertheless, this influence is small and is usually neglected in soil–structure interaction when "smooth footing" conditions are assumed.

In the present example, the computed results are compared with those by Jakub [25], whose solution takes the form

$$K = K_0(k + i\, a_0\, d)(1 + 2i\, D),$$

where K_0 is the static stiffness; k and d are frequency-dependent coefficients, $a_0 = \omega B/C_2$ is the dimensionless frequency; C_2 is the shear wave celerity of the soil, and D is the internal soil damping.

The dimension of the foundation (strip surface footing) and properties of the soil medium are

half-width of footing	$B = 1$,
shear modulus	$G = 1 + 0.1\,i$ (unless otherwise specified),
poisson's ratio	$\nu = 1/3$,
mass density	$\varrho = 1$.

To set a criterion of discretization, Fig. 11.9 shows the rocking stiffness versus the amount of free field discretized for a very low dimensionless frequency, $a_0 = 0.01$. The soil–footing interface was discretized into eight elements (Fig. 11.10) and as

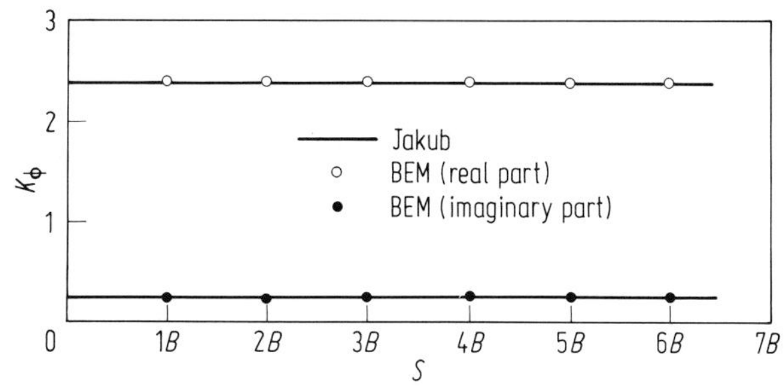

Fig. 11.9. Rocking stiffness K_φ versus extent of free-field discretized S ($a_0 = 0.01$)

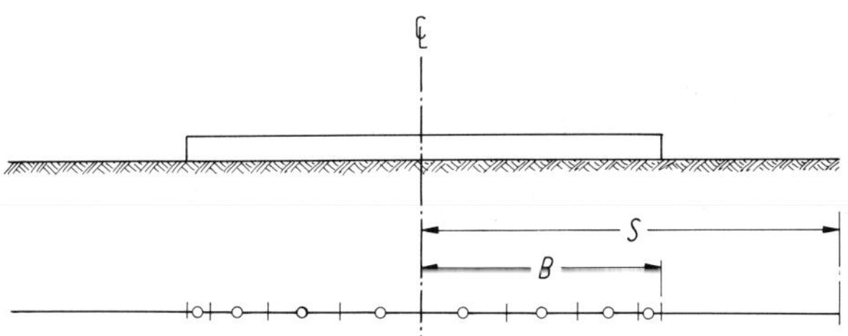

Fig. 11.10. Discretization of the half-space

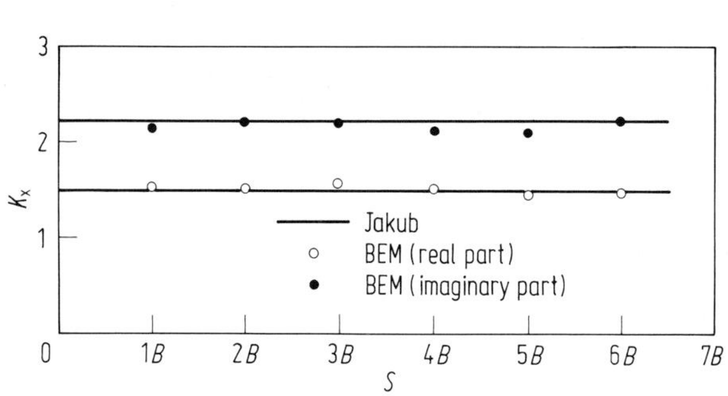

Fig. 11.11. Horizontal stiffness K_x versus extent of free-field discretized S ($a_0 = 0.9$)

expected, the influence of the extent of the free-field discretized is negligible. The same is true for higher frequencies. An analogous study for horizontal stiffness at frequency $a_0 = 0.9$ is shown in Fig. 11.11.

Figure 11.12 presents the real part of the free-surface motion when a unit harmonic motion (horizontal, rocking, and vertical) of frequency $a_0 = 1$ is applied to the footing. All these results are in complete agreement with those presented by Dominguez and Alarcón [9].

Figure 11.13 gives the variation of the rocking stiffness versus frequency for $G=1$. Although internal damping was neglected in this case, imaginary parts appear; this is due to the geometric or radiation damping, i.e., the energy radiated outwards and downwards towards the boundaries at infinity. The real parts decrease with an increase in a_0, which implies a reduction of the stiffness; while the imaginary parts increase, indicating an increase of the damping.

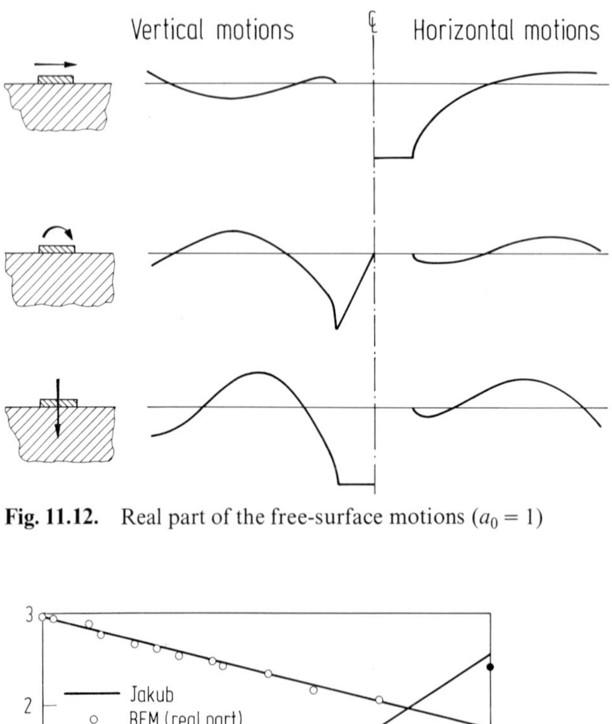

Fig. 11.12. Real part of the free-surface motions ($a_0 = 1$)

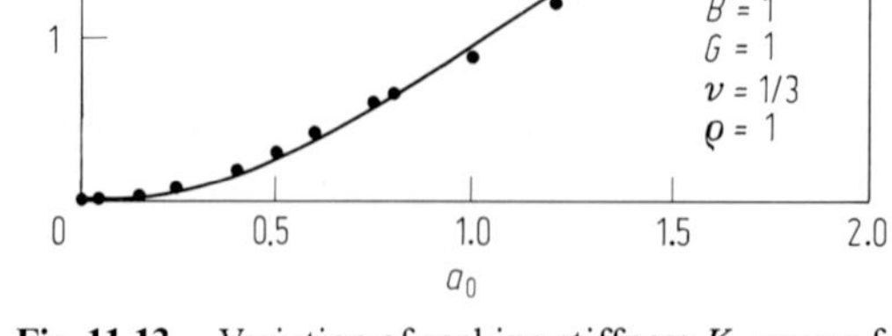

Fig. 11.13. Variation of rocking stiffness K_φ versus frequency a_0

11.6. Free Vibrations

In the formulation described in the last section the fundamental solution employed was found to be frequency dependent. Consequently, if our main interest is only the natural frequencies and modes, it is necessary to carry out the analysis by successively applying different forcing frequencies to the undamped system until resonance occurs. Since the complete set of equations has to be reassembled and solved for each frequency, the procedure is normally found to be time-consuming. The main disadvantage being that once the fundamental solution is itself frequency dependent, the analysis cannot be transformed into an algebraic eigenvalue problem.

An interesting alternative procedure for obtaining the natural frequencies and modes of structures has been presented by Nardini and Brebbia [26]. The formulation has the advantage of reducing the problem of free vibrations to an algebraic eigenvalue problem and is consequently more straightforward. The basic idea consists of simply adopting the static Kelvin fundamental solutions presented in Chapter 5, instead of the frequency-dependent ones introduced in the last section. This leads to the following boundary integral equation which corresponds to the governing equation (11.33) an absence of body forces [26]:

$$c_{ij}\,U_j + \int_\Gamma p^*_{ij}\,U_j\,d\Gamma = \int_\Gamma u^*_{ij}\,P_j\,d\Gamma + \omega^2\varrho \int_\Omega u^*_{ij}\,U_j\,d\Omega\,, \tag{11.37}$$

where the starred field corresponds to the well-known static Kelvin fundamental solution.

As it is seen, Eq. (11.37) involves not only the amplitudes of boundary displacements and tractions, but also the unknown displacements U_j within the domain appearing in the inertia term. Therefore, in order to formulate the problem in terms of boundary values only, a further approximation must be used for the internal displacement amplitudes. To this end, one can make use of a group of functions f^k (k denotes each member of the group; $k=1, N$), which can be multiplied by a set of unknown coefficients α^k_j, such that

$$U_j = \sum_{k=1}^{N} \alpha^k_j\, f^k, \tag{11.38}$$

leading to the following representation of the domain integral of Eq. (11.37):

$$\int_\Omega u^*_{ij}\,U_j\,d\Omega = \sum_{k=1}^{N} \alpha^k_j \int_\Omega u^*_{ij}\,f^k\,d\Omega\,. \tag{11.39}$$

Since our final objective is the transformation of Eq. (11.39) into equivalent boundary integrals, one can associate the group f^k of functions with the displacement fields ψ^k_{lj} and their corresponding stress fields τ^k_{ljm} in the form

$$\tau^k_{ljm,m} = \delta_{lj}\, f^k\,. \tag{11.40}$$

This allows the transformation of Eq. (11.39) into boundary integrals by following the procedures already presented for standard static analysis and gives as a final result

$$c_{ij} U_j + \int_\Gamma p_{ij}^* U_j \, d\Gamma - \int_\Gamma u_{ij}^* P_j \, d\Gamma$$

$$= \omega^2 \varrho \sum_{k=1}^{N} \left(- c_{ij} \psi_{lj}^k + \int_\Gamma u_{ij}^* t_{lj}^k \, d\Gamma - \int_\Gamma p_{ij}^* \psi_{lj}^k \, d\Gamma \right) \alpha_l^k, \tag{11.41}$$

where $t_{lj}^k = \tau_{ljm}^k n_m$ represents the tractions on the boundary corresponding to the displacement fields ψ_{lj}^k.

Equation (11.41) can now be discretized in standard fashion, the boundary integrals corresponding to the inertia term only include known expressions and could be computed analytically or numerically as usual. However, such a computation would represent a substantial effort since it requires integration over the entire boundary for each unknown coefficient α_l^k. Therefore, in order to minimize the computer time, the same interpolation functions already adopted to interpolate U_j and P_j can be used to represent the boundary variation of ψ_{lj}^k and t_{lj}^k, and consequently the same standard matrices $\boldsymbol{H}$ and $\boldsymbol{G}$ will be obtained. In addition, if the total number N of functions f^k is chosen to be the number of nodal points, the unknown coefficients α_l^k can be obtained as functions of the boundary displacement amplitudes in the form (see Eq. (11.38))

$$\boldsymbol{\alpha} = \boldsymbol{F}^{-1} \boldsymbol{U} \tag{11.42}$$

in which matrix $\boldsymbol{F}$ contains the values of f^k at the nodal points.

Equation (11.41) in discretized form, together with Eq. (11.42), leads to a generalized algebraic eigenvalue problem which, despite the approximation in the inertia terms, produce accurate results. In what follows one of the two-dimensional examples presented in Ref. [26] will be shown, this example was solved by using a group of functions of the form

$$f^k(x) = c - r(\xi_k, x),$$

where r is the distance between node ξ_k and the field point x, and c is a suitably chosen constant.

Example 11.4. – Shear Wall. Figure 11.14 shows a shear wall with four openings. The boundary element discretization consisted of 29 quadratic elements with 58 nodes. The results for the free vibration periods for the first eight natural modes are

Table 11.2. Periods of free vibrations for the two methods.

Mode	1	2	3	4	5	6	7	8
BEM	3.022	0.875	0.822	0.531	0.394	0.337	0.310	0.276
FEM	3.029	0.885	0.824	0.526	0.409	0.342	0.316	0.283

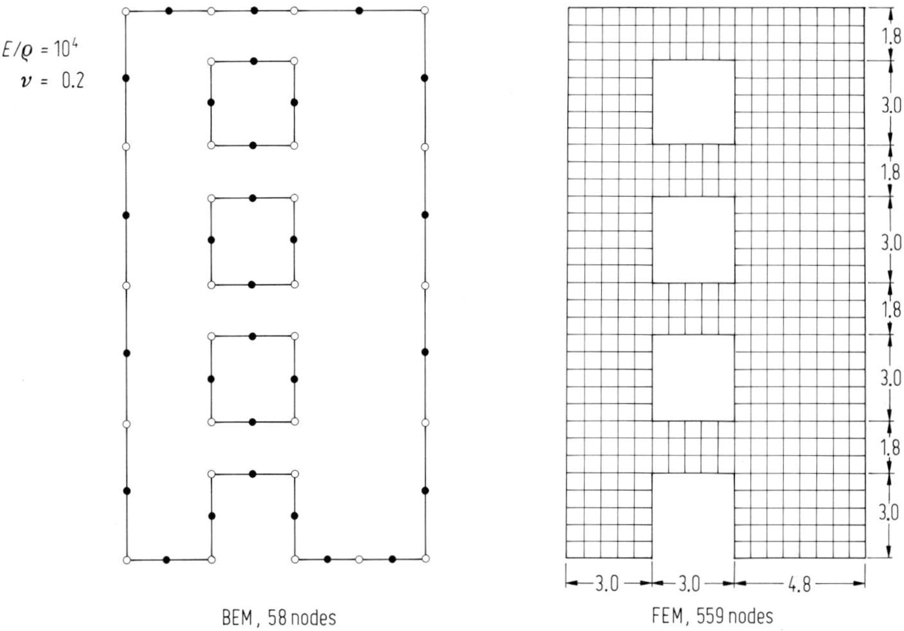

Fig. 11.14. Boundary element and finite element discretizations for shear wall

given in Table 11.2; also included are the finite element results obtained by using the SAP IV program with the mesh presented in Fig. 11.14 (559 nodes). In spite of the complicated geometry and the rather small number of boundary elements employed, the agreement of the solutions is very good.

References

1. Eringen, A. C., and Suhubi, E. S., *Elastodynamics Vol. II: Linear Theory*, Academic Press, London, 1975.
2. Achenbach, J. D., *Wave Propagation in Elastic Solids*, North–Holland, Amsterdam, 1973.
3. Miklowitz, J., *The Theory of Elastic Waves and Waveguides*, North–Holland, Amsterdam, 1978.
4. Cruse, T. A., and Rizzo, F. J., A direct formulation and numerical solution of the general transient elastodynamic problem, I, J. Math. Anal. Appl. **22,** 244–259 (1968).
5. Cruse, T. A., A direct formulation and numerical solution of the general transient elastodynamic problem, II, J. Math. Anal. Appl. **22,** 341–355 (1968).
6. Cole, D. M., Kosloff, D. D., and Minster, J. B., A numerical boundary integral equation method for elastodynamics, I, Bull. Seism. Soc. Amer. **68,** 1331–1357 (1978).
7. Manolis, G. D., and Beskos, D. E., Dynamic stress concentration studies by boundary integrals and Laplace transform, Int. J. Numerical Methods. Engng. **17,** 573–599 (1981).
8. Bellman, R. E., Kalaba, R. E., and Lockett, J., *Numerical Inversion of the Laplace Transform*, American Elsevier, New York, 1966.
9. Dominguez, J., and Alarcón, E., Elastodynamics, in *Progress in Boundary Element Methods Vol. 1*, (C. A. Brebbia, Ed.), Pentech Press, London, Halstead Press, N.Y., 1981.
10. Dominguez, J., Dynamic stiffness of rectangular foundations, M.I. T. Research Report No. R-78-20, Civil Engineering Department, 1978.

11. Hutchinson, R., Determination of membrane vibrational characteristics by the boundary integral equation method, in *Recent Advances in Boundary Element Methods* (C. A. Brebbia, Ed.), pp. 301–315, Southampton, Pentech Press, London, 1978.
12. Niwa, Y., Kobayashi, S., and Kitahara, M., Eigenfrequency analysis of a plate by the integral equation method, in *Theoretical and Applied Mechanics*, Vol. 29, pp. 287–306, University of Tokyo Press, Tokyo, 1981.
13. Rayleigh, J. W. S., *The Theory of Sound*, Dover, New York, 1945.
14. Love, A. E. H., *A Treatise on the Mathematical Theory of Elasticity*, Dover, New York, 1944.
15. Mansur, W. J., Time stepping scheme to solve transient wave propagation problems using the Boundary Element Method. Ph.D. Thesis, University of Southampton, 1983.
16. Doyle, J. M., Integration of the Laplace transformed equations of classical elastokinetics, J. Math. Anal. Appl. **13** (1966).
17. Watson, G. N., *A Treatise on the Theory of Bessel Functions*, MacMillan, New York, 1944.
18. Abramowitz, M., and Stegun, I. R., *Handbook of Mathematical Functions*, Dover, New York, 1965.
19. Papoulis, A., A new method of inversion of the Laplace transform, Quart. Appl. Math. **14,** 405–414 (1957).
20. Durbin, F., Numerical inversion of Laplace transforms: An efficient improvement to Dubner and Abate's method, Computer J. **17,** 371–376 (1974).
21. Bathe, K. J., Wilson, E. L., and Peterson, F. E., SAP IV, A structural analysis program for static and dynamic response of linear systems, Report No. EERC 73-11, University of California, Berkeley, 1973.
22. Chuang, P. H., Application of boundary element methods in elastodynamics, M.Sc. dissertation, University of Southampton, 1981.
23. Clough, R. W., and Chopra, A. K., Earthquake stress analysis in earth dams, Proc. ASCE J. Appl. Mech. **92,** 197–211 (1966).
24. Flügge, W., *Viscoelasticity*, Blaisdell Publ. Co., Waltham, Mass, 1967.
25. Jakub, M., Dynamic stiffness of foundations: 2-D vs 3-D solutions, M.I.T. Research Report No. R 77-36, Civil Engineering Department, 1977.
26. Nardini, D., and Brebbia, C. A., A new approach to free vibration analysis using boundary elements, in *Boundary Element Methods in Engineering* (C. A. Brebbia, Ed.), Springer-Verlag, Berlin, 1982.

Chapter 12

Further Applications in Fluid Mechanics

12.1. Introduction

Some applications of the boundary element method in fluid mechanics have already been discussed in Chapters 2–4 and 9. However, applications of the method in this field are by no means restricted to the cases treated in those chapters. In fact, a wide variety of fluid mechanics problems, some of which involving rather complex features such as nonlinearities, moving boundaries, etc., have been successfully dealt with using boundary elements.

In this chapter, we summarize some of the most important BEM applications in the above-mentioned field, including numerical results that demonstrate the efficiency of the formulations developed.

12.2. Transient Groundwater Flow

Applications of the BEM to transient groundwater flow problems have been carried out by Liggett [1], Liu and Liggett [2], and Lennon *et al.* [3, 4]. These problems are governed by a Laplace equation for the velocity potential u with a kinematic (nonlinear) boundary condition on the free surface as follows [5]:

$$\frac{\partial \eta}{\partial t} = q_1 \frac{\partial \eta}{\partial x_1} - q_2, \tag{12.1}$$

where q_1 and q_2 are the velocities in the x_1 and x_2 directions and η is the elevation of the free surface with relation to an arbitrary plane (Fig. 12.1).

From geometric considerations we have that

$$\frac{\partial \eta}{\partial x_1} = -\tan \beta \tag{12.2}$$

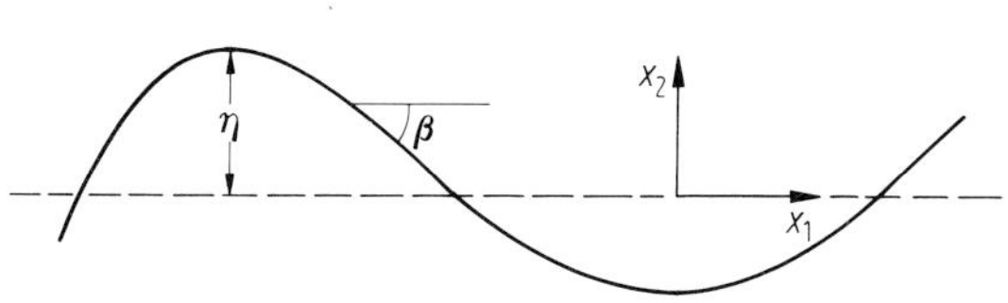

Fig. 12.1. Free-surface elevation with relation to the x_1 axis

in which β is the angle the free surface makes with the x_1 axis. Hence,

$$\frac{\partial \eta}{\partial t} = -\frac{q}{\cos\beta}, \tag{12.3}$$

where $q = \partial u/\partial n$ is the normal velocity. Applying the condition $u = \eta$ at the free surface, Eq. (12.3) becomes

$$\frac{\partial u}{\partial t} = -\frac{q}{\cos\beta}. \tag{12.4}$$

This equation can be written in finite difference form as

$$u^{t+\Delta t} = u^t - \frac{\Delta t}{\cos\beta^t}\left[\theta\, q^{t+\Delta t} + (1-\theta)\, q^t\right], \tag{12.5}$$

where θ is a weighting factor that positions the derivative between the time levels t and $t+\Delta t$. In the equation, the angle β is computed at time t even though the equation is written for the time $t+\Delta t$. Although this problem can be avoided by iteration, the use of a small time step provides sufficient accuracy [1].

As an example of how the free-surface boundary condition in the form of Eq. (12.5) can be introduced into the system of equations (2.81) consider the problem represented by Fig. 3.13 where we assume that there was a drawdown in the downstream water level. System (2.81) can then be rearranged for this problem as

$$[-\boldsymbol{G}_{\mathrm{ABC}} - \boldsymbol{G}_{\mathrm{CD}} - \boldsymbol{G}_{\mathrm{DE}} - \boldsymbol{G}_{\mathrm{EF}}\ \boldsymbol{H}_{\mathrm{AF}}] \begin{Bmatrix} \boldsymbol{q}_{\mathrm{ABC}} \\ \boldsymbol{q}_{\mathrm{CD}}^{t+\Delta t} \\ \boldsymbol{q}_{\mathrm{DE}} \\ \boldsymbol{q}_{\mathrm{EF}} \\ \boldsymbol{u}_{\mathrm{AF}} \end{Bmatrix}$$

$$= [-\boldsymbol{H}_{\mathrm{ABC}} - \boldsymbol{H}_{\mathrm{CD}} - \boldsymbol{H}_{\mathrm{DE}} - \boldsymbol{H}_{\mathrm{EF}}\ \boldsymbol{G}_{\mathrm{AF}}] \begin{Bmatrix} \boldsymbol{u}_{\mathrm{ABC}} \\ \boldsymbol{u}_{\mathrm{CD}}^{t+\Delta t} \\ \boldsymbol{u}_{\mathrm{DE}} \\ \boldsymbol{u}_{\mathrm{EF}} \\ \boldsymbol{q}_{\mathrm{AF}} \end{Bmatrix} \tag{12.6}$$

Substituting $u_{\mathrm{CD}}^{t+\Delta t}$ by its value on Eq. (12.5) yields

$$\left[-\boldsymbol{G}_{\mathrm{ABC}} - \left(\boldsymbol{G}_{\mathrm{CD}} + \boldsymbol{H}_{\mathrm{CD}}\frac{\theta\,\Delta t}{\cos\beta^t}\right) - \boldsymbol{G}_{\mathrm{DE}} - \boldsymbol{G}_{\mathrm{EF}}\ \boldsymbol{H}_{\mathrm{AF}}\right] \begin{Bmatrix} \boldsymbol{q}_{\mathrm{ABC}} \\ \boldsymbol{q}_{\mathrm{CD}}^{t+\Delta t} \\ \boldsymbol{q}_{\mathrm{DE}} \\ \boldsymbol{q}_{\mathrm{EF}} \\ \boldsymbol{u}_{\mathrm{AF}} \end{Bmatrix}$$

$$= [-\boldsymbol{H}_{\mathrm{ABC}} - \boldsymbol{H}_{\mathrm{CD}} - \boldsymbol{H}_{\mathrm{DE}} - \boldsymbol{H}_{\mathrm{EF}}\ \boldsymbol{G}_{\mathrm{AF}}] \begin{Bmatrix} \boldsymbol{u}_{\mathrm{ABC}} \\ \boldsymbol{u}_{\mathrm{CD}}^{t} - \dfrac{(1-\theta)\,\Delta t}{\cos\beta^t}\boldsymbol{q}_{\mathrm{CD}}^{t} \\ \boldsymbol{u}_{\mathrm{DE}} \\ \boldsymbol{u}_{\mathrm{EF}} \\ \boldsymbol{q}_{\mathrm{AF}} \end{Bmatrix}. \tag{12.7}$$

Since all boundary values on the right-hand side of Eq. (12.7) are known, the system of equations can be solved and the normal velocities along the free surface at time $t+\Delta t$ computed. Condition (12.5) is then employed to find the potential values at the free surface and the computation cycle is completed, so that the solution can be advanced on time.

Example 12.1. This example, taken from [1], studies the transient free-surface flow through the block of soil depicted in Fig. 12.2. At time $t=0$, we assume that there was a sudden drawdown from height 10 to height 3 on the downstream water level. Thus, a seepage surface appears on the downstream face and the boundary conditions of the problem are as follows (see Example 3.5): (a) $u=10$ on the upstream face; (b) $q=0$ on the bottom surface; (c) $u=3$ on the downstream face; (d) $u=x_2$ on the seepage surface; and (e) the kinematic boundary condition (12.4) on the free surface.

Results for successive times employing the discretization shown in Fig. 12.2 (24 linear elements) are presented in Fig. 12.3. The line labelled $t=\infty$ results from both a steady-state computation using the procedure described in Example 3.5 and

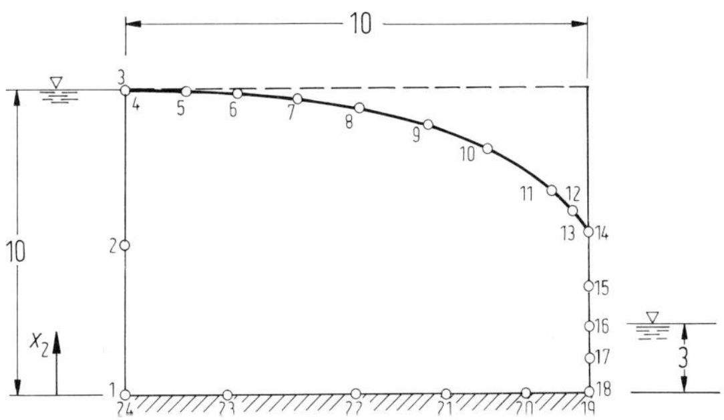

Fig. 12.2. Geometry and discretization of soil block

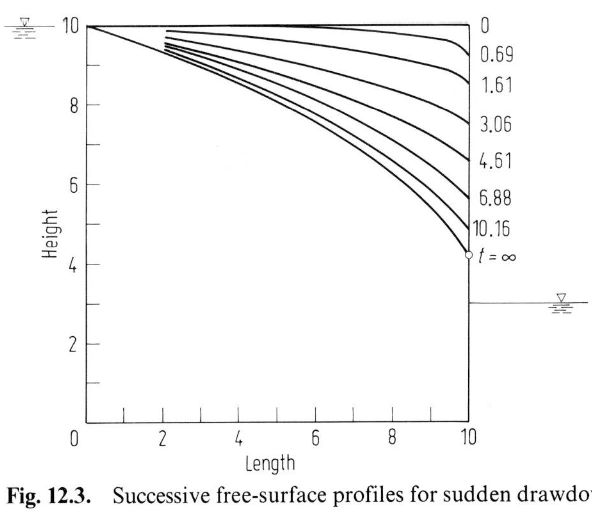

Fig. 12.3. Successive free-surface profiles for sudden drawdown problem

the unsteady computation carried to $t = 30$. The accuracy of the results can be judged from the exact solution given in [6]: the indicated point in Fig. 12.3 shows the exact free-surface–seepage-surface intersection, which agrees well with the computation.

Example 12.2. Groundwater flow problems considering recharge of fluid have been treated using the BEM in [2]. These problems are similar to the previously discussed ones except that now the free-surface boundary condition is written

$$\frac{\partial u}{\partial t} = -\frac{q}{\cos\beta} + W, \tag{a}$$

where W is the recharge intensity. The finite difference analog of this equation is

$$u^{t+\Delta t} = u^t - \frac{\Delta t}{\cos\beta^t}\left[\theta\, q^{t+\Delta t} + (1-\theta)\, q^t\right] + \Delta t\left[\theta\, W^{t+\Delta t} + (1-\theta)\, W^t\right]. \tag{b}$$

The problem of recharge in the Hele–Shaw model of Fig. 12.4, studied experimentally by Marino [7], was solved by the BEM with linear elements and the discretization shown in that figure [2]. The bottom surface and the right and left walls are all impermeable surfaces on which $q = 0$. Initially the fluid was at rest with a level free surface. At time $t = 0$, a uniform recharge rate of 0.056 cm/s was introduced in the leftmost 23.8 cm. The weighting factor θ in equation (b) was taken as 0.7.

Numerical results for the free-surface location at some time levels are plotted in Fig. 12.5. A time step value of $\Delta t = 0.5$ was adopted, this value being a dimen-

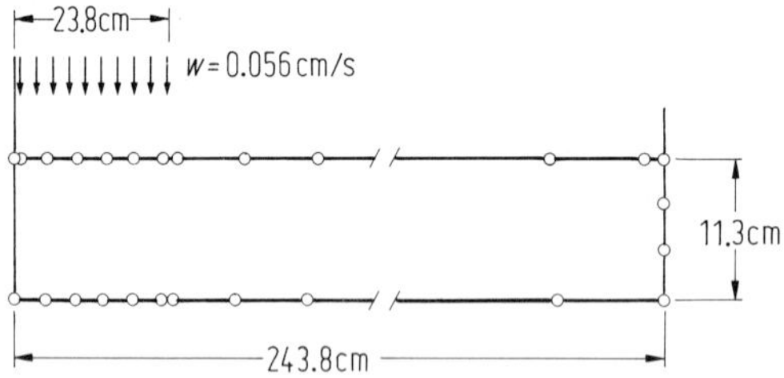

Fig. 12.4. Sketch of the Hele–Shaw model, showing BEM discretization

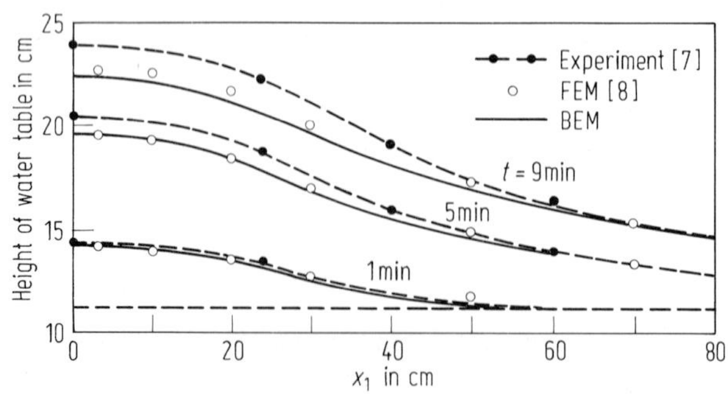

Fig. 12.5. Free-surface profiles at some time levels

sionless one since all variables in the analysis have been nondimensionalized (see [2] for details). Results obtained by Newman *et al.* [8] using the finite element method are also plotted in Fig. 12.5. A good agreement was obtained between the two numerical solutions and also with the experimental curves which indicate a gain in mass (i.e., fluid in the model is excess of the original fluid plus the accumulated recharge) of about 10%, whereas the mass is always conserved in the numerical solutions.

Extension of the previous formulations to axisymmetric and three-dimensional problems have been carried out in [3, 4], respectively. Also, an interesting analysis of recharge to a semi-infinite aquifer, where analytical far-field solutions are utilized as shape functions for an "infinite boundary element" extending from the limit of the significant region up to infinity, was reported in [2].

12.3. Moving Interface Problems

Liu *et al.* [9] employed the BEM to study problems concerning a moving interface between two fluids in porous media, utilizing the so-called "sharp interface model" which assumes that the mixing between the fluids is insignificant and that the thickness of the transition layer is small in comparison with a characteristic length scale in the primary flow direction.

Note that the steady-state interfacial problem (i.e., assuming that the interface has reached a steady position) can be dealt with by using the subregions method described in Section 2.8. The present formulation follows the same idea but with a kinematic interfacial condition which prescribes the movement of the (unknown) interface between the fluids at any instant of time.

In order to formulate the actual problem mathematically, we divide the flow domain into two subregions Ω^1 and Ω^2 occupied by fluids 1 and 2, respectively. The fluid regions are separated by a sharp interface, $x_1 = \eta(x_2, t)$. The boundary-value problem can then be described by Laplace's equation for a potential u (the piezometric head),

$$\nabla^2 u^i = 0 \qquad \text{in } \Omega^i \, (i = 1, 2) \tag{12.8}$$

with boundary conditions of the Dirichlet and Newmann types prescribed along the external surfaces Γ^1 and Γ^2 (Fig. 12.6) and the interfacial boundary conditions at $x_1 = \eta(x_2, t)$:

$$\frac{\partial \eta}{\partial t} = q_2^1 \frac{\partial \eta}{\partial x_2} - q_1^1 = \frac{s}{\alpha}\left(q_2^2 \frac{\partial \eta}{\partial x_2} - q_1^2\right), \tag{12.9}$$

$$s\, u^2 - u^1 = (s - 1)\, x_2 , \tag{12.10}$$

where $\alpha = \mu^2/\mu^1$ and $s = \varrho^2/\varrho^1$, μ^i being the dynamic viscosity and ϱ^i the density of the fluids. To obtain the transient solutions, initial conditions (the initial location of the interface and the initial potential distribution) are also required.

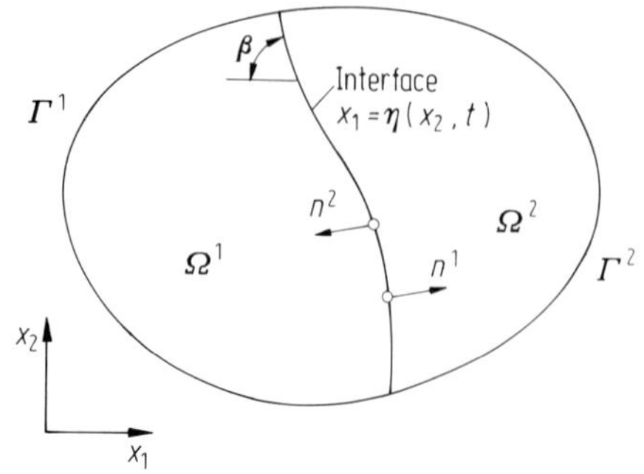

Fig. 12.6. Notation for moving interface problem

For convenience of the numerical computation, the kinematic boundary condition (12.9) can be rewritten

$$\frac{\partial \eta}{\partial t} = \frac{q^1}{\sin \beta} = -\frac{s}{\alpha} \frac{q^2}{\sin \beta}, \tag{12.11}$$

where $\partial \eta / \partial x_2 = \cot \beta$, $q^1 = \partial u^1 / \partial n^1$ and $q^2 = \partial u^2 / \partial n^2$ (see Fig. 12.6).

In order to show how the interface boundary conditions (12.10) and (12.11) can be introduced into the system of equations (2.81), Eqs. (2.117) and (2.118) for subregions 1 and 2, respectively, can be combined in the form

$$\begin{bmatrix} \boldsymbol{H}^1 & \boldsymbol{H}_I^1 & \boldsymbol{0} & \boldsymbol{0} \\ \boldsymbol{0} & \boldsymbol{0} & \boldsymbol{H}^2 & \boldsymbol{H}_I^2 \end{bmatrix} \begin{Bmatrix} \boldsymbol{U}^1 \\ \boldsymbol{U}_I^1 \\ \boldsymbol{U}^2 \\ \boldsymbol{U}_I^2 \end{Bmatrix} = \begin{bmatrix} \boldsymbol{G}^1 & \boldsymbol{G}_I^1 & \boldsymbol{0} & \boldsymbol{0} \\ \boldsymbol{0} & \boldsymbol{0} & \boldsymbol{G}^2 & \boldsymbol{G}_I^2 \end{bmatrix} \begin{Bmatrix} \boldsymbol{Q}^1 \\ \boldsymbol{Q}_I^1 \\ \boldsymbol{Q}^2 \\ \boldsymbol{Q}_I^2 \end{Bmatrix}. \tag{12.12}$$

By imposing the interface boundary conditions (12.10) and (12.11) and remembering that both the potentials and fluxes at the interface are unknowns, the system (12.12) can be reordered as

$$\begin{bmatrix} \boldsymbol{H}^1 & \boldsymbol{H}_I^1 & -\boldsymbol{G}_I^1 & \boldsymbol{0} \\ \boldsymbol{0} & (1/s)\,\boldsymbol{H}_I^2 & (\alpha/s)\,\boldsymbol{G}_I^2 & \boldsymbol{H}^2 \end{bmatrix} \begin{Bmatrix} \boldsymbol{U}^1 \\ \boldsymbol{U}_I^1 \\ \boldsymbol{Q}_I^1 \\ \boldsymbol{U}^2 \end{Bmatrix} = \begin{bmatrix} \boldsymbol{G}^1 & \boldsymbol{0} & \boldsymbol{0} \\ \boldsymbol{0} & \boldsymbol{G}^2 & \boldsymbol{H}_I^2 \end{bmatrix} \begin{Bmatrix} \boldsymbol{Q}^1 \\ \boldsymbol{Q}^2 \\ (1/s - 1)\,\boldsymbol{x}_{2I} \end{Bmatrix}. \tag{12.13}$$

According to the prescribed boundary conditions along Γ^1 and Γ^2, the submatrices corresponding to $\boldsymbol{U}^1$ and $\boldsymbol{Q}^1$, $\boldsymbol{U}^2$ and $\boldsymbol{Q}^2$ may interchange their positions. Notice that the final system matrix in Eq. (12.13) is banded.

Upon solving the above system of linear equations, both the potentials and fluxes at the interface Γ_I^1 become known. Then, values of $\boldsymbol{U}_I^2$ and $\boldsymbol{Q}_I^2$ along Γ_I^2 can be calculated from Eqs. (12.10) and (12.11). Furthermore, the new location of the interface can be found by using the finite difference analog of Eq. (12.11) i.e.,

$$\eta^{t+\Delta t} = \eta^t + \frac{1}{\sin \beta^t} [\theta\, q_I^{1(t)} + (1-\theta)\, q_I^{1(t+\Delta t)}]. \tag{12.14}$$

The procedure can be repeated to find the time history of the interface movement.

Example 12.3. A typical sea water intrusion problem in a confined coastal aquifer was studied by Liu *et al.* [9]. The aquifer was assumed to have constant thickness, a horizontal bottom, and a known fresh water discharge to the sea. At time $t = 0$ the discharge was changed from that of the initial equilibrium value and the movement of the sea water wedge was observed. Bear and Dagan [10] performed experimental studies of such a problem in a Hele-Shaw model; a definition sketch of their model is presented in Fig. 12.7.

The interface between the fluids was discretized into 11–12 linear boundary elements, while 30–35 elements were used to represent the external boundaries. The far end of the salt water aquifer is cut at a distance $x_1 = 200$ cm and either a constant potential condition or the no-flux condition is applied there, as the seaward flow has only a negligible effect at such a distance. The fresh water aquifer is extended 400 cm inland and a constant flux condition is prescribed. The initial steady-state profile was obtained through a full transient analysis [9]. At time $t = 0$ a sudden change of discharge was imposed at the inland end of the

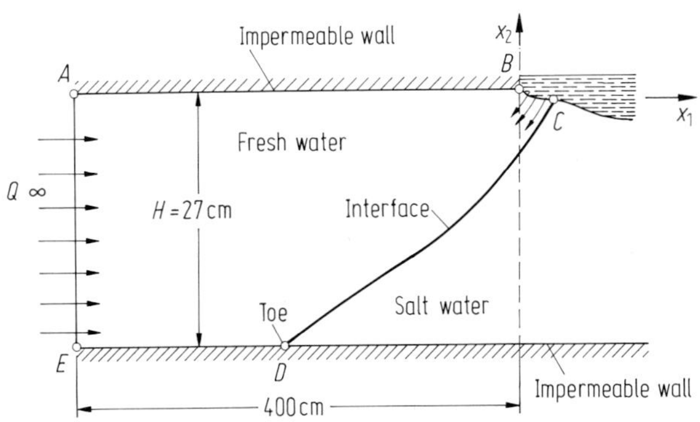

Fig. 12.7. Sketch of the salt water intrusion experiment

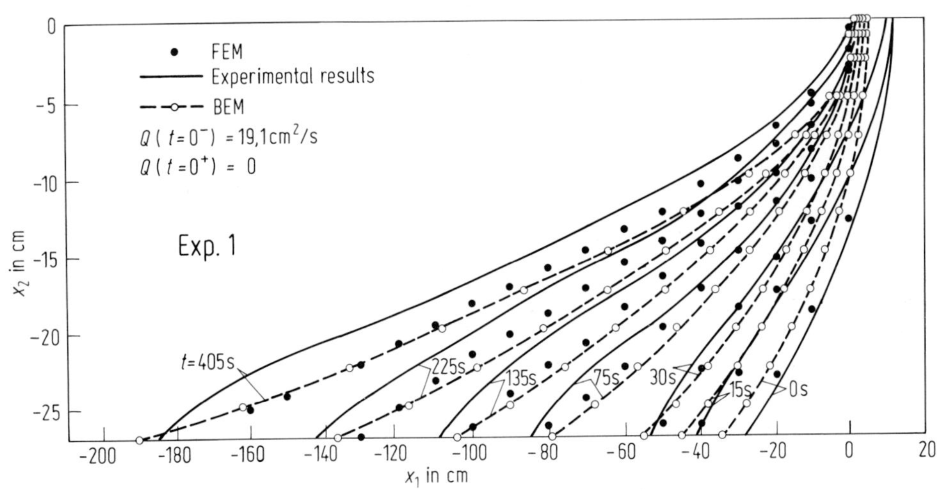

Fig. 12.8. Advancing salt water wedge

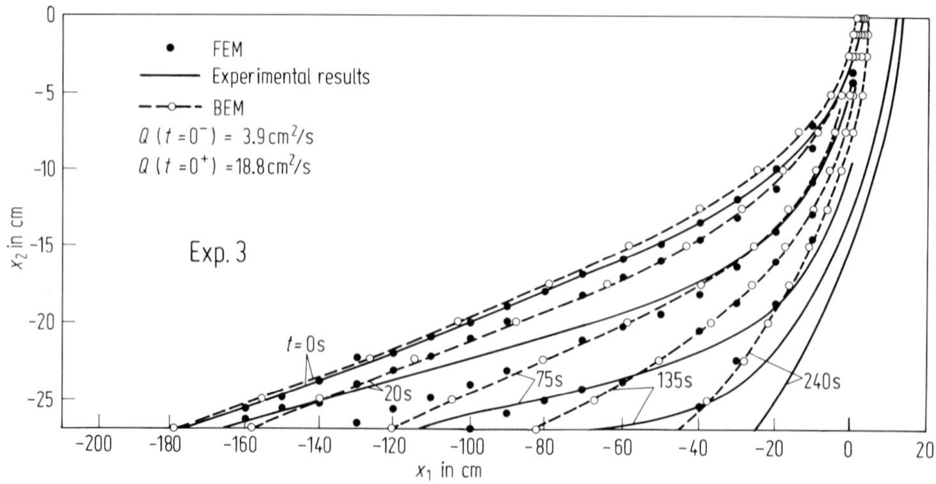

Fig. 12.9. Retreating salt water wedge

aquifer and the subsequent motion of the interface was recorded. Some 40 to 50 time steps were used to cover the duration of Bear and Dagan's experiments [10].

The time histories of the interfaces for two different analyses, one for an advancing salt water wedge (experiment 1 of [10]) and the other for a receding one (experiment 3), are plotted in Figs. 12.8 and 12.9 respectively, compared with the experimental results [10] and a finite element solution [11] which employs the Dupuit–Forchheimer approximation.

Note that a special treatment was devoted to the singular points B and C (Fig. 12.7): singular boundary elements [12] were employed around the point B to improve the accuracy of computation, while a perturbation solution [12, 13] for the region near point C was used to determine its location.

12.4. Axisymmetric Bodies in Cross Flow

The problem of (potential) cross flow about an axisymmetric body immersed in a uniform stream perpendicular to the axis of symmetry of the body was studied by Hess and Smith [14]. In this case, the velocity potential u and its normal derivative q are both proportional to the cosine of the circumferential angle, where this angle is measured from the direction of the uniform stream [15]. The problem of pure cross flow corresponds to the case $n = 1$ (where only an even Fourier expansion is employed) of the general problem of axisymmetric bodies under arbitrary boundary conditions. Combination of this flow with the axisymmetric flow about the same body (the case $n = 0$ discussed in Section 2.13) gives the flow at any angle of attack.

The values of u and q at any point in space may be related to their values at the point in the R-Z plane (assuming the uniform stream to be parallel to the R axis) having the same axial and radial location as the point in question as (refer to

Eq. (2.170) for notation)

$$u(x) = u(\bar{x}) \cos\theta(x)\,,$$
$$q(x) = q(\bar{x}) \cos\theta(x)\,. \tag{12.15}$$

The boundary integral equation for this problem, equivalent to Eq. (2.154) (see Section 2.13), may be written

$$c(\xi)\,u(\xi) + \int_{\bar{\Gamma}} u(\bar{x}) \int_{-\pi}^{\pi} q^*(\xi, x) \cos\theta(x)\, d\theta(x)\, R(x)\, d\bar{\Gamma}(\bar{x})$$
$$= \int_{\bar{\Gamma}} q(\bar{x}) \int_{-\pi}^{\pi} u^*(\xi, x) \cos\theta(x)\, d\theta(x)\, R(x)\, d\bar{\Gamma}(\bar{x})\,. \tag{12.16}$$

The fundamental solution, which can be interpreted as a ring source whose intensity varies as a cosine of the circumferential angle, is given by

$$\bar{u}^*(\bar{\xi}, \bar{x}) = \int_{-\pi}^{\pi} u^*(\xi, x) \cos\theta(x)\, d\theta(x)\,. \tag{12.17}$$

Analogously, we have

$$\bar{q}^*(\bar{\xi}, \bar{x}) = \int_{-\pi}^{\pi} q^*(\xi, x) \cos\theta(x)\, d\theta(x)\,. \tag{12.18}$$

The above integrals can be obtained explicitly in terms of the complete elliptic integrals of the first and second kinds $K(m)$ and $E(m)$ as

$$\bar{u}^*(\bar{\xi}, \bar{x}) = \frac{4}{(a+b)^{1/2}} \left[\frac{2}{m}(K - E) - K\right], \tag{12.19}$$

$$\bar{q}^*(\bar{\xi}, \bar{x}) = \frac{4}{b(a+b)^{1/2}} \left\{\frac{1}{R(x)} \left[\frac{l^2 + c\,R^2(x)}{a-b} E - lK\right] n_R(x)\right.$$
$$\left. + [Z(\xi) - Z(x)] \left(\frac{a}{a-b} E - K\right) n_z(x)\right\}, \tag{12.20}$$

where

$$a = R^2(\xi) + R^2(x) + [Z(\xi) - Z(x)]^2\,,$$
$$b = 2R(\xi)\,R(x)\,,$$
$$c = [Z(\xi) - Z(x)]^2 - R^2(\xi)\,, \tag{12.21}$$
$$l = [Z(\xi) - Z(x)]^2 + R^2(\xi)\,,$$
$$m = \frac{2b}{\mathrm{a} + b}\,.$$

Substituting expressions (12.19) and (12.20) into Eq. (12.16) results in a boundary integral equation whose solution provides the unknown values of u and q on the R-Z plane. Boundary values of u and q at any point along the circumferential direction can then be computed through Eqs. (12.15).

Integration over each boundary element is carried out in the same manner as in the axisymmetric case (Section 2.13). Evaluation of the singular integrals can also be effected in a similar way, i.e., by writing the elliptic integrals in terms of Legendre functions of the second kind and using expansions of these functions for small values of their argument.

12.5. Slow Viscous Flow (Stokes Flow)

The classical Stokes flow problem of determining the steady slow viscous (creeping) flow of an inertialess unbounded incompressible fluid past an obstacle was studied by Youngren and Acrivos [16] using the BEM. The problem is mathematically described by the following equations and boundary conditions [17, 18]

$$u_{i,jj} = p_{,i}\,, \tag{12.22}$$

$$u_{i,i} = 0\,, \tag{12.23}$$

$$u_i = -U_i \quad \text{at the body surface,} \tag{12.24}$$

$$u_i, p \to 0 \quad \text{at infinity,} \tag{12.25}$$

where u_i is the velocity vector, p is pressure, and U_i is the onset velocity vector. In the above equations, all variables are dimensionless and the Cartesian tensor notation defined in Chapter 5 is employed.

Equation (12.22) is the equation of motion which has the form of a vector Poisson's equation, while Eq. (12.23) is the continuity equation. Expressions (12.24) and (12.25) represent the boundary conditions at the body surface (the no-slip condition) and at infinity, respectively. The fundamental solution to this problem is given by [16, 19]

$$u^*_{ij} = \frac{1}{8\pi r}(\delta_{ij} + r_{,i}\,r_{,j})\,, \tag{12.26}$$

$$p^*_j = \frac{r_{,j}}{4\pi r^2} \tag{12.27}$$

in which the notation defined in Eq. (5.57) applies.

By using the divergence theorem [20], we can obtain what might be called Green's formula for the Stokes problem for smooth solenoidal vectors u_i and v_i, and smooth scalars p and q in the bounded domain Ω with boundary Γ,

$$\begin{aligned}&\int_\Omega \{u_i(x)\,[v_{i,jj}(x) - q_{,i}(x)] - v_i(x)\,[u_{i,jj}(x) - p_{,i}(x)]\}\,d\Omega\,(x)\\ &= \int_\Gamma \{u_i(x)\,T_{ij}[v_i(x)]\,n_j(x) - v_i(x)\,T_{ij}[u_i(x)]\,n_j(x)\}\,d\Gamma\,(x)\,,\end{aligned} \tag{12.28}$$

where T_{ij} is the shear stress tensor, defined as

$$T_{ij}[u_i] = -\delta_{ij}\,p + u_{i,j} + u_{j,i}\,,$$
$$T_{ij}[v_i] = -\delta_{ij}\,q + v_{i,j} + v_{j,i}\,. \tag{12.29}$$

By replacing v_i and q with the fundamental solutions u^*_{ki} and p^*_k, identifying u_i and p with the solution to Eqs. (12.22) and (12.23) and using the facts that $u_i(x) = O(|x|^{-1})$ and $p(x) = O(|x|^{-2})$ as $|x| \to \infty$ (see Sections 2.10 and 5.6), we obtain the equivalent of Green's third identity [20] for the Stokes problem, i.e.,

$$u_i(\xi) + \int_\Gamma q^*_{kji}(\xi, x)\, u_k(x)\, n_j(x)\, d\Gamma(x) = \int_\Gamma u^*_{ik}(\xi, x)\, T_{kj}[u_i(x)]\, n_j(x)\, d\Gamma(x)\,, \tag{12.30}$$

$$p(\xi) + \int_\Gamma t^*_{jk}(\xi, x)\, u_k(x)\, n_j(x)\, d\Gamma(x) = \int_\Gamma p^*_k(\xi, x)\, T_{kj}[u_i(x)]\, n_j(x)\, d\Gamma(x)\,, \tag{12.31}$$

where

$$q^*_{ijk} = \frac{-3}{4\pi}\,\frac{r_{,i}\,r_{,j}\,r_{,k}}{r^2}\,, \tag{12.32}$$

$$t^*_{jk} = \frac{1}{2\pi r^3}\,(\delta_{jk} - 3 r_{,j}\, r_{,k})\,. \tag{12.33}$$

Note that the integral equations (12.30) and (12.31) can also be derived through integration by parts. It is also interesting to point out that expressions (12.26) and (12.32) are identical to the fundamental solution and stresses (5.55) and (5.60) for an incompressible solid.

Taking the point ξ in Eq. (12.30) to the boundary, accounting for the jump of the integral in q^* and the no-slip boundary condition (12.24), we obtain the following boundary integral equation:

$$c_{ij}(\xi)\, U_j(\xi) + \int_\Gamma q^*_{kji}(\xi, x)\, U_k(x)\, n_j(x)\, d\Gamma(x) = -\int_\Gamma u^*_{ik}(\xi, x)\, f_k(x)\, d\Gamma(x) \tag{12.34}$$

in which the only unknowns are the local surface stress forces $f_k(x) = T_{kj}[u_i(x)]\, n_j(x)$. Thus, when the Stokes problem is formulated in this manner, the stress force distribution, which is normally the quantity of interest in such calculations, is determined directly. Furthermore, the pressure at any point can then be calculated by using Eq. (12.31).

Equations (12.34) simplify for the case of uniform flow at infinity. Specifically, if $U_i(x)$ is a constant W_i, then it can be shown using Eq. (12.28) that [16]

$$\int_\Gamma q^*_{kji}(\xi, x)\, W_k\, n_j(x)\, d\Gamma(x) = \tfrac{1}{2}\, W_i \tag{12.35}$$

and hence Eq. (12.34) reduces to

$$\tfrac{1}{2}\, W_i + c_{ij}(\xi)\, W_j = -\int_\Gamma u^*_{ik}(\xi, x)\, f_k(x)\, d\Gamma(x)\,. \tag{12.36}$$

The numerical solution of Eqs. (12.34) or (12.36) follows basically the same steps as described in Sections 2.12 and 5.7–5.9. Details of the calculations are provided in [16]. Simplification of Eq. (12.34) for the case of axisymmetric flow is also discussed in [16], where the axisymmetric fundamental solutions are given explicitly in terms of the complete elliptic integrals K and E.

Example 12.4. We reproduce in this example the results obtained by Youngren and Acrivos [16] for the axisymmetric flow problem of a spheroid, given by $x_1^2 + (x_2^2 + x_3^2)/a^2 = 1$, translating in the x_1 direction.

Figure 12.10 presents a comparison between the surface stress forces obtained by discretizing the surface of the spheroid into N equally sized constant boundary elements and the analytical solution. It can be seen that the greatest errors in f_1 occur, as expected, in regions where the gradient of f_1 is the largest (the analytical

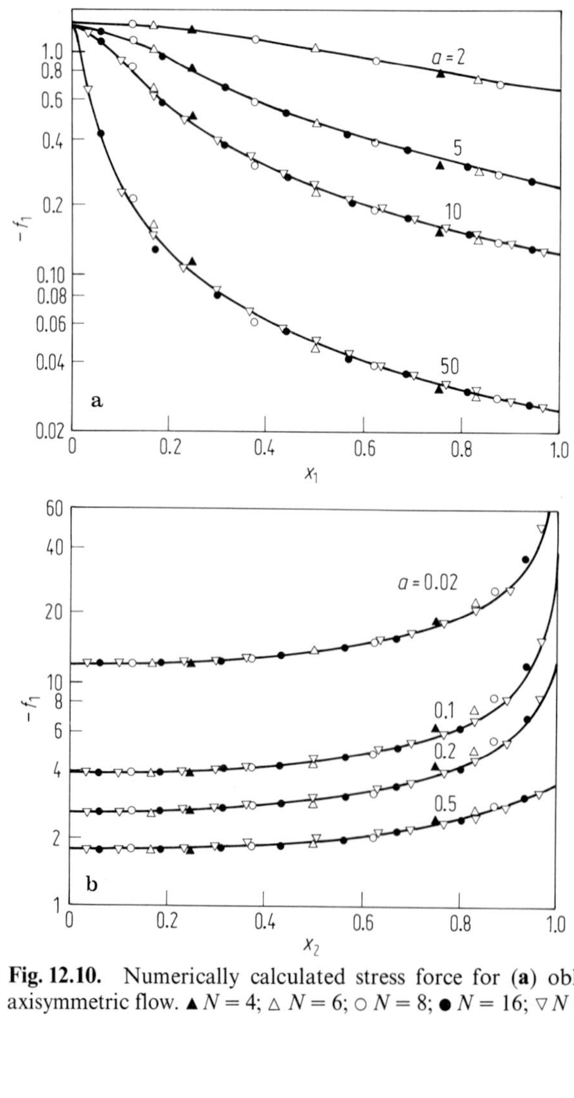

Fig. 12.10. Numerically calculated stress force for (**a**) oblate and (**b**) prolate spheroids in axisymmetric flow. ▲ $N = 4$; △ $N = 6$; ○ $N = 8$; ● $N = 16$; ▽ $N = 30$; (–) analytic solution

value of f_R is identically zero and its numerical value is zero to the accuracy of the calculations). The accuracy of the numerical solution can be further increased by employing a discretization with a larger concentration of boundary elements in regions where high gradients occur.

12.6. General Viscous Flow

Three-dimensional problems of incompressible viscous fluid motion are governed by the Navier–Stokes equations [17–19]

$$\varrho(\dot{u}_i + u_j u_{i,j}) = -p_{,i} + \mu u_{i,jj} \tag{12.37}$$

together with the continuity equation (12.23). The parameters ϱ and μ in the above equation are the mass density and the viscosity of the fluid, respectively, and the dot means time derivative. Note that Eq. (12.22) is a particular case of Eq. (12.37), when the left-hand side of the above equation vanishes.

The prevailing finite difference and finite element methods of solution of such problems usually deal with the Navier–Stokes equations in the above form (in terms of velocity and pressure) or, in the two-dimensional case, rewrite the equations in terms of stream function and vorticity, or stream function alone [21]. The major difficulty experienced by these methods is associated with the fact that implicit numerical procedures are necessary for the kinematic part of the computation. As a consequence, the solution field must comprise the entire flow field, inclusive of the viscous and the inviscid regions. Furthermore, for problems of external flow past finite bodies, satisfaction of boundary conditions prescribed at infinity implies the truncation of the infinite region at a finite distance.

A more convenient form of the Navier–Stokes equations for the development of numerical methods of solution is obtained by using velocity and vorticity as the dependent variables, as proposed by Lighthill [22]. In this way, it is possible to separate the set of equations into a kinetic part which deals with the change of the vorticity field with time and a kinematic part which relates the velocity field at any instant of time to the vorticity field at that instant. The advantages of such approach have already been noticed and, indeed, several formulations employing it in conjunction with the finite difference and the finite element methods have appeared in the literature [23–26]. These formulations recast the kinematic part of the problem into an integral equation for the velocity in terms of the vorticity. The main advantage of doing so is that it permits the explicit point-by-point computation of the velocity.

An immediate consequence of the above feature is that only the vorticity distribution in the viscous region of the flow contributes to the calculation of the velocity anywhere in the flow. Since this viscous region is generally embedded in a much larger inviscid region, a great reduction in the size of the domain involved in the actual computation is achieved. Moreover, for external flow problems, the imposed boundary conditions at infinity are implicitly contained in the integral equation, thus the necessity of truncating the infinite region at a finite distance is avoided. The kinetic part of the flow, however, was kept in the differential form, and some difficulties related to the satisfaction of boundary conditions at solid boundaries still remained, as pointed out by Wu [27].

In this section, we develop efficient boundary element approaches which recast both the kinematic and kinetic aspects of the problem in integral form. In order to introduce these formulations, it is more convenient to replace our tensor notation by a vector one. Thus, Eq. (12.37) is rewritten

$$\varrho \left[\frac{\partial \mathbf{u}}{\partial t} + (\mathbf{u} \cdot \mathbf{\nabla})\, \mathbf{u} \right] = -\, \mathbf{\nabla} p + \mu\, \nabla^2 \mathbf{u}\, . \tag{12.38}$$

By taking the curl of both sides of the above equation, we obtain the vorticity transport equation

$$\frac{\partial \boldsymbol{\omega}}{\partial t} = \mathbf{\nabla} \times (\mathbf{u} \times \boldsymbol{\omega}) + \nu\, \nabla^2 \boldsymbol{\omega} \tag{12.39}$$

in which use has been made of the continuity equation

$$\mathbf{\nabla} \cdot \mathbf{u} = 0 \tag{12.40}$$

and the definition of vorticity

$$\boldsymbol{\omega} = \mathbf{\nabla} \times \mathbf{u}\, . \tag{12.41}$$

The parameter ν in Eq. (12.39) is the kinematic viscosity of the fluid.

The kinematic relation between $\mathbf{u}$ and $\boldsymbol{\omega}$ is described by Eqs. (12.40) and (12.41). For a given distribution of $\boldsymbol{\omega}$, the velocity distribution throughout the flowfield is usually evaluated by using a vector Poisson's equation obtained by taking the curl of Eq. (12.41) accounting for Eq. (12.40),

$$\nabla^2 \mathbf{u} = -\, \mathbf{\nabla} \times \boldsymbol{\omega}\, . \tag{12.42}$$

Comparing this equation with Eq. (2.103) and remembering the derivation of Eq. (2.68), we can write the following integral equation as equivalent to Eq. (12.42):

$$\begin{aligned} &4\pi\, \mathbf{u}(\xi) + \int_\Gamma \mathbf{u}(x) \frac{\partial u^*(\xi, x)}{\partial \mathbf{n}(x)}\, d\Gamma(x) \\ &= \int_\Gamma \frac{\partial \mathbf{u}(x)}{\partial \mathbf{n}(x)}\, u^*(\xi, x)\, d\Gamma(x) + \int_\Omega [\mathbf{\nabla} \times \boldsymbol{\omega}(x)]\, u^*(\xi, x)\, d\Omega(x)\, , \end{aligned} \tag{12.43}$$

where $u^*(\xi, x)$ is the fundamental solution to Laplace's equation defined in Section 2.3 and $\mathbf{n}(x)$ is the unit normal vector.

The correct boundary conditions for the physical problem are prescribed velocities. Therefore, prior to calculate the values of $\mathbf{u}$ throughout the flowfield (for a known vorticity distribution), it is necessary to take the point ξ in the above equation to the boundary, thus yielding a boundary integral equation whose solution produces the values of $\partial \mathbf{u}/\partial \mathbf{n}$ along Γ. These values are then placed in Eq. (12.43) to allow the prediction of the velocity distribution in Ω.

Wu and Thompson [24] contested the validity of the use of Eq. (12.42) to represent the kinematics of the flow. They pointed out that the solution of Eqs. (12.40) and (12.41) for **u** is unique if either u_t (the tangential component of **u**) or u_n (its normal component) is prescribed over the boundary Γ, but the solution of Eq. (12.42) for **u** is unique only if both u_t and u_n are prescribed over Γ. Thus, while solutions of Eqs. (12.40) and (12.41) with prescribed u_t or u_n also satisfy Eq. (12.42), the converse is not necessarily true, i.e., solutions of Eq. (12.42) for prescribed u_t and u_n may not satisfy Eqs. (12.40) and (12.41).

A more convenient integral representation for the kinematic part of the flow was then derived by Wu and Thompson [24] directly from Eqs. (12.40) and (12.41). It follows from an application of Green's theorem for vectors, which can be written [28]

$$\int_\Omega (\mathbf{E}\cdot\nabla^2\mathbf{F} - \mathbf{F}\cdot\nabla^2\mathbf{E})\,d\Omega = \int_\Gamma [\mathbf{E}\times(\nabla\times\mathbf{F}) + \mathbf{E}(\nabla\cdot\mathbf{F}) - \mathbf{F}\times(\nabla\times\mathbf{E}) - \mathbf{F}(\nabla\cdot\mathbf{E})]\cdot\mathbf{n}\,d\Gamma\,, \tag{12.44}$$

where

$$\nabla^2\mathbf{F} = \nabla(\nabla\cdot\mathbf{F}) - \nabla\times\nabla\times\mathbf{F}\,. \tag{12.45}$$

Let $\mathbf{v}^*(\xi, x)$ be a vector fundamental solution to the vector Laplace's equation $\nabla^2\mathbf{F} = \mathbf{0}$ given by

$$\mathbf{v}^*(\xi, x) = \nabla[u^*(\xi, x)]\times\mathbf{a}\,, \tag{12.46}$$

where **a** is a constant unit vector. By direct substitution, it can be seen that

$$\nabla\cdot\mathbf{v}^* = 0\,, \tag{12.47}$$

$$\nabla\times\mathbf{v}^* = \nabla(\mathbf{a}\cdot\nabla u^*) \qquad \text{for } \xi \neq x\,. \tag{12.48}$$

By virtue of Eq. (12.40), there exists a vector potential $\mathbf{\Psi}$ such that [18]

$$\nabla\times\mathbf{\Psi} = \mathbf{u}\,, \tag{12.49}$$

$$\nabla\cdot\mathbf{\Psi} = 0 \tag{12.50}$$

Thus, considering **F** in Eq. (12.44) to be the fundamental solution $\mathbf{v}^*$ and **E** to be the vector potential $\mathbf{\Psi}$, accounting for Eqs. (12.41) and (12.46)–(12.50) and assuming, as in Section 2.2, that a small sphere of radius ε surrounding the point ξ, with surface Γ_ε, is excluded from the domain of integration, Eq. (12.44) becomes

$$\int_\Omega (\nabla u^*\times\mathbf{a})\cdot\boldsymbol{\omega}\,d\Omega = \int_{\Gamma+\Gamma_\varepsilon} \mathbf{\Psi}\times\nabla(\mathbf{a}\cdot\nabla u^*)\cdot\mathbf{n}\,d\Gamma - \int_{\Gamma+\Gamma_\varepsilon} (\nabla u^*\times\mathbf{a})\times\mathbf{u}\cdot\mathbf{n}\,d\Gamma\,. \tag{12.51}$$

The above equation may be rewritten

$$\int_{\Omega} \mathbf{a} \cdot (\boldsymbol{\omega} \times \nabla u^*)\, d\Omega = \int_{\Gamma+\Gamma_\varepsilon} (\mathbf{a} \cdot \nabla u^*)(\mathbf{u} \cdot \mathbf{n})\, d\Gamma - \int_{\Gamma+\Gamma_\varepsilon} \mathbf{a} \cdot [(\mathbf{u} \times \mathbf{n}) \times \nabla u^*]\, d\Gamma\,. \tag{12.52}$$

Taking the limit as $\varepsilon \to 0$, the volume Ω in the integral in the left-hand side of Eq. (12.52) becomes the entire volume bounded by Γ since the volume integral over the interior of Γ_ε goes to zero as $\varepsilon \to 0$. The integrals over Γ_ε in the right-hand side of Eq. (12.52) give

$$\lim_{\varepsilon\to 0} \left\{ \int_{\Gamma_\varepsilon} (\mathbf{a} \cdot \nabla u^*)(\mathbf{u} \cdot \mathbf{n})\, d\Gamma - \int_{\Gamma_\varepsilon} \mathbf{a} \cdot [(\mathbf{u} \times \mathbf{n}) \times \nabla u^*]\, d\Gamma \right\} \tag{12.53}$$

$$= \lim_{\varepsilon\to 0} \left\{ \frac{1}{\varepsilon^2} \int_{\Gamma_\varepsilon} [(\mathbf{a} \cdot \mathbf{n})(\mathbf{u} \cdot \mathbf{n}) - \mathbf{a} \cdot [(\mathbf{u} \times \mathbf{n}) \times \mathbf{n}]]\, d\Gamma \right\} = 4\pi\, \mathbf{a} \cdot \mathbf{u}(\xi)\,.$$

Inserting the above result into Eq. (12.52) and noting that the direction of the vector $\mathbf{a}$ is arbitrary, we obtain

$$4\pi\, \mathbf{u}(\xi) + \int_{\Gamma} (\mathbf{u} \cdot \mathbf{n})\, \nabla u^*\, d\Gamma = \int_{\Gamma} (\mathbf{u} \times \mathbf{n}) \times \nabla u^*\, d\Gamma + \int_{\Omega} \boldsymbol{\omega} \times \nabla u^*\, d\Omega\,. \tag{12.54}$$

A similar expression for two-dimensional flow problems can be obtained by taking $u^*(\xi, x)$ to be the two-dimensional fundamental solution defined in Section 2.3, for which case the result of the limit (12.53) is $[2\pi\, \mathbf{a} \cdot \mathbf{u}(\xi)]$. Thus, the general expression for the velocity $\mathbf{u}$ is of the form [24]

$$\mathbf{u}(\xi) = \frac{1}{2\alpha\pi} \left\{ \int_{\Omega} \frac{\boldsymbol{\omega}(x) \times \mathbf{r}(\xi, x)}{r^d(\xi, x)}\, d\Omega(x) + \int_{\Gamma} \frac{[\mathbf{u}(x) \times \mathbf{n}(x)] \times \mathbf{r}(\xi, x)}{r^d(\xi, x)}\, d\Gamma(x) - \int_{\Gamma} \frac{[\mathbf{u}(x) \cdot \mathbf{n}(x)]\, \mathbf{r}(\xi, x)}{r^d(\xi, x)}\, d\Gamma(x) \right\}, \tag{12.55}$$

where $\alpha = 2$, $d = 3$, and

$$\mathbf{r}(\xi, x) = \{x_1(\xi) - x_1(x) \;\; x_2(\xi) - x_2(x) \;\; x_3(\xi) - x_3(x)\}$$

for three-dimensional problems; $\alpha = 1$, $d = 2$, and

$$\mathbf{r}(\xi, x) = \{x_1(\xi) - x_1(x) \;\; x_2(\xi) - x_2(x)\}$$

for two-dimensional problems.

Note that the use of Eq. (12.55) for the evaluation of $\mathbf{u}$ throughout the flowfield requires the knowledge of both u_t and u_n over Γ. Provided that these values are compatible with each other, i.e., one of them is identical to the value obtained from the solution of Eqs. (12.40) and (12.41) using the other as the prescribed boundary condition, the specification of both in Eq. (12.55) is admissible and does not overspecify the problem.

For problems of external flow past finite bodies, we can consider the region Ω in Eq. (12.55) to be the entire (infinite) region occupied by the fluid. Then, following Section 2.10, the boundary Γ is divided into two parts: the fluid–solid interface on which the no-slip condition ($\mathbf{u}=\mathbf{0}$) applies and a surface infinitely remote from (and enclosing) the body on which the free-stream velocity boundary condition ($\mathbf{u}=\mathbf{u}_\infty$) applies. The surface integrals in (Eq. 12.55) can then be evaluated, giving

$$\mathbf{u}(\xi)=\frac{1}{2\alpha\pi}\int_\Omega \frac{\boldsymbol{\omega}(x)\times\mathbf{r}(\xi,x)}{r^d(\xi,x)}\,d\Omega(x)+\mathbf{u}_\infty\,. \tag{12.56}$$

The above equation can be recognized as the Biot–Savart law of induced velocities [18, 29]. Thus, we can consider the integral equation (12.55) to be an extension of the Biot–Savart law to a region bounded by Γ.

With prescribed values of $\mathbf{u}$ in Γ and known values of $\boldsymbol{\omega}$ in Ω, Eq. (12.55) permits the explicit point-by-point computation of the velocity anywhere in the flow field. Note that the prescribed boundary conditions at infinity are now implicitly contained in Eq. (12.55), as can be seen in Eq. (12.56), and that only the vorticity distribution in the viscous region of the flow contributes to the calculation of the velocity, as the integrand in the domain integral vanishes for $\boldsymbol{\omega}=\mathbf{0}$.

12.6.1. Steady Problems

Before proceeding to the solution of the transient Navier–Stokes equations as depicted in Eq. (12.39), it is convenient to discuss the numerical procedures related to the solution of steady incompressible viscous flow problems. Although many numerical studies of steady flow problems employ the time-dependent equations to obtain the desired steady state asymptotically, this approach introduces an additional dependent variable (the time) into the solution procedure and consequently additional numerical complexities.

As discussed earlier, the kinematic aspect of the flow is described by Eq. (12.55). The kinetic aspect of the flow is represented by the Navier–Stokes equations which, for a steady motion, can be written in the form [18]

$$\nabla\times\boldsymbol{\omega}=\frac{1}{\nu}(\mathbf{u}\times\boldsymbol{\omega}-\nabla h)\,, \tag{12.57}$$

where h is the total head defined by

$$h=\frac{p}{\varrho}+\frac{u^2}{2}\,, \tag{12.58}$$

p being the pressure, ϱ the density of the fluid, and $u^2=\mathbf{u}\cdot\mathbf{u}$.

From the definition of vorticity (Eq. (12.41)), it is clear that

$$\nabla\cdot\boldsymbol{\omega}=0\,. \tag{12.59}$$

Thus, the differential Eqs. (12.57) and (12.58) for the vorticity are analogous to the set of Eqs. (12.40) and (12.41) for the velocity. Therefore, an integral equation for the kinetic part of the flow can be obtained by simply replacing $\mathbf{u}$ by $\boldsymbol{\omega}$ and $\boldsymbol{\omega}$ by the term $(\mathbf{u} \times \boldsymbol{\omega} - \nabla h)/\nu$ in Eq. (12.55).

$$\boldsymbol{\omega}(\xi) = \frac{1}{2\alpha\pi} \left\{ \frac{1}{\nu} \int_{\Omega} \frac{[\mathbf{u}(x) \times \boldsymbol{\omega}(x) - \nabla h(x)] \times \mathbf{r}(\xi, x)}{r^d(\xi, x)} \, d\Omega(x) \right. \tag{12.60}$$

$$\left. + \int_{\Gamma} \frac{[\boldsymbol{\omega}(x) \times \mathbf{n}(x)] \times \mathbf{r}(\xi, x)}{r^d(\xi, x)} \, d\Gamma(x) - \int_{\Gamma} \frac{[\boldsymbol{\omega}(x) \cdot \mathbf{n}(x)]\, \mathbf{r}(\xi, x)}{r^d(\xi, x)} \, d\Gamma(x) \right\} .$$

By applying the divergence theorem [20] to the term in h,

$$\int_{\Omega} \frac{\nabla h(x) \times \mathbf{r}(\xi, x)}{r^d(\xi, x)} \, d\Omega(x) = \int_{\Gamma} \frac{[h(x)\, \mathbf{n}(x)] \times \mathbf{r}(\xi, x)}{r^d(\xi, x)} \, d\Gamma(x) , \tag{12.61}$$

Eq. (12.60) can be rewritten

$$\boldsymbol{\omega}(\xi) = \frac{1}{2\pi\alpha} \left\{ \frac{1}{\nu} \int_{\Omega} \frac{[\mathbf{u}(x) \times \boldsymbol{\omega}(x)] \times \mathbf{r}(\xi, x)}{r^d(\xi, x)} \, d\Omega(x) \right.$$

$$+ \int_{\Gamma} \frac{[\boldsymbol{\omega}(x) \times \mathbf{n}(x)] \times \mathbf{r}(\xi, x) - [\boldsymbol{\omega}(x) \cdot \mathbf{n}(x)]\, r(\xi, x)}{r^d(\xi, x)} \, d\Gamma(x)$$

$$\left. - \frac{1}{\nu} \int_{\Gamma} \frac{[h(x)\, \mathbf{n}(x)] \times \mathbf{r}(\xi, x)}{r^d(\xi, x)} \, d\Gamma(x) \right\} . \tag{12.62}$$

The calculation of $\boldsymbol{\omega}$ throughout the flowfield can be carried out iteratively using the above equation for known values of $\mathbf{u}$ in Ω and $\boldsymbol{\omega}$ and h in Γ. The contribution of the inviscid part of the flow to the computation of $\boldsymbol{\omega}$ anywhere in the flow is zero, as it was in the calculation of $\mathbf{u}$; see Eq. (12.55). Thus, only the values of $\mathbf{u}$ in Γ (which are the prescribed boundary conditions) and in the viscous region of the flow are needed in the calculation of $\boldsymbol{\omega}$. The solution field can then be confined to the viscous region of the flow.

A numerical formulation for the solution of steady incompressible viscous flow problems using Eqs. (12.55) and (12.62) was presented by Wu and Wahbah and is discussed in detail in [30, 31].

Example 12.5. Wu and Wahbah [30] presented a BEM solution for a flow past a square cavity, at a Reynolds number of 600. They employed linear boundary elements and triangular cells with linear interpolation functions to discretize the boundaries and domain of the region under consideration.

The flow configuration for the problem is shown in Fig. 12.11 and represents an open square cavity with a moving plate located one-sixth the cavity width above the cavity. The velocity boundary conditions are also indicated in the figure. The entrance and exit velocity boundary conditions are linear, corresponding to Couette flow profiles without pressure gradients, and are applied at channel sections one-third the width beyond the edges of the cavity.

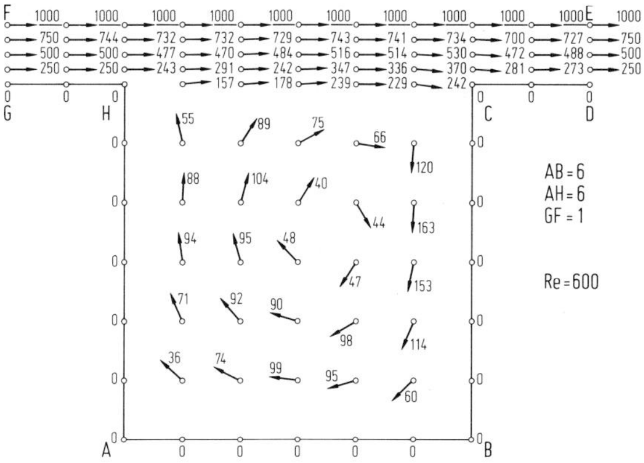

Fig. 12.11. Steady incompressible viscous flow past a cavity

Results for the computed velocity vectors (Fig. 12.11) present the same flow pattern with a central vortex in the cavity as obtained in [32] by using the FEM, and in [33, 34] using the FDM.

12.6.2. Transient Problems

Transient problems of incompressible viscous fluid motion are governed by the set of equations (12.39)–(12.41). An integral equation equivalent to the kinematic part of the flow as represented by Eqs. (12.40) and (12.41) has already been derived (Eq. (12.55)). In what follows, we also derive an integral representation for the kinetic aspect of the flow [35–38].

Rewriting the vorticity transport equation (12.39) as

$$\nabla^2 \boldsymbol{\omega} - \frac{1}{\nu}\frac{\partial \boldsymbol{\omega}}{\partial t} = -\frac{1}{\nu}\,\boldsymbol{\nabla}\times(\mathbf{u}\times\boldsymbol{\omega}) \tag{12.63}$$

and comparing to Eq. (4.1), we note that Eq. (12.63) can be interpreted as a (nonlinear) nonhomogenous diffusion equation, the nonlinearity being included through the convective term in the right-hand side. Thus, an integral equation equivalent to Eq. (12.63) can be readily obtained as (see Eq. (4.29))

$$\begin{aligned}
&\boldsymbol{\omega}(\xi, t_F) + \nu \int_{t_0}^{t_F}\int_{\Gamma} \boldsymbol{\omega}(x, t)\,[\boldsymbol{\nabla} u^*(\xi, x, t_F, t)\cdot \mathbf{n}(x)]\, d\Gamma(x)\, dt \\
&= \nu \int_{t_0}^{t_F}\int_{\Gamma} u^*(\xi, x, t_F, t)\,\{[\boldsymbol{\nabla}\times\boldsymbol{\omega}(x, t)]\times \mathbf{n}(x)\}\, d\Gamma(x)\, dt \\
&\quad + \int_{\Omega} \boldsymbol{\omega}_0(x, t_0)\, u^*(\xi, x, t_F, t_0)\, d\Omega(x) \\
&\quad + \int_{t_0}^{t_F}\int_{\Omega} \{\boldsymbol{\nabla}\times[\mathbf{u}(x, t)\times\boldsymbol{\omega}(x, t)]\}\, u^*(\xi, x, t_F, t)\, d\Omega(x)\, dt\,,
\end{aligned} \tag{12.64}$$

where $u^*(\xi, x, t_F, t)$ is the fundamental solution to the diffusion equation, defined in Eq. (4.26).

The calculation of $\boldsymbol{\omega}$ throughout the flowfield at any time t can be carried out iteratively using the above equations. For simplicity, let us consider only the case of a two-dimensional flow past an obstacle immersed in an infinite fluid. The inner integral in the last term in the above equation can be integrated by parts, giving

$$-\int_{\Omega} [\nabla \cdot (\omega\, \mathbf{u})]\, u^*\, d\Omega = -\int_{\Gamma} u^*\, \omega\, (\mathbf{u} \cdot \mathbf{n})\, d\Gamma + \int_{\Omega} \omega\, (\mathbf{u} \cdot \nabla u^*)\, d\Omega \tag{12.65}$$

and since the first integral in the right-hand side of Eq. (12.65) is zero due to the no-slip condition, Eq. (12.64) becomes

$$\begin{aligned} \omega(\xi, t_F) &+ \nu \int_{t_0}^{t_F} \int_{\Gamma} \omega(x, t)\, [\nabla u^*(\xi, x, t_F, t) \cdot \mathbf{n}(x)]\, d\Gamma(x)\, dt \\ &= \nu \int_{t_0}^{t_F} \int_{\Gamma} u^*(\xi, x, t_F, t)\, [\nabla \omega(x, t) \cdot \mathbf{n}(x)]\, d\Gamma(x)\, dt \\ &\quad + \int_{\Omega} \omega_0(x, t_0)\, u^*(\xi, x, t_F, t_0)\, d\Omega(x) \\ &\quad + \int_{t_0}^{t_F} \int_{\Omega} \omega(x, t)\, [\mathbf{u}(x, t) \cdot \nabla u^*(\xi, x, t_F, t)]\, d\Omega(x)\, dt\,. \end{aligned} \tag{12.66}$$

For a fluid in contact with a solid in motion, the no-slip condition provides a mechanism for the generation (or depletion) of vorticity at the solid surface. In the case where the fluid is initially at rest, the (irrotational) flow set up by the motion of the solid has a nonzero tangential velocity relative to the solid. A discontinuity in tangential velocity therefore results, at $t = 0$, due to the no-slip condition, representing a sheet of vorticity at the boundary [27, 29]. For $t > 0$, the vorticity, which is concentrated at the boundary at $t = 0$, spreads into the interior of the fluid domain by diffusion, and, once there, is carried away by both convection and diffusion.

Equation (12.66) mathematically represents this process. According to a discussion following Eq. (4.33), we note that the third integral in Eq. (12.66) represents the effects of an initial vorticity distribution. Since a stationary fluid cannot coexist with a nonzero vorticity field, the vorticity distribution changes as a result of a convective process, represented by the last term in Eq. (12.66). Finally, the boundary integrals in Eq. (12.66) include the effects of generation (or depletion) of vorticity at the surface Γ, this process being a result of the no-slip condition as previously discussed.

As for the steady-state problem the contribution of the inviscid region of the flow to the computation of $\boldsymbol{\omega}$ anywhere in the flow is zero. Thus, only the values of $\mathbf{u}$ in the viscous region of the flow are needed in the calculation of $\boldsymbol{\omega}$, and the solution field can be confined to this region. Efficient algorithms for the numerical solution of this problem, following the procedures discussed in Chapter 4, are presented in [35, 37, 38].

Extensions of the formulation derived in this section have been carried out by Wu and co-workers for the cases of compressible [39] and turbulent [40, 41] flows.

Example 12.6. BEM results for the classical problem of flow past a circular cylinder, at a Reynolds number of 40, were presented by Wu and Rizk [35].

Computed stream lines and equivorticity contours are shown in Fig. 12.12, in the upper and lower halves of the figure, respectively. The flow patterns plotted in the figure are for an asymptotically obtained steady solution.

In order to compute forces acting on solid bodies in viscous flows, it is important to accurately calculate the vorticity and its normal gradient on the solid boundary. In conventional methods, these quantities are calculate by using one-sided difference formulas, by extrapolating known values on nonboundary points. The results of these computations are sensitive to the grid spacing near the wall and to the order of accuracy of the difference formulas. In the BEM formulation, these

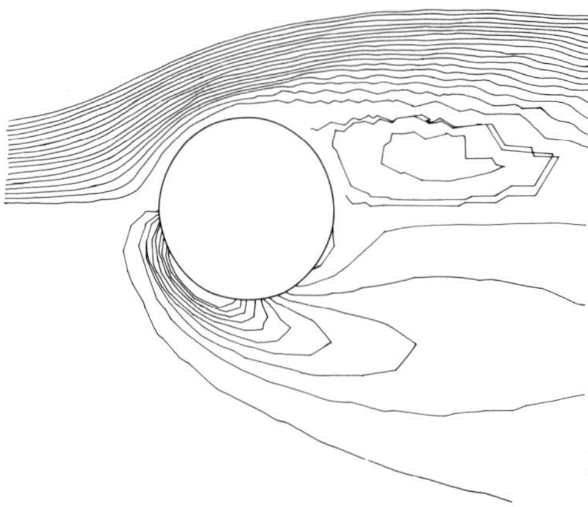

Fig. 12.12. Flow pattern around circular cylinder

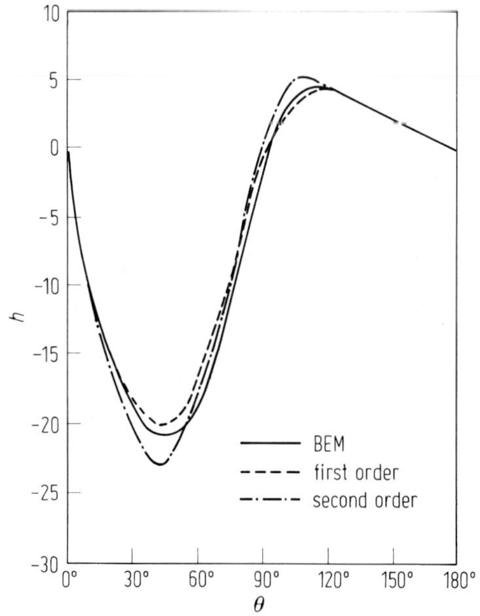

Fig. 12.13. Normal gradient of vorticity on circular cylinder

quantities are obtained directly from the solution of integral equations, thus providing a higher degree of accuray. Values of the normal vorticity gradient obtained by using the BEM and difference formulas of different order are compared in Fig. 12.13. The polar angle θ in the figure is measured from the forward stagnation point on the cylinder.

References

1. Liggett, J. A., Location of free surface in porous media, J. Hydraulics Div. ASCE **103,** 353–365 (1977).
2. Liu, P. L. F., and Liggett, J. A., Boundary solutions to two problems in porous media, J. Hydraulics Div. ASCE **105,** 171–183 (1979).
3. Lennon, G. P., Liu, P. L. F., and Liggett, J. A., Boundary integral equation solution to axisymmetric potential flows: Part 1 – Basic formulation; Part 2 – Recharge and well problems in porous media, Water Resources Res. **15,** 1102–1115 (1979).
4. Lennon, G. P., Liu, P. L. F., and Liggett, J. A., Boundary integral solutions to three-dimensional unconfined Darcy's flow, Water Resources Res. **16,** 651–658 (1980).
5. Connor, J. J., and Brebbia, C. A., *Finite Element Techniques for Fluid Flow,* Butterworths, London, 1976.
6. Polubarinova-Kochina, P. Ya., *Theory of Groundwater Movement,* Princeton University Press, Princeton, 1962.
7. Marino, M. A., Hele-Shaw model study of the growth and decay of groundwater ridges, J. Geophys. Res. **72,** 1195–1205 (1967).
8. Newman, S. P., Narasimhan, T. N., and Witherspoon, P. A., Application of mixed explicit-implicit finite element method to nonlinear diffusion-type problems, in *Proc. Finite Elements in Water Resources Vol. 1,* Pentech Press, London, 1976.
9. Liu, P. L. F., Cheng, A. H. D., Liggett, J. A., and Lee, J. H., Boundary integral equation solutions to moving interface between two fluids in porous media, Water Resources Res. **17,** 1445–1452 (1981).
10. Bear, J., and Dagan, G., Moving interface in coastal aquifers, J. Hydraulics Div. ASCE **90,** 193–216 (1964).
11. Sa da Costa, A. A., and Wilson, J. L., A numerical model of sea water intrusion in aquifers, Report 247, Ralph M. Parsons Lab., M.I.T., Cambridge, 1979.
12. Lafe, O. E., Montes, J. A., Cheng, A. H. D., Liggett, J. A., and Liu, P. L. F., Singularities in Darcy flow through porous media, J. Hydraulics Div. ASCE **106,** 977–997 (1980).
13. Jaswon, M. A. and Symm, G. T., *Integral Equation Methods in Potential Theory and Elastostatics,* Academic Press, New York, 1977.
14. Hess, J. L., and Smith, A. M. O., Calculation of potential flow about arbitrary bodies, Progress in Aeronautical Sciences, Vol. 8 (D. Küchemman, Ed.), Pergamon Press, London, 1967.
15. Hess, J. L., Calculation of potential flow about bodies of revolution having axes perpendicular to the free stream direction, J. Aerospace Sci. **29,** 726 (1962).
16. Youngren, G. K., and Acrivos, A., Stokes flow past a particle of arbitrary shape: A numerical method of solution, J. Fluid Mech. **69,** Part 2, 377–403 (1975).
17. Lamb, H., *Hydrodynamics,* 6th ed., Cambridge University Press, London, 1932.
18. Batchelor, G. K., *An Introduction to Fluid Dynamics,* Cambridge University Press, Cambridge, 1967.
19. Ladyzhenskaya, O. A., *The Mathematical Theory of Viscous Incompressible Flow,* Gordon and Breach, New York, 1963.
20. Kellogg, O. D., *Foundations of Potential Theory,* Springer-Verlag, Berlin, 1929.
21. Huebner, K. H., *The Finite Element Method for Engineers,* Wiley-Interscience, New York, 1975.
22. Lighthill, M. J., Introduction. Boundary layer theory, in *Laminar Boundary Layer,* (L. Rosenhead, Ed.), Oxford University Press, London, 1963.

23. Payne, R. B., Calculations of unsteady viscous flow past a circular cylinder, J. Fluid Mech. **4,** 81–86 (1958).
24. Wu, J. C., and Thompson, J. F., Numerical solutions of time-dependent incompressible Navier–Stokes equations using an integro-differential formulation, Comput. Fluids **1,** 197–215 (1973).
25. Schmall, R. A., and Kinney, R. B., Numerical study of unsteady viscous flow past a lifting plate, AIAA J. **12,** 1566–1573 (1974).
26. Bratanow, T., and Spehert, T., Computational flow development for unsteady viscous flow, NASA CR-2995, 1978.
27. Wu, J. C., Numerical boundary conditions for viscous flow problems, AIAA J. **14,** 1042–1049 (1976).
28. Morse, P. M., and Feshbach, H., *Methods of Theoretical Physics,* McGraw–Hill, New York, 1953.
29. Milne-Thompson, L. M., *Theoretical Aerodynamics,* 4th ed., Dover, New York, 1958.
30. Wu, J. C., and Wahbah, M. M., Numerical solution of viscous flow equations using integral representations, Lecture Notes in Physics Vol. 59, pp. 448–453, Springer-Verlag, New York, 1976.
31. Wahbah, M. M., Computation of internal flows with arbitrary boundaries using the integral representation method, Report, Georgia Institute of Technology, Atlanta, 1978.
32. Taylor, C., and Hood, P., A numerical solution of the Navier–Stokes equations using the finite element technique, Comput. Fluids **1,** 73–100 (1973).
33. Mills, R. D., Numerical solution of the viscous flow equations for a class of closed flows, J. Aeronaut. Soc. **69,** 714–718 (1965).
34. Burggraf, O. R., Analytical and numerical studies of the structure of steady separated flows, J. Fluid Mech. **24,** 113–151 (1966).
35. Wu, J. C., and Rizk, Y. M., Integral-representation approach for time-dependent viscous flow, Lecture Notes in Physics, Vol. 90, pp. 558–564, Springer-Verlag, New York, 1978.
36. Brebbia, C. A., and Wrobel, L. C., The boundary element method, in *Computer Methods in Fluids* (K. Morgan, C. Taylor, and C. A. Brebbia, Eds.), Pentech Press, London, 1980.
37. Rizk, Y., An integral-representation approach for time-dependent viscous flows, Ph.D. Thesis, Georgia Institute of Technology, Atlanta, 1980.
38. Wrobel, L. C., Potential and viscous flow problems using the boundary element method, Ph.D. thesis, Southampton University, Southampton, 1981.
39. El-Refaee, M. M., A numerical study of laminar unsteady compressible flow over airfoils, Ph.D. thesis, Georgia Institute of Technology, Atlanta, 1981.
40. Wu, J. C., Wahbah, M. M., and Sugavanam, A., Some numerical solutions of turbulent flow problems by the use of integral representations, in *Proc. Symposium on Application of Computer Methods in Engineering,* University of Southern California, Los Angeles, 1977.
41. Wu, J. C., and Sugavanam, A., Method for the numerical solution of turbulent flow problems, AIAA J. **16,** 948–955 (1978).

Chapter 13

Coupling of Boundary Elements with Other Methods

13.1. Introduction

Many engineering problems present a certain amount of coupling or interaction between different parts or systems. For instance, systems representing structure, fluid, and soil can be coupled within the same problem, each part represented by a physical region over which a particular numerical solution can be applied. Fluids such as water, air, or lubricants may be interacting against structural elements such as buildings, dams, offshore structures, mechanical components, pressure vessels, etc. Surface structures interact with the soil through their foundations and the behavior of buried structures is strongly coupled with the surrounding rock or soil strata. In many cases it is possible to assume that for all practical purposes, the effect of one system upon the other does not occur concurrently. Typical examples of this uncoupled behavior are wind forces on stiff buildings and hydrodynamic forces on massive off-shore gravity platforms. For these cases the forces on the structure can be computed assuming that the structure is rigid and neglecting the interaction with the surrounding fluid. Boundary elements are recommended to solve these problems due to their ability to model domains extending to infinity. Problems such as wave diffraction, harbor resonance, fluid flow, etc., have been frequently solved using boundary elements. For these cases the boundary element technique offers a very simple data input by comparison with methods such as finite elements of finite differences.

For structures such as flexible masts or compliant off-shore structures, the systems representing solid and fluid may need to be considered at the same moment in time. Systems such as these are said to be coupled "concurrently" as at any time the behavior of one influences the other and vice versa. In many cases we will need to define a part of the total system using boundary elements and another using finite elements, for instance. Coupling may also be necessary to obtain more accurate results in a part of the problem. For instance, the boundary element method generally gives more accurate results than finite elements in regions of stress or flux concentrations. One could, for example, define special boundary elements for regions with singularities and combine them with standard finite elements.

Boundary elements are frequently applied to problems extending to infinity because they satisfy the radiation conditions which are difficult to represent using finite elements. One of the obvious drawbacks of the finite element method is its inability to model domains extending to infinity. Boundary elements, on the contrary, use fundamental solutions which naturally obey the radiation condition. An extensive number of papers now exists on the coupling between boundary and finite elements [1 – 16]. It may be difficult or inconvenient, however, to use the exact fundamental solution in all cases, but some approximations may equally well be applied with negligible losses of accuracy in the numerical solution.

A special type of approximation is the use of Sommerfeld's radiation condition. A related idea is to employ special finite elements – sometimes called "infinite elements" – for the approximate satisfaction of the radiation condition. The use of these and other types of formulations are reviewed in Refs. [17 – 24].

In what follows we will discuss methods of combining boundary element and finite element regions and develop approximate solutions. The advantage of using approximate boundary or domain solutions to represent the radiation conditions is that they do not require coupling of all boundary nodes but only of a node with its neighbors.

13.2. Coupling of Finite Element and Boundary Element Solutions

The idea of combining both numerical techniques is of great interest in many practical problems such as those with unbounded domains or regions of high stress concentration, both of which can be better represented using boundary integral solutions. Finite elements, on the other hand, may be easier to apply in those parts of the domain which present anisotropic or nonlinear behavior.

The use of linear or higher-order functions for boundary elements permits the combination of finite element and boundary element regions without loss of continuity. One approach is to convert the finite element region into a boundary element and this is especially attractive when using mixed finite element formulations, as in these cases the degrees of freedom for both parts are simple to match. More commonly the boundary element region is transformed into an equivalent finite element. This operation generally gives nonsymmetric element matrices although experience shows that these matrices can be symmetrized without substantial losses of accuracy. Symmetric matrices are attractive because they can be easily implemented into existing finite element codes and are computationally more efficient.

Finite element matrices can be deduced from the weighted residual statement previously studied. The starting point for a potential model is

$$\int_{\Omega} q_i \frac{\partial \delta u}{\partial x_i} \, d\Omega = \int_{\Gamma_2} \bar{q} \, \delta u \, d\Gamma + \int_{\Omega} b \, \delta u \, d\Omega \,, \tag{13.1}$$

where δu indicates a variation (or virtual increment) of the potential and q_i is the flux in the x_i direction. $\bar{q}$ is the flux normal to the boundary, known on Γ_2; b are the distributed sources within the domain. Notice that the function identically satisfies the essential boundary conditions of the problem, i.e., $u = \bar{u}$ and $\delta u \equiv 0$ on Γ_1.

Next, an interpolation function is chosen for u over each element. The same function is assumed to be valid for δu which results in symmetric matrices. The procedure is well known and will not be repeated here. (The interested reader is referred to [26].)

For elasticity problems the starting statement is the principle of virtual displacements which can be written

$$\int_{\Omega} \sigma_{ij}\, \delta\varepsilon_{ij}\, d\Omega = \int_{\Gamma_2} \bar{p}_i\, \delta u_i\, d\Gamma + \int_{\Omega} b_i\, \delta u_i\, d\Omega\,. \tag{13.2}$$

Notice that σ_{ij} and ε_{ij} are the components of the stress and strain tensors. $\bar{p}_i$ are the tractions applied on Γ_2, b_i the body force components, and u_i the components of the displacement vector. The conditions on Γ_1 are identically satisfied, i.e., $u_i = \bar{u}_i$ and $\delta u_i \equiv 0$ on Γ_1.

Applying a finite element discretization to Eq. (13.1) or (13.2), we will obtain a final system of equations which can be expressed in matrix form as

$$\boldsymbol{K}\boldsymbol{U} = \boldsymbol{F} + \boldsymbol{D}\,. \tag{13.3}$$

$\boldsymbol{K}$ is the global or stiffness matrix for the system. $\boldsymbol{U}$ are the unknown nodal displacements or potentials. $\boldsymbol{F}$ is an equivalent nodal force or integrated flux vector resulting from the first right-hand side integral of Eq. (13.1) or (13.2). $\boldsymbol{D}$ is a vector due to the distributed sources or body forces.

Notice that the vector $\boldsymbol{F}$ has been obtained by weighting the applied tractions – or fluxes – with the interpolation functions used for the virtual displacements – or potentials. Hence, we can always find a distribution matrix $\boldsymbol{N}$ such that $\boldsymbol{F}$ can now be written

$$\boldsymbol{F} = \boldsymbol{N}\boldsymbol{Q}\,, \tag{13.4}$$

where $\boldsymbol{Q}$ contains the actual nodal values of tractions or fluxes. The coefficients of $\boldsymbol{N}$ will depend on the type of interpolation functions used for the displacements and tractions – or potentials and fluxes. It is important to point out that the symmetry of $\boldsymbol{K}$ is due to the symmetry of the left-hand side integrals in Eqs. (13.1) and (13.2), i.e.,

$$\int_{\Omega} \sigma_{ij}\, \delta\varepsilon_{ij}\, d\Omega = \int_{\Omega} \varepsilon_{ij}\, \delta\sigma_{ij}\, d\Omega\,. \tag{13.5}$$

The finite element matrices can now be written

$$\boldsymbol{K}\boldsymbol{U} = \boldsymbol{N}\boldsymbol{Q} + \boldsymbol{D}\,. \tag{13.6}$$

For a boundary element region, the governing boundary integral equations can be written as follows: For potential problems,

$$c\,u + \int_{\Gamma} q^*\, u\, d\Gamma = \int_{\Gamma} u^*\, q\, d\Gamma + \int_{\Omega} b\, u^*\, d\Omega\,, \tag{13.7}$$

and for elastostatics

$$c_{ij}\, u_j + \int_{\Gamma} p_{ij}^*\, u_j\, d\Gamma = \int_{\Gamma} u_{ij}^*\, p_j\, d\Gamma + \int_{\Omega} b_j\, u_{ij}^*\, d\Omega\,. \tag{13.8}$$

For both cases the resulting boundary integral equations can be written in matrix form as

$$\boldsymbol{H}\boldsymbol{U} = \boldsymbol{G}\boldsymbol{Q} + \boldsymbol{B}, \tag{13.9}$$

where $\boldsymbol{U}$ and $\boldsymbol{Q}$ are the vectors of nodal displacements (or potentials) and tractions (or fluxes), respectively. $\boldsymbol{B}$ is the term due to the body forces (or distributed sources). Notice that $\boldsymbol{H}$ and $\boldsymbol{G}$ have been obtained by using the fundamental solution and the shape functions for the boundary unknowns u and q; a point collocation process is then applied to each boundary node in turn.

Let us now analyze a problem consisting of two regions Ω^1 and Ω^2 as shown in Fig. 13.1. The two regions are joined together at the interface Γ_I; Ω^1 is studied using finite elements and Ω^2 by applying boundary elements. In order to join the two parts, compatibility and equilibrium on Γ_I need to be satisfied, i.e.,

(i) *Compatibility.* The displacements (or potentials) at the Γ_I interface for region 1 ($\boldsymbol{U}_I^1$) and region 2 ($\boldsymbol{U}_I^2$) should be equal, i.e.,

$$\boldsymbol{U}_I^1 = \boldsymbol{U}_I^2 \quad \text{on } \Gamma_I . \tag{13.10}$$

(ii) *Equilibrium.* The tractions (or fluxes) at the Γ_I interface for region 1 ($\boldsymbol{Q}_I^1$) and region 2 ($\boldsymbol{Q}_I^2$) should add to zero, i.e.,

$$\boldsymbol{Q}_I^1 + \boldsymbol{Q}_I^2 = \boldsymbol{0} \quad \text{on } \Gamma_I . \tag{13.11}$$

We can now transform region Ω^2 into an equivalent finite element and then assemble the resulting global matrices together with the finite element matrices of Eq. (13.6).

Let us transform Eq. (13.9) by inverting $\boldsymbol{G}$, i.e.,

$$\boldsymbol{G}^{-1}(\boldsymbol{H}\boldsymbol{U} - \boldsymbol{B}) - \boldsymbol{Q} \tag{13.12}$$

and premultiply the result by the distribution matrix $\boldsymbol{N}$ defined in Eq. (13.4). The result is

$$(\boldsymbol{N}\boldsymbol{G}^{-1}\boldsymbol{H})\,\boldsymbol{U} - (\boldsymbol{N}\boldsymbol{G}^{-1}\boldsymbol{B}) = \boldsymbol{N}\boldsymbol{Q}. \tag{13.13}$$

One can now define

$$\begin{aligned} \boldsymbol{K}' &= \boldsymbol{N}\boldsymbol{G}^{-1}\boldsymbol{H}, \\ \boldsymbol{D}' &= \boldsymbol{N}\boldsymbol{G}^{-1}\boldsymbol{B}, \\ \boldsymbol{F}' &= \boldsymbol{N}\boldsymbol{Q}. \end{aligned} \tag{13.14}$$

Hence Eq. (13.13) presents the following finite element form:

$$\boldsymbol{K}'\boldsymbol{U} = \boldsymbol{F}' + \boldsymbol{D}'. \tag{13.15}$$

The main disadvantage that arises from the above formulation is the fact that the matrix $\boldsymbol{K}'$ is generally asymmetric unlike the finite element matrices which – for symmetric operators – are generally symmetric. The asymmetry arises due to the approximations involved in the discretization and collocation processes as well as the special shapes used for the weighting functions, i.e., the fundamental solution. There is no reason why the matrix $\boldsymbol{K}'$ should be symmetric other than computational convenience and efficiency. The symmetrization of $\boldsymbol{K}'$ is justified using simple "error minimization" techniques or energy-type approaches, but it basically consists of discarding the unsymmetric part of $\boldsymbol{K}'$ by adding it to its transpose and taking the average. To justify this symmetrization by the reciprocal theorem is incorrect.

Georgiou [27] has also explained that the nodal forces in a stiffness relationship are equivalent to some traction distribution and are obtained by the appropriate weighting of this traction distribution around the boundary. Hence, applying a nodal load of 1, say, at any point does not physically represent a point source, but some localized traction distribution which depend on the local geometry and shape functions. Therefore, by applying a load equal to 1 at different points, the physical interpretation of the total loads is the same, but the way in which these loads are distributed is different. It is this difference in the physical interpretation of the exact form of the load states which does not allow a simple interpretation of the reciprocal theorems to be applied and explains the lack of symmetry of $\boldsymbol{K}'$. This effect, however, will tend to be generally not very pronounced and will be negligible at a certain distance. Under this assumption the matrix $\boldsymbol{K}'$ can be symmetrized.

One method of symmetrizing the matrix is by defining an "error" in the nondiagonal terms of $\boldsymbol{K}'$. The error of any coefficient k_{ij} can be written as the difference between k'_{ij} and k'_{ji} and the still unknown coefficient, i.e.,

$$\varepsilon_{ij} = \tfrac{1}{2}\left[(k_{ij} - k'_{ij}) + (k_{ij} - k'_{ji})\right]. \tag{13.16}$$

The square of the error function ε_{ij} can be minimized with respect to k_{ij}, which gives

$$\frac{\partial(\varepsilon_{ij})^2}{\partial k_{ij}} = 2k_{ij} - k'_{ij} - k'_{ji} = 0. \tag{13.17}$$

Hence, the new symmetric coefficients are

$$k_{ij} = \tfrac{1}{2}\left[k'_{ij} + k'_{ji}\right]. \tag{13.18}$$

The equivalent finite element matrix for region 2 (Fig. 13.1) can now be written

$$\boldsymbol{K}^2 = \tfrac{1}{2}\left[\boldsymbol{K}' + \boldsymbol{K}'^{,T}\right] \tag{13.19}$$

and Eq. (13.15) can be expressed as

$$\boldsymbol{K}^2 \boldsymbol{U} = \boldsymbol{F}^2 + \boldsymbol{D}^2, \tag{13.20}$$

where $\boldsymbol{F}^2 = \boldsymbol{F}'$ and $\boldsymbol{D}^2 = \boldsymbol{D}'$ of Eqs. (13.15).

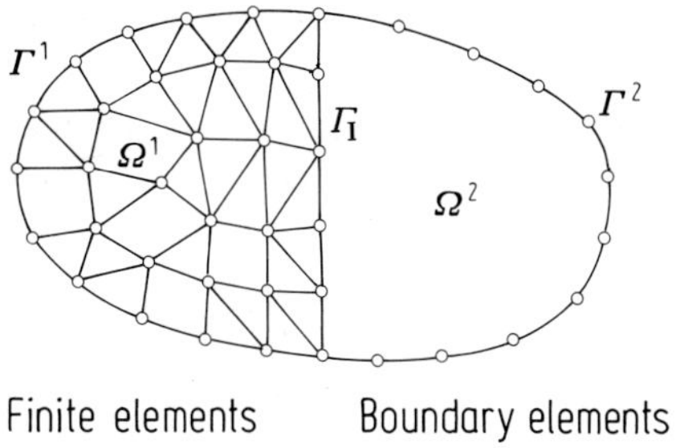

Fig. 13.1. Domain divided into finite element and boundary element regions

Equations (13.20) can now be assembled with the original finite element matrices for region 1, i.e.,

$$\boldsymbol{K}^1 \boldsymbol{U} = \boldsymbol{F}^1 + \boldsymbol{D}^1 \tag{13.21}$$

in the standard manner, which ensures compatibility and equilibrium.

As Georgiou [27] has pointed out, special care should be taken when having corner nodes for which extra conditions may be required (see Section 5). One method of solving this difficulty is by displacing the nodes from the corner, i.e., using discontinuous elements everywhere or just for the corner elements. This procedure has been successfully applied in many programs (see Section 5 and Ref. [13] and other programs in [36]). If the nodes are left at the corners and geometric discontinuities arise, i.e., the tractions have different values at either side of the node considered, the procedure proposed by Chaudonneret [28] can be applied (see Section 5). It was found by Georgiou [27], however, that the straightforward inclusion of the extra equations proposed by Chaudonneret would destroy the "almost" symmetric character of $\boldsymbol{K}'$ and he proposed a method of introducing these "corner conditions" as a set of rotation matrices which are then introduced in the overall formulation in much the same manner as a set of linearly independent constraints is introduced in a global finite element matrix. The technique has been implemented in several applications and found to work very well.

13.2.1. The Energy Approach

It has also become fashionable to attempt to justify the symmetrization of element matrices obtained using boundary elements by energy considerations [6]. This approach, although appealing because of its simplicity, is not strictly correct. The argument centers in the use of energy considerations to prove that the element matrices have to be symmetric even if, as discussed earlier, they are not. The fallacy of the argument rests in using energy principles which are based on symmetry considerations but which do not strictly apply for integral operators. The boundary element statements should be interpreted in the same manner as the mixed formulations of finite elements, many of which give nonsymmetric matrices.

The energy approach starts by asserting that the energy of the boundary element domain can be written for a potential problem as

$$\Pi = \tfrac{1}{2}\int_{\Omega} q_i \frac{\partial u}{\partial x_i}\, d\Omega - \int_{\Gamma_2} \bar{q}\, u\, d\Gamma - \int_{\Omega} b\, u\, d\Omega\,, \tag{13.22}$$

where the first term is the internal energy of the system and the second, the loss of energy due to the external sources. For the case of elastostatics, the expression is the well-known minimum potential energy, i.e.,

$$\Pi = \tfrac{1}{2}\int_{\Omega} \sigma_{ij}\, \varepsilon_{ij}\, d\Omega - \int_{\Gamma_2} \bar{p}_i\, u_i\, d\Gamma - \int_{\Omega} b_i\, u_i\, d\Omega\,. \tag{13.23}$$

Next, one can convert the first integral in Eqs. (13.22) and (13.23) into a boundary integral by using the property that the functions for u (and consequently q or σ) exactly satisfy the equilibrium equations as they are based on the fundamental solutions used in the integral equation analysis. The values of u, p, and q used in the above equations are supposed to be those obtained after the solution of the integral equation problem.

Equation (13.22) and (13.23) can then be written

$$\Pi = \tfrac{1}{2}\int_{\Gamma} q\, u\, d\Gamma - \int_{\Gamma_2} \bar{q}\, u\, d\Gamma - \int_{\Omega} b\, u\, d\Omega \tag{13.24}$$

and

$$\Pi = \tfrac{1}{2}\int_{\Gamma} p_i\, u_i\, d\Gamma - \int_{\Gamma_2} \bar{p}_i\, u_i\, d\Gamma - \int_{\Omega} b_i\, u_i\, d\Omega\,. \tag{13.25}$$

Next the variation of Π is carried out to find the equilibrium requirement. For Eq. (13.24)

$$\delta\Pi = \tfrac{1}{2}\int_{\Gamma} (q\, \delta u + \delta q\, u)\, d\Gamma - \int_{\Gamma_2} \bar{q}\, \delta u\, d\Gamma - \int_{\Omega} b\, \delta u\, d\Omega = 0 \tag{13.26}$$

and for Eq. (13.25)

$$\delta\Pi = \tfrac{1}{2}\int_{\Gamma} (p_i\, \delta u_i + \delta p_i\, u_i)\, d\Gamma - \int_{\Gamma_2} \bar{p}_i\, \delta u_i\, d\Gamma - \int_{\Omega} b_i\, \delta u_i\, d\Omega = 0\,. \tag{13.27}$$

These equations can be written in matrix form using the same notation for both cases as;

$$\delta\Pi = \tfrac{1}{2}\int_{\Gamma} (\delta \boldsymbol{u}^T \boldsymbol{q} + \delta \boldsymbol{q}^T \boldsymbol{u})\, d\Gamma - \int_{\Gamma_2} \delta \boldsymbol{u}^T \bar{\boldsymbol{q}}\, d\Gamma - \int_{\Omega} \delta \boldsymbol{u}^T \boldsymbol{b}\, d\Omega = 0\,, \tag{13.28}$$

where $\boldsymbol{u}$ represents now the displacements or potential functions and $\boldsymbol{q}$ the tractions or fluxes. Upon substitution of the usual interpolation functions for the $\boldsymbol{u}$ and $\boldsymbol{q}$ values, i.e.,

$$\begin{aligned} \boldsymbol{u} &= \boldsymbol{\Phi}^T \boldsymbol{u}^n\,, \\ \boldsymbol{q} &= \boldsymbol{\Phi}^T \boldsymbol{q}^n\,, \end{aligned} \tag{13.29}$$

Eq. (13.28) can be written in the form

$$\delta \Pi = \tfrac{1}{2}\{\delta \boldsymbol{U}^T \boldsymbol{N} \boldsymbol{Q} + \delta \boldsymbol{Q}^T \boldsymbol{N}^T \boldsymbol{U}\} - \delta \boldsymbol{U}^T \boldsymbol{F} - \delta \boldsymbol{U}^T \boldsymbol{D} = 0\,, \tag{13.30}$$

where $\boldsymbol{U}$ and $\boldsymbol{Q}$ are the nodal unknowns. The matrix $\boldsymbol{N}$ is formed by the integrals of the shape functions as in usual finite elements.

One may now substitute the values of $\boldsymbol{Q}$ by those obtained from the integral equation solution (Eq. (13.12)), i.e.,

$$\boldsymbol{Q} = \boldsymbol{G}^{-1}(\boldsymbol{H}\boldsymbol{U}) = \boldsymbol{C}\boldsymbol{U}\,. \tag{13.31}$$

Substituting Eq. (13.31) into Eq. (13.30) gives a symmetric form for the resulting matrices, i.e.,

$$\delta \boldsymbol{U}^T \{\tfrac{1}{2}(\boldsymbol{N}\boldsymbol{C}\boldsymbol{U} + \boldsymbol{C}^T \boldsymbol{N}^T \boldsymbol{U}) - \boldsymbol{F} - \boldsymbol{D}\} = 0 \tag{13.32}$$

which can now be written as in Eq. (13.15), i.e.,

$$\boldsymbol{K}\boldsymbol{U} = \boldsymbol{F} + \boldsymbol{D}\,. \tag{13.33}$$

Notice that now $\boldsymbol{K}$ is symmetric and equal to

$$\boldsymbol{K} = \tfrac{1}{2}\{\boldsymbol{N}\boldsymbol{C} + \boldsymbol{C}^T \boldsymbol{N}^T\}\,. \tag{13.34}$$

This gives the same results as shown previously, i.e. equation (13.19).

Although the final result is the same, the above deduction is incorrect. The error is committed when substituting for $\boldsymbol{Q}$ in terms of the integral equation solution. The $\boldsymbol{Q}$ have to be generated from the displacement or potential fields corresponding to the fundamental solution.

Example 13.1. *Static Interaction* [27]. Consider the example of a building standing on an infinite half-space representing the soil (Fig. 13.2). The finite element domain represents the structure and the infinite half-space is represented by a boundary element region. Notice that since the plane representing the soil extends to infinity, it is in principle necessary to introduce an infinite number of boundary elements. This difficulty can be eliminated in practice in either of the two ways;

(i) By truncating the plane at a finite distance – approximate method – as shown in Fig. 13.2 and used in the present example. Notice that in the present case the domain has been closed with a few large boundary elements. These are not generally needed but have been used here to better compare the BE solution with a FE solution obtained with the same closed domain.

(ii) Using a fundamental solution appropriate to the half-space problem rather than a free-space Green's function. With this last approach, boundary elements are only required at the interface between the structure and the soil, yet the remainder of the soil domain is rigorously taken into account if the soil is homogeneous, isotropic, and elastic (for other cases, more complex types of fundamental solutions are required).

Regions 1 and 2 shown in Fig. 13.2 were then combined, where region 2 could be a boundary element (type (i)) or a finite element domain (Fig. 13.3). The

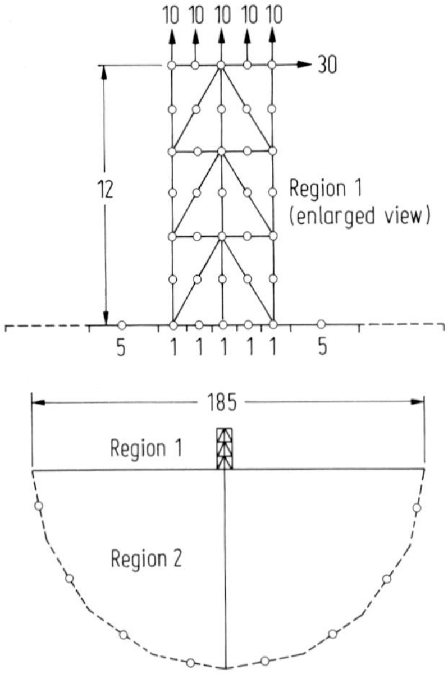

Fig. 13.2. Finite element and boundary element combination mesh

Table 13.1. Vertical displacements along loaded top ($\times 10^{-6}$)

F E solution	Combination solution
− 339	− 355
− 97	− 105
135	135
361	370
600	617

combination of the boundary element region 2 with the finite element region 1 was carried out by finding the equivalent symmetric stiffness matrix for the boundary element domain (see Eq. (13.19)). Five concentrated vertical loads along the top and an additional horizontal load acting at a corner were considered. The results obtained are shown in Table 13.1.

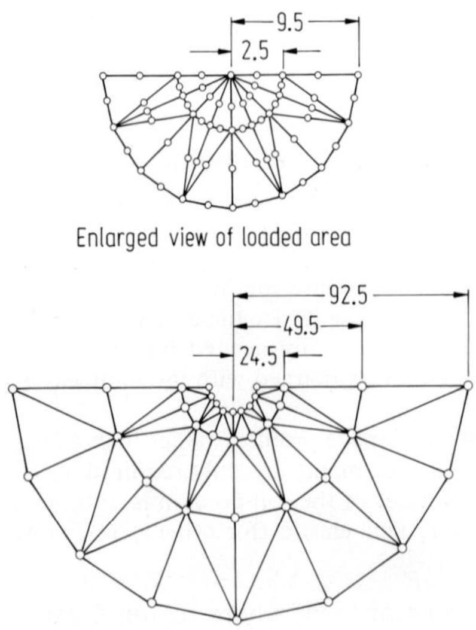

Fig. 13.3. Finite element mesh

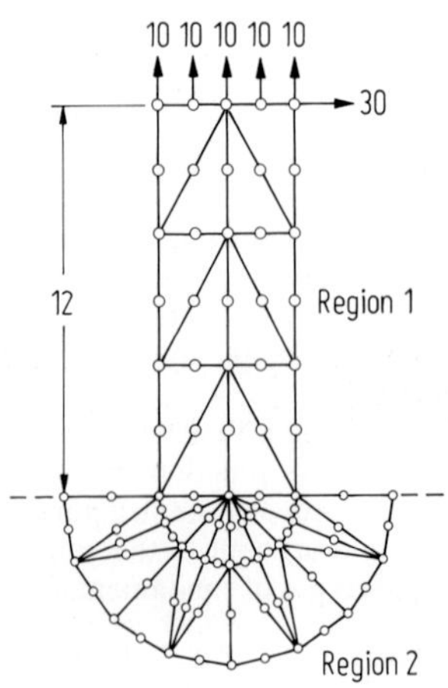

The combination solution is in good agreement with the results obtained using finite elements for the whole domain, but it is interesting to note that the foundation was adequately represented using only 37 boundary element nodes as opposed to 163 for the finite element case. Savings are more pronounced for three-dimensional applications.

13.3. Alternative Approach

This approach consists in using Galerkin's method, weighted residuals or energy considerations to find the equivalent finite element matrices of a boundary element region. Having found the boundary integral statements for region 1 (Fig. 13.1) we can write – neglecting domain integrals for simplicity –

$$c\,u + \int_{\Gamma} q^*\,u\,d\Gamma = \int_{\Gamma} u^*\,q\,d\Gamma \tag{13.35}$$

for potential problems, and the following expression for elastostatics

$$c_{ij}\,u_j + \int_{\Gamma} p_{ij}^*\,u_j\,d\Gamma = \int_{\Gamma} u_{ij}^*\,p_j\,d\Gamma\,. \tag{13.36}$$

As before let us consider that the 'potential energy' for the element can be written as in Section 13.2, i.e.

$$\Pi = \tfrac{1}{2}\int_{\Gamma} q\,u\,d\Gamma - \int_{\Gamma_2} \bar{q}\,u\,d\Gamma \tag{13.37}$$

and

$$\Pi = \tfrac{1}{2}\int_{\Gamma} p_i\,u_i\,d\Gamma - \int_{\Gamma_2} \bar{p}_i\,u_i\,d\Gamma\,. \tag{13.38}$$

Equation (13.35) can be substituted into Eq. (13.37), which gives for points inside the region

$$\begin{aligned}\Pi = \tfrac{1}{2}\int_{\Gamma} q(x)\left\{-\int_{\Gamma} q^*(\xi,x)\,u(\xi)\,d\Gamma(\xi) + \int_{\Gamma} u^*(\xi,x)\,q(\xi)\,d\Gamma(\xi)\right\} d\Gamma(x)\\ - \int_{\Gamma_2} \bar{q}\left\{-\int_{\Gamma} q^*(\xi,x)\,u(\xi)\,d\Gamma(\xi) + \int_{\Gamma} u^*(\xi,x)\,q(\xi)\,d\Gamma(\xi)\right\} d\Gamma(x)\,.\end{aligned} \tag{13.39}$$

Substitution of Eq. (13.36) into Eq. (13.38) produces the potential energy for an elastostatics problem:

$$\begin{aligned}\Pi = \tfrac{1}{2}\int_{\Gamma} p_i(x)\left\{-\int_{\Gamma} p_{ij}^*(\xi,x)\,u_j(\xi)\,d\Gamma(\xi) + \int_{\Gamma} u_{ij}^*(\xi,x)\,p_j(\xi)\,d\Gamma(\xi)\right\} d\Gamma(x)\\ - \int_{\Gamma_2} \bar{p}_i\left\{-\int_{\Gamma} p_{ij}^*(\xi,x)\,u_j(\xi)\,d\Gamma(\xi) + \int_{\Gamma} u_{ij}^*(\xi,x)\,p_j(\xi)\,d\Gamma(\xi)\right\} d\Gamma(x)\,.\end{aligned} \tag{13.40}$$

One can now express u and q (or u_j and p_j) in terms of the usual interpolation functions. This approach involves a double integration, i.e., first with respect to ξ and then with respect to x.

The next steps are similar to those of the energy approach discussed in Section 13.2. Margulies [7] has proposed using the domain integral expressions for the internal energy – see Eqs. (13.22) and (13.23) – rather than their boundary integral expressions (13.37) and (13.38). He points out that the matrices obtained are symmetric and that the procedure is not restricted to the variational formulation but can also be obtained with Galerkin's approach.

Margulies [7] also proposes using a bounded type of Green's function rather than the one for free space. In this case the function u^* can be defined such that

$$u^*(\xi, x) = 0 \qquad \text{on } \Gamma . \tag{13.41}$$

This type of function is such that Eqs. (13.35) and (13.36) simplify to

$$c\,u + \int_\Gamma q^*\,u\,d\Gamma = 0 \tag{13.42}$$

and

$$c_{ij}\,u_j + \int_\Gamma p_{ij}^*\,u_j\,d\Gamma = 0 . \tag{13.43}$$

Using these expressions, the energy formulation without the Γ_2 term can now be written

$$\Pi = \tfrac{1}{2}\int_\Gamma q(x)\left\{-\int_\Gamma q^*(\xi, x)\,u(\xi)\,d\Gamma(\xi)\right\} d\Gamma(x) \tag{13.44}$$

and

$$\Pi = \tfrac{1}{2}\int_\Gamma p_i(x)\left\{-\int_\Gamma p_{ij}^*(\xi, x)\,u_j(\xi)\,d\Gamma(\xi)\right\} d\Gamma(x) . \tag{13.45}$$

This approach also gives symmetric matrices after the unknowns are expressed in term of the interpolation functions.

A similar approach has been frequently used with indirect boundary element formulations [2, 30, 31]. In this case one can start with the following relationship for potential cases:

$$u = \int_\Gamma u^*\,\sigma\,d\Gamma \tag{13.46}$$

and

$$q = \int_\Gamma \frac{\partial u^*}{\partial n}\,\sigma\,d\Gamma . \tag{13.47}$$

We can now define the following energy functional:

$$\Pi(\sigma) = -\tfrac{1}{2}\int_\Gamma \sigma\,q\,d\Gamma . \tag{13.48}$$

Substituting Eq. (13.47), one can write

$$\Pi(\sigma) = -\tfrac{1}{2} \int_{\Gamma} \sigma(x) \left\{ \int_{\Gamma} \frac{\partial u^* (\xi, x)}{\partial n} \sigma(\xi)\, d\Gamma(\xi) \right\} d\Gamma(x) . \tag{13.49}$$

This formulation also gives symmetric matrices. Similar considerations apply for elasticity problems.

13.4. Internal Fluid Problems

Consider now a type of fluid–structure interaction problem, when the fluid is enclosed within a vessel as represented in Fig. 13.4. The vessel may be represented using standard finite elements and the fluid using finite or boundary elements. It seems more appropriate for this type of problems to represent the fluid by means of boundary elements as we are not interested in the internal nodes but wish to construct the fluid matrices and combine them with the structural system.

Let us first assume that the fluid is incompressible and inviscid and that its behavior can be represented by a potential that we will now call φ. The boundary element influence matrices for this case can be written

$$\boldsymbol{H}\boldsymbol{\varphi} = \boldsymbol{G}\boldsymbol{Q} , \tag{13.50}$$

where $\boldsymbol{\varphi}$ and $\boldsymbol{Q}$ are the nodal potential and normal velocity vectors, respectively.

Consider for simplicity that the movement is harmonic. Hence, the velocity at any point k can be written in terms of the displacement U_k as follows:

$$Q_k = i\,\omega\, U_k . \tag{13.51}$$

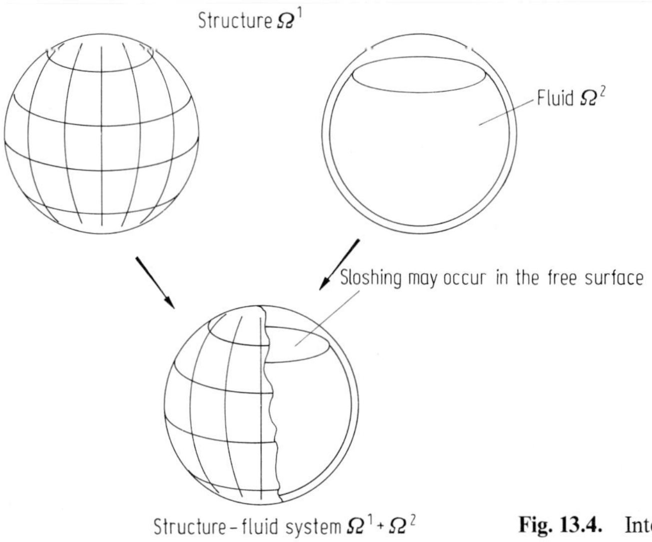

Fig. 13.4. Internal fluid problem

The velocity potential can be related to the pressure P_k through Bernoulli's equation. For harmonic motion this gives

$$\Phi_k = -\frac{1}{i\,\omega\,\varrho} P_k . \tag{13.52}$$

Hence, we can now write Eq. (13.50) as

$$\boldsymbol{HP} = \omega^2 \varrho\, \boldsymbol{GU} . \tag{13.53}$$

Inverting $\boldsymbol{H}$, gives

$$\boldsymbol{P} = \omega^2 \varrho\, \boldsymbol{H}^{-1} \boldsymbol{GU} . \tag{13.54}$$

In order to combine Eq. (13.54) with the finite element system describing the structure, one needs to work in terms of equivalent nodal forces. The relationship between node pressures and equivalent forces can be written in terms of the distribution matrix $\boldsymbol{N}$ such that the nodal forces are

$$\boldsymbol{F} = \boldsymbol{NP} . \tag{13.55}$$

Equation (13.55) now becomes

$$\boldsymbol{F} = \omega^2 \varrho (\boldsymbol{NH}^{-1} \boldsymbol{G})\, \boldsymbol{U} = \boldsymbol{M}_F \boldsymbol{U} . \tag{13.56}$$

The form of $\boldsymbol{M}_F$ is the same as for a finite element formulation. Hence, one can now superimpose it to the mass of the vessel and obtain the final matrices for the fluid–structure system. Notice that the $\boldsymbol{H}$ and $\boldsymbol{G}$ matrices of equation (13.50) have the same form as for a static potential problem. $\boldsymbol{M}_F$ will generally be asymmetric but can be symmetrized by applying the procedure shown in Eq. (13.19).

13.4.1. Free-Surface Boundary Condition

In order to represent sloshing on the free surface of the fluid, one needs to take into consideration the kinematic condition, i.e.,

$$\frac{\partial \varphi}{\partial z} = -\frac{1}{g} \frac{\partial^2 \varphi}{\partial t^2} , \tag{13.57}$$

where z is the normal to the free surface. Notice that $\partial \varphi / \partial z$ is the velocity of the fluid in the vertical direction.

For harmonic problems we can write

$$Q = \frac{\partial \Phi}{\partial z} = \frac{\omega^2}{g} \Phi . \tag{13.58}$$

This normal velocity can be written as a function of the $\boldsymbol{U}$ displacements and the potential as a function of the pressure as shown earlier, i.e., for any point k we have

$$Q_k = i\,\omega\, U_k = -\frac{\omega}{g} \frac{1}{i\,\varrho} P_k . \tag{13.59}$$

Thus,

$$P_k = g\,\varrho\, U_k . \tag{13.60}$$

Once again these pressures can be converted into equivalent nodal forces using the distribution matrix $\boldsymbol{N}$. The equivalent nodal forces $\boldsymbol{F}_s$ on the free surface are defined as

$$\boldsymbol{F}_s = g\,\varrho\,\boldsymbol{N}\boldsymbol{U}_s\,, \tag{13.61}$$

where $\boldsymbol{F}_s$ and $\boldsymbol{U}_s$ represent the forces and displacements on the nodes of the sloshing or free surface. Relationship (13.61) applies only on the nodes on the free surface and can be included in the left-hand side vector of formula (13.56). The nodes on the sloshing surface are then usually eliminated by a matrix condensation technique as they are not directly related to the nodes of the structure.

Example 13.2. *Sloshing in a Rigid Container.* Komatsu [34] has considered several interesting examples involving sloshing, starting with a fluid in a rigid container. In this case the equations of motion under harmonic vibrations can be written

$$(\boldsymbol{K} - \omega^2 \boldsymbol{M})\,\boldsymbol{U} = \boldsymbol{0}\,, \tag{a}$$

where the $\boldsymbol{K}$ matrix is due to the contribution of the free-surface sloshing only and $\boldsymbol{M}$ is the fluid mass matrix. The $\boldsymbol{U}$ vector applies to the free-surface nodes only. By varying ω, the eigenvalues and corresponding eigenvectors can be found.

Table 13.2 presents the sloshing eigenvalues for the fluid in a rigid circular cylinder container with a flat bottom. The results were obtained using a ring boundary element and different meshes. The mesh $10 \times 10 \times 10$ in the table (see Fig. 13.5), for instance, represents 10 elements each for the bottom, wall, and free surface. The eigenvalues are compared with an exact solution [35] with which they are in close agreement.

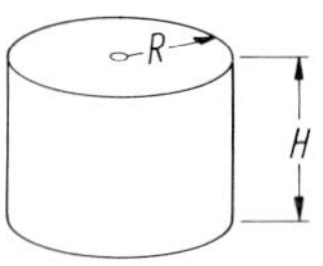

Fig. 13.5. Circular cylindrical tank

Table 13.2. Nondimensional frequencies of sloshing in a circular cylindrical tank ($\omega^2 R/g$; $H/R = 1.0$)

n	m	Mesh				Exact [35]
		$10 \times 10 \times 10$	$15 \times 15 \times 15$	$20 \times 20 \times 20$	$25 \times 25 \times 25$	
0	1	3.86	3.85	3.84	3.84	3.83
	2	7.07	7.08	7.07	7.06	7.02
	3	10.19	10.27	10.27	10.25	10.17
1	1	1.81	1.79	1.77	1.76	1.75
	2	5.48	5.43	5.39	5.38	5.33
	3	8.84	8.74	8.69	8.65	8.54
2	1	3.16	3.12	3.09	3.07	3.04
	2	6.95	6.85	6.81	6.78	6.71
	3	10.42	10.25	10.17	10.13	9.97
3	1	4.40	4.32	4.27	4.25	4.20
	2	8.39	8.23	8.16	8.12	8.02
	3	12.02	11.74	11.61	11.55	11.35

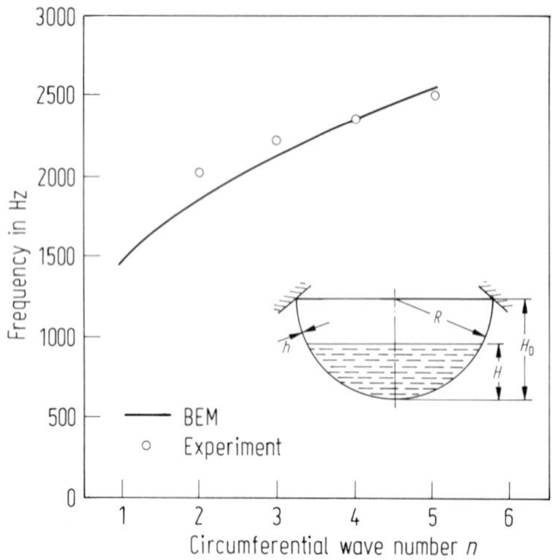

Fig. 13.6. Frequencies of elastic modes of a hemispherical shell partially filled with water. $E = 7000\ \mathrm{kg/mm^2}$, $\nu = 0.3$, $\varrho = 2.8 \times 10^{-6}\ \mathrm{kg\ s^2/cm^4}$, $h = 0.5$ mm, $R = 10$ cm, $H/R = 0.59$, 15 shell elements and 21 ring boundary elements

Example 13.3. *Sloshing in a Flexible Container.* This example is also due to Komatsu [34] and represents the frequencies of a hemispherical shell partially filled with water as shown in Fig. 13.6. The numerical and experimental results shown by Komatsu are in reasonable agreement.

13.4.2. Extension to Compressible Fluid

The extension of the above theory to compressible fluid has important applications in engineering such as problems involving explosions inside pressure vessels, compressive wave propagation, etc. Mathematically it follows the lines of the wave diffraction problems already discussed (Chapter 10). Compressibility of the fluid is important when the frequencies become higher but can usually be neglected at low frequencies [33].

The governing equation for compressible potential flow is

$$\nabla^2 \varphi = \frac{1}{c^2}\frac{\partial^2 \varphi}{\partial t^2}, \tag{13.62}$$

where c is the speed of sound in the fluid. For harmonic motion Eq. (13.62) becomes

$$\nabla^2 \Phi + \varkappa^2 \Phi = 0, \tag{13.63}$$

where $\varkappa$ is the wave number, $\varkappa = \omega/c$.

The Green's function corresponding to Eq. (13.63) is

$$\Phi^* = \frac{i}{4} \mathrm{H}_0^{(1)}(\varkappa r) \qquad \text{for the two-dimensional case} \tag{13.64}$$

and

$$\Phi^* = \frac{1}{4\pi r}\exp(-i\varkappa r) \qquad \text{for three dimensions.} \tag{13.65}$$

The rest of the development follows the same lines as before (Chapter 10).

13.5. Approximate Boundary Elements

We have already pointed out that approximate boundary solutions can be used to form boundary elements. These are of practical interest when the fundamental solution is difficult to obtain or cumbersome to use. They also reduce the size of the resulting matrices as the boundary nodes are now only related to their neighbors rather than to all the other nodes [20, 21, 31].

To illustrate the technique, consider the case of the Laplace equation for simplicity, i.e.,

$$\int_{\Omega} (\nabla^2 u^*)\, u \, d\Omega = \int_{\Gamma} u\, q^* \, d\Gamma - \int_{\Gamma} q\, u^* \, d\Gamma \, , \tag{13.66}$$

where Γ is the total boundary. Let us assume that the above equation refers to the external region of Fig. 13.7 which extends to infinity. The fundamental solution for the three-dimensional Laplace equation is

$$u^* = \frac{1}{4\pi r} \, . \tag{13.67}$$

For any point in the internal domain – including those near but not on the interface Γ_I – one can write

$$\int_{\Gamma} u\, q^* \, d\Gamma = \int_{\Gamma} u^*\, q \, d\Gamma \, . \tag{13.68}$$

Notice that the boundary integrals on the Γ part of the boundary at infinity will disappear when $r \to \infty$.

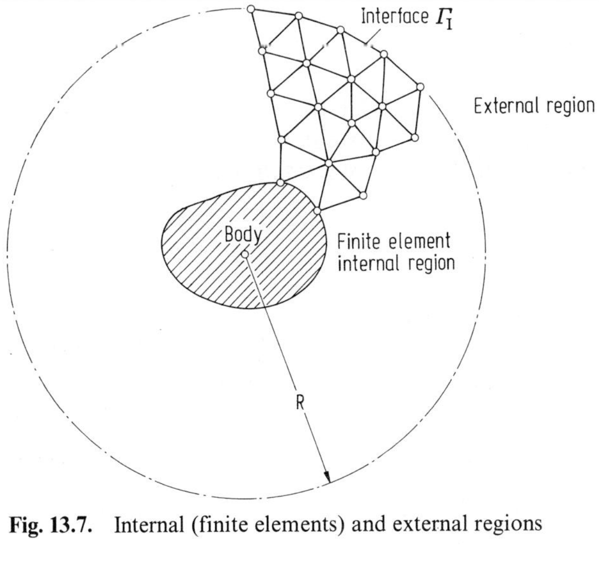

Fig. 13.7. Internal (finite elements) and external regions

Substituting the fundamental solution into Eq. (13.68) leads to

$$\int_{\Gamma_I} \left(\frac{1}{r}\right) q\, d\Gamma = \int_{\Gamma_I} u \frac{\partial}{\partial n}\left(\frac{1}{r}\right) d\Gamma . \tag{13.69}$$

This expression has been obtained by taking the reference point outside the domain – i.e., in the internal region – and because of this, Eq. (13.68) does not present the c term. The boundary values are still interrelated, resulting in a nonbanded system of equations. We can, however, simplify Eq. (13.69) if Γ_I is considered to be a sphere of sufficiently large radius. As R is constant on the sphere, we can write $d\Gamma = R^2 \sin\varphi\, d\theta\, d\varphi$ (θ and φ are the angular coordinates). Notice that $n \equiv r$ on the sphere. Hence, Eq. (13.69) becomes

$$\int_0^{2\pi}\int_0^{2\pi} \left\{\frac{1}{R}\frac{\partial u}{\partial R} + u\left(\frac{1}{R^2}\right)\right\} R^2 \sin\varphi\, d\theta\, d\varphi = 0 . \tag{13.70}$$

This is a special form of Sommerfeld's radiation condition which can also be written approximately on Γ_I as

$$r\frac{\partial u}{\partial r} + u = \text{constant}. \tag{13.71}$$

Equation (13.71) can be applied at each node independently of any others but its neighbors, which gives a banded system of equations as in finite elements.

The importance of the above procedure is that it can be generalized to find approximate radiation conditions in cases where the fundamental solution is complicated. Consider, for instance, the case of the wave or Helmholtz equation in two dimensions, i.e.,

$$\nabla^2 u + \varkappa^2 u = 0 \qquad \text{in } \Omega , \tag{13.72}$$

where $\varkappa$ is the wave number and u the potential function. Referring again to Fig. 13.7 – but assuming now that the problem is two-dimensional – the boundary element region extends to infinity and the fundamental solution is

$$u^* = \frac{i}{4} H_0^{(1)}(\varkappa r) , \tag{13.73}$$

where H is a Hankel function. r is the distance from the point under consideration to any other point on Γ_I.

The inverse weighted residual statement for Eq. (13.72) with similar boundary conditions as before (i.e., essential and natural) is

$$\int_\Omega (\nabla^2 u^* + \varkappa^2 u^*)\, u\, d\Omega = \int_{\Gamma_I} q^*\, u\, d\Gamma - \int_{\Gamma_I} u^*\, q\, d\Gamma . \tag{13.74}$$

As the function u^* satisfies the Helmholtz equation, one can write for any point external to the boundary element region

$$\int_{\Gamma_I} u^* \, q \, d\Gamma = \int_{\Gamma_I} u \, q^* \, d\Gamma \tag{13.75}$$

or in terms of the fundamental solution

$$\int_{\Gamma_I} H_0^{(1)}(\varkappa r) \, q \, d\Gamma = \int_{\Gamma_I} \frac{\partial H_0^{(1)}(\varkappa r)}{\partial n} u \, d\Gamma \, . \tag{13.76}$$

One can simplify this formulation if the observation or reference point is assumed to be at a distance from the region of interest. This means that one can use an asymptotic expansion for the Hankel function and its derivative (note $n \equiv r$ over a circle):

$$\begin{aligned} H_0^{(1)}(\varkappa r) &\simeq \sqrt{\frac{2}{\pi \varkappa r}} \exp\left\{i\left(\varkappa r - \frac{\pi}{4}\right)\right\}, \\ \frac{\partial H_0^{(1)}(\varkappa r)}{\partial n} &= \frac{\partial H_0^{(1)}(\varkappa r)}{\partial r} = -\varkappa H_1^{(1)}(\varkappa r) \simeq i \varkappa \sqrt{\frac{2}{\pi \varkappa r}} \exp\left\{i\left(\varkappa r - \frac{\pi}{4}\right)\right\}. \end{aligned} \tag{13.77}$$

If Γ_I is considered to be a circle of large radius R, one obtains

$$\sqrt{\frac{2}{\pi \varkappa}} e^{-i\pi/4} \int_{\Gamma_I} \frac{1}{\sqrt{R}} \left(\frac{\partial u}{\partial R} - i \varkappa u\right) e^{i\varkappa R} \, d\Gamma = 0 \, . \tag{13.78}$$

As R is constant over Γ, one can write $d\Gamma = R \, d\theta$, i.e.,

$$\int_0^{2\pi} \sqrt{R} \left(\frac{\partial u}{\partial R} - i \varkappa u\right) d\theta = 0 \, . \tag{13.79}$$

This is yet another form of Sommerfeld's condition which can be written on Γ_I as

$$\frac{\partial u}{\partial r} - i \varkappa u = 0 \, . \tag{13.80}$$

Similarly one can deduce the appropriate condition for stress analysis, soil dynamics, fluid flow, and many other complex problems.

Example 13.4. *Wave Diffraction around a Cylindrical Obstruction* [19]. The case of a single vertical column subject to an incident harmonic wave was studied to determine how the application of the radiation condition affects the accuracy of the solution.

The column was surrounded by a finite element mesh as shown in Fig. 13.8 and wave diffraction results were found for different wavelengths, with the application of the radiation condition on the external boundary.

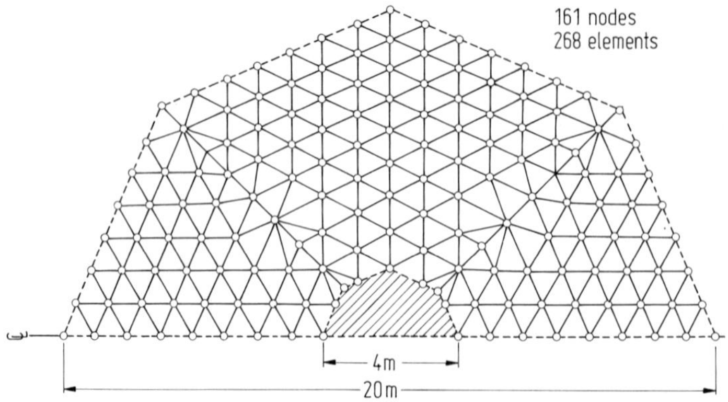

Fig. 13.8. Finite element mesh

As a test to determine the adequacy of the radiation condition in representing a train of plane harmonic waves, the case with no solid cylinder was studied first. For long waves (λ, wavelength $\simeq$ 30 m) the results were very accurate, within 3% of the exact solution. When the wavelength was reduced, the errors tended to increase, which was to be expected as the element mesh became too coarse. It was found that for linear elements such as the ones used here, ideally eight elements per wavelength should be used.

A further test was run to compare our results with those obtained by Mei [20] using a finite element mesh and a series of Hankel functions. The results are for an incident wave of $\lambda = 2\pi$ m and unit incident surface elevation of frequency $\omega = 3.1321$. In Fig. 13.9 the results of the exact solution, the finite element solution due to Mei, and the solution obtained in [19] are compared. The last solution compares favorably taking into consideration the coarseness of the mesh around the cylinder. Mei's solution, for instance, uses 18 elements around the cylinder and represents the geometry of the obstruction better.

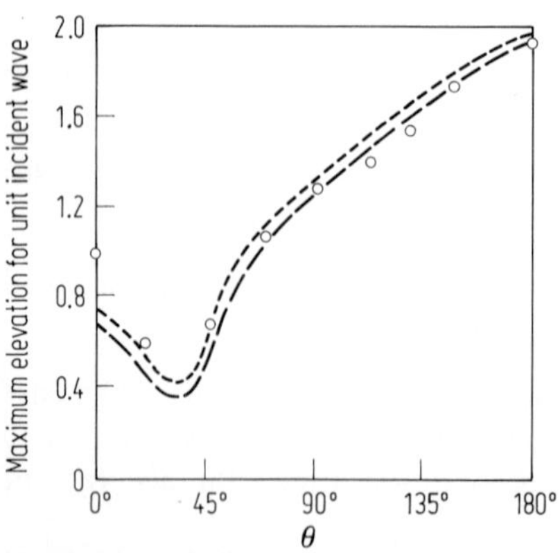

Fig. 13.9. Results for maximum surface elevation around a circular cylinder; radius = 2 m, wavelength = 2π m; (- - -) from Mei [20], (—) analytical; ○ – Ref. [19] results (mesh in Fig. 13.8)

Example 13.5. *Harbor Resonance* [21]. As another example of the application of the radiation condition, we will consider the harbor resonance equation, i.e.,

$$\frac{\partial}{\partial x}\left(h\,\frac{\partial u}{\partial x}\right)+\frac{\partial}{\partial y}\left(h\,\frac{\partial u}{\partial y}\right)+\varkappa^2\,h\,u=0 \tag{a}$$

with boundary condition

$$h\,\frac{\partial u}{\partial n}=q \qquad \text{on } \Gamma_2, \tag{b}$$

where u is the wave elevation referred to the still water level with in-plane coordinates x and y, h is the depth, q is a given value of flux on Γ_2 boundaries, and $\varkappa$ is the wave number.

We have the situation shown in Fig. 13.10. One needs to apply a boundary condition on the fictitious ocean boundary Γ_1 which will allow wave energy to pass across Γ_1 without distortion. In order to do this we need to split the wave elevation into three parts:

(i) the incident field u_0;
(ii) the reflected field calculated from the plane wave reflections by the coastline and called u_R (see Fig. 13.10).
(iii) The scattered field u_S originating from inside the harbor.

Hence the total wave field is

$$u=u_0+u_R+u_S \tag{c}$$

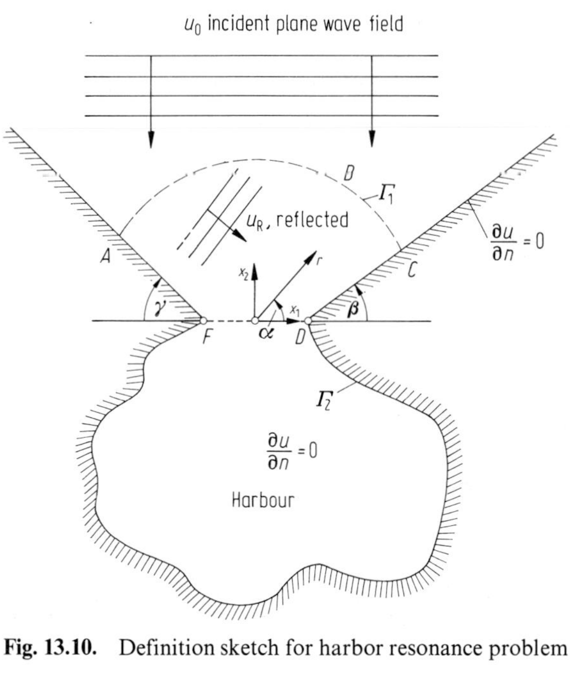

Fig. 13.10. Definition sketch for harbor resonance problem

with the incident field given by

$$u_0 = a \exp\,(-\,i\,\varkappa\,r\cos\,(\theta-\alpha))\,. \tag{d}$$

We know that the scattered wave field u_S will be travelling away from the harbor mouth on Γ_1. This fact is represented by the Sommerfeld boundary condition which may be written

$$\lim_{r\to\infty}\left[\sqrt{r}\left(\frac{\partial u_S}{\partial r}+i\,\varkappa\,u_S\right)\right]=0 \tag{e}$$

or approximately

$$\frac{\partial u_S}{\partial r}+i\,\varkappa\,u_S=0 \qquad \text{on } \Gamma_1\,. \tag{f}$$

Using equation (f) we may write in terms of the total field u

$$\frac{\partial u}{\partial r}+i\,\varkappa\,u=\frac{\partial}{\partial r}(u_0+u_R)+i\,\varkappa(u_0+u_R)=f\,. \tag{g}$$

Note that the right-hand side is a known function f of the horizontal coordinates, the angles determining the local geometry and the angle of incidence of the wave field u_0.

We may now incorporate this boundary condition into our variational-type statement to obtain

$$\begin{aligned}&\int_\Omega\left[\frac{\partial}{\partial x}\left(h\,\frac{\partial u}{\partial x}\right)+\frac{\partial}{\partial y}\left(h\,\frac{\partial u}{\partial y}\right)+\varkappa^2 h\,u\right]u^*\,d\Omega\\ &=\int_{\Gamma_2}\left(h\,\frac{\partial u}{\partial n}-q\right)u^*\,d\Gamma+\int_{\Gamma_1}h\left(\frac{\partial u}{\partial n}+i\,\varkappa\,u-f\right)u^*\,d\Gamma\end{aligned} \tag{h}$$

which after integration by parts gives

$$\begin{aligned}&\int_\Omega\left(h\,\frac{\partial u}{\partial x}\,\frac{\partial u^*}{\partial x}+h\,\frac{\partial u}{\partial y}\,\frac{\partial u^*}{\partial y}-\varkappa^2 hu\,u^*\right)d\Omega\\ &=\int_{\Gamma_2}q\,u^*\,d\Gamma+\int_{\Gamma_1}h\,(f-i\,\varkappa\,u)\,u^*\,d\Gamma\end{aligned} \tag{i}$$

which is the starting expression for our finite element part formulation incorporating the boundary condition (f) on Γ_1. Numerical results with and without the use of the radiation condition were obtained for the classical case of the old Duncan Basin within Table Bay in South Africa.

To simulate deterministically the effect of waves entering the harbor one discretizes it into finite elements (Fig. 13.11). Along the contour ABCDEF the usual zero normal flux condition $\partial u/\partial n = 0$ is applied. At the harbor entrance AF one applies the radiation condition for an incident plane wave of unit amplitude entering the harbor in the negative y direction, for the moment ignoring reflections from the neighboring coastline. The resulting elevations at node 70 for unit incident waves of different frequencies are plotted in Fig. 13.12. The first peaks occur at the same frequencies as those predicted by the harmonic analysis given by Brebbia and Adey [32]. For instance, the first peak represents the first significant period of about 11.45 min which is the mode corresponding to water flowing in and out of the basin and is called the "pumping mode". The second peak also corresponds to a simple motion, the so-called "sloshing mode". The results have been plotted up to $\omega = 0.4$ rad s^{-1} which is the lower bound of the wind generated part of the wave spectrum.

It is interesting to note that after a series of mainly longitudinal modes, the transverse modes start to play a more significant part. The effect of these modes is considerable especially from the probabilistic response point of view. The application of the radiation condition at the harbor mouth adequately represents the input of wave energy into the system as well as energy passing out of the harbor. A more extensive discussion of this example can be found in Walker and Brebbia [21].

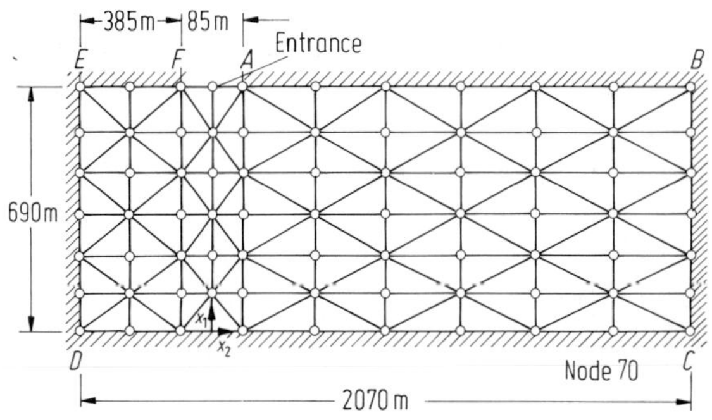

Fig. 13.11. Mesh for the old Duncan Basin

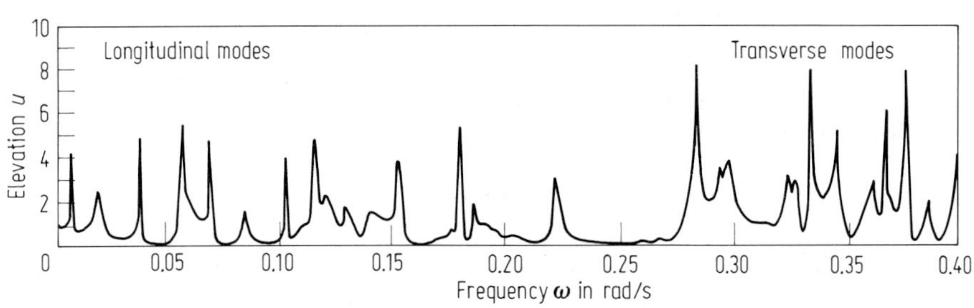

Fig. 13.12. Maximum surface elevation at node 70 with radiation

13.6. Approximate Finite Elements

Applications of finite elements are usually restricted to finite domains and to comparatively smooth variations of the interpolation functions, although sometimes they can be combined with boundary elements as seen in the previous sections. For problems such as fracture mechanics or infinite domains special finite elements can also be developed. These elements may simply be based on including appropriate singular functions – of order $r^{1/2}$ for instance – in fracture mechanics or developing "infinite" elements to take radiation into consideration. Finite elements extending to infinity are especially important in coupled problems and have received a great deal of attention. Although they may be used instead of boundary elements in certain problems, it is important to understand that they are essentially different as they are based on the classical finite element approach, which is a "domain" technique. Assume, for instance, that one has a finite element potential problem, but instead of using the standard basis functions for u and δu one now looks for decaying functions to model the behavior of the potential and flux when the domain tends to infinity. This is better explained by considering the interface shown in Fig. 13.13. The dashed lines represent planes normal to the interface emanating from the corners of the element – which for simplicity is assumed to have straigth sides. The ζ coordinate emanating from the center of the element can be used as a reference axis.

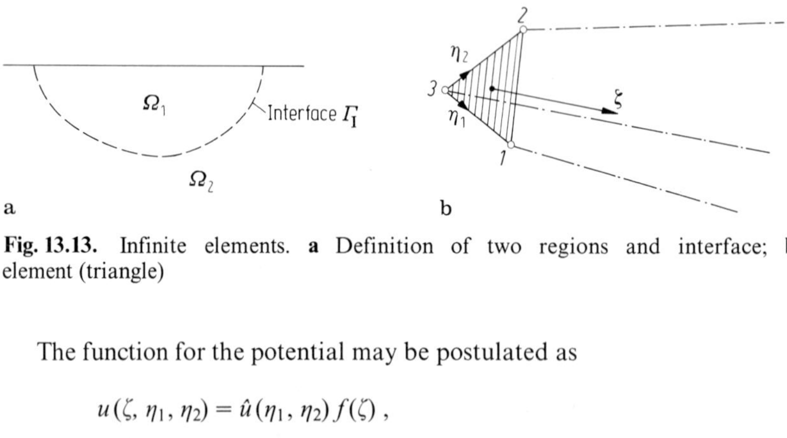

Fig. 13.13. Infinite elements. **a** Definition of two regions and interface; **b** Interface element (triangle)

The function for the potential may be postulated as

$$u(\zeta, \eta_1, \eta_2) = \hat{u}(\eta_1, \eta_2) f(\zeta) , \tag{13.81}$$

where $\hat{u}(\eta_1, \eta_2)$ is the standard two-dimensional finite element interpolation function with η_1 and η_2 the homogeneous coordinates for the triangular element. $f(\zeta)$ is a decay function that reduces the magnitude of u as ζ increases. It is unity at $\zeta = 0$ and tends to zero at infinity. The function proposed by Ungless [17, 24] was simply

$$f(\zeta) = \frac{1}{1 + \zeta/L} . \tag{13.82}$$

We can now substitute this expression into the finite element equations and work as usual. The difference now is that the integration over Ω is carried out over the infinite domain of the element and one must resort to numerical integration.

Instead of function (13.82), Bettess [18] has used exponential decay. His shape functions in the ζ direction are defined as Lagrange polynomials multiplied by exponential decay terms. In this case a set of internal nodes is needed because of the type of polynomials used, the simplest of which is a linear function.

The form of the solution will depend on the coefficients taken for the exponential decay, i.e., on the λ parameter of the equation

$$f(\zeta) = F(\zeta)\, e^{-\zeta/\lambda}. \tag{13.83}$$

The advantage of infinite elements is that they can be used for a series of problems for which the fundamental solution required for boundary elements may be difficult to determine. The method does not give an accurate indication of the behavior of the system towards infinity, but in this way the effect of the far region on the domain of interest can be introduced easily.

Example 13.6 *Diffraction and Refraction of Waves by a Parabolic Shoal, Surmounted by a Cylindrical Island* [22]. Figure 13.14 shows the geometry of the

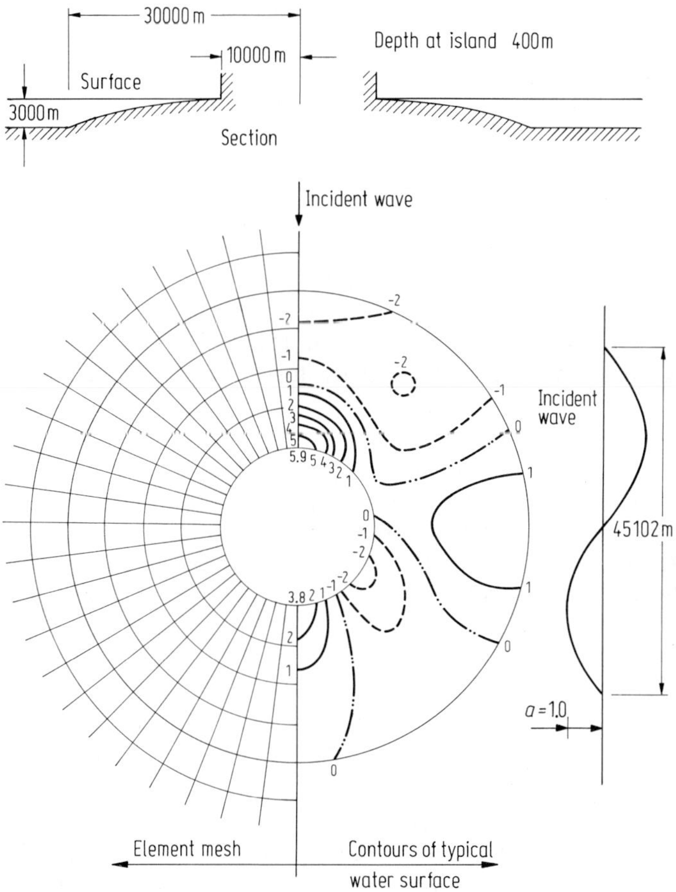

Fig. 13.14. Wave diffraction by parabolic shoal and cylinder

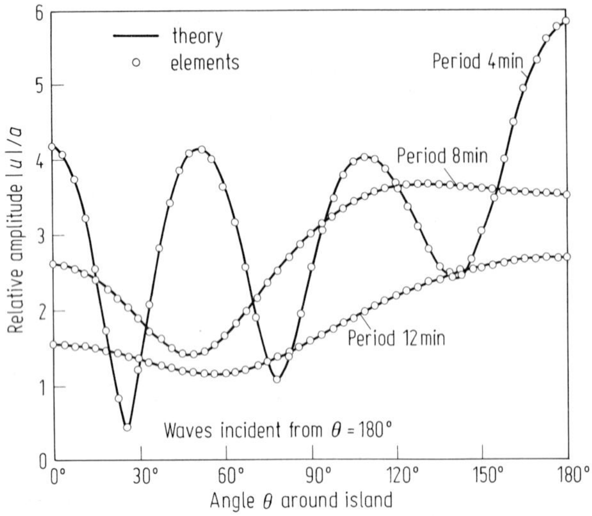

Fig. 13.15. Relative amplitudes on cylinder

problem and the element mesh used. The finite element mesh is enclosed by a number of *infinite* elements; the shape function in the radial direction is given by

$$P(r)\, e^{-r/L}\, e^{-i\varkappa r}\,. \tag{a}$$

(for time dependence insert $e^{i\omega t}$). Here $P(r)$ is a polynomial in r, L is the so-called decay length, and $\varkappa$ is the wavenumber corresponding to the frequency ω. This shape function satisfies the Sommerfeld radiation condition. L was chosen so that near the island the decay $e^{-r/L}$ roughly matched the decay of the first term of the general series solution, $H_0^{(1)}(\varkappa r)$, where $H_0^{(1)}$ is the Hankel function of zero order of the first kind. See Bettess and Zienkiewicz for a fuller explanation of the above theory [22].

Figure 13.15 shows a comparison of the elevations at the island itself compared with the analytical solution given by Homma [37] and Vastano and Reid [38]. The agreement is very good, especially at moderate wavelengths.

The weakness of this method is that to determine the parameter L "some" knowledge of the exact solution is required before it can be used. However, the method is efficient if this information is available.

References

1. Shaw, R. P., and Falby, W., FEBIE: A combined finite element-boundary integral equations method, in *Proc. 1st Symposium on Innovative Numerical Analysis in Applied Engineering Science,* Versailles, CETIM, 1977.
2. McDonald, B. H., and Wexler, A., Finite element solution of unbounded field problems, IEEE Trans. Microwave Theory Tech. **MTT-20,** 841–847 (1972).
3. Osias, J. R., Wilson, R. B., and Seitelman, L. A., Combined boundary integral equation finite element analysis of solids, in *Proc. 1st Symp. on Innovative Numerical Analysis in Applied Engineering Science,* Versailles, CETIM, 1977.

4. Brebbia, C. A., Basis of boundary elements, in *Recent Advances in Boundary Element Methods* (C. A. Brebbia, Ed.), Pentech Press, London, 1978.
5. Brebbia, C. A., and Georgiou, P., Combination of boundary and finite elements for elastostatics, Appl. Math. Modelling **3,** 212–220 (1979).
6. Zienkiewicz, O. C., Kelly, D. W., and Bettess, P., The coupling of the finite element method and boundary solution procedures, Int. J. Numerical Meth. Engng. **11,** 355–375 (1977).
7. Margulies, M., Combination of the boundary element and finite element methods, in *Progress in Boundary Element Methods, Vol. 1* (C. A. Brebbia, Ed.), Pentech Press, London; Halstead Press, New York, 1981.
8. Silvester, P., and Hsieh, M. S., Finite element solution of two dimensional exterior field problems, Proc. IEE **118,** 1743–1747 (1971).
9. Chari, M. V. K., and Silvester, P. (Eds.), *Finite Elements in Electrical and Magnetic Field Problems,* Wiley, New York, 1978.
10. Margulies, M., Exact treatment of the exterior problem in the combined FEM-BEM, in *New Developments in Boundary Element Methods* (C. A. Brebbia, Ed.), CML Publications, Southampton; Butterworths, London, 1980.
11. Felippa, C. A., Interfacing finite elements and boundary element discretizations, in *Boundary Element Methods* (C. A. Brebbia, Ed.), Springer-Verlag, Berlin, 1981.
12. Hartman, F., The derivation of stiffness matrices from integral equations, in *Boundary Element Methods* (C. A. Brebbia, Ed.), Springer-Verlag, Berlin, 1981.
13. Volait, F., Three dimensional super-element by the boundary integral equation method for elastostatics, in *Boundary Element Methods* (C. A. Brebbia, Ed.), Springer-Verlag, Berlin, 1981.
14. Beer, G., and Meek, J. L., The coupling of boundary and finite element methods for infinite domain problems in elasto-plasticity, in *Boundary Element Methods* (C. A. Brebbia, Ed.), Springer-Verlag, Berlin, 1981.
15. Dendrou, B. A., and Dendrou, S. A., A finite element-boundary integral scheme to formulate rock-effects on the lines of an underground intersection, in *Boundary Element Methods* (C. A. Brebbia, Ed.), Springer-Verlag, Berlin, 1981.
16. Katz, C., The use of Green's functions in the numerical analysis of potential, elastic and plate bending problems, in *Boundary Element Methods* (C. A. Brebbia, Ed.), Springer-Verlag, Berlin, 1981.
17. Ungless, R. F., An infinite finite element, MSc. thesis, Dept. of Civil Eng., University of British Columbia, Canada, 1973.
18. Bettess, P., Infinite elements, Int. J. Numerical Meth. Engng. **11,** 53–64 (1977).
19. Brebbia, C. A., and Walker, S., Simplified boundary elements for radiation problems, Appl. Math. Modelling **2,** 135–137 (1978).
20. Chen, H. S., and Mei, C. C., Oscillations and wave forces in a man-made harbor in the open sea, in *Proc. 10th Symp. on Naval Hydrodynamics,* pp. 573–594, London, 1974.
21. Walker, S., and Brebbia, C. A., Harbour resonance problems using finite elements, Adv. Water Res. **1,** 205–211 (1978).
22. Bettess, P., and Zienkiewicz, O. C., Diffraction and refraction of surface waves using finite and infinite elements, Int. J. Numerical Meth. Engng. **11,** 1271–1290 (1977).
23. Beer, G., and Meek, J. L., A boundary finite element for underground mining applications, in *New Developments in Boundary Element Methods* (C. A. Brebbia, Ed.), CML Publications, Southampton; Butterworths, London, 1980.
24. Anderson, D. L., and Ungless, R. L., Infinite finite elements, in *Proc. Int. Symp. on Innovative Numerical Analysis in Applied Engineering Science,* Versailles, CETIM, 1977.
25. Gartling, D. K., and Becker, E. B., Finite element analysis of viscous incompressible fluid flow, Comp. Meth. Appl. Mech. Engng. **8,** 51–60 (1976).
26. Brebbia, C. A., and Connor, J. J., *Finite Element Techniques for Fluid Flow,* 2nd ed., Butterworths, London, 1977.
27. Georgiou, P., The coupling of the direct boundary element method with the finite element displacement technique in elastostatics, Ph.D. thesis, Southampton University, 1981.
28. Chaudonneret, M., On the discontinuity of the stress vector in the boundary integral equation method for elastic analysis, in *Recent Advances in Boundary Element Methods* (C. A. Brebbia, Ed.), Pentech Press, London, 1978.

29. Jeng, G., and Wexler, A., Self-adjoint variational formulation of problems having non-self adjoint operators, IEEE Trans. Microwave Theory Tech. **MTT-26,** 91–94 (1978).
30. Jaswon, M. A., and Symm, G. T., *Integral Equation Methods in Potential Theory and Elastostatics,* Academic Press, London, 1977.
31. Alliney, S., An approximate numerical model for radiation problems, Appl. Math. Modelling **6,** 192–196 (1982).
32. Brebbia, C. A., and Adey, R., Circulation problems, in *Proc. Finite Element Symp. Atlas Computer,* Rutherford Laboratory, Didcot, UK, HMSO, London, 1974.
33. Walker, S., Boundary elements in fluid/structure interaction problems, in *New Developments in Boundary Element Methods* (C. A. Brebbia, Ed.), CML Publications, Southampton; Butterworths, London, 1980.
34. Komatsu, T., Fluid-structure interaction, in *Progress in Boundary Element Methods, Vol. 2,* Pentech Press, London; Springer-Verlag, New York, 1982.
35. Abramson, H. N. (Ed.), The dynamic behavior of fluids in moving containers, NASA-SP-106, 1966.
36. Brebbia, C. A. (Ed.), *Boundary Element Methods in Engineering,* Proceedings of the 4th International Conference on Boundary Element Methods, Springer-Verlag, Berlin, 1982.
37. Homma, S., On the behavior of seismic sea waves around circular island, Geophys. Magazine **21,** 199–208 (1950).
38. Vastano, A. C., and Reid, R. O., Tsunami response for islands: Verification of a numerical procedure, J. Marine Res. **25,** 129–139 (1967).

Chapter 14

Computer Program for Two-Dimensional Elastostatics

14.1. Introduction

We will now describe a simple Fortran computer program for the solution of two-dimensional elastostatic problems using linear boundary elements, i.e., elements with linear variations of displacements and tractions.

The program requires less steps than a finite element code as it is unnecessary to have a special assembler subroutine. The number of unknowns is substantially smaller as only nodes on the boundary are required. Figure 14.1 compares the main steps of the two methods. Not only is the number of steps reduced for boundary elements but the data input is very much easier than for finite elements. Internal results are computed only at required points and not everywhere in the domain, as in the case of finite elements.

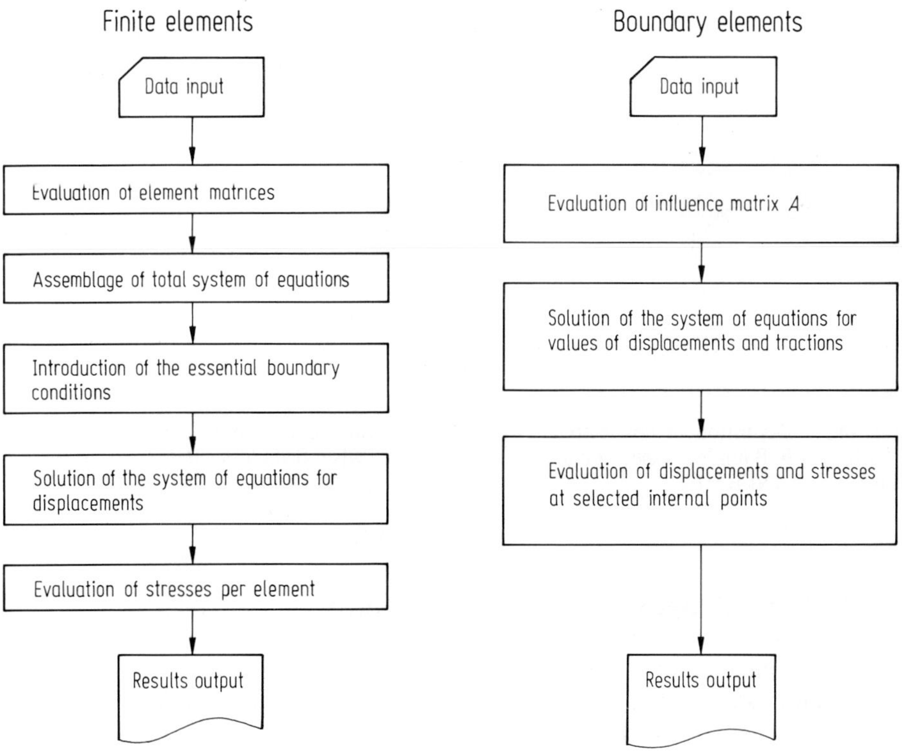

Fig. 14.1. Finite element versus boundary element program

The program has the following capabilities:

1. Solution of elastostatic plane stress – plane strain problems with isotropic material properties.
2. Computation of surface tractions and displacements plus stresses and displacements at any internal points. In addition, stresses on the boundary are also calculated.
3. Planes of symmetry normal to the X and Y axis can be taken into consideration without need to define boundary nodes on these planes (see Fig. 14.2).
4. The program admits boundary at infinity without any special difficulty (Fig. 14.3).
5. The code allows multisurfaces.
6. Traction discontinuities are allowed to occur by using double nodes in the manner described in Section 5.12. The only limitation is that at least one traction in each direction has to be prescribed in one of the double nodes.

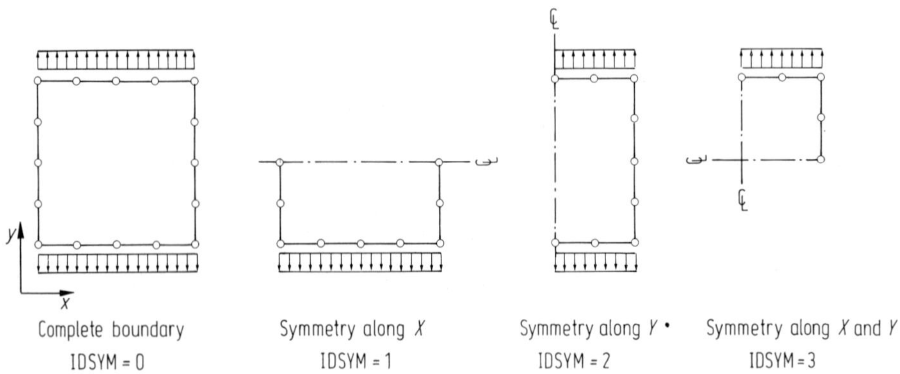

Fig. 14.2. Symmetry planes

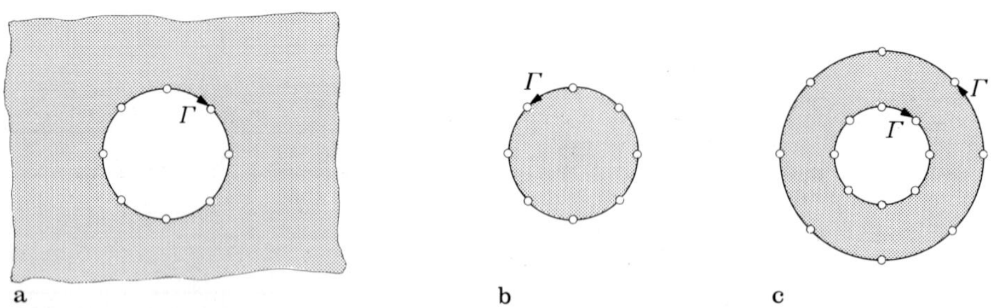

Fig. 14.3. Definition of connectivity. **a** Unbounded case; connectivity in clockwise sense, INFB = 1. **b** Bounded case; connectivity in anticlockwise direction, INFB = 0. **c** Multi-surface case, INFB = 0

14.2. Main Program and Data Structure

The macro flow chart for the boundary element program can be seen in Fig. 14.4. The main program defines the maximum dimensions of the system of equations (in this case 100 or 50 nodes with 2 unknowns per node). It also allocates channel 5 for the input (IRE = 5) and 6 for the output (IWR = 6).

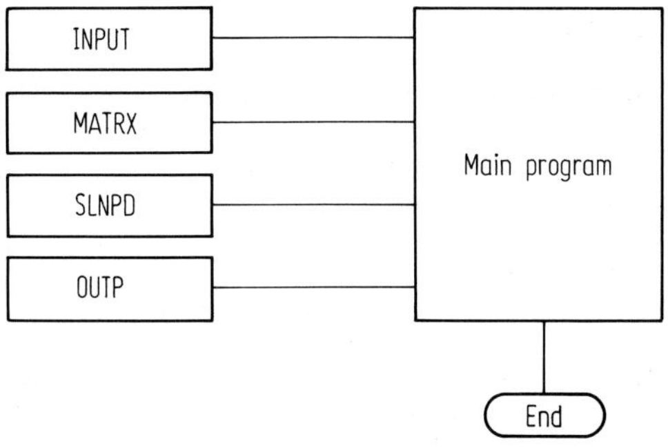

Fig. 14.4. Main flow chart

The main program calls the following subroutines:

INPUT	This routine reads the program input.
MATRX	Computes matrix ***A*** and the right-hand side vector ***F***. They are called A (100, 100) and XM (100).
SLNPD	Solves the system of equations using Gauss elimination. This routine allows for row interchange if required.
OUTPT	Outputs the boundary solution including computation and printout of boundary stresses and internal displacements and stresses.

The variables used by the program, together with their meaning are given below:

NE	Number of elements.
NN	Number of nodes.
NP	Number of internal points.
NN2	Number of nodes multiplied by 2 (NN2 = 2*NN).
NT	Total number of boundary nodes plus internal points (NT = NN + NP).
E	Young's modulus.
PO	Poisson's ratio.
IPL	Index to indicate type of problem (1 for plane stress, 2 for plane strain).
C1 to C11	Constants including material parameters needed for the fundamental solution and other expressions.
IDSYM	Index indicating symmetry (0 for no-symmetry case; 1 for plane of symmetry normal to *Y*; 2 for plane of symmetry normal to *X*; 3 when there are two planes of symmetry normal to *Y* and *X*).
XSYM and YSYM	Coordinates *X* and *Y*, respectively, that define the position of the planes of symmetry.
INFB	Infinite boundary index (1 means that there is a boundary at infinity; 0 if the body is bounded).
IFA and NIF	Parameters needed for symmetry loop.
IFAIL	Index to indicate if matrix ***A*** is singular.

The following arrays are used by the program, the problem-dependent ones are indicated with the required dimensions for possible modifications. (Note that herein the listings are presented for NN = 50, NE = 50, and NP = 50).

D(2, 2)	Kronecker delta matrix.
XI(6, 3)	The abcissas of Gaussian integration points.
W(6, 3)	The weights of Gaussian integration points.
IDUP(NN)	Pointer vector which indicates the double node (= 0 no double node; = number of node with the same coordinates).
INC(NE, 2)	Element connectivity.

ISYM(NT)	Vector that indicates if the node or internal point is located on any symmetry plane (= 0 if not; = 1 for plane normal to *Y*; = 2 for plane normal to *X*; = 3 if located at the intersection of the symmetry planes).
C(NE)	Element length vector.
S(NN, 3)	Matrix that will contain the stresses at boundary nodes.
X(NT)	*X* coordinates of boundary nodes and internal points.
Y(NT)	*Y* coordinates of boundary nodes and internal points.
IFIP(NN2)	Boundary condition index vector (= 0 if traction is prescribed; = 1 if displacement is prescribed).
A(NN2, NN2)	System matrix.
P(NN2)	Auxiliary vector to read prescribed tractions and displacements. At the end it contains the nodal tractions.
XM(NN2)	Right-hand side vector in system of equations. After subroutine SLNPD is called, it contains the boundary solution and at the end the nodal displacements.

Listing of Main Program

```
C.... PROGRAM FOR SOLUTION OF TWO DIMENSIONAL
C.... ELASTOSTATIC PROBLEMS BY THE
C.... BOUNDARY ELEMENT METHOD
      COMMON /RW/ IRE,IWR
      COMMON /A/ D(2,2),XI(6,3),W(6,3),IDUP(50),INC(50,2),C(50),
     *S(50,3),ISYM(100),X(100),Y(100),IFIP(100),A(100,100),P(100),
     *XM(100)
C.... INPUT
      IRE=5
      IWR=6
      CALL INPUT(NE,NN,NP,IPL,PO,NN2,NT,C1,C2,
     *C3,C4,C5,C6,C7,C8,C9,C10,C11,IDSYM,XSYM,YSYM,INFB)
C.... COMPUTE MATRIX A AND INDEPENDENT TERM XM
      CALL MATRX(NE,NN,NN2,NT,C1,C2,C3,C4,C5,C6,C7,C8,
     *C9,C10,C11,PO,IDSYM,XSYM,YSYM,INFB,IFA,NIF)
C.... SOLVE SYSTEM OF EQUATIONS
      CALL SLNPD(NN2,C8,IFAIL)
      IF(IFAIL.NE.0)GO TO 4
C.... OUTPUT RESULTS
      CALL OUTPT(NN,NT,NN2,NE,IFA,NIF,C1,C2,C3,C4,C5,C6,
     *C7,C8,C9,C10,C11,PO,XSYM,YSYM)
    4 STOP
      END
```

14.3. Subroutine INPUT

The input required by the program is read in this subroutine. The input data consists of the following group of cards;

1. *Title Card.* One card containing the title of the problem. The title should start in column 16.
2. *Basic Parameter Card.* One card containing the following information:
   INFB, NE, NN, NP, IPL, IDSYM, E, PO
   FORMAT (I1, I4, 4 I5, 2 F10.0).
3. *Coordinates of Boundary Nodes and Internal Points.* NT cards indicating for each node or point K (internal points are numbered in sequence to boundary nodes; i.e., NN + 1 ≦ K ≦ NT):
   K, X(K), Y(K), IDUP(K), ISYM(K)
   FORMAT (I5, 2F10.0, 2I5).

Note: In case of two nodes with the same coordinates (double nodes), the first one to be given must have the coordinates X and Y and IDUP has to be left blank. The second one may have the coordinates left blank but must have IDUP equal to the number K of the former. The program will automatically generate the coordinates and identify the two nodes as double.

4. *Element Connectivity.* NE cards containing the element number K, the first node and the second node in the form:

   K, INC(K, 1), INC(K, 2)
   FORMAT (3I5).

5. *Boundary Values Prescribed.* One card indicating the number of nodes with displacement prescribed (NFIP) and the number of nodes with nonzero tractions prescribed (NDFIP):

   NFIP, NDFIP
   FORMAT (2I5)

Note: The program initially assumes all tractions to be prescribed with zero value in both directions. Therefore, only the nonzero prescribed tractions have to be given.

6. *Displacements Prescribed.* NFIP cards containing for each node K:

   K, P(2∗K − 1), P(2∗K), IFIP(2∗K − 1), IFIP(2∗K)
   FORMAT (I5, 2F10.0, 2I5).

Note: Nodes located on symmetry planes must not have the displacement in the normal direction to the corresponding plane prescribed. The condition of zero displacement in this direction is automatically accomplished by the system of equations and consequently this displacement is calculated zero.

7. *Tractions Prescribed.* NDFIP cards indicating for each node K:

   K, P(2∗K − 1), P(2∗K)
   FORMAT (I5, 2F10.0).

Listing of Subroutine INPUT

```
      SUBROUTINE INPUT(NE,NN,NP,IPL,PO,NN2,NT,C1,C2,
     *C3,C4,C5,C6,C7,C8,C9,C10,C11,IDSYM,XSYM,YSYM,INFB)
      COMMON /RW/ IRE,IWR
      COMMON /A/ D(2,2),XI(6,3),W(6,3),IDUP(50),INC(50,2),C(50),
     *S(50,3),ISYM(100),X(100),Y(100),IFIP(100),A(100,100),P(100),
     *XM(100)
      WRITE(IWR,1)
    1 FORMAT(//////////,24X,'* * *  B O U N D A R Y  E L E M E N T  M E
     *T H O D  A P P L I E D  T O  * * *',//,24X,'* * *  P L A N E  E L
     *A S T O S T A T I C  P R O B L E M S                 * * *',///)
C.... TITLE OF PROBLEM
      READ(IRE,2)
    2 FORMAT('                                                            ')
      WRITE(IWR,2)
C.... GENERAL INFORMATION ABOUT THE PROBLEM
      READ(IRE,3)INFB,NE,NN,NP,IPL,IDSYM,E,PO
    3 FORMAT(I1,I4,4I5,2F10.0)
      IF(INFB.EQ.0)GO TO 60
      WRITE(IWR,61)
   61 FORMAT(//,13X,'* INFINITE BOUNDARY *')
   60 WRITE(IWR,4)NE,NN,NP,IPL,IDSYM,E,PO
    4 FORMAT(//,15X,'NO. ELEMENTS =',I5,//,15
     *X,'NO.    NODES =',I5,//,15X,'NO.   POINTS =',I5,//,15X,'PROBL.  T
     *YPE =',I5,//,15X,'SYMME.  TYPE =',I5,///,15X,'MATERIAL PROPERTIES'
     *,//,15X,'E          =',F10.0,
     *//,15X,'POISSON =',F15.5,///,30X,'COORDINATES OF BOUNDARY NODES',
     *//,12X,'NODE',14X,'X',15X,'Y',12X,'DOUBLE',/)
      NN2=NN*2
      NT=NN+NP
C.... NODES AND POINTS COORDINATES
      DO 5 I=1,NN
      READ(IRE,6)K,X(K),Y(K),IDUP(K),ISYM(K)
    6 FORMAT(I5,2F10.0,2I5)
      IF(IDUP(K).EQ.0)GO TO 5
      J=IDUP(K)
      IDUP(J)=K
      X(K)=X(J)
      Y(K)=Y(J)
```

```
    5 CONTINUE
      DO 63 K=1,NN
      IF(IDUP(K).NE.0)GO TO 62
      WRITE(IWR,7)K,X(K),Y(K)
      GO TO 63
   62 WRITE(IWR,16)K,X(K),Y(K),IDUP(K)
   16 FORMAT(10X,I5,5X,F15.4,1X,F15.4,7X,I5)
   63 CONTINUE
    7 FORMAT(10X,I5,5X,F15.4,1X,F15.4)
      IF(NP.EQ.0)GO TO 9
      WRITE(IWR,8)
    8 FORMAT(//,30X,'COORDINATES OF INTERNAL POINTS',//,11X,'POINT',14X,
     *'X',15X,'Y',/)
      K=NN+1
      READ(IRE,14)(J,X(J),Y(J),ISYM(J),JJ=K,NT)
   14 FORMAT(I5,2F10.0,5X,I5)
      WRITE(IWR,7)(J,X(J),Y(J),J=K,NT)
C.... NODES AND POINTS AT SYMMETRY LINES
    9 IF(IDSYM.EQ.0)GO TO 49
      WRITE(IWR,42)
   42 FORMAT(//,30X,'BOUNDARY NODES AND INTERNAL POINTS AT SYMMETRY LINE
     *(S)',//,12X,'L. X',12X,'L. Y',/)
      DO 43 K=1,NT
      IF(ISYM(K).EQ.0)GO TO 43
      IZZ=ISYM(K)
      GO TO (44,45,46),IZZ
   44 YSYM=Y(K)
      WRITE(IWR,47)K
   47 FORMAT(10X,I5)
      GO TO 43
   45 XSYM=X(K)
      WRITE(IWR,48)K
   48 FORMAT(26X,I5)
      GO TO 43
   46 WRITE(IWR,50)K,K
   50 FORMAT(10X,I5,11X,I5)
   43 CONTINUE
C.... ELEMENT CONNECTIVITY
   49 WRITE(IWR,10)
   10 FORMAT(//,30X,'ELEMENT CONNECTIVITY',//,13X,'EL',13X,'N. 1',12X,'N
     *. 2',14X,'L',/)
      DO 11 I=1,NE
      READ(IRE,12)K,INC(K,1),INC(K,2)
   12 FORMAT(3I5)
      II=INC(K,1)
      IF=INC(K,2)
   11 C(K)=SQRT((X(IF)-X(II))**2+(Y(IF)-Y(II))**2)
      WRITE(IWR,13)(I,INC(I,1),INC(I,2),C(I),I=1,NE)
   13 FORMAT(10X,I5,11X,I5,11X,I5,5X,F15.4)
C.... CONSTANTS
      G=E/(2.*(1.+PO))
      C11=PO
      IF(IPL-1)40,40,41
   40 PO=PO/(1.+PO)
      C11=0.
   41 C2=3.-4.*PO
      C3=1./((1.-PO)*12.56637062)
      C4=1.-2.*PO
      C6=2.*C3*G
      C7=1.-4.*PO
      C1=C3/(2.*G)
      C5=C1/2.
      C8=2.*G/(1.-PO)
      C9=PO/(1.-PO)
      C10=(2.-PO)/(1.-PO)
C.... BOUNDARY VALUES PRESCRIBED
      DO 19 I=1,NN2
      P(I)=0.
   19 IFIP(I)=0
      READ(IRE,20)NFIP,NDFIP
   20 FORMAT(2I5)
      WRITE(IWR,21)NFIP,NDFIP
   21 FORMAT(//,15X,'NO. DISPL. PRESC. =',I5,//,15X,'NO. TRACT. PRESC. =
     *',I5,///,15X,'DISPLACEMENTS
     *',//,12X,'NODE',14X,'U',15X,'V',/)
      IF(NFIP.EQ.0)GO TO 22
```

```
      DO 23 I=1,NFIP
      READ(IRE,24)K,P(2*K-1),P(2*K),IFIP(2*K-1),IFIP(2*K)
   24 FORMAT(I5,2F10.0,2I5)
      IND=IFIP(2*K-1)+2*IFIP(2*K)
      GO TO (25,26,27),IND
   25 WRITE(IWR,28)K,P(2*K-1)
   28 FORMAT(10X,I5,5X,F15.4)
      GO TO 23
   26 WRITE(IWR,29)K,P(2*K)
   29 FORMAT(10X,I5,21X,F15.4)
      GO TO 23
   27 WRITE(IWR,30)K,P(2*K-1),P(2*K)
   30 FORMAT(10X,I5,5X,F15.4,1X,F15.4)
   23 CONTINUE
   22 IF(NDFIP.EQ.0)GO TO 31
      WRITE(IWR,34)
   34 FORMAT(//,15X,'TRACTIONS',//,12X,'NODE',13X,'PX',14X,'PY',/)
      DO 32 I=1,NDFIP
      READ(IRE,33)K,P(2*K-1),P(2*K)
   33 FORMAT(I5,2F10.0)
   32 WRITE(IWR,30)K,P(2*K-1),P(2*K)
C.... INTEGRATION POINTS
   31 XI(1,3)=-0.932469514203152
      XI(2,3)=-0.661209386466265
      XI(3,3)=-0.238619186083197
      XI(4,3)=-XI(3,3)
      XI(5,3)=-XI(2,3)
      XI(6,3)=-XI(1,3)
      W(1,3)=0.171324492379170
      W(2,3)=0.360761573048139
      W(3,3)=0.467913934572691
      W(4,3)=W(3,3)
      W(5,3)=W(2,3)
      W(6,3)=W(1,3)
      XI(1,2)=-0.861136311594053
      XI(2,2)=-0.339981043584856
      XI(3,2)=-XI(2,2)
      XI(4,2)=-XI(1,2)
      W(1,2)=0.347854845137454
      W(2,2)=0.652145154862546
      W(3,2)=W(2,2)
      W(4,2)=W(1,2)
      XI(1,1)=-0.577350269189626
      XI(2,1)=-XI(1,1)
      W(1,1)=1.
      W(2,1)=1.
      RETURN
      END
```

14.4. Subroutine MATRX

This subroutine computes matrix $\boldsymbol{A}$ and vector $\boldsymbol{F}$ of Eq. (5.103). The element influence coefficients are computed calling subroutine FUNC.

FUNC Computes the element influence matrices $\boldsymbol{h}$ and $\boldsymbol{g}$ necessary to form system matrix $\boldsymbol{A}$ and independent term $\boldsymbol{F}$ (now in XM).

Subroutine MATRX generates the system of equations by assembling directly matrix $\boldsymbol{A}$ without going to the global $\boldsymbol{H}$ and $\boldsymbol{G}$ matrices. This is done by considering the prescribed boundary conditions for the node under consideration before assembling. The leading diagonal submatrices corresponding to $\boldsymbol{H}$ are calculated using rigid-body translations. Consequently, when the body is unbounded a different type of rigid body consideration needs to be applied (see Eq. (5.108)).

Symmetry is taken into consideration by instead of reflecting the elements, reflecting the singular nodes. This procedure is very simple and completely general.

Listing of Subroutine MATRX

```
      SUBROUTINE MATRX(NE,NN,NN2,NT,C1,C2,C3,C4,C5,C6,C7,C8,
     *C9,C10,C11,PO,IDSYM,XSYM,YSYM,INFB,IFA,NIF)
      COMMON /A/ D(2,2),XI(6,3),W(6,3),IDUP(50),INC(50,2),C(50),
     *S(50,3),ISYM(100),X(100),Y(100),IFIP(100),A(100,100),P(100),
     *XM(100)
      COMMON /A4/ H(3,4),G(3,4),HL(3,4),GL(3,4)
C.... KRONECKER DELTA
      D(1,1)=1.
      D(2,2)=1.
      D(1,2)=0.
      D(2,1)=0.
C.... CLEAR ARRAYS
      DO 1 I=1,NN2
      XM(I)=0.
      DO 1 J=1,NN2
    1 A(I,J)=0.
C.... COMPUTE PARAMETERS FOR SYMMETRY LOOP
      IFA=1
      NIF=1
      IF(IDSYM.EQ.1)IFA=2
      IF(IDSYM.NE.2)GO TO 60
      IFA=3
      NIF=2
   60 IF(IDSYM.EQ.3)IFA=4
C.... TEST FOR INFINITE BOUNDARY
      IF(INFB.EQ.0)GO TO 90
      DO 91 I=1,NN2
      IF(IFIP(I).NE.0)GO TO 92
      A(I,I)=1.
      GO TO 91
   92 XM(I)=-P(I)
   91 CONTINUE
C.... SYMMETRY LOOP
   90 DO 2 ISY=1,IFA,NIF
C.... COMPUTE CHANGE SIGN CONTROLLING PARAMETERS
      GO TO (70,71,71,73),ISY
   71 IIS=4-ISY
      IFS=IIS
      GO TO 70
   73 IIS=1
      IFS=2
C.... LOOP OVER BOUNDARY NODES
   70 DO 2 I=1,NN
      XS=X(I)
      YS=Y(I)
      IF(ISY.EQ.2.OR.ISY.EQ.4)YS=2.*YSYM-YS
      IF(ISY.GE.3)XS=2.*XSYM-XS
C.... GENERATE MATRIX A AND INDEPENDENT TERM XM
C.... NOTE: COLUMNS OF  A  THAT CORRESPOND TO  G  ARE MULTIPLIED
C.... BY  C8  TO AVOID ROUNDOFF ERRORS IN THE SOLUTION.
      DO 10 J=1,NE
      II=INC(J,1)
      IF=INC(J,2)
      ICOD=1
      IF(ISY.NE.1.AND.ISYM(I).NE.(ISY-1))GO TO 6
      IF(I.EQ.II.OR.I.EQ.IDUP(II))ICOD=2
      IF(I.EQ.IF.OR.I.EQ.IDUP(IF))ICOD=3
    6 CALL FUNC(ICOD,J,C1,C2,C3,C4,C5,C6,C7,PO,II,IF,XS,YS,ISY,IIS,
     *IFS)
      DO 10 K=1,2
      JJ=2*(I-1)+K
      M=0
      DO 10 NX=1,2
      DO 10 NV=1,2
      M=M+1
      IC=2*INC(J,NX)+NV-2
      IF(IFIP(IC).NE.0)GO TO 67
      A(JJ,IC)=A(JJ,IC)+H(K,M)
      XM(JJ)=XM(JJ)+G(K,M)*P(IC)
      GO TO 68
   67 A(JJ,IC)=A(JJ,IC)-G(K,M)*C8
      XM(JJ)=XM(JJ)-H(K,M)*P(IC)
C..... COMPUTE REMAINING COEFFICIENTS BY APPLYING RIGID BODY TRANSLATIONS
```

```
68 GO TO (61,62,63,64),ISY
62 IF(NV-2)61,64,61
63 IF(NV-1)61,64,61
64 H(K,M)=-H(K,M)
61 IF(IFIP(JJ+NV-K).NE.0)GO TO 69
   A(JJ,JJ+NV-K)=A(JJ,JJ+NV-K)-H(K,M)
   GO TO 10
69 XM(JJ)=XM(JJ)+H(K,M)*P(JJ+NV-K)
10 CONTINUE
 2 CONTINUE
   RETURN
   END
```

14.5. Subroutine FUNC

This subroutine computes all the element integrals required for the system of equations, internal displacements, and internal stresses. Numerical integrals are performed over nonsingular elements by using two, four, and six Gauss integration points. The number of points is selected according to the relative distance from the singular node or point to the center of the element under consideration.

For elements with the singularity at one of its extremities the required integrals are computed analytically to obtain more accurate results. Notice that at the end of the subroutine a symmetry test is included to determine if it is necessary to change the sign of some rows of the element submatrices.

Listing of Subroutine FUNC

```
      SUBROUTINE FUNC(ICOD,JA,C1,C2,C3,C4,C5,C6,C7,PO,II,IF,XS,YS,ISY,
     *IIS,IFS)
C.... INTEGRALS OVER BOUNDARY ELEMENTS
      COMMON /A/ D(2,2),XI(6,3),W(6,3),IDUP(50),INC(50,2),C(50),
     *S(50,3),ISYM(100),X(100),Y(100),IFIP(100),A(100,100),P(100),
     *XM(100)
      COMMON /A4/ H(3,4),G(3,4),HL(3,4),GL(3,4)
      DIMENSION DXY(2),BN(2),B(2),DR(2),UL(2,2),PL(2,2),ULL(2,2,2),
     *PLL(2,2,2)
      DO 5 KK=1,3
      DO 5 L=1,4
      GL(KK,L)=0.
      HL(KK,L)=0.
      G(KK,L)=0.
    5 H(KK,L)=0.
      DXY(1)=X(IF)-X(II)
      DXY(2)=Y(IF)-Y(II)
      GO TO (1,2,2,1),ICOD
    1 BN(1)=DXY(2)/C(JA)
      BN(2)=-DXY(1)/C(JA)
C.... SELECT NO. INTEGRATION POINTS
      SEL=0.5*SQRT((2.*XS-X(II)-X(IF))**2+(2.*YS-Y(II)-Y(IF))**2)/C(JA)
      NPI=4
      IF(SEL.LE.1.5)NPI=6
      IF(SEL.GT.5.5)NPI=2
      INP=NPI/2
C.... COMPUTE MATRICES NUMERICALLY
      DO 50 KK=1,NPI
      XMXI=0.5*(1.+XI(KK,INP))*DXY(1)+X(II)-XS
      YMYI=0.5*(1.+XI(KK,INP))*DXY(2)+Y(II)-YS
      R=SQRT(XMXI**2+YMYI**2)
      B(1)=-0.25*(XI(KK,INP)-1.)*C(JA)
      B(2)=0.25*(XI(KK,INP)+1.)*C(JA)
      DR(1)=XMXI/R
      DR(2)=YMYI/R
      DRDN=DR(1)*BN(1)+DR(2)*BN(2)
```

```
C.... COMPUTE MATRICES H AND G
      DO 6 I=1,2
      DO 6 J=1,2
      UL(I,J)=-C1*(C2*ALOG(R)*D(I,J)-DR(I)*DR(J))
    6 PL(I,J)=-C3*((C4*D(I,J)+2.*DR(I)*DR(J))*DRDN+C4*(DR(J)*BN(I)-DR(I)
     **BN(J)))/R
      DO 7 LA=1,2
      IC=0
      DO 7 LL=1,2
      DO 7 JJ=1,2
      IC=IC+1
      G(LA,IC)=G(LA,IC)+UL(LA,JJ)*B(LL)*W(KK,INP)
    7 H(LA,IC)=H(LA,IC)+PL(LA,JJ)*B(LL)*W(KK,INP)
      IF(ICOD.NE.4)GO TO 50
C.... COMPUTE MATRICES HL AND GL (STRESSES AT INTERNAL POINTS)
   10 DO 11 I=1,2
      DO 11 J=I,2
      DO 11 K=1,2
      ULL(I,J,K)=C3*(C4*(DR(J)*D(K,I)+DR(I)*D(K,J)-DR(K)*D(I,J))+2.*DR(I
     *)*DR(J)*DR(K))/R
      B1=2.*DRDN*(C4*DR(K)*D(I,J)+PO*(DR(J)*D(I,K)+DR(I)*D(J,K))-4.*DR(I
     *)*DR(J)*DR(K))
      B2=2.*PO*(BN(I)*DR(J)*DR(K)+BN(J)*DR(I)*DR(K))
      B3=C4*(2.*BN(K)*DR(I)*DR(J)+BN(J)*D(I,K)+BN(I)*D(J,K))
   11 PLL(I,J,K)=C6*(B1+B2+B3-C7*BN(K)*D(I,J))/R**2
      IL=0
      DO 12 I=1,2
      DO 12 J=I,2
      IL=IL+1
      IC=0
      DO 12 IAA=1,2
      DO 12 JAA=1,2
      IC=IC+1
      GL(IL,IC)=GL(IL,IC)+B(IAA)*ULL(I,J,JAA)*W(KK,INP)
   12 HL(IL,IC)=HL(IL,IC)+B(IAA)*PLL(I,J,JAA)*W(KK,INP)
   50 CONTINUE
      GO TO 18
C.... COMPUTE MATRICES H AND G ANALYTICALLY (BOUNDARY CONSTRAINT EQ.)
    2 AL=C5*C2*C(JA)
      AA=AL*(0.5-ALOG(C(JA)))
      DO 15 I=1,2
      DO 15 J=1,4
      IT=(J/2)*2+2-J
      G(I,J)=C5*DXY(I)*DXY(IT)/C(JA)
      IF(IT.EQ.I)G(I,J)=G(I,J)+AA
   15 CONTINUE
      IAA=-2
      IF(ICOD.EQ.3)IAA=0
      G(1,3+IAA)=G(1,3+IAA)+AL
      G(2,4+IAA)=G(2,4+IAA)+AL
      H(1,2-IAA)=C3*C4*(1.+IAA)
      H(2,1-IAA)=-H(1,2-IAA)
C.... SYMMETRY TEST
   18 IF(ISY.EQ.1)GO TO 8
      DO 24 I=IIS,IFS
      DO 24 J=1,4
      H(I,J)=-H(I,J)
   24 G(I,J)=-G(I,J)
      IF(ICOD.NE.4.OR.ISY.EQ.4)GO TO 8
      DO 25 J=1,4
      HL(2,J)=-HL(2,J)
   25 GL(2,J)=-GL(2,J)
    8 RETURN
      END
```

14.6. Subroutine SLNPD [1]

This is a standard subroutine for the solution of nonpositive definite matrices using Gauss elimination. If the matrix $\boldsymbol{A}$ has a zero in the diagonal, it will interchange rows, deciding that the system matrix is singular only when no row interchange will produce a nonzero diagonal coefficient for a given row position.

The final result is a vector with the unknown boundary displacements and tractions and is overwritten in B(NN2) which locally corresponds to XM(NN2) in the main program.

Listing of Subroutine SLNPD

```
      SUBROUTINE SLNPD(N,C8,IFAIL)
C.... SOLVE SYSTEM OF EQUATIONS BY GAUSS ELIMINATION PROCESS AND
C.... RETURN RESULT IN VECTOR XM
      COMMON /RW/ IRE,IWR
      COMMON /A/ D(2,2),XI(6,3),W(6,3),IDUP(50),INC(50,2),C(50),
     *S(50,3),ISYM(100),X(100),Y(100),IFIP(100),A(100,100),P(100),
     *B(100)
      N1=N-1
      DO 100 K=1,N1
      K1=K+1
      CC=A(K,K)
      IF(ABS(CC))1,1,3
    1 DO 7 J=K1,N
      IF(ABS(A(J,K)))7,7,5
    5 DO 6 L=K,N
      CC=A(K,L)
      A(K,L)=A(J,L)
    6 A(J,L)=CC
      CC=B(K)
      B(K)=B(J)
      B(J)=CC
      CC-A(K,K)
      GO TO 3
    7 CONTINUE
    8 WRITE(IWR,2)K
    2 FORMAT(/////,20X,'* * * SINGULARITY IN ROW',I5,' * * *')
      IFAIL=1
      GO TO 300
    3 DO 4 J=K1,N
    4 A(K,J)=A(K,J)/CC
      B(K)=B(K)/CC
      DO 10 I=K1,N
      CC=A(I,K)
      DO 9 J=K1,N
    9 A(I,J)=A(I,J)-CC*A(K,J)
   10 B(I)=B(I)-CC*B(K)
  100 CONTINUE
      IF(ABS(A(N,N)))8,8,101
  101 B(N)=B(N)/A(N,N)
      DO 200 L=1,N1
      K=N-L
      K1=K+1
      DO 200 J=K1,N
  200 B(K)=B(K)-A(K,J)*B(J)
C.... MULTIPLY CALCULATED TRACTIONS BY  C8  TO OBTAIN ACTUAL VALUES
      DO 70 I=1,N
      IF(IFIP(I).EQ.0)GO IO 70
      B(I)=B(I)*C8
   70 CONTINUE
      IFAIL=0
  300 RETURN
      END
```

14.7. Subroutine OUTPT

This subroutine prints the results for boundary displacements and tractions. In addition, it calculates and prints the boundary stresses and internal displacements and stresses.

The boundary stresses are computed using subroutine FENC; these stresses are important in most practical applications. The internal displacements and stresses are computed by integrating over the boundary elements using subroutine FUNC.

The total output consists of

(i) displacements and tractions at boundary nodes,
(ii) displacements and stresses at boundary nodes and internal points.

Listing of Subroutine OUTPT

```
      SUBROUTINE OUTPT(NN,NT,NN2,NE,IFA,NIF,C1,C2,C3,C4,C5,C6,
     *C7,C8,C9,C10,C11,PO,XSYM,YSYM)
C.... OUTPUT RESULTS
      COMMON /RW/ IRE,IWR
      COMMON /A/ D(2,2),XI(6,3),W(6,3),IDUP(50),INC(50,2),C(50),
     *S(50,3),ISYM(100),X(100),Y(100),IFIP(100),A(100,100),P(100),
     *XM(100)
      COMMON /A4/ H(3,4),G(3,4),HL(3,4),GL(3,4)
      DIMENSION U(2),SA(4)
      WRITE(IWR,6)
    6 FORMAT(
     *///,30X,'BOUNDARY DISPLACEMENTS AND TRACTIONS',//,12X,'NODE',14
     *X,'U',15X,'V',14X,'PX',14X,'PY',/)
C.... BOUNDARY DISPLACEMENTS AND TRACTIONS
      DO 8 I=1,NN2
      IF(IFIP(I).EQ.0)GO TO 8
    9 PA=XM(I)
      XM(I)=P(I)
      P(I)=PA
    8 CONTINUE
      WRITE(IWR,11)(I,XM(2*I-1),XM(2*I),P(2*I-1),P(2*I),I=1,NN)
   11 FORMAT(10X,I5,5X,F15.4,1X,F15.4,1X,F15.4,1X,F15.4)
C.... DISPLACEMENTS AND STRESSES AT NODES AND INT. POINTS
      WRITE(IWR,12)
   12 FORMAT(//,15X,'DISPLACEMENTS AND STRESSES AT NODES AND INTERNAL PO
     *INTS',//,11X,'NO/PT',14X,'U',15X,'V',14X,'SX',14X,'SXY',13X,'SY',
     *14X,'SZ',/)
C.... COMPUTE BOUNDARY STRESSES
      DO 14 I=1,NN
      DO 14 J=1,3
   14 S(I,J)=0.
C.... LOOP OVER ALL BOUNDARY ELEMENTS
      DO 30 I=1,NE
      II=INC(I,1)
      IF=INC(I,2)
      CC1=(Y(IF)-Y(II))/C(I)
      CC2=(X(II)-X(IF))/C(I)
      CALL FENC(C8,C9,C10,CC1,CC2,I)
      DO 30 JP=1,2
      IIF=INC(I,JP)
      XFAC=2.
      IF(IDUP(IIF).NE.0.OR.ISYM(IIF).NE.0)XFAC=1.
      DO 30 IR=1,3
      M=0
      DO 30 IP=1,2
      IO=2*IIF+IP-2
      S(IIF,IR)=S(IIF,IR)+G(IR,IP)*P(IO)/XFAC
      DO 30 JR=1,2
      M=M+1
      IO=2*INC(I,IP)+JR-2
   30 S(IIF,IR)=S(IIF,IR)-H(IR,M)*XM(IO)/XFAC
```

```
C.... PRINT VALUES ON THE BOUNDARY
      DO 13 I=1,NN
      SA(4)=C11*(S(I,1)+S(I,3))
      IF(ISYM(I).NE.0)S(I,2)=0.
   13 WRITE(IWR,15)I,XM(2*I-1),XM(2*I),S(I,1),S(I,2),S(I,3),SA(4)
   15 FORMAT(10X,I5,4X,6(1X,F15.4))
      IF(NN.EQ.NT)GO TO 5
C.... COMPUTE INTERNAL VALUES
      NNI=NN+1
      ICOD=4
C.... LOOP OVER ALL INTERNAL POINTS
      DO 16 I=NNI,NT
      U(1)=0.
      U(2)=0.
      DO 17 J=1,3
   17 SA(J)=0.
C.... SYMMETRY LOOP
      DO 20 ISY=1,IFA,NIF
      XS=X(I)
      YS=Y(I)
      IF(ISY.EQ.2.OR.ISY.EQ.4)YS=2.*YSYM-YS
      IF(ISY.GE.3)XS=2.*XSYM-XS
      GO TO (70,71,71,73),ISY
   71 IIS=4-ISY
      IFS=IIS
      GO TO 70
   73 IIS=1
      IFS=2
C.... INTEGRATE OVER THE BOUNDARY
   70 DO 20 J=1,NE
      II=INC(J,1)
      IF=INC(J,2)
      CALL FUNC(ICOD,J,C1,C2,C3,C4,C5,C6,C7,PO,II,IF,XS,YS,ISY,
     *IIS,IFS)
      DO 20 K=1,3
      M=0
      DO 20 NX=1,2
      DO 20 NV=1,2
      M=M+1
      ICA=2*INC(J,NX)+NV-2
      IF(K.LT.3)U(K)=U(K)-H(K,M)*XM(ICA)+G(K,M)*P(ICA)
   20 SA(K)=SA(K)-HL(K,M)*XM(ICA)+GL(K,M)*P(ICA)
      SA(4)=C11*(SA(1)+SA(3))
   16 WRITE(IWR,15)I,U(1),U(2),SA(1),SA(2),SA(3),SA(4)
    5 RETURN
      END
```

14.8. Subroutine FENC

This subroutine computes the expressions for stresses at boundary nodes using the formulas presented in (5.114).

Listing of Subroutine FENC

```
      SUBROUTINE FENC(C8,C9,C10,CC1,CC2,I)
C.... EXPRESSIONS FOR STRESSES AT BOUNDARY NODES
      COMMON /A/ D(2,2),XI(6,3),W(6,3),IDUP(50),INC(50,2),C(50),
     *S(50,3),ISYM(100),X(100),Y(100),IFIP(100),A(100,100),P(100),
     *XM(100)
      COMMON /A4/ H(3,4),G(3,4),HL(3,4),GL(3,4)
C.... MATRIX H
      CO=-C8/C(I)
      H(1,1)=CO*CC2**3
      H(1,2)=-CO*CC1*CC2**2
      H(1,3)=-H(1,1)
      H(1,4)=-H(1,2)
```

```
      H(2,1)=H(1,2)
      H(2,2)=CO*CC2*CC1**2
      H(2,3)=-H(1,2)
      H(2,4)=-H(2,2)
      H(3,1)=H(2,2)
      H(3,2)=-CO*CC1**3
      H(3,3)=-H(2,2)
      H(3,4)=-H(3,2)
C.... MATRIX G
      G(1,1)=CC1**3+C10*CC1*CC2**2
      G(1,2)=-CC2*CC1**2+C9*CC2**3
      G(2,1)=CC2**3-C9*CC2*CC1**2
      G(2,2)=CC1**3-C9*CC1*CC2**2
      G(3,1)=-CC1*CC2**2+C9*CC1**3
      G(3,2)=CC2**3+C10*CC2*CC1**2
      RETURN
      END
```

14.9. Examples

In order to illustrate the data preparation for the computer program, two simple examples are presented below.

14.9.1. Square Plate

This example consists of a square plate under biaxial load. The discretization is presented in Fig. 14.5 and the corresponding input data is given in Table 14.1. In this analysis four boundary elements with six boundary nodes and four internal points are employed.

The complete printout corresponding to the solution of the problem is presented in what follows. As can be easily checked, the analytical solution is obtained.

Table 14.1. Data file for square plate.

```
                        SQUARE PLATE PROBLEM
  4          6         4         1         3        5.                    0.3
  1 2.                                                        1
  2 2.                      1.
  3 2.                      2.
  4                                                  3
  5 1.                      2.
  6                         2.                                2
  7 1.                      1.
  8                                                           3
  9                         1.                                2
 10 1.                                                        1
  1          1         2
  2          2         3
  3          4         5
  4          5         6
  3          3
  4                                                           1
  5                                                           1
  6                                                           1
  1 2.
  2 2.
  3 2.
```

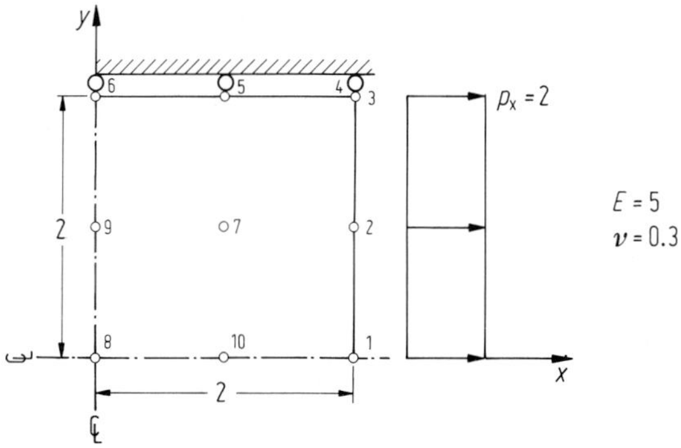

Fig. 14.5. Definition of square plate problem

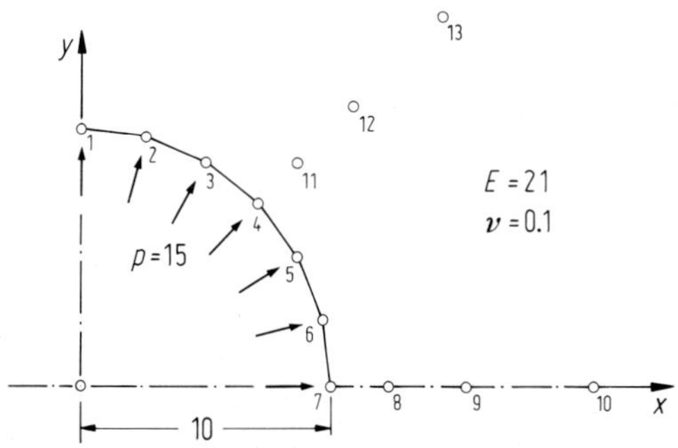

Fig. 14.6. Definition of cylindrical cavity problem

Table 14.2. Data file for cylindrical cavity problem.

```
                      CYLINDRICAL CAVITY PROBLEM
1    6      7     6       2     3    21.          0.1
     1                   10.               2
     2  2.5882            9.6593
     3  5.                8.6603
     4  7.0711            7.0711
     5  8.6603            5.
     6  9.6593            2.5882
     7 10.                                 1
     8 12.                                 1
     9 15.                                 1
    10 20.                                 1
    11  8.4853            8.4853
    12 10.6066           10.6066
    13 14.1421           14.1421
     1      1     2
     2      2     3
     3      3     4
     4      4     5
     5      5     6
     6      6     7
            7
     1                   15.
     2  3.8823           14.4889
     3  7.5              12.9904
     4 10.6066           10.6066
     5 12.9904            7.5
     6 14.4889            3.8823
     7 15.
```

Line Printer Output

```
*** BOUNDARY ELEMENT METHOD APPLIED TO ***
*** PLANE ELASTOSTATIC PROBLEMS        ***

SQUARE PLATE PROBLEM

NO. ELEMENTS =   4
NO.    NODES =   6
NO.   POINTS =   4
PROBL.  TYPE =   1
SYMME.  TYPE =   3

MATERIAL PROPERTIES
E        =       5.
POISSON  =       0.30000

               COORDINATES OF BOUNDARY NODES
NODE           X            Y            DOUBLE

  1            2.0000       0.0
  2            2.0000       1.0000
  3            2.0000       2.0000       4
  4            2.0000       2.0000       3
  5            1.0000       2.0000
  6            0.0          2.0000

               COORDINATES OF INTERNAL POINTS
POINT          X            Y

  7            1.0000       1.0000
  8            0.0          0.0
  9            0.0          1.0000
 10            1.0000       0.0

               BOUNDARY NODES AND INTERNAL POINTS AT SYMMETRY LINE(S)
L. X           L. Y

  1
               6
  8            8
               9
 10
```

ELEMENT CONNECTIVITY

EL	N. 1	N. 2	L
1	1	2	1.0000
2	2	3	1.0000
3	4	5	1.0000
4	5	6	1.0000

NO. DISPL. PRESC. = 3

NO. TRACT. PRESC. = 3

DISPLACEMENTS

NODE	U	V
4		0.0
5		0.0
6		0.0

TRACTIONS

NODE	PX	PY
1	2.0000	0.0
2	2.0000	0.0
3	2.0000	0.0

BOUNDARY DISPLACEMENTS AND TRACTIONS

NODE	U	V	PX	PY
1	0.7280	-0.0000	2.0000	0.0
2	0.7280	-0.0000	2.0000	0.0
3	0.7280	-0.0000	2.0000	0.0
4	0.7280	0.0	0.0	0.6000
5	0.3640	0.0	0.0	0.6000
6	0.0000	0.0	0.0	0.6000

DISPLACEMENTS AND STRESSES AT NODES AND INTERNAL POINTS

NO/PT	U	V	SX	SXY	SY	SZ
1	0.7280	-0.0000	2.0000	0.0	0.6000	0.0
2	0.7280	-0.0000	2.0000	0.0	0.6000	0.0
3	0.7280	-0.0000	2.0000	0.0	0.6000	0.0
4	0.7280	0.0	2.0000	0.0	0.6000	0.0
5	0.3640	0.0	2.0000	0.0	0.6000	0.0
6	0.0000	0.0	2.0000	0.0	0.6000	0.0
7	0.3640	-0.0000	2.0000	-0.0000	0.6000	0.0
8	-0.0000	-0.0000	2.0000	-0.0000	0.6000	0.0
9	-0.0000	-0.0000	2.0000	0.0000	0.6000	0.0
10	0.3640	-0.0000	2.0000	0.0000	0.6000	0.0

Line Printer Output

```
*** BOUNDARY ELEMENT METHOD APPLIED TO ***
*** PLANE ELASTOSTATIC PROBLEMS        ***

   CYLINDRICAL CAVITY PROBLEM

 * INFINITE BOUNDARY *

   NO. ELEMENTS =    6
   NO.    NODES =    7
   NO.   POINTS =    6
   PROBL.  TYPE =    2
   SYMME.  TYPE =    3

   MATERIAL PROPERTIES
   E       =       21.
   POISSON =        0.10000

                COORDINATES OF BOUNDARY NODES
 NODE               X               Y            DOUBLE

    1             0.0            10.0000
    2             2.5882          9.6593
    3             5.0000          8.6603
    4             7.0711          7.0711
    5             8.6603          5.0000
    6             9.6593          2.5882
    7            10.0000          0.0

                COORDINATES OF INTERNAL POINTS
POINT               X               Y

    8            12.0000          0.0
    9            15.0000          0.0
   10            20.0000          0.0
   11             8.4853          8.4853
   12            10.6066         10.6066
   13            14.1421         14.1421

                BOUNDARY NODES AND INTERNAL POINTS AT SYMMETRY LINE(S)
 L. X              L. Y

                    1
    7
    8
    9
   10
```

ELEMENT CONNECTIVITY

EL	N. 1	N. 2	L
1	1	2	2.6105
2	2	3	2.6105
3	3	4	2.6106
4	4	5	2.6106
5	5	6	2.6105
6	6	7	2.6105

NO. DISPL. PRESC. = 0

NO. TRACT. PRESC. = 7

DISPLACEMENTS

NODE	U	V

TRACTIONS

NODE	PX	PY
1	0.0	15.0000
2	3.8823	14.4889
3	7.5000	12.9904
4	10.6066	10.6066
5	12.9904	7.5000
6	14.4889	3.8823
7	15.0000	0.0

BOUNDARY DISPLACEMENTS AND TRACTIONS

NODE	U	V	PX	PY
1	0.0000	7.7353	0.0	15.0000
2	2.0020	7.4717	3.8823	14.4889
3	3.8676	6.6990	7.5000	12.9904
4	5.4697	5.4697	10.6066	10.6066
5	6.6990	3.8675	12.9904	7.5000
6	7.4717	2.0020	14.4889	3.8823
7	7.7353	0.0000	15.0000	0.0

DISPLACEMENTS AND STRESSES AT NODES AND INTERNAL POINTS

NO/PT	U	V	SX	SXY	SY	SZ
1	0.0000	7.7353	14.7577	0.0	-14.8737	-0.0116
2	2.0020	7.4717	12.7729	-7.4078	-12.8888	-0.0116
3	3.8676	6.6990	7.3499	-12.8308	-7.4658	-0.0116
4	5.4697	5.4697	-0.0580	-14.8157	-0.0580	-0.0116
5	6.6990	3.8675	-7.4658	-12.8308	7.3499	-0.0116
6	7.4717	2.0020	-12.8888	-7.4078	12.7729	-0.0116
7	7.7353	0.0000	-14.8737	0.0	14.7577	-0.0116
8	6.4100	0.0000	-10.2220	-0.0000	10.2136	-0.0008
9	5.1274	0.0000	-6.5259	-0.0000	6.5259	-0.0000
10	3.8455	0.0000	-3.6707	-0.0000	3.6707	-0.0000
11	4.5325	4.5325	-0.0042	-10.2178	-0.0042	-0.0008
12	3.6256	3.6255	-0.0000	-6.5259	-0.0000	-0.0000
13	2.7192	2.7192	0.0000	-3.6707	0.0000	0.0000

14.9.2. Cylindrical Cavity Problem

In this example the analysis of a cylindrical cavity under internal pressure is carried out. The problem domain is the infinite medium and, consequently, INFB = 1 in this case. The discretization is depicted in Fig. 14.6 and the input data is presented in Table 14.2. For the solution presented six boundary elements and seven boundary nodes are used. Additional information is obtained at six internal points.

The solution obtained by the program is presented. These results are in good agreement with the analytical solution [2], and the maximum error is around 2%.

References

1. Brebbia, C. A., and Ferrante, A. J., *Computational Methods for the Solution of Engineering Problems*, Pentech Press, London, 1978.
2. Timoshenko, S. P., and Goodier, J. N., *Theory of Elasticity*, 3rd ed., McGraw–Hill, Tokyo, 1970.
3. Brebbia, C. A., *The Boundary Element Method for Engineers*, Pentech Press, London and Halstead Press, New York, 1978.

Appendix A

Numerical Integration Formulas

A.1. Introduction

In this appendix guidelines for the numerical computation of the element and cell integrals are presented. Since Gaussian integration formulas are the ones that present the best accuracy for a given number of points, emphasis will be given to this sort of numerical integration procedure.

In what follows, the numerical integration formulas are divided into two groups. The first one corresponds to standard integration and should be used when the integrals involved do not present any singularities. The second group is concerned with integration over elements or cells in which the singular (source or load) point is located at the extremity of the integration domain and, therefore, should be applied in these cases only (e.g., computation of leading diagonal submatrices of $\boldsymbol{H}$ and $\boldsymbol{G}$).

A.2. Standard Gaussian Quadrature

A.2.1. One-Dimensional Quadrature [1]

$$I = \int_{-1}^{1} f(x)\, dx \cong \sum_{i=1}^{n} f(x_i)\, w_i\,, \tag{A.1}$$

where x_i is the coordinate of the ith integration point, w_i is the associated weighting factor, and n is the total number of integration points; they are listed in Table A.1. The error associated is $E_n = O\,(d^{2n}f/dx^{2n})$.

A.2.2. Two- and Three-Dimensional Quadrature for Rectangles and Rectangular Hexahedra

Two- and three-dimensional formulas are obtained by combining expression (A.1) in the form

$$\int_{-1}^{1}\int_{-1}^{1} f(x, y)\, dx\, dy \cong \sum_{j=1}^{n}\sum_{i=1}^{n} f(x_i, y_j)\, w_i\, w_j \tag{A.2}$$

and

$$\int_{-1}^{1}\int_{-1}^{1}\int_{-1}^{1} f(x, y, z)\, dx\, dy\, dz \cong \sum_{k=1}^{n}\sum_{j=1}^{n}\sum_{i=1}^{n} f(x_i, y_j, z_k)\, w_i\, w_j\, w_k\,, \tag{A.3}$$

Table A.1.

$\pm x_i$	w_i	$\pm x_i$	w_i
$n=2$		$n=8$	
0.57735 02691 89626	1.00000 00000 00000	0.18343 46424 95650	0.36268 37833 78362
		0.52553 24099 16329	0.31370 66458 77887
		0.79666 64774 13627	0.22238 10344 53374
$n=3$		0.96028 98564 97536	0.10122 85362 90376
0.00000 00000 00000	0.88888 88888 88889		
0.77459 66692 41483	0.55555 55555 55556	$n=9$	
		0.00000 00000 00000	0.33023 93550 01260
$n=4$		0.32425 34234 03809	0.31234 70770 40003
0.33998 10435 84856	0.65214 51548 62546	0.61337 14327 00590	0.26061 06964 02935
0.86113 63115 94053	0.34785 48451 37454	0.83603 11073 26636	0.18064 81606 94857
		0.96816 02395 07626	0.08127 43883 61574
$n=5$			
0.00000 00000 00000	0.56888 88888 88889	$n=10$	
0.53846 93101 05683	0.47862 86704 99366	0.14887 43389 81631	0.29552 42247 14753
0.90617 98459 38664	0.23692 68850 56189	0.43339 53941 29247	0.26926 67193 09996
		0.67940 95682 99024	0.21908 63625 15982
$n=6$		0.86506 33666 88985	0.14945 13491 50581
0.23861 91860 83197	0.46791 39345 72691	0.97390 65285 17172	0.06667 13443 08688
0.66120 93864 66265	0.36076 15730 48139		
0.93246 95142 03152	0.17132 44923 79170	$n=12$	
		0.12523 34085 11469	0.24914 70458 13403
$n=7$		0.36783 14989 98180	0.23349 25365 38355
0.00000 00000 00000	0.41795 91836 73469	0.58731 79542 86617	0.20316 74267 23066
0.40584 51513 77397	0.38183 00505 05119	0.76990 26741 94305	0.16007 83285 43346
0.74153 11855 99394	0.27970 53914 89277	0.90411 72563 70475	0.10693 93259 95318
0.94910 79123 42759	0.12948 49661 68870	0.98156 06342 46719	0.04717 53363 86512

Table A.2.

n	i	η_1^i	η_2^i	η_3^i	w_i
1 (linear)	1	1/3	1/3	1/3	1
2	1	1/2	1/2	0	1/3
(quadratic)	2	0	1/2	1/2	1/3
	3	1/2	0	1/2	1/3
4 (cubic)	1	1/3	1/3	1/3	− 9/16
	2	3/5	1/5	1/5	25/48
	3	1/5	3/5	1/5	25/48
	4	1/5	1/5	3/5	25/48
7 (quintic)	1	0.333 333 33	0.333 333 33	0.333 333 33	0.225 000 00
	2	0.797 426 99	0.101 286 51	0.101 286 51	0.125 939 18
	3	0.101 286 51	0.797 426 99	0.101 286 51	0.125 939 18
	4	0.101 286 51	0.101 286 51	0.797 426 99	0.125 939 18
	5	0.059 715 87	0.470 142 06	0.470 142 06	0.132 394 15
	6	0.470 142 06	0.059 715 87	0.470 142 06	0.132 394 15
	7	0.470 142 06	0.470 142 06	0.059 715 87	0.132 394 15

where the integration point coordinates and weighting factors are listed in Table A.1.

A.2.3. Triangular Domain

Numerical integration over a triangle can be performed in terms of the triangular coordinates η_1, η_2, and η_3 defined in Chapter 3 as follows (see Fig. A.1):

$$I=\int_0^1\left(\int_0^{1-\eta_2} f(\eta_1,\eta_2,\eta_3)\,d\eta_1\right)d\eta_2 \cong \sum_{i=1}^{n} f(\eta_1^i,\eta_2^i,\eta_3^i)\,w_i \tag{A.4}$$

in which the triangular coordinates and associated weighting factors are due to Hammer *et al.* [2] and given in Table A.2.

By combining expression (A.4) with (A.1), numerical integration formulas for three-dimensional pentahedral cells can be obtained as before.

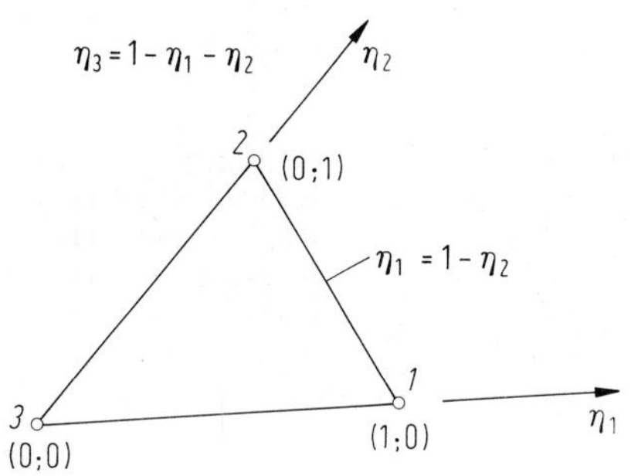

Fig. A.1. Definition of triangular coordinates (η_1, η_2)

A.3. Computation of Singular Integrals

A.3.1. One-Dimensional Logarithmic Gaussian Quadrature Formulas [1]

$$I=\int_0^1 \ln\frac{1}{x}f(x)\,dx \cong \sum_{i=1}^{n} f(x_i)\,w_i\,, \tag{A.5}$$

where Table A.3 presents the required points and weights. Note that the above expression is useful for two-dimensional boundary element applications where a logarithmic singularity often occurs.

A.3.2. Numerical Integration over Triangles and Squares with $1/r$ Singularity

Numerical quadrature formulas for these cases have been presented by Cristescu and Loubignac [3] and also later by Pina *et al.* [4]. The latter have been adopted

Table A.3.

n	x_i	w_i	n	x_i	w_i
2	0.11200880	0.71853931	8	0.13320243 (– 1)	0.16441660
	0.60227691	0.28146068		0.79750427 (– 1)	0.23752560
3	0.63890792 (– 1)	0.51340455		0.19787102	0.22684198
	0.36899706	0.39198004		0.35415398	0.17575408
	0.76688030	0.94615406 (–1)		0.52945857	0.11292402
4	0.41448480 (– 1)	0.38346406		0.70181452	0.57872212 (– 1)
	0.24527491	0.38687532		0.84937932	0.20979074 (– 1)
	0.55616545	0.19043513		0.95332645	0.36864071 (– 2)
	0.84898239	0.39225487 (– 1)	9	0.10869338 (– 1)	0.14006846
5	0.29134472 (– 1)	0.29789346		0.64983682 (– 1)	0.20977224
	0.17397721	0.34977622		0.16222943	0.21142716
	0.41170251	0.23448829		0.29374996	0.17715622
	0.67731417	0.98930460 (– 1)		0.44663195	0.12779920
	0.89477136	0.18911552 (– 1)		0.60548172	0.78478879 (– 1)
6	0.21634005 (– 1)	0.23876366		0.75411017	0.39022490 (– 1)
	0.12958339	0.30828657		0.87726585	0.13867290 (– 1)
	0.31402045	0.24531742		0.96225056	0.24080402 (– 2)
	0.53865721	0.14200875	10	0.90425944 (– 2)	0.12095474
	0.75691533	0.55454622 (– 1)		0.53971054 (– 1)	0.18636310
	0.92266884	0.10168958 (– 1)		0.13531134	0.19566066
7	0.16719355 (– 1)	0.19616938		0.24705169	0.17357723
	0.10018568	0.27030264		0.38021171	0.13569597
	0.24629424	0.23968187		0.52379159	0.93647084 (– 1)
	0.43346349	0.16577577		0.66577472	0.55787938 (– 1)
	0.63235098	0.88943226 (– 1)		0.79419019	0.27159893 (– 1)
	0.81111862	0.33194304 (– 1)		0.89816102	0.95151992 (– 2)
	0.94084816	0.59327869 (– 2)		0.96884798	0.16381586 (– 2)

Note: Numbers are to be multiplied by the power of 10 in parentheses.

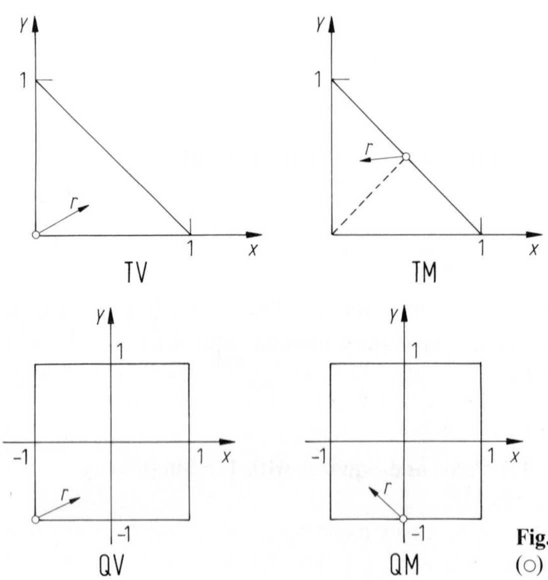

Fig. A.2. Description of configurations, (○) denotes position of singularity.

here and are to be applied as follows:

$$\int_{\omega} \frac{1}{r} f(x, y)\, d\omega \cong \sum_{i=1}^{n} f(x_i, y_i)\, w_i, \tag{A.6}$$

where ω represents any of the integration domains depicted in Fig. A.2. The points and weights are grouped in Table A.4 in which the notation TV k or TM k designates a formula of degree k for the configuration TV or TM shown in Fig. A.2; the same pattern is followed for QV k and QM k. When more than one formula is given for the same degree, they are distinguished by the use of primes.

Note that all the formulas exhibit positive weights and integration points inside the region of integration with the exception of formula TV 2″. Point 1 of this rule lies slightly outside the triangle.

A.3.3. Numerical Evaluation of Cauchy Principal Values

It is known that the computation of Cauchy principal value integrals can be accomplished by using the finite parts of the integrals involved [5]. To illustrate the present matter, consider the Cauchy principal value integral

$$I = \int_a^b \frac{1}{(x-s)} f(x)\, dx = \lim_{\varepsilon \to 0} \left\{ \int_a^{s-\varepsilon} \frac{f(x)}{x-s}\, dx + \int_{s+\varepsilon}^{b} \frac{f(x)}{x-s}\, dx \right\}, \qquad a < s < b, \tag{A.7}$$

where $f(x)$ is supposed to satisfy a Hölder condition at s and $f(s) \neq 0$ so that the singularity is actually of order one.

Clearly, each of the two integrals on the right of Eq. (A.7) can be evaluated by finite part integration, i.e.,

$$\begin{aligned} I' &= \oint_a^s \frac{f(x)}{x-s}\, dx, \\ I'' &= \oint_s^b \frac{f(x)}{x-s}\, dx, \\ I &= I' + I'', \end{aligned} \tag{A.8}$$

where $\oint$ denotes a finite part integral [5–7].

A finite part integral is a linear and continuous functional which includes the regular integrals; consequently, many properties of standard integrals also apply here, but not all of them are generally valid. For instance, in the case of expressions (A.8), translation and reflection of the interval of integration are permitted in the standard form, but not scaling. If one scales the second expression (A.8) to a unit length, the following expression must be used [7]:

$$\oint_s^b \frac{f(x)}{x-s}\, dx = \oint_0^1 \frac{f[(b-s)\,t+s]}{t}\, dt + f(s) \ln |b-s| . \tag{A.9}$$

Table A.4.

Formula	Coordinates x_i	Coordinates y_i	Weights w_i
TV 1	0.25	$y_1 = x_1$	1.24645048
TV 2′	0.16666667	$y_1 = x_1$	0.93483790
	0.81742619	0.18257381	0.15580629
	$x_3 = y_2$	$y_3 = x_2$	$w_3 = w_2$
TV 2″	0.53764799	$y_1 = x_1$	0.25071485
	0.0	0.35514705	0.49786782
	$x_3 = y_2$	$y_3 = x_2$	$w_3 = w_2$
TV 3	0.16385495	0.04756957	0.31161231
	0.61114353	0.17753138	0.31161293
	$x_3 = y_1$	$y_3 = x_1$	$w_3 = w_1$
	$x_4 = y_2$	$y_4 = x_2$	$w_4 = w_2$
TM 1	0.125	$y_1 = x_1$	1.76274717
TM 2′	0.28295366	$y_1 = x_1$	1.01519055
	0.24706775	0.75293225	0.37377831
	$x_3 = y_2$	$y_3 = x_2$	$w_3 = w_2$
TM 2″	0.46805571	$y_1 = x_1$	1.00578450
	0.50271173	0.0	0.37848134
	$x_3 = y_2$	$y_3 = x_2$	$w_3 = w_2$
TM 3	0.073328962	0.33623395	0.25051549
	0.26640495	0.61878355	0.63085810
	$x_3 = y_1$	$y_3 = x_1$	$w_3 = w_1$
	$x_4 = y_2$	$y_4 = x_2$	$w_4 = w_2$
QV 1	−0.26501817	$y_1 = x_1$	3.52549435
QV 2′	−0.58105530	$y_1 = x_1$	2.37881900
	1.0	−0.21877566	0.57333767
	$x_3 = y_2$	$y_3 = x_2$	$w_3 = w_2$
QV 2″	0.39666491	$y_1 = x_1$	0.13250102
	−0.15632872	−1.0	1.19649666
	$x_3 = y_2$	$y_3 = x_2$	$w_3 = w_2$
QV 3	−0.37512304	−0.92928746	1.02276580
	0.69629093	−0.15602536	0.73998134
	$x_3 = y_1$	$y_3 = x_1$	$w_3 = w_1$
	$x_4 = y_2$	$y_4 = x_2$	$w_4 = w_2$
QM 1	0.0	−0.34313433	4.81211825
QM 2′	0.0	0.13130626	2.79404031
	0.76138824	−1.0	1.00903897
	$x_3 = x_2$	$y_3 = y_2$	$w_3 = w_2$
QM 2″	0.0	−0.65952349	3.64221415
	1.0	0.64186697	0.58495205
	$x_3 = -x_2$	$y_3 = y_2$	$w_3 = w_2$
QM 3	0.44855808	−0.73405341	1.61441436
	0.57322583	0.45407346	0.79164476
	$x_3 = x_1$	$y_3 = y_1$	$w_3 = w_1$
	$x_4 = -x_2$	$y_4 = y_2$	$w_4 = w_2$

Table A.5.

x_i	w_i
$n = 2$	
− 0.1318813079 1298666721 5670599477	(1) −0.1309307341 4159542875 9658491249
0.6318813079 1298666721 5670599477	(1) 0.1309307341 4159542875 9658491249
$n = 3$	
(− 1) −0.3969686527 5763972364 5602577574	(1) −0.2051668519 3485338763 3677477371
0.3333333333 3333333333 3333333333	(1) 0.1588235294 1176470588 2352941176
0.8396968652 7576397236 4560257757	0.4634332252 3088681751 3245361943
$n = 4$	
(− 1) −0.1835380883 6002397574 3306030308	(1) −0.2593628689 4206353141 1986911649
0.1997381810 4998674591 7610636251	(1) 0.1672029801 3533497561 5800624534
0.5839530474 6507837000 6565400789	0.6754334170 2765424584 0955338683
0.9099872556 4561260632 5479241316	0.2461654710 3963131212 0907532468
$n = 5$	
(− 1) −0.1034864035 5610450391 3216228982	(1) −0.3020220188 0167489499 1983779419
0.1318792969 2542320444 8930810801	(1) 0.1706705519 0523283409 5676514538
0.4110445071 4059390565 4861696346	0.7655771077 6620893412 7999254496
0.7206100604 8734965175 5922829110	0.3940320876 1668993050 9861730341
0.9423703313 5779924408 7161842197	0.1539054735 8152174432 5211663977
$n = 6$	
(− 2) −0.6556081602 2445314564 8046162307	(1) −0.3371785154 0107025091 6620840281
(− 1) 0.9325015355 0834953341 4314714990	(1) 0.1723519097 4105612509 3829789834
0.3004462888 0792320156 8234438420	0.8125117342 9233737746 8770569286
0.5594481971 0305566728 1833365623	0.4683097364 8467610133 5742374651
0.8008212350 0722462324 6583676282	0.2618278247 3224660162 0697426204
0.9599558342 0686235675 4960217165	0.1056167610 9088117780 2700134329
$n = 7$	
(− 2) −0.4486017976 3859046160 9844627776	(1) −0.3670695942 0541972514 7338842722
(− 1) 0.6930868259 3891477203 9902359773	(1) 0.1732436531 9568001126 4627034186
0.2276496031 4328665070 4908345661	0.8400787344 9383995169 0577594652
0.4390637610 4952800515 8487273996	0.5112251766 0748204799 5326075488
0.6616489750 8666059405 0408569646	0.3222444519 0458395756 1663455934
0.8513209482 8767406448 2920243734	0.1876463072 3129246428 5563208552
0.9705646914 5741732651 6168706619	(− 1) 0.7706473986 0198717293 9877507356
$n = 8$	
(− 2) −0.3242501597 2971718707 6226447723	(1) −0.3930636810 0685511256 7343685901
(− 1) 0.5349076607 2145996351 6325759777	(1) 0.1737406310 9002152856 2172435669
0.1778273263 9269532029 1476101206	0.8576345422 7306884948 2376287755
0.3507178785 5254952494 1759499784	0.5383607734 8249803899 5953311080
0.5458195204 6848788049 7113433351	0.3597575915 9203737679 5700571782
0.7334270806 5718625775 7460821281	0.2372676990 9455128730 7264145652
0.8849830533 7011672211 4798993637	0.1414622854 9686852832 8034959954
0.9774543783 7979958378 2452886991	(− 1) 0.5874760722 9311759142 3832261022

Note: Numbers are to be multiplied by the power of 10 in parentheses.

It is interesting to note that higher-order singularities can also be integrated in the finite part sense, but a complete treatment of this theory is far beyond the scope of the present appendix. The interested reader should report to Refs. [5–7] for more detailed explanations.

Numerical integration formulas for the computation of finite part integrals have been presented by Kutt [8]. The table corresponding to expressions (A.8) is partially reproduced here for completeness. This table was computed for unit lengths and consequently expression (A.9) must be used such that

$$I' = \oint_a^s \frac{f(x)}{x-s}\,dx \cong -\sum_{i=1}^{n} f[(a-s)\,x_i + s]\,w_i - f(s)\ln|a-s|\,, \tag{A.10}$$

$$I'' = \oint_s^b \frac{f(x)}{x-s}\,dx \cong \sum_{i=1}^{n} f[(b-s)\,x_i + s]\,w_i + f(s)\ln|b-s|\,, \tag{A.11}$$

where x_i and w_i are the coordinates and weights of the Gaussian type integration formulas listed in Table A.5. Note that the first integration point lies slightly outside the integration limits, but since $f(x)$ is normally known analytically, this is of no consequence.

The above formulas can be applied for the computation of Cauchy principal values in two-dimensional boundary elements (i.e., leading diagonal submatrices of $\boldsymbol{H}$). For three-dimensional applications and integration of inelastic cells (e.g., plasticity, creep, etc.), they can still be applied by adopting a polar coordinate system (r, φ, θ) which allows for one-dimensional finite part integration with respect to r and standard quadrature formulas with respect to the angles φ and θ.

References

1. Stroud, A. H., and Secrest, D., *Gaussian Quadrature Formulas,* Prentice–Hall, New York, 1966.
2. Hammer, P. C., Marlowe, O. J., and Stroud, A. H., Numerical integration over simplexes and cones, Math. Tables Other Aids Comput. **10,** 130–139 (1956).
3. Cristescu, M., and Loubignac, G., Gaussian quadrature formulas for functions with singularities in 1/r over triangles and quadrangles, in *Recent Advances in Boundary Element Methods* (C. A. Brebbia, Ed.), 375–390, Pentech Press, London, 1978.
4. Pina, H. L. G., Fernandes, J. L. M., and Brebbia, C. A., Some numerical integration formulae over triangles and squares with a $1/r$ singularity, Appl. Math. Modelling **5,** 209–211 (1981).
5. Gel'fand, I. M., and Shilov, G. E., *Generalized Functions,* Vol. 1, Academic Press, New York, 1964.
6. Kutt, H. R., The numerical evaluation of principal value integrals by finite part integration, Numer. Math. **24,** 205–210 (1975).
7. Kutt, H. R., On the numerical evaluation of finite part integrals involving an algebraic singularity, Report WISK 179, The National Research Institute for Mathematical Sciences, Pretoria, 1975.
8. Kutt, H. R., Quadrature formulae for finite part integrals, Report WISK 178, The National Research Institute for Mathematical Sciences, Pretoria, 1975.

Appendix B

Semi-Infinite Fundamental Solutions

In this appendix the complementary part of the fundamental solutions for half-space and half-plane problems are presented. These expressions, when added to the corresponding Kelvin solution, produce the desired fundamental solutions (see expression (5.67)) for three- and two-dimensional problems.

B.1. Half-Space [1]

With reference to Fig. B.1, the complementary expressions for the displacements due to unit point loads applied within the half-space are given by

$$u_{11}^c = K_d \left\{ \frac{8(1-\nu)^2 - (3-4\nu)}{R} + \frac{(3-4\nu)R_1^2 - 2c\bar{x}}{R^3} + \frac{6c\bar{x}R_1^2}{R^5} \right\},$$

$$u_{12}^c = K_d r_2 \left\{ \frac{(3-4\nu)r_1}{R^3} - \frac{4(1-\nu)(1-2\nu)}{R(R+R_1)} + \frac{6c\bar{x}R_1}{R^5} \right\},$$

$$u_{13}^c = \frac{r_3}{r_2} u_{12}^c,$$

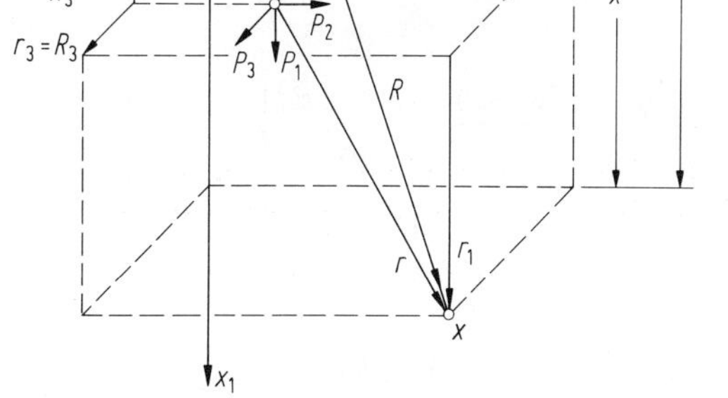

Fig. B.1. Unit point loads applied within the half-space ($|P_1| = |P_2| = |P_3| = 1$)

$$u_{21}^c = K_d r_2 \left\{ \frac{(3-4v)\,r_1}{R^3} + \frac{4(1-v)(1-2v)}{R(R+R_1)} - \frac{6c\,\bar{x}\,R_1}{R^5} \right\},$$

$$u_{22}^c = K_d \left\{ \frac{1}{R} + \frac{(3-4v)\,r_2^2}{R^3} + \frac{2c\,\bar{x}}{R^3}\left(1 - \frac{3r_2^2}{R^2}\right) + \frac{4(1-v)(1-2v)}{R+R_1}\left(1 - \frac{r_2^2}{R(R+R_1)}\right) \right\}, \tag{B.1}$$

$$u_{23}^c = K_d r_2 r_3 \left\{ \frac{3-4v}{R^3} - \frac{4(1-v)(1-2v)}{R(R+R_1)^2} - \frac{6c\,\bar{x}}{R^5} \right\},$$

$$u_{31}^c = \frac{r_3}{r_2} u_{21}^c,$$

$$u_{32}^c = u_{23}^c,$$

$$u_{33}^c = K_d \left\{ \frac{1}{R} + \frac{(3-4v)\,r_3^2}{R^3} + \frac{2c\,\bar{x}}{R^3}\left(1 - \frac{3r_3^2}{R^2}\right) + \frac{4(1-v)(1-2v)}{R+R_1}\left(1 - \frac{r_3^2}{R(R+R_1)}\right) \right\},$$

where $(i = 1, 2, 3)$

$$\begin{aligned} R &= (R_i R_i)^{1/2}, \\ r_i &= x_i(x) - x_i(\xi), \\ R_i &= x_i(x) - x_i(\xi'), \\ c &= x_1(\xi) \geqq 0, \\ \bar{x} &= x_1(x) \geqq 0, \\ K_d &= \frac{1}{16\pi(1-v)\,G}. \end{aligned} \tag{B.2}$$

The complementary expressions for the corresponding stresses are as follows:

$$\sigma_{111}^c = K_s \left\{ \frac{(1-2v)\,r_1}{R^3} - \frac{3(3-4v)\,\bar{x}\,R_1^2 - 3c\,R_1(5\bar{x}-c)}{R^5} - \frac{30c\,\bar{x}\,R_1^3}{R^7} \right\},$$

$$\sigma_{121}^c = K_s r_2 \left\{ \frac{1-2v}{R^3} - \frac{3(3-4v)\,\bar{x}\,R_1 - 3c(3\bar{x}+c)}{R^5} - \frac{30\,c\,\bar{x}\,R_1^2}{R^7} \right\},$$

$$\sigma_{131}^c = K_s r_3 \left\{ \frac{1-2v}{R^3} - \frac{3(3-4v)\,\bar{x}\,R_1 - 3c(3\bar{x}+c)}{R^5} - \frac{30c\,\bar{x}\,R_1^2}{R^7} \right\},$$

$$\sigma_{221}^{c} = K_s \left\{ \frac{(1-2\nu)(3r_1 - 4\nu R_1)}{R^3} - \frac{3(3-4\nu)r_2^2 r_1 - 6cR_1[(1-2\nu)\bar{x} - 2\nu c]}{R^5} \right.$$

$$\left. - \frac{30c\, r_2^2 \bar{x} R_1}{R^7} - \frac{4(1-\nu)(1-2\nu)}{R(R+R_1)} \left(1 - \frac{r_2^2}{R(R+R_1)} - \frac{r_2^2}{R^2} \right) \right\},$$

$$\sigma_{231}^{c} = K_s r_2 r_3 \left\{ -\frac{3(3-4\nu)r_1}{R^5} + \frac{4(1-\nu)(1-2\nu)}{R^2(R+R_1)} \left(\frac{1}{R+R_1} + \frac{1}{R} \right) - \frac{30c\,\bar{x}R_1}{R^7} \right\},$$

$$\sigma_{331}^{c} = K_s \left\{ \frac{(1-2\nu)[3r_1 - 4\nu R_1]}{R^3} - \frac{3(3-4\nu)r_3^2 r_1 - 6cR_1[(1-2\nu)\bar{x} - 2\nu c]}{R^5} \right.$$

$$\left. - \frac{30c\, r_3^2 \bar{x} R_1}{R^7} - \frac{4(1-\nu)(1-2\nu)}{R(R+R_1)} \left(1 - \frac{r_3^2}{R(R+R_1)} - \frac{r_3^2}{R^2} \right) \right\},$$

$$\sigma_{112}^{c} = K_s r_2 \left\{ -\frac{1-2\nu}{R^3} - \frac{3(3-4\nu)R_1^2}{R^5} + \frac{6c}{R^5} \left(c + (1-2\nu)R_1 + \frac{5\bar{x}R_1^2}{R^2} \right) \right\},$$

$$\sigma_{122}^{c} = K_s \left\{ \frac{(1-2\nu)r_1}{R^3} - \frac{3(3-4\nu)r_2^2 R_1}{R^5} - \frac{6c}{R^5} \left[\bar{x}R_1 - (1-2\nu)r_2^2 - \frac{5r_2^2 \bar{x} R_1}{R^2} \right] \right\},$$

$$\sigma_{132}^{c} = K_s r_2 r_3 \left\{ -\frac{3(3-4\nu)R_1}{R^5} + \frac{6c}{R^5} \left(1 - 2\nu + \frac{5\bar{x}R_1}{R^2} \right) \right\}, \tag{B.3}$$

$$\sigma_{222}^{c} = K_s r_2 \left\{ \frac{(1-2\nu)(5-4\nu)}{R^3} - \frac{3(3-4\nu)r_2^2}{R^5} \right.$$

$$\left. - \frac{4(1-\nu)(1-2\nu)}{R(R+R_1)^2} \left[3 - \frac{r_2^2(3R+R_1)}{R^2(R+R_1)} \right] + \frac{6c}{R^5} \left[3c - (3-2\nu)R_1 + \frac{5r_2^2 \bar{x}}{R^2} \right] \right\},$$

$$\sigma_{232}^{c} = K_s r_3 \left\{ \frac{1-2\nu}{R^3} - \frac{3(3-4\nu)r_2^2}{R^5} - \frac{4(1-\nu)(1-2\nu)}{R(R+R_1)^2} \left[1 - \frac{r_2^2(3R+R_1)}{R^2(R+R_1)} \right] \right.$$

$$\left. - \frac{6c\,\bar{x}}{R^5} \left(1 - \frac{5r_2^2}{R^2} \right) \right\},$$

$$\sigma_{332}^{c} = K_s r_2 \left\{ \frac{(1-2\nu)(3-4\nu)}{R^3} - \frac{3(3-4\nu)r_3^2}{R^5} \right.$$

$$\left. - \frac{4(1-\nu)(1-2\nu)}{R(R+R_1)^2} \left[1 - \frac{r_3^2(3R+R_1)}{R^2(R+R_1)} \right] + \frac{6c}{R^5} \left[c - (1-2\nu)R_1 + \frac{5r_3^2 \bar{x}}{R^2} \right] \right\},$$

$$\sigma_{113}^{c} = \frac{r_3}{r_2} \sigma_{112}^{c},$$

$$\sigma_{123}^{c} = \sigma_{132}^{c},$$

$$\sigma_{133}^c = K_s \left\{ \frac{(1-2\nu)\, r_1}{R^3} - \frac{3(3-4\nu)\, r_3^2 R_1}{R^5} - \frac{6c}{R^5} \left[\bar{x} R_1 - (1-2\nu)\, r_3^2 - \frac{5 r_3^2 \bar{x} R_1}{R^2} \right] \right\},$$

$$\sigma_{223}^c = K_s r_3 \left\{ \frac{(1-2\nu)(3-4\nu)}{R^3} - \frac{3(3-4\nu)\, r_2^2}{R^5} \right.$$
$$\left. - \frac{4(1-\nu)(1-2\nu)}{R(R+R_1)^2} \left[1 - \frac{r_2^2(3R+R_1)}{R^2(R+R_1)} \right] + \frac{6c}{R^5} \left[c - (1-2\nu)\, R_1 + \frac{5 r_2^2 \bar{x}}{R^2} \right] \right\},$$

$$\sigma_{233}^c = K_s r_2 \left\{ \frac{1-2\nu}{R^3} - \frac{3(3-4\nu)\, r_3^2}{R^5} - \frac{4(1-\nu)(1-2\nu)}{R(R+R_1)^2} \left[1 - \frac{r_3^2(3R+R_1)}{R^2(R+R_1)} \right] \right.$$
$$\left. - \frac{6c\,\bar{x}}{R^5} \left(1 - \frac{5 r_3^2}{R^2} \right) \right\},$$

$$\sigma_{333}^c = K_s r_3 \left\{ \frac{(1-2\nu)(5-4\nu)}{R^3} - \frac{3(3-4\nu)\, r_3^2}{R^5} \right.$$
$$\left. - \frac{4(1-\nu)(1-2\nu)}{R(R+R_1)^2} \left[3 - \frac{r_3^2(3R+R_1)}{R^2(R+R_1)} \right] + \frac{6c}{R^5} \left[3c - (3-2\nu)\, R_1 + \frac{5 r_3^2 \bar{x}}{R^2} \right] \right\},$$

in which

$$K_s = \frac{1}{8\pi(1-\nu)}. \tag{B.4}$$

The fundamental tractions can be computed from expressions (B.3) by using the relation

$$p_{ij}^c = \sigma_{jki}^c\, n_k\,. \tag{B.5}$$

B.2. Half-Plane [2]

With reference to Fig. B.2, the complementary part of the plane strain fundamental displacements can be written as

$$u_{11}^c = \bar{K}_d \left\{ -[8(1-\nu)^2 - (3-4\nu)] \ln R + \frac{[(3-4\nu)\, R_1^2 - 2c\,\bar{x}]}{R^2} + \frac{4c\,\bar{x} R_1^2}{R^4} \right\},$$

$$u_{12}^c = \bar{K}_d \left\{ \frac{(3-4\nu)\, r_1 r_2}{R^2} + \frac{4c\,\bar{x} R_1 r_2}{R^4} - 4(1-\nu)(1-2\nu)\,\theta \right\},$$

$$u_{21}^c = \bar{K}_d \left\{ \frac{(3-4\nu)\, r_1 r_2}{R^2} - \frac{4c\,\bar{x} R_1 r_2}{R^4} + 4(1-\nu)(1-2\nu)\,\theta \right\}, \tag{B.6}$$

$$u_{22}^c = \bar{K}_d \left\{ -[8(1-\nu)^2 - (3-4\nu)] \ln R + \frac{[(3-4\nu)\, r_2^2 + 2c\,\bar{x}]}{R^2} - \frac{4c\,\bar{x}\, r_2^2}{R^4} \right\},$$

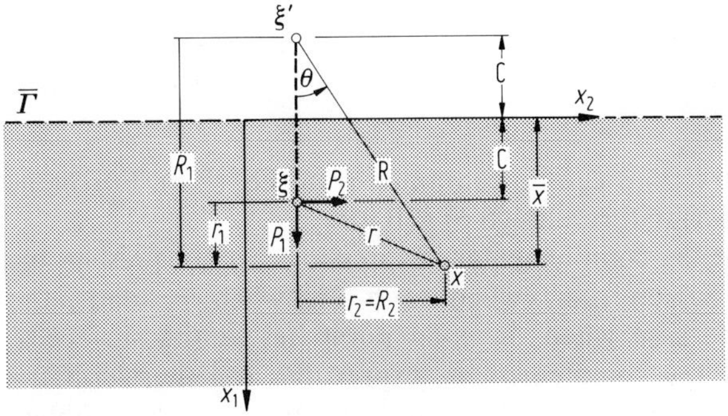

Fig. B.2. Unit point loads applied within the half-plane. ξ is the load point, x is the field point, ξ' is the image of ξ, and $|P_1| = |P_2| = 1$

where the notation presented in (B.2) is used ($i = 1, 2$) and

$$\begin{aligned} \theta &= \arctan\left(\frac{R_2}{R_1}\right), \\ \bar{K}_d &= \frac{1}{8\pi(1-\nu)\,G}. \end{aligned} \tag{B.7}$$

The complementary stresses due to unit forces acting inside the half-plane are given by (plane strain)

$$\begin{aligned}
\sigma^c_{111} &= -\bar{K}_s\left\{\frac{(3\bar{x}+c)(1-2\nu)}{R^2} + \frac{2[R_1(R_1^2+2c\,\bar{x}) - 2\bar{x}\,r_2^2(1-2\nu)]}{R^4} - \frac{16c\,\bar{x}\,R_1 r_2^2}{R^6}\right\}, \\
\sigma^c_{121} &= -\bar{K}_s\, r_2\left\{-\frac{(1-2\nu)}{R^2} + \frac{2[\bar{x}^2 - 2c\,\bar{x} - c^2 + 2\bar{x}\,R_1(1-2\nu)]}{R^4} + \frac{16c\,\bar{x}\,R_1^2}{R^6}\right\}, \\
\sigma^c_{221} &= -\bar{K}_s\left\{\frac{(\bar{x}+3c)(1-2\nu)}{R^2} + \frac{2[R_1(r_2^2+2c^2) - 2c\,r_2^2 + 2\bar{x}\,r_2^2(1-2\nu)]}{R^4}\right. \\
&\qquad \left. + \frac{16c\,\bar{x}\,R_1 r_2^2}{R^6}\right\}, \\
\sigma^c_{112} &= -\bar{K}_s\, r_2\left\{\frac{(1-2\nu)}{R^2} - \frac{2[c^2 - \bar{x}^2 + 6c\,\bar{x} - 2\bar{x}\,R_1(1-2\nu)]}{R^4} + \frac{16c\,\bar{x}\,r_2^2}{R^6}\right\}, \\
\sigma^c_{122} &= -\bar{K}_s\left\{\frac{(3\bar{x}+c)(1-2\nu)}{R^2} + \frac{2[(2c\,\bar{x}+r_2^2)\,R_1 - 2\bar{x}\,R_1^2(1-2\nu)]}{R^4} - \frac{16c\,\bar{x}\,R_1 r_2^2}{R^6}\right\}, \\
\sigma^c_{222} &= -\bar{K}_s\, r_2\left\{\frac{3(1-2\nu)}{R^2} + \frac{2[r_2^2 - 4c\,\bar{x} - 2c^2 - 2\bar{x}\,R_1(1-2\nu)]}{R^4} + \frac{16c\,\bar{x}\,R_1^2}{R^6}\right\},
\end{aligned} \tag{B.8}$$

where

$$\bar{K}_s = \frac{1}{4\pi(1-\nu)}. \tag{B.9}$$

and again the corresponding tractions can be computed from Eq. (B.5). For plane stress ν is replaced by $\bar{\nu} = \nu/(1+\nu)$ in formulas (B.6)–(B.9).

It is interesting to note that the complementary expressions do not present any singularities within the actual region $x_1 \geqq 0$ when $c > 0$ (i.e., when the load point is located inside Ω^*). For the case when the load point lies at the surface $\bar{\Gamma}$ ($c \to 0$), it is easily seen that the complementary expressions together with the Kelvin solution (see relation (5.67)) produce the complete solution to the problem of Boussinesq-Cerruti [1, 3] in three dimensions or Flamant's problem [3, 4] in two dimensions. In the latter case, for instance, the fundamental displacements and tractions become ($\xi \in \bar{\Gamma}$),

$$\begin{aligned} u_{11}^* &= -K_d'\{2(1-\nu)\ln r - r_{,1}^2\}, \\ u_{12}^* &= -K_d'\{(1-2\nu)\,\theta - r_{,1}\,r_{,2}\}, \\ u_{21}^* &= -K_d'\{-(1-2\nu)\,\theta - r_{,2}\,r_{,1}\}, \\ u_{22}^* &= -K_d'\{2(1-\nu)\ln r - r_{,2}^2\}, \end{aligned} \tag{B.10}$$

and

$$p_{ij}^* = -\frac{2}{\pi r}\left\{r_{,i}\,r_{,j}\frac{\partial r}{\partial n}\right\}, \tag{B.11}$$

where

$$K_d' = \frac{1}{2\pi G}. \tag{B.12}$$

The above expressions clearly indicate that as $c \to 0$ the half-plane fundamental solution still produces singularities of the same order as the corresponding Kelvin fundamental solution; the same argument is valid for three dimensions. It is important to note that the traction-free condition over the surface of the half-plane is now provided by the occurrence of $\partial r/\partial n$ in expression (B.11) (i.e., $\partial r/\partial n = 0$ for $\xi, x \in \bar{\Gamma}$).

References

1. Mindlin, R. D., Force at a point in the interior of a semi-infinite solid, Physics **7**, 195–202 (1936).
2. Telles, J. C. F., and Brebbia, C. A., Boundary element solution for half-plane problems, Int. J. Solids Structures **17**, 1149–1158 (1981).
3. Love, A. E. H., *A Treatise on the Mathematical Theory of Elasticity*, Dover, New York, 1944.
4. Timoshenko, S. P., and Goodier, J. N., *Theory of Elasticity*, 3rd ed., McGraw–Hill, Tokyo, 1970.

Appendix C

Some Particular Expressions for Two-Dimensional Inelastic Problems

With reference to the initial strain formulation of Section 7.2, the following expressions are valid for two-dimensional plasticity problems:

$$\begin{aligned} \sigma_z &= \nu(\sigma_x+\sigma_y)+E(\varepsilon_x^p+\varepsilon_y^p+\Delta\varepsilon_x^p+\Delta\varepsilon_y^p) && \text{(plane strain)} \\ &= 0 && \text{(plane stress)}\,, \end{aligned} \tag{C.1}$$

$$\begin{aligned} \varepsilon_x' &= \frac{1}{E}[\sigma_x-\nu(\sigma_y+\sigma_z)]+\Delta\varepsilon_x^p\,, \\ \varepsilon_{xy}' &= \frac{\sigma_{xy}}{2G}+\Delta\varepsilon_{xy}^p\,, \\ \varepsilon_y' &= \frac{1}{E}[\sigma_y-\nu(\sigma_x+\sigma_z)]+\Delta\varepsilon_y^p\,, \\ \varepsilon_z' &= \varepsilon_x^p+\varepsilon_y^p && \text{(plane strain)} \\ &= \left[\frac{\nu}{E}(\sigma_x+\sigma_y)+\Delta\varepsilon_x^p+\Delta\varepsilon_y^p\right] && \text{(plane stress)}\,, \end{aligned} \tag{C.2}$$

$$\varepsilon_{et}=\frac{\sqrt{2}}{3}\sqrt{(\varepsilon_x'-\varepsilon_y')^2+(\varepsilon_y'-\varepsilon_z')^2+(\varepsilon_z'-\varepsilon_x')^2+6(\varepsilon_{xy}')^2}\,, \tag{C.3}$$

$$\begin{aligned} \Delta\varepsilon_x^p &= \frac{\Delta\varepsilon_e^p}{3\varepsilon_{et}}(2\varepsilon_x'-\varepsilon_y'-\varepsilon_z')\,, \\ \Delta\varepsilon_{xy}^p &= \frac{\Delta\varepsilon_e^p}{\varepsilon_{et}}\varepsilon_{xy}'\,, \\ \Delta\varepsilon_y^p &= \frac{\Delta\varepsilon_e^p}{3\varepsilon_{et}}(2\varepsilon_y'-\varepsilon_z'-\varepsilon_x')\,. \end{aligned} \tag{C.4}$$

For the initial stress formulation of Chapter 7 it is convenient to write a_{ij} in vector form as follows:

$$\boldsymbol{a}=\begin{Bmatrix} a_{11} \\ 2a_{12} \\ a_{22} \\ a_{33} \end{Bmatrix}\,. \tag{C.5}$$

In addition, $\bar{d}_{ij}$ defined in Chapter 8 (expression (8.18)) can be represented for plane strain in the form

$$\bar{\boldsymbol{d}} = 2G \begin{Bmatrix} a_{11}+\omega \\ a_{12} \\ a_{22}+\omega \\ a_{33}+\omega \end{Bmatrix}, \qquad \omega = \frac{\nu}{1-2\nu}(a_{11}+a_{22}+a_{33}), \tag{C.6}$$

whereas for plane stress

$$\bar{\boldsymbol{d}} = 2G \begin{Bmatrix} a_{11}+\omega \\ a_{12} \\ a_{22}+\omega \\ 0 \end{Bmatrix}, \qquad \omega = \frac{\bar{\nu}}{1-2\bar{\nu}}(a_{11}+a_{22}). \tag{C.7}$$

The above vectors allow expressions (7.52) and (7.56) to be written as

$$\gamma' = \boldsymbol{a}^T \bar{\boldsymbol{d}} + \frac{d\psi}{d\varepsilon_e^p} \tag{C.8}$$

and

$$\begin{Bmatrix} d\sigma_x \\ d\sigma_{xy} \\ d\sigma_y \\ d\sigma_z \end{Bmatrix} = \begin{Bmatrix} d\sigma_x^e \\ d\sigma_{xy}^e \\ d\sigma_y^e \\ d\sigma_z^e \end{Bmatrix} - \frac{1}{\gamma'} \bar{\boldsymbol{d}} \boldsymbol{a}^T \begin{Bmatrix} d\sigma_x^e \\ d\sigma_{xy}^e \\ d\sigma_y^e \\ d\sigma_z^e \end{Bmatrix} \tag{C.9}$$

in which

$$\begin{aligned} d\sigma_z^e &= \nu(d\sigma_x^e + d\sigma_y^e) \qquad \text{(plane strain)} \\ &= 0 \qquad \text{(plane stress)}. \end{aligned} \tag{C.10}$$

Also, expression (8.17) is now of the form

$$\begin{Bmatrix} \dot{\sigma}_x^a \\ \dot{\sigma}_{xy}^a \\ \dot{\sigma}_y^a \\ \dot{\sigma}_z^a \end{Bmatrix} = \gamma \langle \Phi \rangle \bar{\boldsymbol{d}}, \tag{C.11}$$

where it should be noted that σ_z is computed by the relation

$$\begin{aligned} \sigma_z &= \nu(\sigma_x + \sigma_y + \sigma_x^a + \sigma_y^a) - \sigma_z^a \qquad \text{(plane strain)} \\ &= 0 \qquad \text{(plane stress)}. \end{aligned} \tag{C.12}$$

Subject Index